Heat Transfer and Fluid Flow in Microstructures

Heat Transfer and Fluid Flow in Microstructures

Guest Editors

Xuan Zhang
Steven Wang
Jinyao Ho
Bingqiang Ji
Long Zhang

MDPI

Basel • Beijing • Wuhan • Barcelona • Belgrade • Novi Sad • Cluj • Manchester

Guest Editors

Xuan Zhang
Department of Energy and Power Engineering
Beijing Institute of Technology
Beijing
China

Steven Wang
Department of Mechanical Engineering
City University of Hong Kong
Hong Kong
China

Jinyao Ho
School of Mechanical and Aerospace Engineering
Nanyang Technological University
Singapore
Singapore

Bingqiang Ji
School of Astronautics
Beihang University
Beijing
China

Long Zhang
Department of Energy and Power Engineering
Beijing Institute of Technology
Beijing
China

Editorial Office
MDPI AG
Grosspeteranlage 5
4052 Basel, Switzerland

This is a reprint of the Special Issue, published open access by the journal *Micromachines* (ISSN 2072-666X), freely accessible at: https://www.mdpi.com/journal/micromachines/special_issues/49D8FFA8H4.

For citation purposes, cite each article independently as indicated on the article page online and as indicated below:

Lastname, A.A.; Lastname, B.B. Article Title. *Journal Name* **Year**, *Volume Number*, Page Range.

ISBN 978-3-7258-7825-3 (Hbk)
ISBN 978-3-7258-7826-0 (PDF)
https://doi.org/10.3390/books978-3-7258-7826-0

Contents

About the Editors

Xuan Zhang

Xuan Zhang is an Associate Professor at the Beijing Institute of Technology. He received a Ph.D. degree from Tsinghua University in 2019 and then worked at Nanyang Technological University as a Postdoctoral Research Fellow until 2022. He focuses on the mechanisms and applications of heat/mass transfer and fluid flow coupling processes, especially phase change problems (e.g., icing/frosting/freezing, condensation/evaporation/boiling, and solidification/melting). He has published/co-authored 110+ SCI journal articles, with 10 recognized as ESI Highly Cited/Hot, Featured, Editor's Pick, Front Cover, and CNKI Highly Cited papers, and 5 books/chapters; he has also been invited to deliver 10 reports at conferences and applied for/authorized 20+ patents and 10+ software systems. His works have been cited 3200+ times in Google Scholar with a H-index of 32 and have been reported 50+ times by famous science and technology media, such as *People's Daily*, *China Science Daily*, *The New York Times*, and *EurekAlert*. Prof. Zhang has been listed among the World's Top 2% Scientists by Stanford University/Elsevier (since 2023) and the 0.5% Top Scholars by ScholarGPS Rankings (since 2025). He has received 10+ grants from the National Natural Science Foundation of China, Ministry of Science and Technology of China, etc., and has served as Associate Editor for *Particulate Science and Technology*; an Editorial Board Member for *Scientific Reports*; a Youth Editor Member for *Coatings*, *Exploration*, etc.; a Guest Editor for *Applied Thermal Engineering*, *Micromachines*, etc.; as a Conference Committee Member for Frosting-Icing 2025, MSTP-2024, etc.; and as a Reviewer for 70+ academic journals. He has been recognized with the following awards: the First Prize of Science and Technology Progress Award of Shandong Province, Wu Zhonghua Outstanding Graduate Student (Chinese Society of Engineering Thermophysics), Outstanding Doctoral Thesis (Tsinghua University), Highly Cited Doctoral Thesis (CNKI), etc.

Steven Wang

Steven Wang is an Associate Professor and serves as the Associate Vice President (Resources Planning) at the City University of Hong Kong and is the Deputy Director of the Centre for Nature-Inspired Engineering. He obtained his PhD from Monash University (Supervisor Prof David Boger, FRS) and his Bachelor of Engineering from the University of Melbourne. Before assuming his current position, he conducted postdoctoral research at ETH Zurich from 2015 to 2016 and at the Massachusetts Institute of Technology from 2014 to 2015. His research interests include energy and environmental engineering, fluid mechanics, advanced materials for engineering applications, indoor air and ventilation, and multiphase flows for engineering applications. He has published extensively in prestigious journals, including *Nature*, *Nature Chemical Engineering*, *Nature Communications*, *Science Advances*, and *Physical Review Fluids*. His contributions to the field were recognized when he received the Research Project of the Year: STEM award at the Times Higher Education (THE) Awards Asia 2024.

Jinyao Ho

Jinyao Ho is an Assistant Professor at Nanyang Technological University. He received his PhD degree from Nanyang Technological University in 2019 and was subsequently appointed as a Research Fellow from July 2019 to February 2020. Between March 2020 and December 2021, he undertook his postdoctoral training at the Energy Transport Research Laboratory at the University of Illinois at Urbana-Champaign. His research focuses on the development of advanced thermal devices to enhance condensation and anti-frosting, boiling and evaporation, and anti-corrosion and anti-fouling. The research in his group also focuses on the development of new surface engineering strategies to promote interfacial transport of heat and mass and the development of new organic micro/nanocomposite materials for the thermal management of lithium-ion batteries.

Bingqiang Ji

Bingqiang Ji is a professor at the School of Astronautics at Beihang University. He received his B.S. and Ph.D. degrees from Huazhong University of Science and Technology and Tsinghua University in 2014 and 2019, respectively, and then worked at the University of Illinois at Urbana-Champaign and City University of Hong Kong as a postdoctoral research fellow. His research interests include multiphase flow, interface phenomena, and flow instability. He has published extensively in prestigious journals, including *Nature Physics*, *PNAS*, *Nature Communications*, *Physical Review Letters*, *Journal of Fluid Mechanics*, and *Advanced Materials*.

Long Zhang

Long Zhang is an Assistant Professor at the Beijing Institute of Technology. His research focuses on the coupled mechanisms of heat and mass transfer and fluid flow, specifically phase change problems, such as frosting, icing, and solidification, alongside key technologies in heat pump and refrigeration systems. He has led over 10 projects funded by the National Natural Science Foundation of China, the Ministry of Industry and Information Technology, the Beijing Municipal Science and Technology Commission, and key state laboratories. Zhang has published over 90 SCI/EI journal papers, including more than 40 as the first or corresponding author with two ESI highly cited papers, and authored three English books. As of March 2026, he has over 2,000 Google Scholar citations and an H-index of 25. He holds over 20 patents and software copyrights, has co-authored three industry standards, and has delivered eight invited academic presentations. His honors include the First Prize of Shandong Provincial Science and Technology Progress, the First Prize from the Architectural Society of China, and the Outstanding Doctoral Dissertation Nomination Award from Hong Kong Polytechnic University; he has been listed among the Elsevier/Stanford World's Top 2% Scientists for three consecutive years.

Preface

Heat transfer and fluid flow are seen in many mechanical and electronic equipment and systems and thus have attracted a great deal of attention. With advances in manufacturing technologies, an increasing number of microstructures have been used to control heat transfer and fluid flow in recent decades. For example, micro-/nano-structured surfaces are applied to enhance boiling heat transfer and to achieve a potentially higher critical heat flux, superhydrophobic surfaces with two-stage structures show promising applications in anti-icing/frosting, and microchannels are used in microfluidic chips. Significant effort has been devoted to deepening our understanding of heat transfer and fluid flow and advancing the development of relevant applications. The aim of this Special Issue of *Micromachines*, "Heat Transfer and Fluid Flow in Microstructures", is to present recent advances in heat transfer and fluid flow in microstructures, including but not limited to associated theoretical analyses, experimental measurements, numerical simulations, and practical applications.

Xuan Zhang, Steven Wang, Jinyao Ho, Bingqiang Ji, and Long Zhang
Guest Editors

micromachines

Editorial

Editorial for the Special Issue on Heat Transfer and Fluid Flow in Microstructures

Xuan Zhang [1,*], Steven Wang [2], Jin Yao Ho [3], Bingqiang Ji [2,4] and Long Zhang [1,*]

1 Department of Energy and Power Engineering, School of Mechanical Engineering, Beijing Institute of Technology, Beijing 100081, China
2 Department of Mechanical Engineering, City University of Hong Kong, Hong Kong 999077, China
3 School of Mechanical and Aerospace Engineering, Nanyang Technological University, Singapore 639798, Singapore
4 School of Astronautics, Beihang University, Beijing 100191, China
* Correspondence: xuan.zhang@bit.edu.cn (X.Z.); long.zhang@bit.edu.cn (L.Z.)

1. Introduction to the Special Issue "Heat Transfer and Fluid Flow in Microstructures"

By constructing precise micro-/nano-scale topological features on solid interfaces, microstructured surface technology offers a revolutionary approach to actively regulating flow and heat transfer processes [1–4]. It overcomes the performance limitations of conventional smooth surfaces, transforming passive interfaces into active functional units and thereby enabling precise control over fluid dynamics, phase change behavior, and energy transport. This technology plays a critical role in fields such as high-efficiency cooling of electronic chips [5], aerospace thermal protection [6], biomedical microfluidics [7], and enhanced heat transfer in energy systems [8]. Its core problems lie in endowing surfaces with novel physical functions and system-level performance through structural design.

The strategic significance of this field extends beyond improving local heat and mass transfer efficiency; it also reveals and exploits a series of unique physical mechanisms dominated by microscopic surface geometry. These mechanisms include inducing controlled flow separation and vortex generation through designed micro-ribs, dimples, and other features [9–11] to enhance mixing and convection; precisely regulating the pinning and slipping behaviors of the solid–liquid–gas three-phase contact line [12] for active management of wettability and phase change processes; and utilizing interfacial forces such as capillary forces and Marangoni stresses generated by microstructures to drive or stabilize fluid motion. A deeper understanding and wider application of these mechanisms can allow microstructured surfaces to become a key enabling technology in addressing engineering challenges such as high heat flux dissipation, enhanced phase change heat transfer, flow drag reduction, and surface anti-icing/anti-corrosion.

However, translating the significant potential of microstructured surfaces into stable, reliable, and scalable engineering applications involves complex multi-scale and multi-physics coupling challenges [13]. This requires research not only to establish structure–performance relationships between surface topography and local heat/mass transfer characteristics at the microscale but also to evaluate overall thermal performance, mechanical durability, and long-term service stability at the macroscale system level. Specific scientific issues include the optimal design and performance prediction of microstructure shape, size, and arrangement; flow and heat transfer behavior of complex fluids (e.g., non-Newtonian fluids, nanofluids) over structured surfaces [14]; retention and failure mechanisms of structural functions in multi-phase flow environments [15]; and efficient,

 https://doi.org/10.3390/mi17020203

low-cost manufacturing processes and integration technologies for large-scale microstructure fabrication [16,17].

Therefore, research on microstructured surfaces has become a comprehensive interdisciplinary frontier integrating surface engineering, micro-/nano-fabrication, fluid mechanics, heat transfer, and materials science. The ultimate goal is to achieve "designed functions" through "designed surfaces", driving innovation in next-generation high-performance thermal management devices, efficient energy conversion systems, advanced microfluidic chips, and aerospace power systems. This calls for full-chain innovation spanning fundamental mechanisms, material systems, manufacturing processes, and system integration, achieving deep synergy among structure and function, material and process, and design and application.

This Special Issue focuses on this interdisciplinary field, bringing together a range of research achievements covering fundamental studies, numerical simulations, experimental characterizations, and application explorations. Based on their research directions and content characteristics, the contributions can be grouped into the following three topics: fundamental studies on the mechanisms of single-phase flow [contributions 4, 5, 13–15], fundamental studies on the mechanisms of multi-phase flow [contributions 1, 6, 8, 10, 11], and extended research on the applications for multi-phase flow [contributions 2, 3, 7, 9, 12], as illustrated in Figure 1. This classification framework systematically presents a complete research spectrum, from fundamental flow principles to engineering solutions.

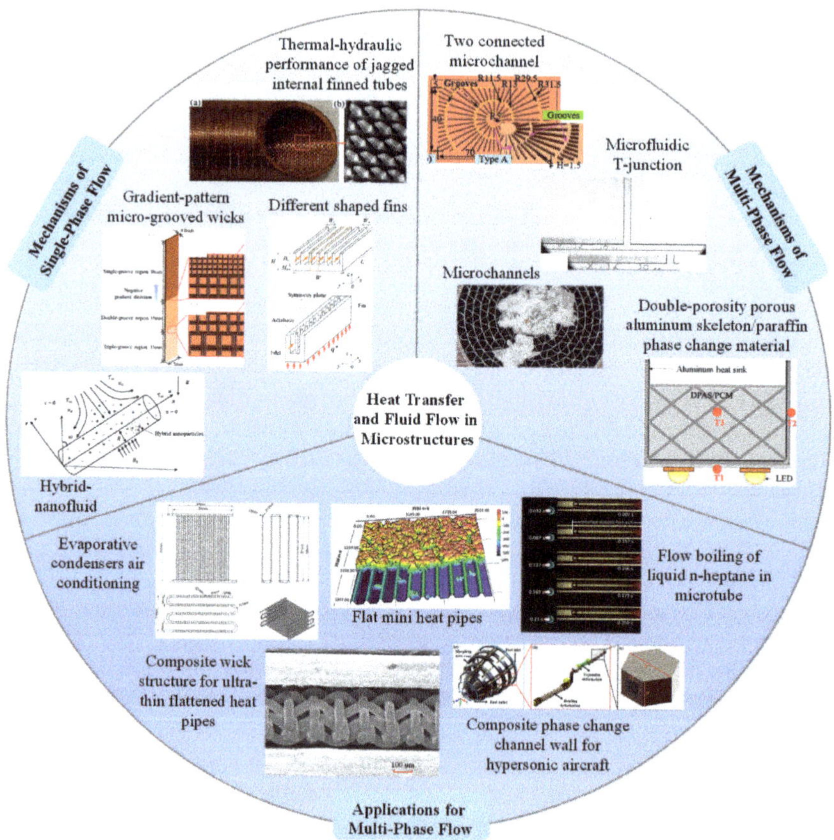

Figure 1. Topics covered in the Special Issue titled "Heat Transfer and Fluid Flow in Microstructures".

https://doi.org/10.3390/mi17020203

Overall, this Special Issue reflects the latest advances in microstructured surface technology in both fundamental research and engineering applications. It not only enhances our understanding of the flow and heat transfer mechanisms influenced by micro-/nano-structures but also provides theoretical guidance and technical solutions for the design and system integration of functional surfaces orientated toward practical needs. Future research is expected to further advance toward intelligent responsive surfaces, multifunctional integrated structures, cross-scale synergistic design, and reliable applications in extreme environments, continuously driving this technology toward a more efficient, reliable, and intelligent future.

2. Mechanisms of Single-Phase Flow

This section explores the enhancement of single-phase heat transfer through engineered microstructures and advanced fluids. Key studies investigate the performance of novel 3D jagged finned tubes [contribution 4], laser-fabricated capillary-gradient wicks [contribution 5], liquid metal-cooled wavy microchannels [contribution 13], hybrid nanofluids in shaped tube exchangers [contribution 14], and magneto-convective hybrid nanofluid flow [contribution 15]. Collectively, they demonstrate how deliberate structural design and fluid engineering can disrupt flow, augment mixing, and significantly improve thermal transport.

3. Mechanisms of Multi-Phase Flow

This section delves into the role of microstructure in governing phase change processes. Research examines thermal management using gradient porosity phase change materials [contribution 1], flow boiling pattern evolution in T-shaped microchannels [contribution 6], the pseudo-desublimation of AdBlue in SCR systems [contribution 8, 11], and the performance of interconnected microchannel heat sinks [contribution 10]. These works highlight how micro-scale geometry and system architecture critically influence boiling dynamics, flow stability, and overall thermal performance in multi-phase systems.

4. Applications for Multi-Phase Flow

This section bridges fundamental research and practical thermal management solutions. It covers the development of composite wicks for ultra-thin heat pipes [contribution 2], bio-inspired phase change cooling channels for hypersonic aircraft [contribution 3], evaporative condensers for CO_2 air conditioning [contribution 7], flow boiling of fuel in microtubes [contribution 9], and vaporization in flat mini heat pipes with porous structures [contribution 12]. These studies provide actionable design insights and validate performance in high-heat-flux cooling applications in electronics, aerospace, and energy systems.

To conclude, we would like to acknowledge all the authors for their contributions to the success of this Special Issue, "Heat Transfer and Fluid Flow in Microstructures", as well as the reviewers, whose feedback helped to improve the quality of the published papers.

Acknowledgments: The Guest Editors are very grateful to Special Issue Editors from the *Micromachines* publishing office for their great guidance, assistance, help, and support, which led to the success of this Special Issue.

Conflicts of Interest: The authors declare no conflicts of interest.

List of Contributions:

1. Huang, S.; Long, C.; Hu, Z.; Xu, Y.; Zhang, B.; Zhi, C. Thermal Performance of Heat Sink Filled with Double-Porosity Porous Aluminum Skeleton/Paraffin Phase Change Material. *Micromachines* **2024**, *15*, 806. https://doi.org/10.3390/mi15060806.

2. Zhou, W.; Yang, Y.; He, J.; Chen, R.; Jian, Y.; Shao, D.; Wu, A. An Experimental Study of a Composite Wick Structure for Ultra-Thin Flattened Heat Pipes. *Micromachines* **2024**, *15*, 764. https://doi.org/10.3390/mi15060764.
3. Li, W.; Zhao, J.; Wu, X.; Liang, L.; Wang, W.; Yan, S. Structure Design and Heat Transfer Performance Analysis of a Novel Composite Phase Change Active Cooling Channel Wall for Hypersonic Aircraft. *Micromachines* **2024**, *15*, 623. https://doi.org/10.3390/mi15050623.
4. Huang, S.; Deng, M.; Chen, Z.; Yang, D.; Xu, Y.; Lan, N. Experimental and Numerical Investigations on Thermal-Hydraulic Performance of Three-Dimensional Overall Jagged Internal Finned Tubes. *Micromachines* **2024**, *15*, 513. https://doi.org/10.3390/mi15040513.
5. Huang, G.; Liao, J.; Fan, C.; Liu, S.; Miao, W.; Zhang, Y.; Ta, S.; Yang, G.; Cui, C. Gradient-Pattern Micro-Grooved Wicks Fabricated by the Ultraviolet Nanosecond Laser Method and Their Enhanced Capillary Performance. *Micromachines* **2024**, *15*, 165; https://doi.org/10.3390/mi15010165.
6. Bao, X.; Yang, F.; Zhang, X. Experimental Study of Flow Boiling Regimes Occurring in a Microfluidic T-Junction. *Micromachines* **2023**, *14*, 2235. https://doi.org/10.3390/mi14122235.
7. Dang, T.; Nguyen, H. A Study on the Simulation and Experiment of Evaporative Condensers in an R744 Air Conditioning System. *Micromachines* **2023**, *14*, 1826. https://doi.org/10.3390/mi14101826.
8. Picus, C.; Mihai, I.; Suciu, C. Pseudo-Desublimation of AdBlue Microdroplets through Selective Catalytic Reduction System Microchannels and Surfaces. *Micromachines* **2023**, *14*, 1807. https://doi.org/10.3390/mi14091807.
9. Rashid, M.; Ahmad, N.; Swati, R.; Khan, M. Flow Boiling of Liquid n-Heptane in Microtube with Various Fuel Flow Rate: Experimental and Numerical Study. *Micromachines* **2023**, *14*, 1760. https://doi.org/10.3390/mi14091760.
10. Jiang, Z.; Song, M.; Shen, J.; Zhang, L.; Zhang, X.; Lin, S. Experimental Investigation on the Flow Boiling of Two Microchannel Heat Sinks Connected in Parallel and Series for Cooling of Multiple Heat Sources. *Micromachines* **2023**, *14*, 1580. https://doi.org/10.3390/mi14081580.
11. Picus, C.; Mihai, I.; Suciu, C. Experimental Investigations upon Ultrasound Influence on Calefaction of AdBlue in Selective Catalytic Reduction Systems (SCR). *Micromachines* **2023**, *14*, 1488. https://doi.org/10.3390/mi14081488.
12. Mihai, I.; Suciu, C.; Picus, C. Assessment of Vapor Formation Rate and Phase Shift between Pressure Gradient and Liquid Velocity in Flat Mini Heat Pipes as a Function of Internal Structure. *Micromachines* **2023**, *14*, 1468. https://doi.org/10.3390/mi14071468.
13. Yu, T.; Guo, X.; Tang, Y.; Zhang, X.; Wang, L.; Wu, T. Numerical Investigation of Fluid Flow and Heat Transfer in High-Temperature Wavy Microchannels with Different Shaped Fins Cooled by Liquid Metal. *Micromachines* **2023**, *14*, 1366. https://doi.org/10.3390/mi14071366.
14. Bouselsal, M.; Mebarek-Oudina, F.; Biswas, N.; Ismail, A. Heat Transfer Enhancement Using Al_2O_3-MWCNT Hybrid-Nanofluid inside a Tube/Shell Heat Exchanger with Different Tube Shapes. *Micromachines* **2023**, *14*, 1072. https://doi.org/10.3390/mi14051072.
15. Sohut, F.; Khan, U.; Ishak, A.; Soid, S.; Waini, I. Mixed Convection Hybrid Nanofluid Flow Induced by an Inclined Cylinder with Lorentz Forces. *Micromachines* **2023**, *14*, 982. https://doi.org/10.3390/mi14050982.

References

1. Chen, J.; Tian, H.; Xu, W.; Guo, H.; Wang, C.; Gu, J.; Cao, Y. Experimental Research Progress on Gas–Liquid Flow and Heat Transfer Characteristics in Micro Pulsating Heat Pipes. *Micromachines* **2026**, *17*, 37. [CrossRef] [PubMed]
2. Chen, Z.; Zhou, Y.; Yang, Z.; Song, Z.; Li, J.; Zhang, G.; Han, F. A New Preparation Method of the Dual Scale Surface Microstructures with Self-Cleaning and Efficient Boiling Heat Transfer Based on Wire Electrical Discharge Machining and Laser Etching. *Appl. Therm. Eng.* **2025**, *279*, 127822. [CrossRef]
3. Gao, J.; Zhang, J.; Li, K.; Lu, L. Effect of Superhydrophobic Microstructures on the Heat Transfer Performance of Surgical Electrode: Droplet and Bubble Dynamics Investigation. *Appl. Phys. Lett.* **2025**, *126*, 141601. [CrossRef]
4. Ouyang, Z.; Yan, Y.; Long, Y.; Luo, B.; Su, Z.; Long, J. Fabrication of Copper-Based Two-Tiered Surface Microstructures by Picosecond Laser Micromachining in Combination with Electrodeposition for Enhanced Two-Phase Heat Transfer. *Appl. Surf. Sci.* **2025**, *686*, 162096. [CrossRef]

5. Fang, W.; Sang, J.; Li, H.; Cao, X.; Yin, S.; Huang, Y. Energy-Efficient Cooling of Silicon-Based Microfluidic Chips Enabled by Integrated Microstructural and Surface Modification Strategies. *Int. J. Heat Mass Transf.* **2026**, *254*, 127632. [CrossRef]
6. Zhou, K.; Xiao, G.; Huang, Y. Fabricating Physicochemical Microstructures with Super Hydrophilicity on Cf/SiC Composites Surface via Picosecond-Laser Induced Ablation. *Ceram. Int.* **2023**, *49*, 34291–34302. [CrossRef]
7. Yang, X.; Li, C.; Jiao, Y.; Wang, L.; Wang, F. Microfluidic Construction of Heterogeneous Micro-Nano-Structured Materials for Biomedical Applications. *ACS Mater. Lett.* **2025**, *7*, 1462–1487. [CrossRef]
8. Ni, S.; Li, Y.; Ji, D.; Zhang, X.; Zhao, J. Effect of Surfactants on Flow Boiling Heat Transfer and Bubble Dynamic: An Experimental Study. *Int. J. Heat Mass Transf.* **2026**, *255*, 127851. [CrossRef]
9. Jiang, Z.; Song, M.; Dang, C.; Xu, Y.; Azizifar, S.; Zhen, H. Experimental Study on Flow Boiling of Two Rectangular Expanding Microchannel Heat Sinks in Series Connection and Flow Patterns Analysis. *Int. Commun. Heat Mass Transf.* **2024**, *155*, 107470. [CrossRef]
10. Jiang, Z.; Song, M.; Dang, C.; Jiang, Y.; Wan, M.P.; Chao, C.Y.H. Two Mathematical Models of Flow Boiling and Flow Instability in Rectangular Expanding Microchannel Heat Exchangers and Structure Optimization. *Appl. Therm. Eng.* **2023**, *228*, 120483. [CrossRef]
11. Li, F.; Han, J.; Wang, Z.; Dai, Y.; Chen, P. Numerical Study on Heat Transfer Performance of Turbulence Enhancement Configurations for Galinstan Based Mini-Channel Cooling. *Micromachines* **2026**, *17*, 83. [CrossRef] [PubMed]
12. Chen, J.; Yang, Q.; Liu, Z.; Zhang, R.; Dong, Y.; Chen, J.; Chen, H. Dynamic Behavior Regulation of Droplet Transfer Based on Arc-Bubble Control by Ultrasound-Induced during Underwater Wet Welding. *J. Manuf. Process.* **2024**, *129*, 147–160. [CrossRef]
13. Zhang, X.; Wu, X.M. Time and Frequency Characteristics of Pressure Fluctuations during Subcooled Nucleate Flow Boiling. *Heat Transf. Eng.* **2018**, *39*, 642–653. [CrossRef]
14. Abas, S.A.; Ullah, H.; Fiza, M.; Akgul, A.; Jan, A.U.; El-Rahman, M.A.; Al-Mekhlafi, S.M. Second Order Slip Micropolar MHD Hybrid Nanofluid Flow over a Stretching Surface with Uniform Heat Source and Activation Energy: Numerical Computational Approach. *Results Eng.* **2025**, *25*, 104060. [CrossRef]
15. Cui, J.; Cui, Y.; Cui, J.; Cui, Y. Effects of Surface Wettability and Roughness on the Heat Transfer Performance of Fluid Flowing through Microchannels. *Energies* **2015**, *8*, 5704–5724. [CrossRef]
16. Zhou, F.; Fang, W.; Sun, W.; Huang, J. Self-Adaptive Microchannel Heat Exchanger Enabled by Deformable Alloys for Heat Transfer Characteristics. *Appl. Therm. Eng.* **2026**, *288*, 129516. [CrossRef]
17. Pereira, J.; Souza, R.; Lima, R.; Moreira, A.; Moita, A. An Overview of the Recent Advances in Pool Boiling Enhancement Materials, Structure, and Devices. *Micromachines* **2024**, *15*, 281. [CrossRef]

https://doi.org/10.3390/mi17020203

micromachines

Article

Thermal Performance of Heat Sink Filled with Double-Porosity Porous Aluminum Skeleton/Paraffin Phase Change Material

Shufeng Huang [1,2,*], Chuanshun Long [1], Zhihan Hu [1,*], Yingshuai Xu [1], Bin Zhang [1] and Changjian Zhi [1]

1 School of Mechanical and Electronic Engineering, East China University of Technology, Nanchang 330013, China
2 Jiangxi Engineering Province Engineering Research Center of New Energy Technology and Equipment, East China University of Technology, Nanchang 330013, China
* Correspondence: author: shufeng3030@163.com (S.H.); hld112121@126.com (Z.H.)

Abstract: Phase change materials (PCMs) are used to cool high-power-density electronic devices because of their high latent heat and chemical stability. However, their low thermal conductivity limits the application of PCMs. To solve this problem, a double-porosity porous aluminum skeleton/paraffin phase change materials (DPAS/PCM) was prepared via additive manufacturing and the water-bath method. The thermal performance of the DPAS/PCM heat sink (HS) was experimentally investigated to examine the effects of the positive- and reverse-gradient porosity structures of the DPAS/PCM. The results show that a positive-gradient porosity arrangement is more conducive to achieving a low-temperature cooling target for LED operation. In particular, the temperature control time for the positive gradient porosity structure increased by 4.6–13.7% compared with the reverse gradient porosity structure. Additionally, the thermal performances of uniform porous aluminum skeleton/paraffin (UAS) and DPAS/PCMs were investigated. The temperature control effect of the DPAS/PCM was better than that of the UAS/PCM HS at high critical temperatures. Compared with the UAS/PCM HS, the temperature control time of the DPAS/PCM HS is increased by 7.8–12.5%. The results of this work show that the prepared DPAS/PCM is a high-potential hybrid system for thermal management of high-power electronic devices.

Keywords: double-porosity porous aluminum skeleton; phase change material; heat sink; thermal management

1. Introduction

With the development trend of high performance, miniaturization and functional integration of modern optoelectronic devices such as light emitting diodes (LED) [1–3] and lasers [4], a large amount of heat is generated inside electronic components, which increase the operating temperature that affects the life and reliability of the modern electronic devices [5]. Different efficient heat dissipation techniques for electronic devices have been proposed, including substrates with high thermal conductivity [6,7], pool boiling [8], jet impingement cooling [9], and heat pipes [10]. Phase change material (PCM) has significant potential to cool high heat-generating electronic components due to being readily available, inexpensive, compatible with other materials, chemically inert, recyclable, and thermally stable [11]. Therefore, phase change materials are widely used in LED [12], energy-saving buildings [13], intelligent equipment [14,15], aerospace [16] and other fields. However, due to the inherently low thermal conductivity of PCM, it is a challenge to achieve efficient heat dissipation in temperature management, which limits their further engineering applications [17,18]. To address this drawback, various methods have been developed to improve the thermal conductivity of PCM, including inserting metal fins [19,20], adding high thermal conductivity nanoparticles [21], embedding metal foams [22,23] and microencapsulating [24].

Among the many methods for enhancing thermal conductivity, the introduction of inner fins has attracted attention in recent years. Fok et al. [25] conducted thermal management on portable electronic devices using heat sinks (HSs) comprising fins of n-eicosane PCM; the results showed that the HSs with PCM fins were more effective than those without fins. Other studies have also been conducted that employed inner fins to improve the thermal conductivity of PCMs [26,27]. Although inner fins can improve the thermal performance of a PCM-based thermal control system, they lead to an increase in the size, weight, and manufacturing cost of the system.

Accordingly, another reinforcement technique involving the addition of nanoparticles has been attempted to solve the problem of low thermal conductivity. Owing to their large specific surface area, nanoparticles have been used to increase heat dissipation in asphalt [28] as well as enhance the thermal properties of cement mortar [29] and nanofluids [30]. Although the introduction of nanoparticles into a PCM reduces the weight of the thermal control system, the nanoparticles can agglomerate, which reduces the cooling effect of the PCM. Further, nanoparticles increase the viscosity of PCMs, which limits convective heat transfer.

In addition to the abovementioned methods, embedding PCM into porous materials with high thermal conductivity has proven to be the most effective way to solve the low thermal conductivity of PCMs [31]. Baby and Balaji [32] proposed a copper foam (CuF)-PCM HS to dissipate heat from electronic components. The results showed that the maximum enhancement ratio is 3 when the set point temperature is 52 °C; the effect of orientation on increasing heat transfer performance is not significant. Rehman and Ali [33] investigated the addition of copper foam and iron-nickel foam to PCM to cool electronic devices and chose RT-35HC as the PCM. As a result, the temperature of the HS with copper foam PCM was 5–6 °C lower than that with the Fe-Ni foam PCM. The same authors [34] studied the effect of the PCM volume on thermal performance. Compared with pure nickel foam, the temperature of CoF-HS with 0.8 PCM volume fraction was reduced by 26%. Wang et al. [12] investigated an HS without a PCM, an HS with a pure PCM, and an HS with a porous metal sintered felt (PMFSF)-PCM. The addition of the PMFSF enhanced the heat transfer of the PCM and reduced the heat source temperature. It can be clearly seen from the literature that metal foams (such as copper, iron, aluminum, and nickel) and non-metallic foams (such as carbon and graphene) can be used with PCM to improve the efficiency of PCM-based thermal control modules [31–34]. However, the limited improvements in thermal conductivity are insufficient to satisfy the increasing thermal storage and dissipation requirements of high-power electronics.

Considering the above situation, it is necessary to develop a new PCM with a high latent that is suitable for cooling high-power electronics. Considering the operating temperatures of high-power electronic devices, very little work has been conducted using double-porosity porous skeleton/paraffin PCM. This study proposes a novel double-porosity porous aluminum skeleton/paraffin PCM (DPAS/PCM) based on additive manufacturing and a water bath method. The preparation and experimental system are described in detail in Section 2. Furthermore, the transient thermal performance of the DPAS/PCMs is experimentally studied in Section 3. Finally, the thermal properties of the double-porosity and uniform HSs were compared.

2. Materials and Methods

2.1. Experimental Platform

As shown in Figure 1, the experimental platform comprised an HS with LEDs (Cree Inc., Durham, NC, USA), an altimeter (Mitutoyo, Shanghai, China), a DC power supply (Guangdong Henghuiyuan Electronics Co., Ltd., Guangdong, China), a temperature acquisition system (Changzhou Jinailian Electronic Technology Co., Ltd., Changzhou, China), a computer (China Hewlett-Packard Co., Ltd., Beijing, China), and a Plexiglas box. The HS was placed in the plexiglass box to prevent heat exchange with its surroundings. The temperature detection unit consisted of K-type thermocouples, a temperature detector, and a computer. Two K-type thermocouples were glued to the LED base plate midpoint (T1) and cavity side midpoint (T2) to analyze the instantaneous temperatures T1 and T2. Another K-type thermocouple was secured and positioned using an altimeter to test the T3 instantaneous temperature. The LEDs were powered by a DC power supply, and four power levels of 11.7, 14.7, 17.7, and 21.3 W were applied to each HS to test the transient performance of the HSs. The specific experimental steps are as follows: (1) Pre-regulate the DC power supply to output the desired power. (2) Start the temperature acquisition system and turn on the power to apply constant power to the LEDs to heat the DPAS/PCM heat sink. (3) Turn off the power to end the heating process when T1 reaches the limit temperature of 90 or 100 °C. (4) Heat sink naturally cools until the system temperature drops below 27 °C and stops the temperature acquisition system. In addition, the temperature control experiment was repeated three times for each sample at each heating power, and the average value was obtained as the experimental result. Specifically, the employed K-type thermocouples had an accuracy of 0.1 °C, and the input power accuracy was ±0.1 W.

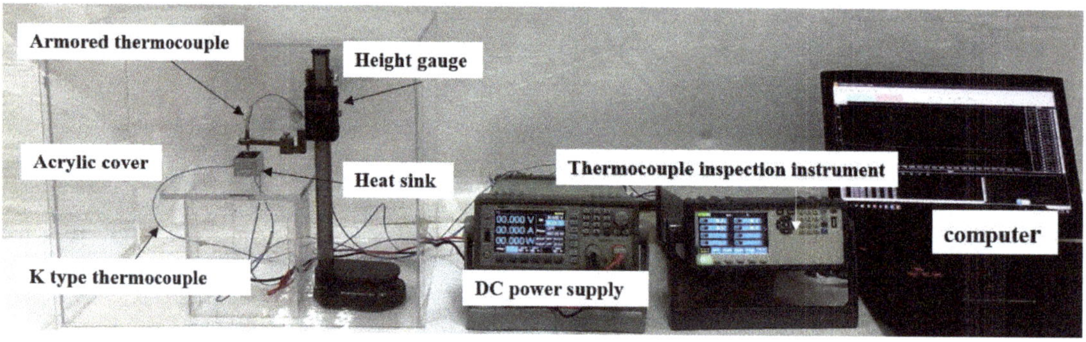

Figure 1. Physical image of experimental platform.

2.2. LED Heat Sink (HS) Cooling Device

A light-emitting diode (LED) HS-cooling device (LED-HD) stores the heat generated by the LED during operation as heat and emits the remaining heat through natural convection on the outer surface of the aluminum cavity and radiation heat dissipation of the HS, ensuring that the LED operates below the limit temperature. As shown in Figure 2, the device consists of a cubic aluminum cavity, a DPAS/PCM, and an LED light heat source. The heat storage of the LED-HD is mainly provided by the inner DPAS/PCM. The dimensions of the aluminum cavity were as follows: 30 mm in length, 30 mm in width, 25 mm in height, and 2 mm in wall thickness. The LED heat source was fixed in the middle of the bottom surface using a thermally conductive two-sided adhesive, and the power of the LED was 8–35 W. As shown in Figure 2, the DPAS/PCM HS has three temperature measuring points: the bottom (T1), side (T2), and HS center (T3). To test the central temperature of the AS/PCM, a hole with a depth of 10 mm and a diameter of 1 mm was drilled at the center of the DPAS/PCM. In addition, the DPAS/PCM HS was equipped with three temperature-measuring points, as shown in Figure 2: the center of the HS (T3), bottom (T1), and side (T2). To obtain the central temperature of the DPAS/PCM, a small hole with a depth of 10 mm and a diameter of 1 mm was drilled in the center of the DPAS/PCM with a drill bit. The safe operating temperature range is 45 °C to 100 °C.

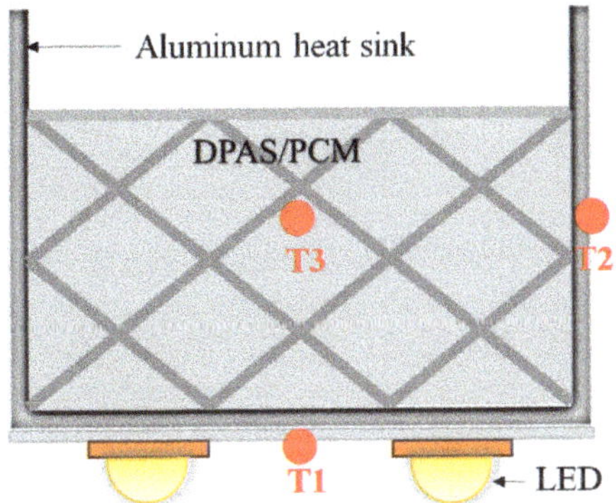

Figure 2. Diagram of a heat sink (HS) constituting a double-porosity porous aluminum skeleton/paraffin phase change material.

2.3. Preparation of Double-Porosity Porous Aluminum Skeleton/Paraffin Phase Change Material

The preparation of the DPAS/PCM was divided into two steps: fabrication of a double-porosity porous aluminum skeleton (DPAS) using the additive manufacturing method and preparation of the DPAS/PCM using the water bath method. Figure 3 shows the four types of DPAS samples based on the AM. Figure 3a shows a DPAS with 65% porosity at the bottom and 85% porosity at the top, named DPAS6585. The arrangement of DPAS6585, with low porosity in the first layer and high porosity in the second layer, was defined as a positive-gradient porosity. Conversely, DPAS8565 is called reverse gradient porosity. Figure 3f shows that the DPAS has a three-dimensional connected pore structure and a rough surface, which is conducive to natural convection heat transfer. For comparative analysis, a uniform porous aluminum skeleton (UAS) with the same overall porosity of 75% was also prepared, as shown in Figure 3e. For DPAS6585 and DPAS8565, the filament size of the 65% porosity layer is 3.01 mm, and the filament size of the 85% porosity layer is 1.86 mm. For DPAS6090 and DPAS9060, the filament size of the 60% porosity layer is 3.25 mm, and the filament size of the 85% porosity layer is 1.50 mm. As shown in Figure 3f, the printed skeleton has a rough surface topography.

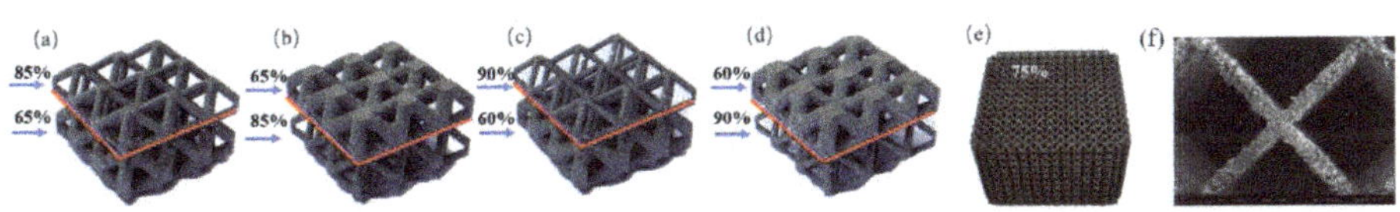

Figure 3. Porous aluminum skeleton samples: **(a)** DPAS6585, **(b)** DPAS8565, **(c)** DPAS6090, **(d)** DPAS9060, **(e)** uniform porous aluminum skeleton, **(f)** the morphology of DPAS6585/PCM.

The DPAS/PCMs were prepared in four steps based on the water bath method. (1) A heat-resistant soft silicone mold was placed in a beaker, an appropriate amount of paraffin was weighed and placed into the silicone mold, and then DPAS was placed on the paraffin. (2) The above beaker was placed in a water bath; the temperature of the water bath was adjusted to about 80 °C and maintained for a period of time, and the DPAS was completely immersed in the paraffin. (3) After being kept at 80 °C for a period of time, the power supply was turned off, and the sample cooled naturally until the paraffin in the beaker solidified. (4) Excess paraffin was removed from the surfaces of the samples to obtain samples with dimensions of 30 × 30 × 20 mm, as shown in Figure 4. In addition, Figure 4e shows that a uniform porous aluminum skeleton/paraffin PCM (UAS/PCM) was prepared for comparison. As shown in Table 1, the thermal conductivity of the AS/PCM has been tested based on the transient plane heat source method.

Figure 4. DPAS/PCM samples: (**a**) DPAS6585/PCM, (**b**) DPAS8565/PCM, (**c**) DPAS6090/PCM, (**d**) DPAS9060/PCM, (**e**) UAS/PCM.

Table 1. Thermal conductivities of AS/PCMs.

AS/PCM	DPAS6585/PCM	DPAS8565/PCM	DPAS6090/PCM	DPAS9060/PCM	UAS/PCM
k (W/m·K)	12.23	10.92	11.74	8.7	11.92

3. Results and Discussion

3.1. Thermal Performance Analysis of a Typical DPAS/PCM HS

Figure 5 displays the transient temperature of the DPAS6585/PCM HS at the input power of 11.7 W. From Figure 5, the transient temperature curve can be divided into four segments: solid, melting, liquid, and solidification. The average velocity V of T1 is defined as $V = \Delta T/\Delta t$, where Δt is the duration of the corresponding interval, and ΔT is the change in temperature T1 during Δt.

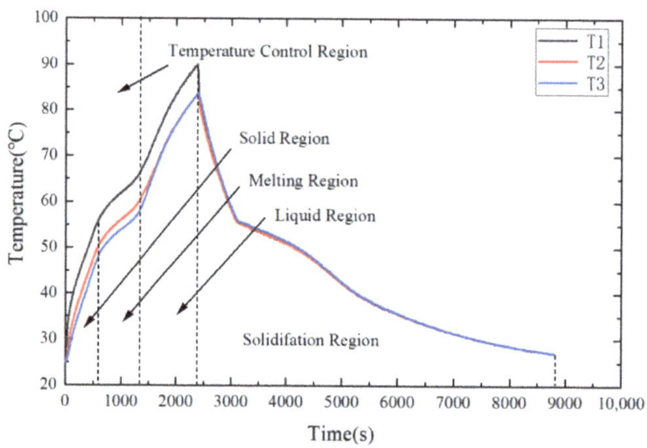

Figure 5. Transient temperature of DPAS/PCM HS at an input power of 11.7 W.

The average temperature increase rates of T1 in the solid and liquid regions were 0.0515 and 0.0225 K/s, respectively. Meanwhile, the average temperature rise rate in the melting region was only 0.0136 K/s, which is 0.604 times that in the liquid-state region and 0.264 times that in the solid-state region. Thus, the latent heat of the phase transition inhibits the temperature increase of the LED in the melting region.

3.2. DPAS/PCM during the Heating Stage

Figure 6 presents a comparison between temperatures T1 of the DPAS/PCM HSs (DPAS6585/PCM, DPAS8565/PCM, DPAS6090/PCM, DPAS9060/PCM) during the heating phase. In the solid-state phase, T1 of the DPAS/PCM HS with a positive gradient porosity was smaller than that of the DPAS/PCM HS with a reverse gradient porosity; specifically, $T1_{60\text{–}90\%} < T1_{90\text{–}60\%}$, $T1_{65\text{–}85\%} < T1_{85\text{–}65\%}$. Furthermore, $T1_{60\text{–}90\%} < T1_{90\text{–}60\%}$, $T1_{65\text{–}85\%} < T1_{85\text{–}65\%}$ is also apparent in the melting stage. This can be attributed to the following reasons: the thermal conductivity of DPAS6585/PCM is 11.99% higher than that of DPAS8565/PCM based on the thermal conductivity results. In addition, the thermal conductivity of DPAS6090/PCM is 34.94% higher than that of DPAS9060/PCM. Therefore, T1 for PCMs with positive-gradient porosity is smaller than that for PCMs with negative-gradient porosity. In the liquid phase, the temperature rise rate of the four DPAS/PCMs is basically the same and ultimately reaches 90 °C.

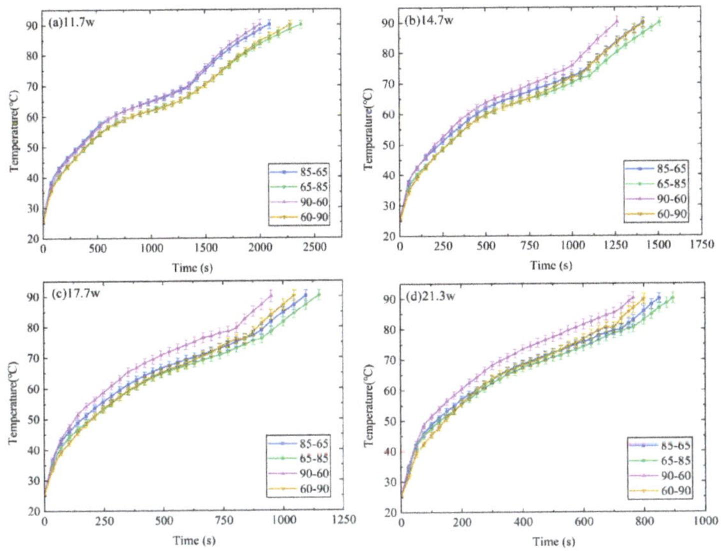

Figure 6. Comparison between T1 of different DPAS/PCM HSs during the heating stage: (**a**) 11.7 W, (**b**) 14.7 W, (**c**) 17.7 W, (**d**) 21.3 W. The T1 of the positive gradient porosity heat sink is smaller than that of the reverse gradient porosity heat sink ($T1_{60\text{–}90\%} < T1_{90\text{–}60\%}$, $T1_{65\text{–}85\%} < T1_{85\text{–}65\%}$).

Figure 7 presents a comparison between temperatures T3 of the different DPAS/PCM HSs during the heating stage. In the solid phase, no significant difference is observed in the transient temperature for all the DPAS/PCMs because the overall porosity of the DPAS/PCMs was the same at 75%, while the heat transfer was dominated by the aluminum skeleton in the solid phase. After entering the melting stage, a significant gap occurs between T3 with the porosity difference of 20% and T3 with the porosity difference of 30%, $T3_{60\text{–}90\%} > T3_{65\text{–}85\%}$, $T3_{90\text{–}60\%} > T3_{85\text{–}65\%}$, which increased as the power increased. As the DPAS/PCM enters the liquid phase, $T3_{70\text{–}80\%}$ gradually approaches $T3_{80\text{–}70\%}$, and the difference in T3 decreases because natural convection dominates the heat transfer. Because high-porosity DPAS/PCMs have larger pores, which are conducive to natural convection.

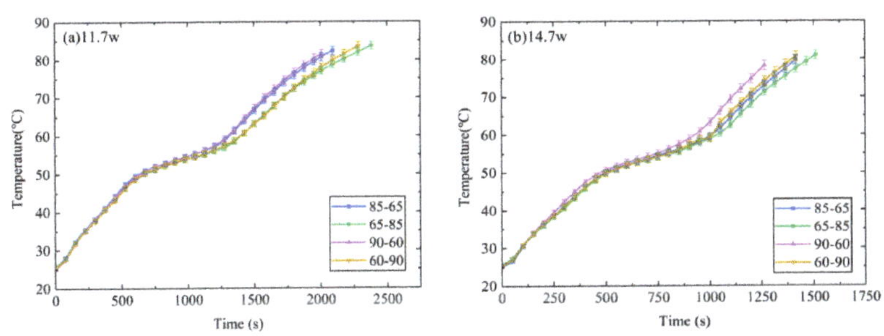

Figure 7. *Cont.*

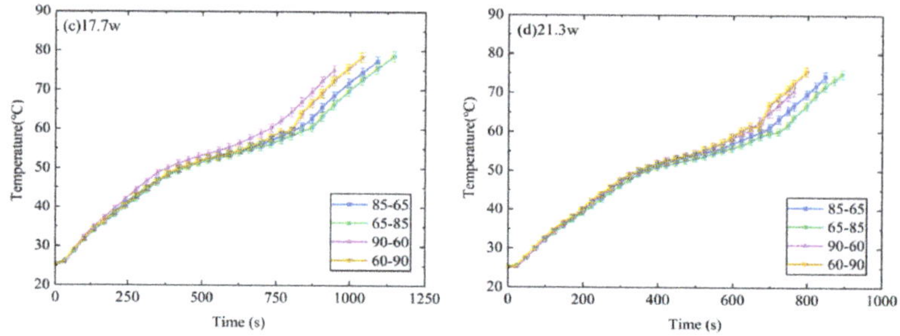

Figure 7. Comparison between T3 of different DPAS/PCM HSs during the heating stage: (**a**) 11.7 W, (**b**) 14.7 W, (**c**) 17.7 W, (**d**) 21.3 W. T3 with a porosity difference of 20% is smaller than that with a porosity difference of 30% at high power.

Figure 8 compares the temperature difference ΔT2 for the various DPAS/PCMs. The ΔT2 value for positive gradient porosity is lower than ΔT2 for reverse gradient porosity at any power. For example, at the power of 11.7 W, $\Delta T2_{65\text{–}85\%}$ is approximately 2 °C lower than $\Delta T2_{85\text{–}65\%}$; $\Delta T2_{60\text{–}90\%}$ and $\Delta T2_{90\text{–}60\%}$ are also 2° lower. The results show that a positive-gradient porosity arrangement is more conducive to achieving a low-temperature cooling target for LED operation.

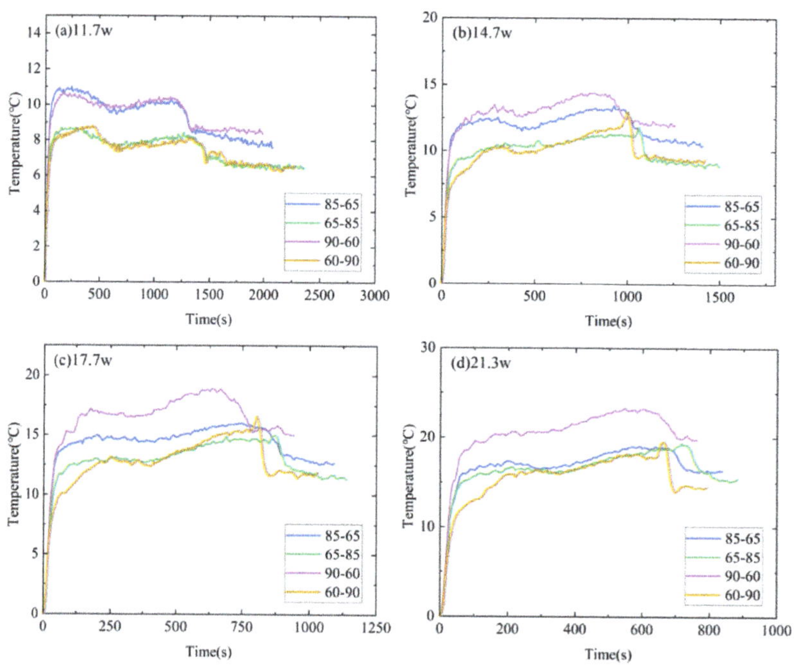

Figure 8. Comparison between ΔT2 of different DPAS/PCM HSs at the heating stage: (**a**) 11.7 W, (**b**) 14.7 W, (**c**) 17.7 W, (**d**) 21.3 W. The ΔT2 of the positive gradient porosity heat sink is smaller than that of the reverse gradient porosity heat sink ($\Delta T2_{60\text{–}90\%} < \Delta T2_{90\text{–}60\%}$, $\Delta T2_{65\text{–}85\%} < \Delta T2_{85\text{–}65\%}$).

3.3. Temperature Control Time Analysis of DPAS/PCM

Figure 9a displays a comparison of the temperature control time for the different DPAS/PCM HSs, with 90 °C as the critical temperature. As shown in Figure 9, all DPAS/PCM HSs exhibit a decrease in temperature control time with increasing power, and the temperature control time for HSs with the porosity difference of 20% (DPAS6585/PCM,

DPAS8565/PCM) is significantly longer than that for HSs with the porosity difference of 30% (DPAS6090/PCM, DPAS9060/PCM). In addition, the HS temperature control time is longer for positive-gradient porosity. From Figure 9a, the temperature control time of DPAS6090 increased by 4.6–13.7% compared with DPAS9060. At the critical temperature of 90 °C, DPAS6585/PCM exhibits the best temperature control effect. Taking the power of 11.7 W as an example, the temperature control times of DPAS6585/PCM were 13.8%, 4.4%, and 18.7% higher than those of DPAS8565/PCM, DPAS6090/PCM, and DPAS9060/PCM, respectively. Meanwhile, at the critical temperature of 60 °C, the DPAS6090/PCM presented the best temperature control at the power of 11.7 W. The temperature control time of the DPAS6090/PCM increased by 0.06%, 27.9% and 27.9% compared to the DPAS6585/PCM, DPAS9565/PCM, and DPAS9060/PCM, respectively. Therefore, to realize the low-temperature operation of LED electronics, a DPAS/PCM HS with a positive gradient porosity and a smaller gradient porosity difference is preferred for temperature control.

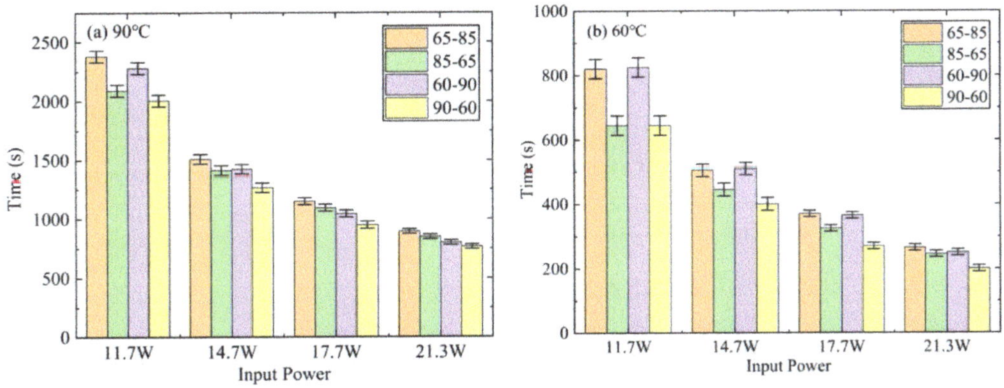

Figure 9. Temperature control time of DPAS/PCM HS at critical temperatures of (**a**) 90 °C and (**b**) 60 °C. At critical temperatures of 90°C, DPAS65-85/PCM has the longest temperature control time.

4. Comparative Study of DPAS/PCM and UAS/PCM

4.1. Comparison at Heating Stage

Figure 10 presents a comparison of T1 for the DPAS/PCM and UAS/PCM HSs during the heating stage. In the solid-state phase, T1 of the DPAS/PCM HS is lower than that of the UAS/PCM HS at a higher power. In addition, T1 of the DPAS/PCM HS is lower than that of the UAS/PCM HS at a higher power during the melting stage. The reason for the lower T1 of the DPAS6585/PCM at high power is because, compared with the UAS/PCM HS with 75% porosity, the positive gradient porosity structure of the DPAS6585/PCM provides high thermal conductivity in the lower part and relatively good natural convection in the upper part. Consequently, the combination of these two factors results in the lower T1 for the DPAS/PCM HSs. In addition, based on the thermal conductivity results, the thermal conductivity of DPAS6585/PCM is 2.6% higher than that of UAS/PCM. Therefore, T1 of the DPAS/PCM HS is lower than that of the UAS/PCM HS. The temperature control time for the DPAS/PCM HS is longer than that for the UAS/PCM HS because a larger volume of paraffin with higher porosity was used in the upper layer of the DPAS/PCM HS.

Figure 11 compares the temperature difference ΔT2 for the DPAS/PCM and UAS/PCM HSs. It can be seen that $\Delta T2_{65\text{–}85\%} < \Delta T2_{75\%}$ at high power; $\Delta T2_{65\text{–}85\%}$ is 1~5 °C lower than $\Delta T2_{75\%}$, which indicates that DPAS/PCM HS is more suitable for LEDs operating at high power.

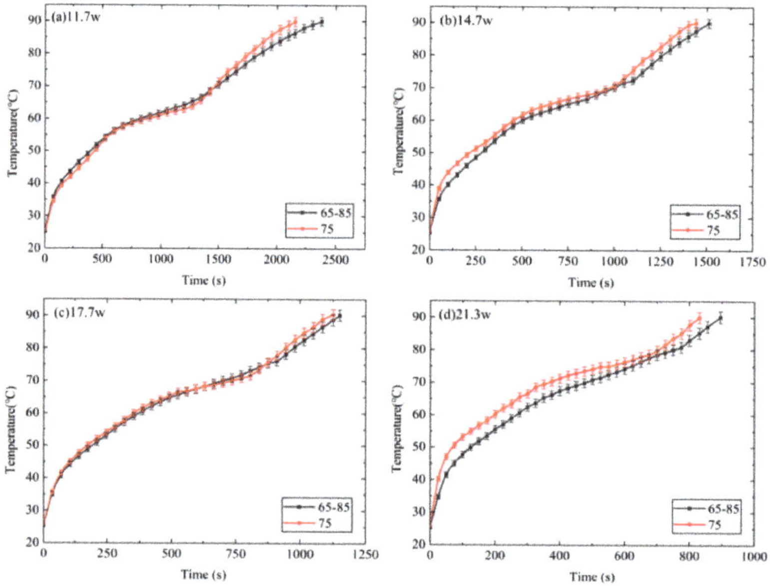

Figure 10. Comparison of T1 for DPAS/PCM and UAS/PCM HSs during the heating stage considering powers of (**a**) 11.7 W, (**b**) 14.7 W, (**c**) 17.7 W, (**d**) 21.3 W. The temperature control time for the DPAS/PCM heat sink was longer than for the UAS/PCM heat sink.

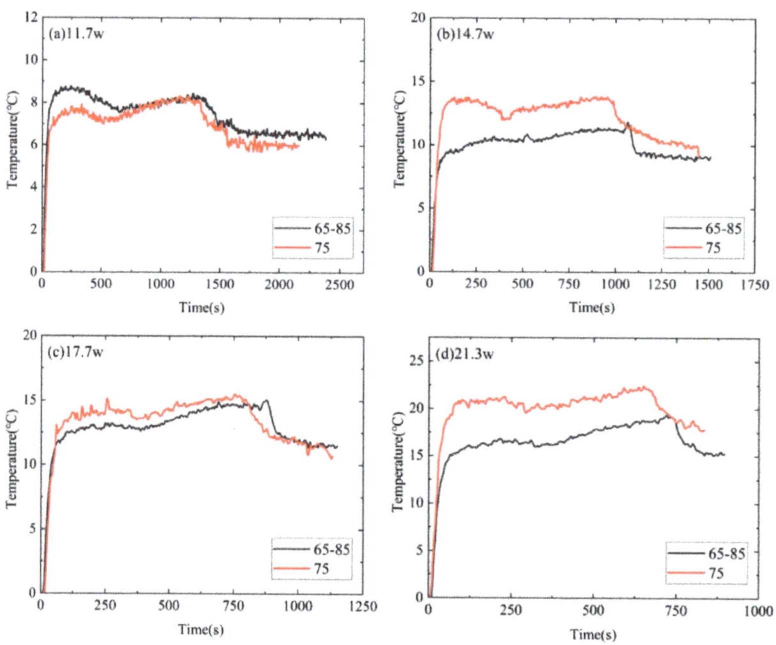

Figure 11. Comparison of ΔT2 for DPAS/PCM and UAS/PCM HSs during the heating stage at powers of (**a**) 11.7 W, (**b**) 14.7 W, (**c**) 17.7 W, (**d**) 21.3 W. The DPAS/PCM heat sink is more suitable for LEDs operating at high power.

4.2. Comparison at Cooling Stage

Figure 12 presents a comparison of the cooling curves (T1) for the DPAS/PCM and UAS/PCM HSs. The cooling time of the UAS/PCM HS with 75% porosity was 7225 s. The cooling times for the DPAS6585/PCM, DPAS8565/PCM, DPAS6090/PCM, and DPAS9060/PCM HSs were 6440, 6330, 6005, and 5690 s, respectively. Moreover, the results show that the cooling process of the DPAS/PCM HS was shorter than that of the UAS/PCM HS. The cooling time of the DPAS/PCM HS is 12.19~26.98% less than that of the UAS/PCM HS. In addition, the cooling time of the HS with reverse-gradient porosity was shorter than that of the HS with positive-gradient porosity. For example, the cooling time of the DPAS8565/PCM HS is 100 s shorter than that of the DPAS6585/PCM HS, while that of the DPAS9060/PCM HS is 315 s shorter than that of the DPAS6090/PCM HS.

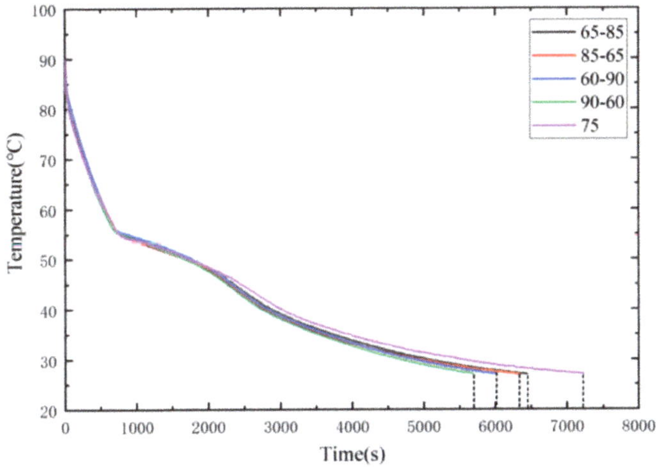

Figure 12. Comparison of cooling curves (T1) for DPAS/PCM and UAS/PCM HS. The cooling process of the DPAS/PCM heat sink is shorter than that of the UAS/PCM heat sink.

4.3. Comparison of Temperature Control Times

Figure 13 presents a comparison of the temperature-control times for the DPAS/PCM and UAS/PCM HSs. As shown in Figure 13, the temperature control time of the DPAS/PCM HS was significantly prolonged compared with that of the UAS/PCM HS. At the critical temperature of 90 °C, the temperature control effect of the DPAS/PCM is better than that of the UAS/PCM HS, but the gap gradually reduces with increasing power. The temperature control time of the DPAS/PCM increased by 12.5% at 11.7 W compared with that of the UAS/PCM HS, whereas it only increased by 7.8% at 21.3 W. Meanwhile, at the critical temperature of 60 °C, the gap between DPAS/PCM and UAS/PCM HSs became larger with increasing power. Uniform HSs are preferred under low-power conditions; i.e., at 11.7 W, the temperature control time of DPAS/PCM HSs is 7.8% shorter than that of UAS/PCM HSs. Under high-power conditions, the DPAS/PCM HS is preferred, and the temperature time of the DPAS/PCM HS is 32.5% longer than that of UAS/PCM HS at 21.3 W, for example. Therefore, UAS/PCM or DPAS/PCM HSs are selected in engineering applications depending on the power and operating temperatures of the electronic devices.

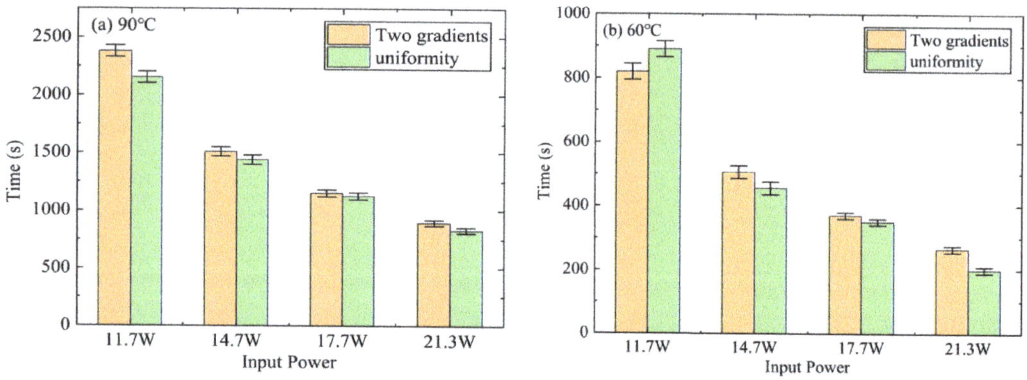

Figure 13. Temperature control time comparison between DPAS/PCM and UAS/PCM at critical temperatures of (**a**) 90 °C and (**b**) 60 °C. The temperature control time of the double-porosity heat sink is better than that of the uniformity heat sink under high-power conditions.

5. Conclusions

(1) A DPAS/PCM is proposed to solve the problem of low thermal conductivity of the PCM, and the positive-gradient porosity structure of the DPAS/PCM is more conducive to achieving low-temperature LED operation, mainly because of the high heat conduction near the heat source and the lowest resistance of natural convection in the upper part;

(2) The positive- and reverse-gradient porosity structures of the DPAS/PCM are very sensitive to the thermal performance. The LED temperature of the DPAS/PCM HSs with a positive-gradient porosity was lower than that of the DPAS/PCM HSs with a reverse-gradient porosity;

(3) Compared with the UAS/PCM HS, the DPAS/PCM HS is preferred under high-power conditions, and the temperature of the DPAS/PCM HS was 32.5% higher than that of the UAS/PCM HS at 21.3 W.

The results will provide some reference for the optimal design of the porous skeleton, which is the research to be focused on for future work.

Author Contributions: S.H. was responsible for writing the research results and conducting the primary experiments; C.L. and Z.H. contributed to the article review and provided the experimental conditions; Y.X., B.Z. and C.Z. assisted in completing the necessary experimental investigations. All authors have read and agreed to the published version of the manuscript.

Funding: This work was financially supported by the National Natural Science Foundation of China (52065003); Jiangxi Provincial Natural Science Foundation (20202BAB204021, 20212BAB204001); Jiangxi Engineering Province Engineering Research Center of New Energy Technology and Equipment, East China University of Technology (JXNE2020-04, JXNE2021-07); Postdoctoral Research Funding project of Jiangxi, China (2021KY50); and Engineering Research Center of Nuclear Technology Application, Ministry of Education, China (HJSJYB2021-2, HJSJYB2021-3).

Data Availability Statement: All data supporting the results of this study have been included in this article.

Conflicts of Interest: The authors declare no conflicts of interest.

References

1. Amano, H. Development of GaN-based blue LEDs and metalorganic vapor phase epitaxy of GaN and related materials. *Prog. Cryst. Growth Charact. Mater.* **2016**, *62*, 126–135. [CrossRef]
2. Qi, Z.; Wang, L.; Liang, Y.; Liu, P.; Zhu, H.; Wang, Y. Deep-ultraviolet light communication in sunlight using 275-nm LEDs. *Appl. Phys. Lett.* **2023**, *123*, 161109. [CrossRef]
3. Zhang, S.; Li, W.; Wang, K.; Guo, Y.; Gong, Z.; Zhang, Z.; Zhang, Y.; Zhou, X.; Guo, T.; Wu, C. Noncarrier Injection Mode for Realizing One Line-to-Three LEDs Driving Method. *IEEE Trans. Electron Devices* **2024**, *71*, 2486–2490. [CrossRef]

4. Liu, J.; Tang, M.; Deng, H.; Shutts, S.; Wang, L.; Smowton, P.; Jin, C.-Y.; Chen, S.; Seeds, A.; Liu, H. Theoretical Analysis and Modelling of Degradation for III-V Lasers on Si. *J. Phys. D Appl. Phys.* **2022**, *55*, 404006. [CrossRef]
5. Hosseinizadeh, S.F.; Tan, F.L.; Moosania, S.M. Experimental and Numerical Studies on Performance of PCM-Based Heat Sink with Different Configurations of Internal Fins. *Appl. Therm. Eng.* **2011**, *31*, 3827–3838. [CrossRef]
6. Mikulics, M.; Kordoš, P.; Fox, A.; Kočan, M.; Lüth, H.; Sofer, Z.; Hardtdegen, H. Efficient heat dissipation in AlGaN/GaN heterostructure grown on silver substrate. *Appl. Mater. Today* **2017**, *7*, 134–137. [CrossRef]
7. Zhao, M.; Tang, X.; Huo, W.; Han, L.; Deng, Z.; Jiang, Y.; Wang, W.; Chen, H.; Du, C.; Jia, H. Characteristics of AlGaN/GaN high electron mobility transistors on metallic substrate. *Chin. Phys. B* **2020**, *29*, 048104. [CrossRef]
8. Peng, Y.; Mou, Y.; Wang, T.; Wang, H.; Liang, R.; Wang, X.; Chen, M.; Luo, X. Effective Heat Dissipation of QD-Based WLEDs by Stacking QD Film on Heat-Conducting Phosphor-Sapphire Composite. *IEEE Trans. Electron Devices* **2019**, *66*, 2637–2642. [CrossRef]
9. Sajjad, U.; Sadeghianjahromi, A.; Ali, H.M.; Wang, C.C. Enhanced pool boiling of dielectric and highly wetting liquids—A review on surface engineering. *Appl. Therm. Eng.* **2021**, *195*, 117074. [CrossRef]
10. McGlen, R.J.; Jachuck, R.; Lin, S. Integrated thermal management techniques for high power electronic devices. *Appl. Therm. Eng.* **2004**, *24*, 11431156. [CrossRef]
11. Yin, Z.; Zhang, X.; Huang, Z.; Liu, S.; Zhang, W.; Liu, Y.; Wu, X.; Fang, M.; Min, X. Paraffin/expanded graphite phase change composites with enhanced thermal conductivity prepared by implanted β-SiC nanowires with chemical vapor deposition method. *Mater. Res. Express* **2018**, *5*, 025503. [CrossRef]
12. Wang, H.; Wang, F.; Li, Z.; Tang, Y.; Yu, B.; Yuan, W. Experimental Investigation on the Thermal Performance of a Heat Sink Filled with Porous Metal Fiber Sintered Felt/Paraffin Composite Phase Change Material. *Appl. Energy* **2016**, *176*, 221–232. [CrossRef]
13. Meng, E.; Cai, R.; Sun, Z.; Yang, J.; Wang, J. Experimental Study of the Passive and Active Performance of Real-Scale Composite PCM Room in Winter. *Appl. Therm. Eng.* **2021**, *185*, 116418. [CrossRef]
14. Samimi, F.; Babapoor, A.; Azizi, M.; Karimi, G. Thermal Management Analysis of a Li-Ion Battery Cell Using Phase Change Material Loaded with Carbon Fibers. *Energy* **2016**, *96*, 355–371. [CrossRef]
15. Mousavi, S.; Siavashi, M.; Zadehkabir, A. A New Design for Hybrid Cooling of Li-Ion Battery Pack Utilizing PCM and Mini Channel Cold Plates. *Appl. Therm. Eng.* **2021**, *197*, 117398. [CrossRef]
16. Raj, C.R.; Suresh, S.; Singh, V.K.; Bhavsar, R.R.; Vasudevan, S.; Archita, V. Experimental Investigation on Nanoalloy Enhanced Layered Perovskite PCM Tamped in a Tapered Triangular Heat Sink for Satellite Avionics Thermal Management. *Int. J. Therm. Sci.* **2021**, *167*, 107007. [CrossRef]
17. Li, Z.; Wu, Y.; Zhuang, B.; Zhao, X.; Tang, Y.; Ding, X.; Chen, K. Preparation of Novel Copper-Powder-Sintered Frame/Paraffin Form-Stable Phase Change Materials with Extremely High Thermal Conductivity. *Appl. Energy* **2017**, *206*, 1147–1157. [CrossRef]
18. Fang, X.; Fan, L.-W.; Ding, Q.; Yao, X.-L.; Wu, Y.-Y.; Hou, J.-F.; Wang, X.; Yu, Z.-T.; Cheng, G.-H.; Hu, Y.-C. Thermal Energy Storage Performance of Paraffin-Based Composite Phase Change Materials Filled with Hexagonal Boron Nitride Nanosheets. *Energy Convers. Manag.* **2014**, *80*, 103–109. [CrossRef]
19. Mahdi, J.M.; Lohrasbi, S.; Ganji, D.D.; Nsofor, E.C. Accelerated Melting of PCM in Energy Storage Systems via Novel Configuration of Fins in the Triplex-Tube Heat Exchanger. *Int. J. Heat Mass Transf.* **2018**, *124*, 663–676. [CrossRef]
20. Tamraparni, A.; Hoe, A.; Deckard, M.; Zhang, C.; Elwany, A.; Shamberger, P.J.; Felts, J.R. Design and Optimization of Lamellar Phase Change Composites for Thermal Energy Storage. *Adv. Eng. Mater.* **2021**, *23*, 2001052. [CrossRef]
21. Zhou, J.; Zhao, J.; Li, H.; Cui, Y.; Li, X. Enhanced Thermal Properties for Nanoencapsulated Phase Change Materials with Functionalized Graphene Oxide (FGO) Modified PMMA. *Nanotechnology* **2020**, *31*, 295704. [CrossRef] [PubMed]
22. Ghalambaz, M.; Zhang, J. Conjugate Solid-Liquid Phase Change Heat Transfer in Heatsink Filled with Phase Change Material-Metal Foam. *Int. J. Heat Mass Transf.* **2020**, *146*, 118832. [CrossRef]
23. Zhu, M.; Wang, Z.; Zhang, H.; Sun, X.; Dou, B.; Wu, W.; Zhang, G.; Jiang, L. Experimental Investigation of the Comprehensive Heat Transfer Performance of PCMs Filled with CMF in a Heat Storage Device. *Int. J. Heat Mass Transf.* **2022**, *188*, 122582. [CrossRef]
24. Yu, S.; Wang, X.; Wu, D. Microencapsulation of n-octadecane Phase Change Material with Calcium Carbonate Shell for Enhancement of Thermal Conductivity and Serving Durability: Synthesis, Microstructure, and Performance Evaluation. *Appl. Energy* **2014**, *114*, 632–643. [CrossRef]
25. Sahoo, S.K.; Das, M.K.; Rath, P. Application of TCE-PCM Based Heat Sinks for Cooling of Electronic Components: A Review. *Renew. Sustain. Energy Rev.* **2016**, *59*, 550–582. [CrossRef]
26. Ali, H.M.; Arshad, A.; Jabbal, M.; Verdin, P.G. Thermal Management of Electronics Devices with PCMs Filled Pin-Fin Heat Sinks: A Comparison. *Int. J. Heat Mass Transf.* **2018**, *117*, 1199–1204. [CrossRef]
27. Arshad, A.; Ali, H.M.; Yan, W.-M.; Hussein, A.K.; Ahmadlouydarab, M. An Experimental Study of Enhanced Heat Sinks for Thermal Management Using n-eicosane as Phase Change Material. *Appl. Therm. Eng.* **2018**, *132*, 52–66. [CrossRef]
28. Qureshi, F.A.; Ahmad, N.; Ali, H.M. Heat Dissipation in Bituminous Asphalt Catalyzed by Different Metallic Oxide Nanopowders. *Constr. Build. Mater.* **2021**, *276*, 122220. [CrossRef]
29. Khan, M.A.; Imam, M.K.; Irshad, K.; Ali, H.M.; Hasan, M.A.; Islam, S. Comparative Overview of the Performance of Cementitious and Non-cementitious Nanomaterials in Mortar at Normal and Elevated Temperatures. *Nanomaterials* **2021**, *11*, 911. [CrossRef]

30. Mishra, S.R.; Mathur, P.; Ali, H.M. Analysis of Homogeneous–Heterogeneous Reactions in a Micropolar Nanoflid past a Nonlinear Stretching Surface: Semi-analytical Approach. *J. Therm. Anal. Calorim.* **2021**, *144*, 2247–2257. [CrossRef]
31. Zehri, A.; Samani, M.K.; Latorre, M.G.; Nylander, A.; Nilsson, T.; Fu, Y.; Wang, N.; Ye, L.; Liu, J. High Porosity and Light Weight Graphene Foam Heat Sink and Phase Change Material Container for Thermal Management. *Nanotechnology* **2020**, *31*, 424003. [CrossRef] [PubMed]
32. Baby, R.; Balaji, C. Experimental Investigations on Thermal Performance Enhancement and Effect of Orientation on Porous Matrix Filled PCM Based Heat Sink. *Int. Commun. Heat Mass Transf.* **2013**, *46*, 27–30. [CrossRef]
33. Rehman, T.; Ali, H.M. Experimental Study on the Thermal Behavior of RT-35HC Paraffi Within Copper and Iron-Nickel Open Cell Foams: Energy Storage for Thermal Management of Electronics. *Int. J. Heat Mass Transf.* **2020**, *146*, 118852. [CrossRef]
34. Rehman, T.; Ali, H.M. Thermal Performance Analysis of Metallic Foam-Based Heat Sinks Embedded with RT-54HC Paraffi: An Experimental Investigation for Electronic Cooling. *J. Therm. Anal. Calorim.* **2020**, *140*, 979–990. [CrossRef]

micromachines

Article

An Experimental Study of a Composite Wick Structure for Ultra-Thin Flattened Heat Pipes

Wenjie Zhou [1,2,*], Yong Yang [2], Junfeng He [2], Ri Chen [2], Yue Jian [2], Dan Shao [1,*] and Aihua Wu [1]

1 Guangdong Key Laboratory of Battery Safety, Guangzhou Institute of Energy Testing, Guangzhou 511447, China; wuahgz@163.com
2 School of Mechatronic Engineering, Guangdong Polytechnic Normal University, Guangzhou 510450, China; yy2008@gpnu.edu.cn (Y.Y.); hejunfeng@gpnu.edu.cn (J.H.); rchen01@gpnu.edu.cn (R.C.); jianyue@gpnu.edu.cn (Y.J.)
* Correspondence: zhouwj@gpnu.edu.cn (W.Z.); shaod1005@163.com (D.S.)

Abstract: As the thickness of an ultra-thin flattened heat pipe (UTHP) decreases, the fabrication difficulty increases exponentially, and the thermal performance deteriorates rapidly. In this study, three types of composite wicks were developed for UTHPs with a 0.6 mm thickness: copper foam and mesh wick (CFMW), two layers of different mesh wick (TDMW), and three layers of the same mesh wick (TSMW). The startup and steady-state performances of the UTHPs with liquid filling ratios of 60% to 120% were investigated. The findings indicated that the CFMW UTHP with a filling ratio of 100% exhibited the best startup performance, with the highest equilibrium temperature of 58.37 °C. The maximum heat transport capacities of the CFMW, TDMW, and TSMW UTHP samples were 9, 8, and 8.5 W, respectively, at their corresponding optimum filling ratios of 110%, 90%, and 100%. The CFMW UTHP exhibited the lowest evaporation and condensation thermal resistances of 0.151 and 0.189 K/W, respectively, which were 24.67% and 41.85% lower than those of the TSMW UTHP. CFMW can be used to improve the thermal performance of UTHPs. This study provides important guidelines for the structural design, fabrication technology, and performance improvement of high-performance UTHPs used in portable electronic devices.

Keywords: heat pipe; wick; filling ratio; thermal performance

1. Introduction

As portable electronic devices have become smaller and more powerful, the operating chips generate more and more heat, making it difficult to dissipate in a timely manner. This results in elevated temperatures within these devices, which in turn constrains their further development [1–3]. Ultra-thin flattened heat pipes (UTHPs) exhibit a number of advantages, including small size, lightweight, quick startup performance, good temperature uniformity, and customizable shape [4,5]. Therefore, the rational use of high-performance UTHPs can effectively solve the cooling issue in these devices, reduce the operating temperature, and enhance reliability and safety [6,7]. The heat generated by chips is rapidly transferred to the device's bracket via UTHPs, and then it is distributed to the surrounding environment through natural convection, thus eliminating the local overheating. Consequently, the thickness and performance of UTHPs are key factors in determining the effectiveness of heat dissipation in devices.

The thickness of a flattened heat pipe is determined by the parameters of the copper tube (e.g., diameter, wall thickness, and length), the manufacturing process, and the application requirements [8]. As the thickness of the UTHP decreases, the difficulty of manufacturing increases exponentially, and the heat transfer performance deteriorates rapidly. Li et al. [9] developed a UTHP with a bilateral arch-shaped sintered wick. The impact of the wick structure, flattened thickness, and particle size on its thermal performance was investigated. The findings indicated that a reduction in the internal cavity results in a notable increase in the flow resistance of the liquid and vapor, accompanied by a corresponding decrease in the thermal performance. Lin and Wong [10] conducted an investigation into the causes of performance degradation in flattened heat pipes. The findings revealed that a liquid accumulation inside the condensation section was the primary factor leading to the deterioration in the performance of the powder-sintered heat pipe, while the main reason for the grooved heat pipe was the reduction in liquid supply to the evaporation section due to the vapor–liquid interfacial shear [11,12], which resulted in the tendency for the heat pipe to dry out. Lurie et al. [13] analyzed a flat heat pipe with a sintered wick using a topology optimization method. The findings demonstrated that the utilization of the optimal wick can enhance its operational performance. In particular, the heat transport capacity was found to be twice that of the heat pipe with a conventional wick.

The wick structure is one of the most critical factors influencing the thermal performance of UTHPs. Yang et al. [14] developed a 1 mm thick UTHP with a copper wire braided wick, which was composed of small-diameter and large-diameter copper wires. Tang et al. [15] studied a novel mesh wick made from copper fine lines. The findings indicated that the wick exhibited a greater capillary force than the conventional wicks. Zhou et al. [16–18] developed a variety of spiral woven mesh wick (SWM) structures for use in UTHPs, with the objective of meeting the heat dissipation requirements of a variety of electronic devices. Specifically, a band-shaped SWM structure was designed for UTHPs with a thickness of 0.4 mm to enhance the cooling capacity of smartphones [16]. Additionally, a hybrid SWM wick structure with two diameters of copper wires in strands was fabricated to improve the thermal performance [17], and the effect of the combination of copper wires with two diameters on the thermal performance of UTHPs was studied [18]. The findings indicated that the optimized hybrid SWM wick structure can enhance the heat transport capacity of UTHPs by 53.85%.

The mass of the working fluid exerts a considerable influence on thermal performance. It can be demonstrated that heat pipes with a similar wick structure will only exhibit optimal performance when the liquid filling ratio is appropriate. Li et al. [19] investigated the thermal performance of UTHPs with different filling ratios using a visualization method. The findings showed that a filling ratio of 15% was optimal. If the filling ratio was excessive, the liquid would boil in the evaporator at a higher heating power, and the liquid bridge phenomenon would occur with the liquid subsequently gathering in the condensation section. Chen and Chou [20] designed a flat heat pipe to study the influence of the filling ratio on its thermal performance. The findings indicated that 25% was the optimal filling ratio. Tharayil et al. [21] developed a loop heat pipe to investigate the optimal filling ratio. The experimental findings showed that the optimal value was 30%.

Considering the aforementioned studies, it is evident that numerous scholars have conducted valuable research utilizing theoretical calculations, visualization studies, and experimental analyses. However, there is a paucity of research on the systematic study of UTHP structural design, manufacturing processes, and heat transfer performance. As the thickness of a UTHP decreases, the fabrication of the material and the enhancement of its thermal performance become increasingly challenging. In particular, for thinner UTHPs, the effects of numerous factors, including the material, structure, processes, wick dimension, and working fluid mass, on thermal performance are relatively more significant than those of thicker UTHPs. Consequently, in this study, the manufacturing process and thermal performance of UTHPs with a thickness of 0.6 mm were investigated. Three types of composite wicks were employed in the fabrication of thin-walled cylindrical heat pipes, which were subsequently flattened to manufacture the UTHPs. The startup and steady-state performance levels of the UTHP samples were tested and discussed, with particular consideration given to the effect of heating power, filling ratio, and wick structure. The findings of this study are of significant value in the fields of structural design and fabrication technology regarding performance improvement in high-performance UTHPs, which is a crucial component in portable electronic devices.

2. Experimental Samples and Test System

2.1. Fabrication of the UTHP Samples

The composite sintered wick structures of the UTHP samples were as follows: copper foam and mesh wick (CFMW), two layers of different mesh wick (TDMW), and three layers of the same mesh wick (TSMW). The three types of composite wicks were constructed using copper foam, 100-mesh, and 200-mesh copper wire meshes, respectively. The thicknesses of the three structures were 0.2, 0.1, and 0.2 mm, respectively. The porosity of the sintered copper foam was 60%. The copper wire meshes were orthogonal woven meshes (OWMs). The CFMW was produced by sintering a layer of copper foam and a layer of 200-mesh copper wire mesh. The TDMW was sintered from a single layer of 100-mesh and a single layer of 200-mesh copper wire mesh. The TSMW was manufactured by sintering three layers of 200-mesh copper wire mesh. Consequently, the thickness of all three composite wicks after sintering was 0.3 mm.

Figure 1 illustrates the cylindrical heat pipe following the shrinking and welding processes. The diameter and length of cylindrical heat pipes were 6 and 130 mm, respectively. The UTHP samples were manufactured through a series of processes, including cutting and shrinking the copper tubes, sintering the composite wicks, welding the tail ends, annealing the copper tubes, filling the working fluids, vacuuming and degassing the copper tubes, welding the head ends, and flattening the cylindrical heat pipes. The tube shrinking process for conventional heat pipes was used in the radial forging process (RFP). During the RFP, a three-phase asynchronous motor drives two- or three-split dies in a circumrotation and radial reciprocating movement, while the dies shape the tube with a high-frequency, short-stroke movement. The RFP is generally suitable for the shaping of copper tubes with a thickness of more than 0.20 mm. However, the copper tube thickness for the experimental UTHP was only 0.15 mm, and the strength was found to be poor, particularly after the annealing process. This is because the hard copper tube becomes soft.

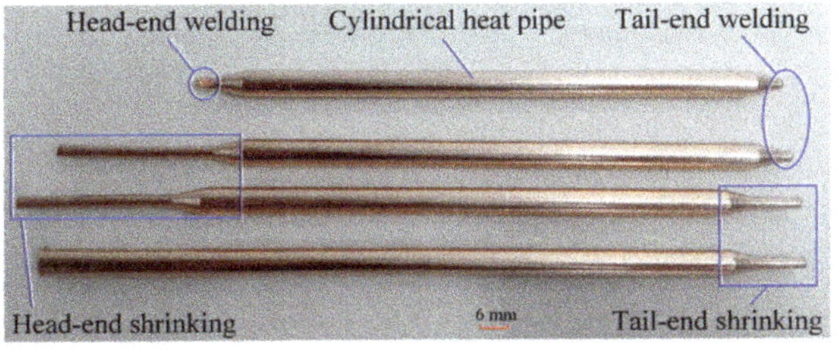

Figure 1. Shrinking and welding processes of cylindrical heat pipes.

The shaping section of the UTHP was susceptible to bending and breaking in the RFP. In order to avoid these serious defects, a novel shrinking process, namely the high-speed spinning process (HSSP), was used to form the experimental copper tube. During the HSSP, a high-speed spindle motor drives an integral die to rotate at a speed of approximately 10,000 rpm, while the die is fed to the tube and the tube is shaped under the synthetic movement of the die. The working area exhibited a high degree of smoothness following the HSSP. The vacuum process was conducted on an automated primary degassing machine. The UTHP samples were vertically inserted into a sealed chamber with a sealing structure that was connected through a tube to a vacuum pump with a limiting vacuum of 0.08 Pa. A digital vacuum gauge was installed on the pipe. As the tube was connected to the internal cavity of the UTHP during the vacuuming process, the vacuum gauge reading could also be used to measure the internal pressure of the UTHP sample. During the actual batch production of UTHP samples, the vacuum gauge reading was 10^{-2}–10^{1} Pa, a vacuum level that fully met the requirements of UTHPs.

Following a three-hour sintering process in a high-temperature reducing atmosphere (950 °C, nitrogen–hydrogen mixture gas), the composite wick was attached to the inner wall of the copper tube. As illustrated in Figure 2, scanning electron microscopy (SEM, Zeiss EVO 18 Special Edition from Oberkochen, Germany) was used to examine the microstructures of the three wicks. The samples for SEM observation were prepared by cutting the UTHPs longitudinally with a precision wire-cutting machine. The width and thickness of the CFMW, TDMW, and TSMW were 4.0 and 0.3 mm, respectively. The porosities of the three wicks were measured as 72%, 77%, and 60%, respectively.

Figure 2. SEM photographs of longitudinal sections of the composite wick structures: (**a**) CFMW, (**b**) TDMW, and (**c**) TSMW.

The working fluids in the experimental UTHP samples were deionized water with volumetric filling ratios of 60%, 70%, 80%, 90%, 100%, 110%, and 120%. A value of 100% signified that the wick was saturated, indicating that all pores were filled with deionized water. Figure 3 depicts the UTHP samples. The length and thickness of the samples were 130 and 0.6 mm, respectively. The UTHP samples were designated as CFMW UTHP, TDMW UTHP, and TSMW UTHP, in accordance with the wick structures.

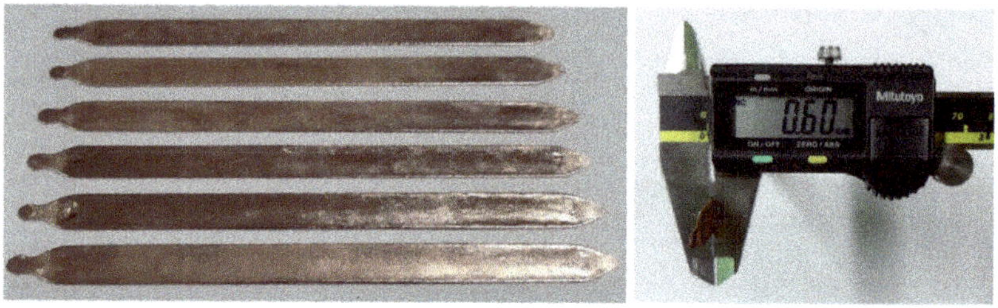

Figure 3. UTHP samples.

2.2. Startup Performance Testing of UTHP

Figure 4 shows the startup performance test system for UTHP samples, which comprised a constant temperature water tank, a positioning block, and a temperature data acquisition system. The sample was positioned vertically during the experiment. The tail end was immersed in the water stored in the tank at a temperature of 60 ± 0.1 °C, with an insertion depth of 30 mm. The head end was maintained in the air. A thermocouple point was established at a distance of 10 mm from the top of the UTHP. The temperature value of the point was collected and recorded using an acquisition system (NI 9213 from Austin, TX, USA). The time from the start of the test to equilibrium was the start time of the UTHP. The equilibrium state was defined as a temperature variation of the point of less than 0.3 °C over a period of 5 s. The ambient environment temperature was 26 ± 1 °C.

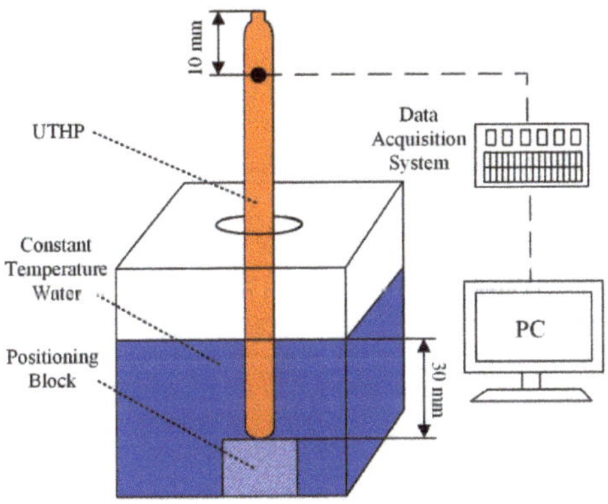

Figure 4. Schematic diagram of the startup performance test system for UTHP samples.

2.3. Steady-State Performance Testing of UTHP

Figure 5 presents the steady-state performance test system for UTHP samples. The system was primarily composed of a heating block, a DC power supply, a cooling block, a water-cooling bath system, and a temperature data acquisition system. A heating rod was inserted into the copper heating block. The heating rod was equipped with a power supply that exhibited an accuracy of 0.5% for current and 0.1% for voltage. The cooling block was cooled by the water at a constant temperature, which was controlled to within a tolerance of 0.2 °C using the water-cooling bath. The water flow rate was regulated with a flow meter with a precision of 2 L per hour. The temperature data were obtained at nine test points by means of T-type thermocouples. Following calibration, the T-type thermocouples had an accuracy of 0.1 °C for temperature. One end of each thermocouple was in contact with the surface of the UTHP, while the other end was connected to an acquisition card with an accuracy of 0.02 °C. During the test, the UTHP sample was positioned horizontally, with its head end and tail end placed on the cooling block and heating block, respectively. The application of thermal grease with a 6.0 W/(m·K) thermal conductivity resulted in a reduction in the contact thermal resistance. A polyurethane block with two test points (T_1 and T_2) was subjected to compression in the evaporation section. The adiabatic section was equipped with four test points, designated T_3, T_4, T_5, and T_6. The condensation section was compressed by the other polyurethane block, which had three test points: T_7, T_8, and T_9. The cooling water was supplied at a rate of 20 L per hour through the cooling block, with a water temperature of approximately 55 °C.

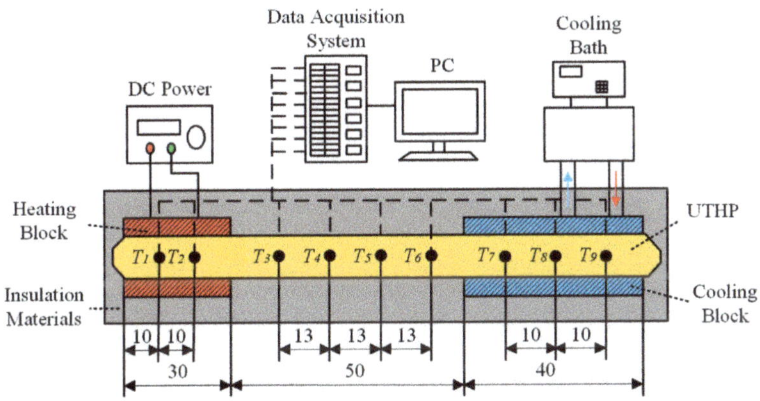

Figure 5. Schematic diagram of the steady-state performance test system (units: mm).

During the test, the UTHPs were insulated with thermal insulation materials to prevent heat loss. The heating power exhibited a range of values between 4.5 and 9 W, with each increment representing a 0.5 W increase. Once the heating power had been adjusted to initiate the test, the test PC recorded the temperature data of all test points until the UTHP sample reached equilibrium. This is defined as a period during which the temperature fluctuation of the points is less than 0.5 °C within 120 s. The maximum heat transport capacity is defined as the maximum heating power. This is the heating power at which the sample continues to operate in a normal state, that is, the evaporation section does not dry out with a rapid rise in temperature, and the temperature difference among the test points is less than 5 °C.

The average temperatures (T_e, T_a, and T_c) of the evaporation, adiabatic, and condensation sections of a UTHP sample were obtained using the following equations:

$$T_e = \frac{1}{2}(T_1 + T_2), \tag{1}$$

$$T_a = \frac{1}{4}(T_3 + T_4 + T_5 + T_6), \tag{2}$$

$$T_c = \frac{1}{3}(T_7 + T_8 + T_9), \tag{3}$$

The thermal resistances (R_c and R_e) of the condensation and evaporation sections of a UTHP sample were obtained as follows:

$$R_c = \frac{T_a + T_c}{P_h}, \tag{4}$$

$$R_e = \frac{T_e + T_a}{P_h}, \tag{5}$$

where P_h is the heating power.

2.4. Analysis of the Testing Errors Associated with the Test System

The testing error of the steady-state performance test system is mainly derived from the heating and temperature acquisition system. The relative testing error can be obtained using the following equations [22]:

$$\frac{E[y(x)]}{y(x)} = \frac{\sqrt{\sum_{i=1}^{n}\left(\frac{\partial y}{\partial x_i}E_{x_i}\right)^2}}{y(x)}, \tag{6}$$

where x is an independent variable; $y(x)$ is a given function of x; and E_x and $E[y(x)]$ are the absolute testing errors of x and $y(x)$, respectively.

According to these calculations, the relative testing error of the heating power was 0.2%, and the maximum relative testing error of the thermal resistance was 6.28%.

3. Results and Discussion

3.1. Startup Performance of the UTHP Samples

Figure 6 illustrates the startup performance of the UTHP samples with CFMW, TDMW, and TSMW, under varying liquid filling ratios. The temperature of the UTHP test point exhibited a rapid increase and subsequent stabilization with the increase in testing time. The startup performance of the UTHP was also influenced by the changes in the filling ratios. The UTHP samples with CFMW, TDMW, and TSMW demonstrated optimal startup performance at the corresponding fill rates of 100%, 80%, and 80%. This result indicates that the test point temperature increased the most rapidly, while the stable temperature value was the highest. The startup performance of the UTHP samples with CFMW was found to be significantly impaired by high or low filling ratios, but samples with TDMW and TSMW exhibited greater sensitivity to high filling ratios than to low filling ratios. The startup performance of the UTHP samples with TDMW and TSMW deteriorated with an increase in the filling ratio, particularly when the value exceeded 90%. The startup performance of the TDMW UTHP samples was found to be inferior when the filling ratio was 90% in comparison to those with the filling ratios of 60% and 70%. Conversely, the performance of TSMW UTHP was superior when the filling ratio was 90% compared to those with filling ratios of 60% and 70%. For UTHPs, a suitable liquid filling ratio is defined as a sufficient quantity of working liquid present in the evaporation section to participate in the phase change, while the heat source is rapidly dissipated. Therefore, an appropriate filling ratio not only facilitates a rapid startup but also ensures a minimal temperature difference between the evaporation and the condensation sections. Figure 6d illustrates the optimal startup performance of the UTHPs with CFMW, TDMW, and TSMW at 100%, 80%, and 80% filling ratios, respectively. The start times of the three structures were approximately 10 s, and the equilibrium temperatures were 58.37, 58.13, and 56.34 °C, respectively. Among the three structures, the CFMW UTHP exhibited the most optimal startup performance, while that of the TSMW UTHP was the least optimal. The observed performance differences can be attributed to the interconnected nature of the pores inside the CFMW and the minimal backflow resistance of the working liquid. Furthermore, the TDMW and TSMW were composed of orthogonal woven meshes, and the liquid backflow resistance along the direction of their length was considerable. Moreover, the TSMW was sintered by three layers of 200-mesh wires with pores smaller than 100-mesh wires, which resulted in greater resistance.

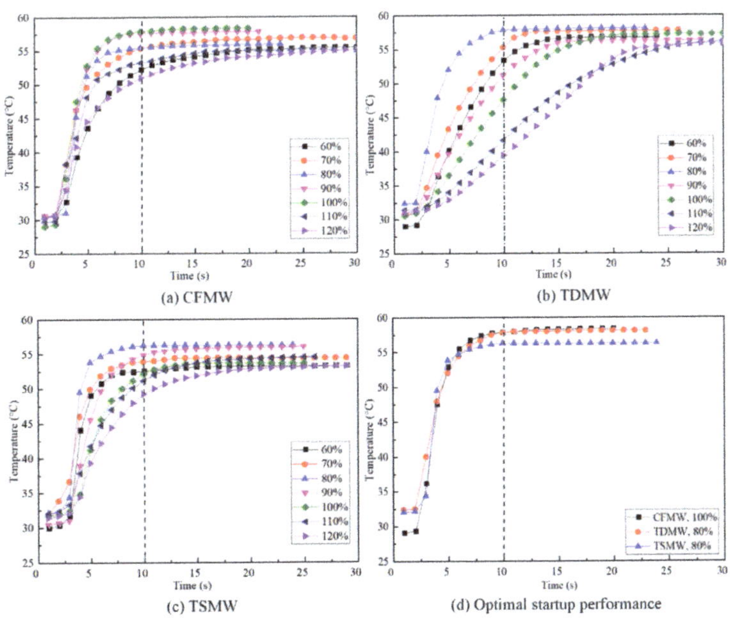

Figure 6. Startup performance of the UTHP samples.

3.2. *Temperature Uniformities of the UTHP Samples*

Figure 7 depicts the temperature uniformities of UTHPs with different wick structures and varying filling ratios at their maximum heating powers. The temperature of the evaporation section exhibited a continuous increase in response to an increase in the heat input. These sections were dried out at the heat inputs that exceeded the maximum heat transport capacity (Q_{max}) values. The UTHP samples with filling ratios below 90% exhibited drying in the evaporation sections with the temperature soaring at a heating power of 7.5 W. This phenomenon could be mitigated if the mass of the working liquid was increased to an appropriate level. However, at a high filling ratio, such as 120%, the liquid would accumulate in the UTHP at a high heating power. This phenomenon would result in a reduction in the effective working length of the sample, accompanied by a shift in the primary mode of heat transfer from phase change to wall conduction. Thus, the heat transfer coefficient was significantly reduced. The performance of the CFMW UTHP exhibited a more pronounced change in response to variations in liquid filling ratios than that of the TDMW and TSMW UTHPs. When the filling ratios were below 80%, the axial temperature differences in the TDMW and TSMW UTHPs were minimal, and these values increased with the heating power. The CFMW UTHP samples demonstrated the optimal overall temperature performance under the appropriate filling ratios in comparison to the TDMW and TSMW UTHPs. The maximum temperature of the evaporation sections was maintained below 57 °C while maintaining a slight difference in temperature. Moreover, the CFMW UTHPs demonstrated the capacity to withstand additional working fluids, which could effectively reduce the dry-out phenomenon. The favorable capillary force of the CFMW facilitated the backflow of the working fluid at high filling ratios. Consequently, the condensation temperature could be maintained above 53 °C at 7.5 W. It was observed that liquid blockages occurred in the condensation sections of the UTHPs with TDMW and TSMW. The temperature differences between the adiabatic and the condensation sections were considerable, with the condensation temperature filling below 52 °C. The Q_{max} values were 9, 8, and 8.5 W, respectively, at the corresponding optimum filling ratios of 110%, 90%, and 100% for the CFMW, TDMW, and TSMW UTHPs.

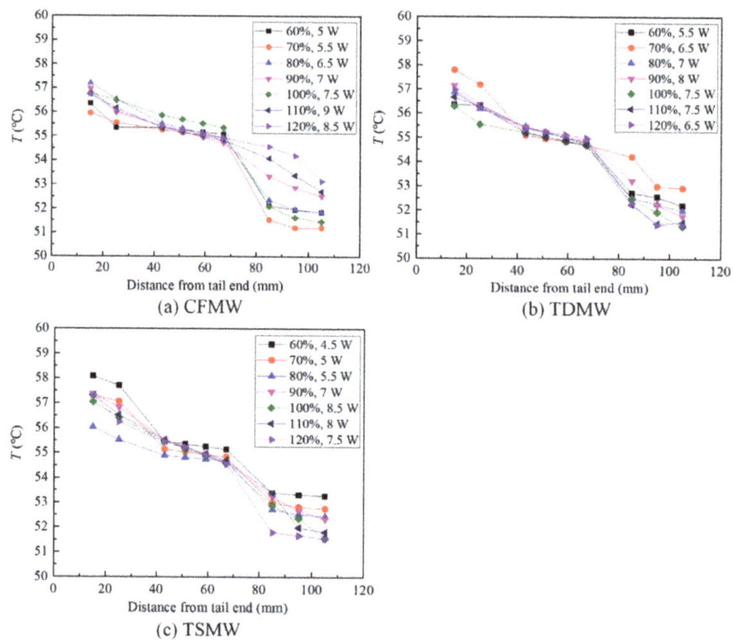

Figure 7. Temperature uniformities of the UTHP samples.

3.3. Thermal Resistances of the UTHP Samples

Figure 8 presents the evaporation thermal resistance (R_e) values of the UTHP samples with CFMW, TDMW, and TSMW at different filling ratios. The overall R_e values in the normal state were found to be stable and below 0.25 K/W. The favorable surface and internal microporous structures of copper foam result in superior capillary properties, which facilitate the replenishment of the evaporation section with working fluid during the phase change process, particularly at high heating powers [23]. Consequently, the CFMW UTHP samples exhibited the lowest R_e of 0.151 K/W. The R_e values exhibited a rapid increase with increasing input power before reaching the dry-out point. The UTHP samples had sufficient working fluid to undergo phase change and absorb additional heat in the evaporation sections, given high liquid filling ratios. Thus, the R_e values exhibited a gradual increase with the input power, maintaining relatively low values. The R_e exhibited a relatively stable profile when the filling ratio was close to the optimal value. The ratio of the wick to vapor spaces was found to be suitable within the UTHP, given the optimal liquid filling ratio. At that time point, neither the phenomenon of drying out nor clogging was observed in the UTHP. The minimal circulation flow resistance of vapor–liquid resulted in relatively low thermal resistance.

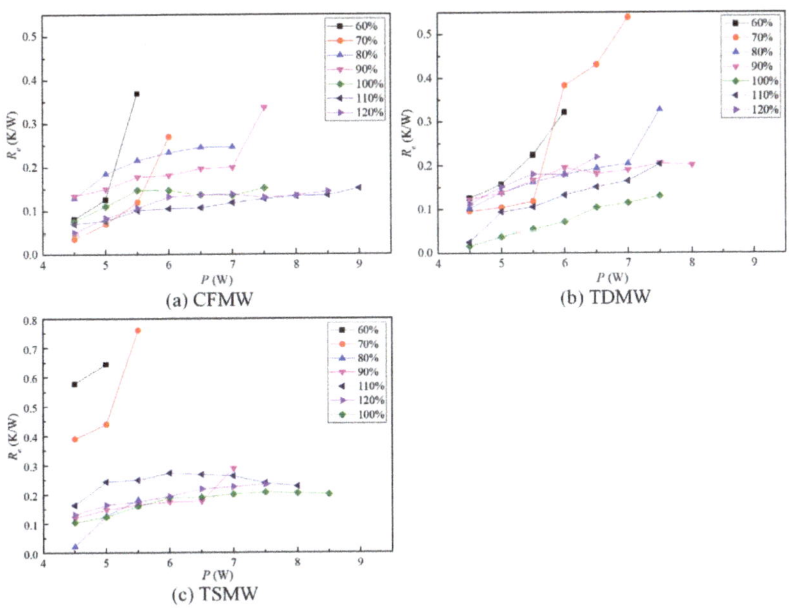

Figure 8. Evaporation thermal resistances of the UTHP samples at various filling ratios.

Figure 9 shows the condensation thermal resistance (R_c) values of the UTHPs with CFMW, TDMW, and TSMW at varying filling ratios. The R_c values of the UTHP samples with appropriate filling ratios were relatively stable and below 0.5 K/W. The CFMW UTHP samples exhibited the lowest R_c value, at 0.189 K/W. The CFMW and TSMW UTHP samples were dried out under low liquid filling ratios (60% and 70%). However, under high liquid filling ratios (110% and 120%), the working liquids within the heat pipes became excessive, resulting in the accumulation of liquids in the condensation sections, thereby increasing the R_c values. The R_c values of the CFMW UTHP samples exhibited a significantly lower level at high filling ratios. This reduction was caused by the CFMW, which exhibited a favorable capillary force, effectively enabling the condensation liquid to flow back to the evaporation section to avoid blockage. The R_c values of the TDMW UTHP samples exhibited an initial increase, followed by a subsequent decrease, with increasing input power. The reason is that the heat transfer mass of the UTHP at a low heating power was relatively small, and the blockage of the vapor–liquid flow reduced the condensation temperature and increased the R_c accordingly. As the heating load increased gradually, the temperature difference increased, and the pressure difference between the two ends also increased. The elevated pressure was sufficient to overcome the flow resistance of the working fluid, thus enabling the working fluid to circulate normally and the R_c to gradually decrease.

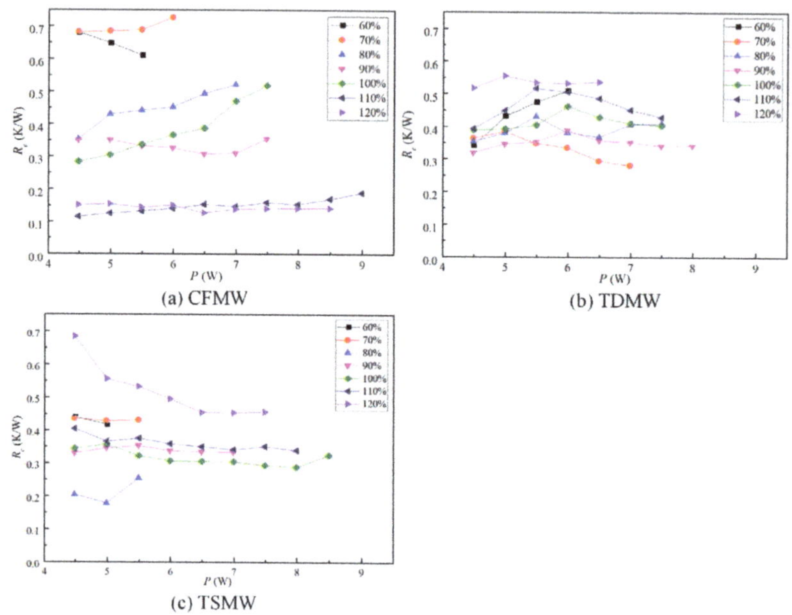

Figure 9. Condensation thermal resistances of the UTHP samples at various filling ratios.

3.4. Comparison of the Maximum Heat Transport Capacity Observed in This Study with Those Reported in Earlier Publications

Table 1 presents the main parameters of the heat pipes with thicknesses ranging from 0.4 to 1.0 mm developed by other researchers and in the present study. The parameters included wick type, total thickness, and Q_{max}. It can be observed that the Q_{max} of the experimental UTHP with a total thickness of 0.6 mm manufactured with the optimized CFMW wick is larger and can reach 9 W, a value that is only slightly lower than the Q_{max} of the UTHP with a total thickness of 1.0 mm designed by Tang et al. [24], with a Q_{max} value of 11 W. The optimized UTHP sample with CFMW demonstrated good startup performance, reaching a steady state in 10 s during the startup performance test. The temperature at the test point was 58.37 °C, with a temperature difference of only 1.63 °C with the test water. In the steady-state performance test, the CFMW UTHP sample exhibited the lowest evaporation and condensation thermal resistances, with corresponding values of 0.151 and 0.189 K/W, respectively, which were 24.67% and 41.85% lower than those of the TSMW UTHP. This indicates that the CFMW can improve the Q_{max} and enhance the thermal performance of the UTHP. The findings of this study provide valuable guidelines for the optimal design of UTHP for cooling portable electronic devices.

Table 1. Q_{max} value comparison between the present study and earlier publications.

Literature	Wick Type	Total Thickness	Q_{max}
Chen et al. [25]	OWM and groove	0.4 mm	4.5 W
Zhou et al. [16]	Band-shape SWM	0.4 mm	5.25 W
Yu et al. [26]	Three SWMs	0.4 mm	6 W
Ahamed et al. [27]	SWM	0.6 mm	5 W
Sun et al. [28]	Multi-layer OWM and micropillars	0.67 mm	8.1 W
Cui et al. [29]	Multi-layer OWM	0.68 mm	8 W
Zhou et al. [30]	Center OWM	0.75 mm	8.5 W
Zhou et al. [31]	Copper foam and OWM	0.8 mm	5 W
Tang et al. [24]	Annular OWM	1.0 mm	11 W
Present study	Optimized CFMW	0.6 mm	9 W

4. Conclusions

In this study, UTHPs with 0.6 mm thickness were fabricated and investigated for cooling portable electronic devices. The composite sintered wick structures of the UTHP samples were as follows: copper foam and mesh wick (CFMW), two layers of different mesh wick (TDMW), and three layers of the same mesh wick (TSMW). The startup and steady-state performance levels of the UTHP samples with liquid filling ratios of 60% to 120% were tested and studied, with particular consideration given to the effect of heating power, the filling ratio, and the wick structure. The main findings were as follows:

The filling ratio of UTHP had a greater effect on its thermal performance than that of the wick structure. The startup performance of the CFMW UTHPs was found to be significantly affected by high or low filling ratios. However, the samples with TDMW and TSMW were more sensitive to high filling ratios than to low filling ratios. The startup time of all UTHPs was 10 s, and the equilibrium temperatures were 58.37, 58.13, and 56.34 °C, respectively, at the corresponding filling ratios of 100%, 80%, and 80% for the three types of UTHP samples. The dry-out phenomenon can be alleviated by increasing the mass of the working liquid accordingly. The maximum heat transport capacities were 9, 8, and 8.5 W, respectively, at the corresponding optimum filling ratios of 110%, 90%, and 100% for the three types of UTHP samples. The CFMW UTHP exhibited the lowest evaporation and condensation thermal resistances of 0.151 and 0.189 K/W, respectively, which were 24.67% and 41.85% lower than those of the TSMW UTHP. CFMW can be used to improve the Q_{max} and enhance the thermal performance of UTHPs. This study provides important guidelines for the structural design, fabrication technology, and performance improvement of high-performance UTHPs used in portable electronic devices.

Author Contributions: W.Z. designed the experiment. J.H., Y.J. and W.Z. carried out the experiments. Y.Y., R.C., D.S. and A.W. analyzed the experimental data. W.Z. wrote the manuscript. All authors have read and agreed to the published version of the manuscript.

Funding: This study was funded by Basic and Applied Basic Research Foundation of Guangdong Province (grant numbers 2022A1515110041 and 2022A1515110691), Guangzhou Science and Technology Plan Project (grant numbers 202201011104, 202201011219, 202201011357, and 2023A04J0290), National Natural Science Foundation of China (grant number U22A2062), Youth Science and Technology Talent Lifting Project of Guangzhou Science and Technology Association (grant number QT2024-014), Guangdong Colleges and Universities Young Innovative Talents Project (grant number 2022KQNCX039), Project of Enhancing Key Construction Disciplines Research Capacity in Guangdong Province (grant number 2022ZDJS012), Major Science and Technology R&D Program of Jiangxi Province (Grant No. 20223AAE02008), and Talent Special Project of Guangdong Polytechnic Normal University (grant numbers 2021SDKYA104, 2021SDKYA030, and 2021SDKYA036).

Data Availability Statement: The original contributions presented in the study are included in the article, further inquiries can be directed to the corresponding author.

Conflicts of Interest: The authors declare no conflicts of interest.

References

1. Tang, H.; Tang, Y.; Wan, Z.P.; Li, J.; Yuan, W.; Lu, L.S.; Li, Y.; Tang, K.R. Review of Applications and Developments of Ultra-Thin Micro Heat Pipes for Electronic Cooling. *Appl. Energy* **2018**, *223*, 383–400. [CrossRef]
2. Deng, D.X.; Zeng, L.; Sun, W. A Review on Flow Boiling Enhancement and Fabrication of Enhanced Microchannels of Microchannel Heat Sinks. *Int. J. Heat Mass Tranf.* **2021**, *175*, 121332. [CrossRef]
3. Mihai, I.; Suciu, C.; Picus, C.M. Assessment of Vapor Formation Rate and Phase Shift between Pressure Gradient and Liquid Velocity in Flat Mini Heat Pipes as a Function of Internal Structure. *Micromachines* **2023**, *14*, 1468. [CrossRef]
4. Jiang, L.L.; Ling, J.T.; Jiang, L.Z.; Tang, Y.; Li, Y.; Zhou, W.; Gao, J.W. Thermal Performance of a Novel Porous Crack Composite Wick Heat Pipe. *Energy Convers. Manag.* **2014**, *81*, 10–18. [CrossRef]
5. Song, W.; Xu, Y.; Xue, L.; Li, H.; Guo, C. Visualization Experimental Study on Silicon-Based Ultra-Thin Loop Heat Pipe Using Deionized Water as Working Fluid. *Micromachines* **2021**, *12*, 1080. [CrossRef]
6. Li, Y.; Zhou, W.J.; He, J.B.; Yan, Y.Y.; Li, B.; Zeng, Z.X. Thermal Performance of Ultra-Thin Flattened Heat Pipes with Composite Wick Structure. *Appl. Therm. Eng.* **2016**, *102*, 487–499. [CrossRef]

7. Zhou, G.H.; Li, J.; Lv, L.C. An Ultra-Thin Miniature Loop Heat Pipe Cooler for Mobile Electronics. *Appl. Therm. Eng.* **2016**, *109*, 514–523. [CrossRef]
8. Zhou, W.J.; Li, Y.; Huang, G.W.; Yang, Y.; Xu, L.Y.; He, J.F.; Jian, Y. Thermal Performance of a Large-diameter Thin Flattened Heat Pipe with Novel Composite Wick Structure. *Case Stud. Therm. Eng.* **2024**, *57*, 104352. [CrossRef]
9. Li, Y.; He, J.B.; He, H.F.; Yan, Y.Y.; Zeng, Z.X.; Li, B. Investigation of Ultra-Thin Flattened Heat Pipes with Sintered Wick Structure. *Appl. Therm. Eng.* **2015**, *86*, 106–118. [CrossRef]
10. Lin, K.T.; Wong, S.C. Performance Degradation of Flattened Heat Pipes. *Appl. Therm. Eng.* **2013**, *50*, 46–54. [CrossRef]
11. Do, K.H.; Kim, S.J.; Garimella, S.V. A Mathematical Model for Analyzing the Thermal Characteristics of a Flat Micro Heat Pipe with a Grooved Wick. *Int. J. Heat Mass Tranf.* **2008**, *51*, 4637–4650. [CrossRef]
12. Lefèvre, F.; Rullière, R.; Pandraud, G.; Lallemand, M. Prediction of the Temperature Field in Flat Plate Heat Pipes with Micro-Grooves–Experimental Validation. *Int. J. Heat Mass Tranf.* **2008**, *51*, 4083–4094. [CrossRef]
13. Lurie, S.A.; Rabinskiy, L.N.; Solyaev, Y.O. Topology Optimization of The Wick Geometry in a Flat Plate Heat Pipe. *Int. J. Heat Mass Transf.* **2019**, *128*, 239–247. [CrossRef]
14. Yang, K.; Tu, C.; Zhang, W.; Yeh, C.; Wang, C. A Novel Oxidized Composite Braided Wires Wick Structure Applicable for Ultra-Thin Flattened Heat Pipes. *Int. Commun. Heat Mass Transf.* **2017**, *88*, 84–90. [CrossRef]
15. Tang, Y.; Tang, H.; Li, J.; Zhang, S.W.; Zhuang, B.S.; Sun, Y.L. Experimental Investigation of Capillary Force in a Novel Sintered Copper Mesh Wick for Ultra-Thin Heat Pipes. *Appl. Therm. Eng.* **2017**, *115*, 1020–1030. [CrossRef]
16. Zhou, W.J.; Li, Y.; Chen, Z.S.; Deng, L.Q.; Gan, Y.H. Ultra-Thin flattened Heat Pipe with a Novel Band-Shape Spiral Woven Mesh Wick for Cooling Smartphones. *Int. J. Heat Mass Tranf.* **2020**, *146*, 118792. [CrossRef]
17. Zhou, W.J.; Li, Y.; Chen, Z.S.; Deng, L.Q.; Gan, Y.H. A Novel Ultra-Thin flattened Heat Pipe with Biporous Spiral Woven Mesh Wick for Cooling Electronic Devices. *Energy Convers. Manag.* **2019**, *180*, 769–783. [CrossRef]
18. Zhou, W.J.; Li, Y.; Chen, Z.S.; Deng, L.Q.; Gan, Y.H. Experimental Study on The Heat Transfer Performance of Ultra-Thin flattened Heat Pipe with Hybrid Spiral Woven Mesh Wick Structure. *Appl. Therm. Eng.* **2020**, *170*, 115009. [CrossRef]
19. Li, D.Q.; Huang, Z.; Liao, X.N.; Zu, S.F.; Jian, Q.F. Heat and Mass Transfer Characteristics of Ultra-Thin Flat Heat Pipe with Different Liquid Filling Rates. *Appl. Therm. Eng.* **2021**, *199*, 117588. [CrossRef]
20. Chen, J.S.; Chou, J.H. Cooling Performance of Flat Plate Heat Pipes with Different Liquid filling Ratios. *Int. J. Heat Mass Transf.* **2014**, *77*, 874–882. [CrossRef]
21. Tharayil, T.; Asirvatham, L.G.; Ravindran, V.; Wongwises, S. Effect of Filling Ratio on The Performance of a Novel Miniature Loop Heat Pipe Having Different Diameter Transport Lines. *Appl. Therm. Eng.* **2016**, *106*, 588–600. [CrossRef]
22. Holman, J.P. *Experimental Methods for Engineers*, 8th ed.; McGraw Hill: New York, NY, USA, 2012; pp. 62–68.
23. Sun, Y.L.; Li, H.M.; Chen, J.Y.; Wu, C.X.; Tang, H.; Zhang, S.W.; Tang, Y. Hierarchical Sintered Porous Surfaces with Enhanced Pool Boiling Heat Transfer Performance for High-Power Cooling Applications. *Appl. Therm. Eng.* **2024**, *249*, 123368. [CrossRef]
24. Tang, Y.L.; Hong, S.H.; Wang, S.F.; Deng, D.W. Experimental Study on Thermal Performances of Ultra-Thin Flattened Heat Pipes. *Int. J. Heat Mass Transf.* **2019**, *134*, 884–894. [CrossRef]
25. Chen, Z.S.; Li, Y.; Zhou, W.J.; Deng, L.Q.; Yan, Y.Y. Design, Fabrication and Thermal Performance of a Novel Ultra-Thin Vapor Chamber for Cooling Electronic Devices. *Energy Convers. Manag.* **2019**, *187*, 221–231. [CrossRef]
26. Yu, J.; Li, Y.; Chen, Z.S.; Luo, Q.L.; Chen, H.Y.; Tang, X.K. Effect of the Passage Area Ratio of Wick on an Ultra-Thin Vapor Chamber with a Spiral Woven Mesh Wick. *Appl. Therm. Eng.* **2021**, *196*, 117282. [CrossRef]
27. Ahamed, M.S.; Saito, Y.; Mashiko, K.; Mochizuki, M. Characterization of a High Performance Ultra-Thin Heat Pipe Cooling Module for Mobile Hand Held Electronic Devices. *Heat Mass Transf.* **2017**, *53*, 3241–3247. [CrossRef]
28. Sun, Y.L.; Liang, F.Y.; Tang, H.; Tang, Y.; Zhang, S.W.; Chua, K.J. Ultrathin flexible Heat Pipes with Microsorum Fortunei Structural-Like Wick for Cooling Fexible Electronic Devices. *Int. J. Heat Mass Tranf.* **2023**, *202*, 123743. [CrossRef]
29. Cui, Z.; Jia, L.; Wang, Z.; Dang, C.; Yin, L.F. Thermal Performance of an Ultra-Thin Flat Heat Pipe with Striped Super-Hydrophilic Wick Structure. *Appl. Therm. Eng.* **2022**, *208*, 118249. [CrossRef]
30. Zhou, W.J.; Li, Y.; Chen, Z.S.; Deng, L.Q.; Gan, Y.H. Effect of the Passage Area Ratio of Liquid to Vapor on an Ultra-Thin Flattened Heat Pipe. *Appl. Therm. Eng.* **2019**, *162*, 114215. [CrossRef]
31. Zhou, W.J.; Xie, P.D.; Li, Y.; Yan, Y.Y.; Li, B. Thermal Performance of Ultra-Thin Flattened Heat Pipes. *Appl. Therm. Eng.* **2017**, *117*, 773–781. [CrossRef]

micromachines

MDPI

Article

Structure Design and Heat Transfer Performance Analysis of a Novel Composite Phase Change Active Cooling Channel Wall for Hypersonic Aircraft

Weichen Li [1], Jieliang Zhao [1,*], Xiangbing Wu [1], Lulu Liang [1], Wenzhong Wang [1] and Shaoze Yan [2]

1 School of Mechanical Engineering, Beijing Institute of Technology, Beijing 100081, China; 3120210421@bit.edu.cn (W.L.); 18801165131@163.com (X.W.); lianglulu12345@163.com (L.L.); wangwzhong@bit.edu.cn (W.W.)
2 Division of Intelligent and Biomechanical Systems, State Key Laboratory of Tribology in Advanced Equipment, Department of Mechanical Engineering, Tsinghua University, Beijing 100084, China; yansz@mail.tsinghua.edu.cn
* Correspondence: jielzhao@bit.edu.cn

Abstract: Efficient and stable heat dissipation structure is crucial for improving the convective heat transfer performance of thermal protection systems (TPSs) for hypersonic aircraft. However, the heat dissipation wall of the current TPS is limited by a single material and structure, inefficiently dissipating the large amount of accumulated heat generated during the high-speed maneuvering flight of hypersonic aircraft. Here, a convection cooling channel structure of TPS is proposed, which is an innovative multi-level structure inspired by the natural honeycomb. An active cooling channel (PCM-HC) is designed by using a variable-density topology optimization method and filled with phase change material (PCM). Numerical simulations are used to investigate the thermal performance of the PCM-HC wall, focusing on the influence of PCM properties, structural geometric parameters, and PCM types on heat transfer characteristics. The results demonstrate that the honeycomb-like convection cooling channel wall, combined with PCM latent heat of phase change, exhibits superior heat dissipation capability. With a heat flux input of 50 kW/m^2, the maximum temperature on the inner wall of PCM-HC is reduced by 12 K to 20 K. Different PCMs have opposing effects on heat transfer performance due to their distinct thermophysical properties. This work can provide a theoretical basis for the design of high-efficiency cooling channel, improving the heat dissipation performance in the TPS of hypersonic aircraft.

Keywords: active cooling thermal protection; honeycomb structure; topological optimization; phase change material; heat transfer performance

1. Introduction

Hypersonic aircraft have gradually become a research hotspot in the field of aerospace due to their high maneuverability [1] and strong penetration capability [2]. However, when the aircraft cruises at speeds exceeding 5 Ma in near space, it will be subjected to severe aerodynamic heating [3] and air viscosity friction [4], resulting in surface temperatures of over one thousand degrees Celsius [5,6]. Although most of the heat can be isolated by the Thermal Protection System (TPS) of the aircraft, the remaining aerodynamic heat transmitted to the inside of the structure will cause internal thermal strain and result in an increase in the temperature of the cabin, which will affect the normal use of the thermally sensitive equipment and change the mechanical properties of the structure's materials, thus affecting its strength and stiffness [7,8]. Therefore, it is necessary to develop a convective cooling structure that can both bear external load and actively cool internally to better dissipate the heat inside the aircraft's structure.

The fundamental structure of thermal protection systems comprises an outer high-temperature structural layer and an internal non-load-bearing thermal insulation layer [9,10].

The structural layer, composed of high-temperature-resistant metals or composite materials, is designed to withstand external aerodynamic loads, while the thermal insulation layer, consisting of insulating materials applied as coatings and skins, serves to isolate the substantial heat generated by aerodynamic heating. Given that hypersonic vehicles endure prolonged exposure to high-heat-flux aerodynamic thermal environments, coupled with stringent requirements for flight speed and structural integrity, thermal protection structures often exhibit considerable thickness, thereby unavoidably introducing issues of increased mass that adversely impact flight velocity and escalate fuel consumption [11]. Existing convective cooling channels have been employed in scenarios characterized by high heat flux density, such as in the supersonic combustion ramjet engines of hypersonic vehicles [12], where adjustments to channel geometry and dimensions enhance heat dissipation capabilities. Similarly, hypersonic vehicles' surfaces confront severe thermal loading environments, prompting the integration of convective cooling channels into TPS to replace bulky thermal insulation materials with high-efficiency heat transfer mechanisms. This approach significantly aids in reducing vehicle mass while preserving structural load-bearing capabilities.

Presently, active TPS are predominantly employed in hypersonic vehicles, leveraging active cooling techniques to remove greater amounts of heat, thereby minimizing the limitations imposed by insufficient material heat resistance and mitigating structural damage to the vehicle [13]. According to different cooling methods, active TPS can be divided into three types: sweat cooling, thin film cooling, and convective cooling [14]. Liu et al. [15] designed a sweating-type TPS, which played a role in reducing the structural temperature to some extent, and the sweating cooling reduced the internal structure temperature of the test model by 50 K. Zhou et al. [16] experimentally verified the influence of low compressibility and the ratio of coolant to heat flux density on the thin film cooling performance of the model. Lee et al. [17] numerically studied the thermal performance of a pin fin heat sink and reduced its thermal resistance by 11% and weight by 30% by optimizing the fin density. Among these active cooling technologies, convective cooling, as a clean and renewable cooling method, has received particular attention and become one of the effective thermal protection methods for hypersonic aircraft in long-term service. In addition to the above features, one of the most attractive features of convective cooling (regenerative cooling) is the use of fuel as a coolant, which not only avoids coolant waste but also achieves efficient and environmentally friendly utilization of residual heat energy [18, 19].

However, traditional convective cooling techniques also have certain problems, including low heat dissipation, large coolant consumption, and susceptibility to failure [20,21]. Eckert et al. [22] studied the coolant consumption of three cooling methods, namely impinging cooling, convective cooling, and thin film cooling, under the same mainstream conditions. It was found that convective cooling has the highest coolant consumption among the three cooling methods: under laminar flow conditions, convective cooling is 3 times higher than sweating cooling, and under turbulent flow conditions, it is 1.5 times higher than sweating cooling. This also results in a larger weight of the convective cooling system, and its application is mainly concentrated on heat sinks for miniature devices such as electronic components [23], posing significant technical challenges for TPS applications in spacecraft [24]. Thus far, researchers have conducted some studies on the application of convective cooling technology in the aerospace field, including the use of cold liquid hydrogen circuits for engine cooling in aerospace vehicles and its application in TPS for SSTO hot zones [25]. These strategies can enhance fluid perturbation and heat transfer in convective cooling structures, but they cannot fundamentally solve the aforementioned issues associated with convective cooling technology.

In order to enhance the heat dissipation capability of convective cooling channels and improve the load-bearing performance of the channels themselves, natural honeycomb structures have come into our focus. The honeycomb structure itself exhibits excellent mechanical properties, and moreover, the temperature of the honeycomb can maintain dynamic stability, varying between 25 and 45 °C according to environmental temperatures [26]. However, existing research has mostly focused on the macroscopic structure of honeycombs, with less attention given to the biomimetic utilization value of the microstructure of honeycombs. Scholars have conducted relevant research on the structural characteristics and component properties of honeycombs. Zhang et al. characterized the multilevel structure of natural honeycombs at different scales in detail [27–29], and used ESEM, AFM, and other methods to observe and analyze the microstructure and composition of the honeycomb, revealing the micro-morphology of the layered composite material composed of beeswax and bee silk. Zhang also studied the principles of anisotropy and temperature control in honeycomb structures through experiments and numerical simulations, proposing a novel porous material design with a surface composite coating. By mimicking the microstructure of honeycombs, this design breaks through the limitation of only biomimicry from a macroscopic perspective. Duan [30,31] proposed honeycomb equivalent mechanical models and expansion performance models, analyzing the mechanical properties of honeycombs at different temperatures. It was found that bees improve the mechanical performance of honeycombs by cyclically using bee silk, which affects the internal heat transfer capacity of the honeycomb and thereby regulates the temperature of the honeycomb.

The honeycomb structure provides an excellent biomimetic object for the design of convective cooling channels. Recently, many scholars have conducted research on the biomimetic heat exchange structure of honeycombs. Lu [32] used a corrugated wall model to evaluate the heat transfer of microporous aluminum honeycomb under convective cooling in a compact heat exchanger, obtaining the optimal honeycomb cell morphology to achieve maximum overall heat transfer efficiency. Wen [33] studied the performance of sandwich metal honeycomb structures under forced air convection conditions through experiments and numerical methods, testing honeycomb structures made of different materials and with different topological structures. The results showed that the total pressure drop was related to the specific surface area density and the honeycomb cell shape, while the total heat transfer rate was related to the specific surface area density, cell shape, and thermal conductivity of the solid material. Pei et al. [34] proposed a honeycomb-shaped ultra-compact plate heat exchanger and studied its flow and heat transfer characteristics, obtaining the optimal geometric structure of the heat exchanger. Therefore, the honeycomb structure has great potential for application in the field of heat exchange, and it is necessary to break through the limitations of traditional porous materials and design convective cooling structures from the perspective of honeycomb microstructure in order to achieve better heat transfer effects.

With the help of microstructure features, the honeycomb achieves a more reasonable heat distribution through heat transfer, thereby improving the heat exchange capacity between the honeycomb walls and the external environment. In order to better apply the honeycomb's ability to convective cooling channel wall heat transfer, the phase change latent heat characteristics of phase change materials (PCM) can effectively delay the heat transfer on the wall surface [35], allowing more heat to accumulate on the aerodynamic heated flow side of the channel and avoiding transfer to the far end, thus improving the heat transfer efficiency with the fuel medium. PCM has been widely used in fields such as aerospace and battery thermal management. Kim et al. [36] proposed a new type of spacecraft thermal control hardware combined with PCM and studied the thermal carrying capacity of PCM through simulation. Wang et al. [37] studied the influence of composite PCMs composed of expanded graphite and paraffin with different melting points on battery thermal management performance at different phase change temperatures, and improved the efficiency of PCM through numerical optimization.

In recent years, topology optimization methods have gradually become a new way of designing heat transfer structures [38–40]. By satisfying conditions such as objective functions and design variables, topology optimization seeks the optimal material layout in the design domain to achieve a certain performance optimum [41]. Tang et al. [38] proposed a topology optimization model for steady-state nonlinear heat conduction structures under large temperature gradients and found that solid material variability affects topology optimization configurations. Zhuang et al. [42] studied the topology optimization problem of transient heat conduction structures with the integration of temperature gradients at fixed time intervals as the objective function. Xia et al. [43] combined the level set method with bi-directional evolutionary structural optimization to perform heat transfer topology optimization on a flat plate. However, the aforementioned studies commonly focus on optimizing thermal parameters to enhance the thermal characteristics of structures, with little consideration given to the load-bearing properties of thermal structures. In the field of statistics, there is a paucity of research related to topology optimization with the objective of minimizing flexibility.

Topology optimization has also found application in the design of biomimetic honeycomb heat transfer structures. Zhou [44] developed a honeycomb structure with cooling channels, employing topology optimization based on parameters such as thermal shielding and dissipation to study its thermomechanical behavior. Li et al. [45] proposed a heterogeneous honeycomb structure and established a single-ring topology optimization model, successfully reducing its thermal elastic deformation. Huang et al. [46], through the MTO optimization method, adjusted the thermal conductivity and convective area to design a honeycomb structure with high cooling performance. While honeycomb structures hold great potential in the field of heat transfer structures, existing topology optimization designs only mimic the macrostructure of natural honeycombs, namely by using homogeneous pore walls without microstructure to imitate their hexagonal structure. It is necessary to break through the limitations of traditional porous materials and conduct topology optimization design from the perspective of honeycomb microstructure to achieve better heat transfer efficiency.

In order to better utilize the biomimetic value of honeycomb microstructure, this study combines biomimetics with topology optimization methods to biomimetically design honeycomb structures at multiple scales, considering both the macroscopic and microscopic features of natural honeycombs. By integrating PCM with the honeycomb structure and utilizing the latent heat storage characteristics of PCM, a novel composite phase change material-honeycomb channel (PCM-HC) is designed. The thermal performance of the PCM-HC's wall is investigated, taking into account its heat transfer properties.

The remaining sections of this paper are organized as follows. Section 2 presents the construction of a biomimetic honeycomb convective cooling channel (HC) and outlines the micro-design principles for HC. Section 3 introduces a density-based topology optimization method based on the SIMP interpolation model, which is used for two-dimensional cross-sectional topology optimization of HC. Based on this, a multi-scale structural design method called PCM-HC is proposed. In Section 4, a transient heat transfer analysis model for HC and PCM-HC is established using the finite element method (FE). Section 5 analyzes the influence of PCM presence, geometric parameter variations, and the use of different PCMs on the wall heat transfer performance of PCM-HC. Finally, Section 6 provides a summary of the main findings of this study.

2. Biomimetic Honeycomb HC Model

Honeycombs possess distinctive structural characteristics that are highly suitable for biomimetic emulation. For instance, they are designed with minimal material usage while maintaining stability, making them an ideal choice for porous structures. Additionally, the inherent insulation capability of honeycombs is worthy of extensive attention. Honeybees employ an ingenious architectural strategy by incorporating propolis as an additive into beeswax, forming the cell walls of honeycombs. This facilitates the regulation of hive temperature in response to a decrease in external environmental temperature. Furthermore, bees cover the surface of their nests with a detachable covering, which serves as passive insulation to regulate temperature. This mechanism helps prevent heat loss from the interior and protects against excessive external heat disturbance [47]. Therefore, at the macroscopic level, the hexagonal shape of the honeycomb is chosen as its overall configuration to enhance the mechanical strength of the convective cooling channel. Building upon this, at the microscopic level, the structural characteristics of biomimetic honeycomb cell walls, resembling layered composite materials, are utilized, with PCM filling the channel interstices, aiming to improve the heat transfer performance of the channel.

2.1. HC Model Construction

As shown in Figure 1c, leveraging the macroscopic hexagonal configuration of natural honeycombs, a design model of a honeycomb-inspired convective cooling channel, denoted as HC, was proposed based on the multi-level structure of honeycomb. The HC has a hexagonal shape with dimensions of $D \times H \times T$ (20 mm × 10 mm × 2 mm) and includes internal cooling channels for fuel flow to achieve convective heat transfer. In the application scenario of hypersonic variant aircraft nose cones, multiple HC units and other channel structures including stretching and bending deformation modules (Figure 1b) are connected to the internal skeleton structure of the aircraft morphing nose cone (Figure 1a) to remove the aerodynamic heat conducted from the skin to the internal structure. This design strategy can effectively improve the heat management performance of the hypersonic aircraft. Subsequent sections primarily focus on the heat transfer performance analysis of the biomimetic honeycomb convective cooling channel. The analysis excludes consideration of the remaining modified duct structures based on variant head cones.

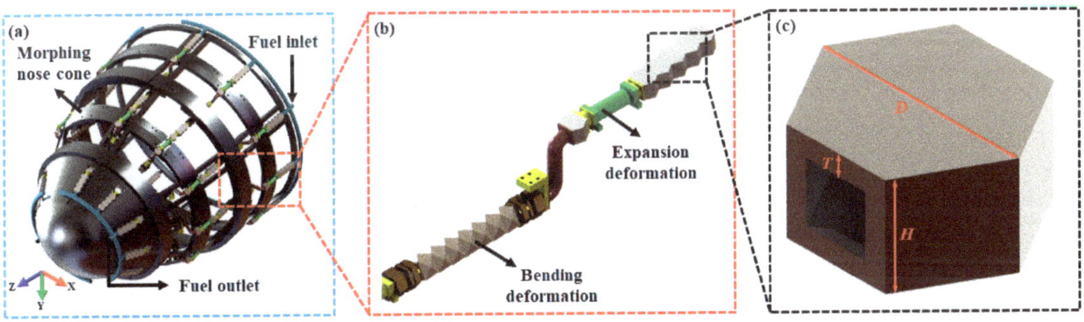

Figure 1. HC model diagram. (**a**) The distribution pattern of HC on the surface of the morphing nose cone. (**b**) The connection method between HC and other functional structures. (**c**) The model structure of HC, with dimensions specified as $D \times H \times T$ (20 mm × 10 mm × 2 mm).

2.2. Micro-Design Principles of HC

The pore walls of natural honeycombs consist of two parts: the initial layer and the adhesive layer. The initial layer corresponds to the new honeycomb constructed during the initial stages of use, while the adhesive layer gradually forms during the honeycomb's usage. Consequently, the cell walls of natural honeycombs are considered as fiber composite materials composed of beeswax and bee silk. Taking into account the microstructure and mechanical properties of natural honeycombs, bees enhance the thermal stability of honeycombs by adding bee silk to the surface of new honeycomb walls [28]. Therefore, the design principles for the microstructure of HC walls are illustrated in Figure 2, mimicking the microstructural characteristics of natural honeycombs for HC wall design. Firstly, suitable metal materials are chosen to fabricate HC channels, representing the base material of natural honeycombs, beeswax. Subsequently, utilizing topology optimization methods, parts of the base material are reasonably removed and replaced with other materials (i.e., PCM) to form a micro-layered composite structure similar to honeycomb pore walls.

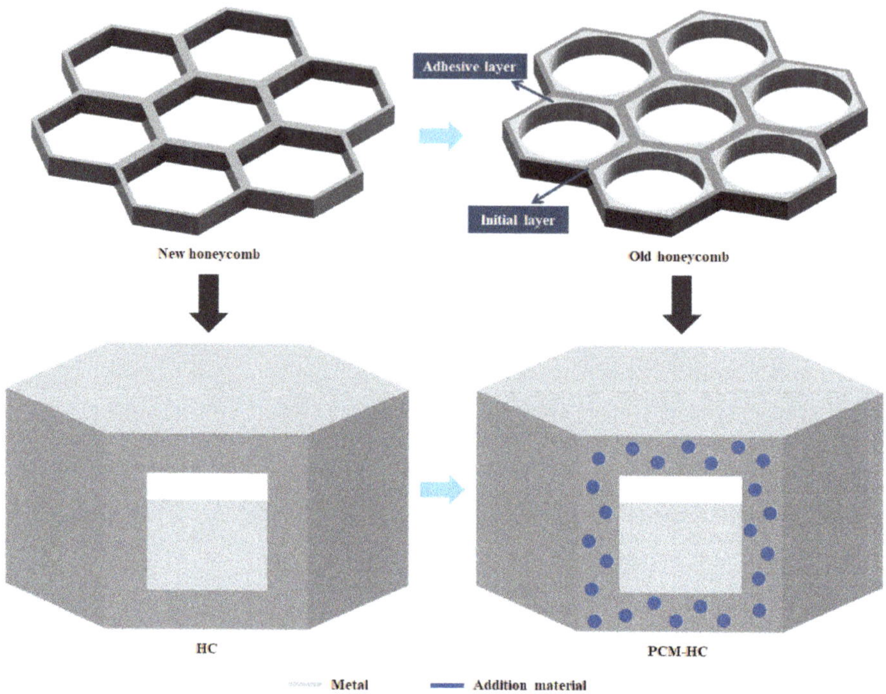

Figure 2. The micro-design method for HC. The HC originates from the new honeycomb, whereas the old honeycomb consists of an initial layer and adhesive layer. Utilizing metal materials as the substrate, and incorporating additional materials, a layered composite structure is designed for PCM-HC, mimicking the layered composite structure of the old honeycomb.

3. Topology Optimization Design of PCM-HC

3.1. Topology Optimization Method

A density-based method for topology optimization of continuum structures was employed. The optimization objective was to maximize stiffness while maintaining the optimal HC bearing characteristics by removing some wall material and filling it with PCM to improve the heat transfer capacity of the wall. The density-based method assumes that the material density of each element is different. Material optimization is achieved through independent variation of the relative density of the material. The mathematical model is given as follows [48]:

$$\begin{aligned} \text{Find}: \rho &= \{\rho_1, \rho_2, \cdots, \rho_n\} \in R, \\ \text{Minimize}: C &= \tfrac{1}{2} U^T K U, \\ \text{Subject to}: KU &= F, \\ \rho_i &\in [0,1], \forall_i, \\ g &= \textstyle\sum_i \rho_i - V_0 \leq 0, \end{aligned} \tag{1}$$

where ρ, R, U, K, F, and V_0 are the design variable, the design domain, the displacement matrix, the stiffness matrix, the load matrix, and the target optimized volume, respectively.

In topology optimization, the objective of maximizing stiffness is translated into minimizing the elastic strain energy expressed by the objective function C using matrix notation.

The material interpolation models can be classified into two methods: the RAMP method and the SIMP method. These two models are quite similar, but the SIMP method exerts slightly stronger penalization on the element density. The material interpolation model based on the SIMP formulation is given as follows [49]:

$$E_i = p(\eta_i)E_{i_0} = \eta_i^p E_{i_0} \tag{2}$$

where E_{i_0} is the elastic modulus of the i-th element in the fully dense state ($\eta_i = 1$), $P(\eta_i)$ denotes the interpolation function, and p is the penalty coefficient, which is set to 3.0 [48].

Under the condition of inertial load being overloaded, the classical SIMP model may lead to distortion in low-density elements due to the mismatch between material mass and stiffness models. To address this issue, a polynomial interpolation model can be employed to mitigate the occurrence of element distortion:

$$E_i = [(1-\omega)\eta_i^p + \omega\eta_i]E_{i_0} \tag{3}$$

where ω is the weight coefficient of the linear term. Therefore, the ratio between the penalty coefficient of element mass and stiffness is defined as follows:

$$r_{mK} = \frac{1}{(1-\omega)\eta_i^{p-1} + \omega} \tag{4}$$

By incorporating ω in the denominator, regardless of the value of η_i within the range of [0, 1], r_{mK} remains finite in size, resolving the issue of element distortion in the SIMP model.

3.2. Topology Optimization Process

The HC channel is a symmetrical structure (Figure 3a) that can be split in half along the symmetry plane, while the 2D cross section gradually lengthens into an arch-like structure along the axial direction. The stress distribution of the 2D cross section is similar, allowing the 3D problem to be simplified to a 2D problem. Therefore, the optimization of the structural topology was carried out on two cross sections of the HC channel, namely at the end and middle sections, and the result was extrapolated to the 3D topology structure (Figure 3b).

Utilizing the topology optimization module (ATOM) in Abaqus for structural topology optimization, a 2D cross-sectional topology optimization model was established. The bottom surfaces of both end sections and the midsection were fully constrained, while the upper surface was subjected to a uniformly distributed load F. Considering the practical application scenario of PCM-HC heat pipes, the model distinguished between the design domain and non-design domain. A boundary layer within 0.2 mm of the outer contour of the wall was frozen (Figure 3c,d). Within ATOM, stress, volume fraction, and weighted strain energy were initially defined as responses to be constrained or targeted. Stress constraints were based on the compressive strength of SUS304 (515 MPa), with an upper limit of 500 MPa. Volume fraction constraints were set as design variables, with volume fraction gradients of 40%, 55%, and 70% (representing the ratio of optimized volume to original volume). The weighted strain energy was selected as the objective function, with the optimization direction set to minimize. Simultaneously, in the geometric constraint module, a minimum size constraint of 0.1 was defined to prevent the generation of overly thin structures, thereby achieving more uniformly distributed material distribution in the optimized results.

The static stress contour maps of the end and middle sections of the HC channel in Figure 3e,f show that the stress concentration is mainly located around the middle region of the upper wall surface and at the diagonal intersection between the upper wall and side walls. These regions should be preserved as the main load-bearing parts in the design of the PCM-HC channel. As the bottom surface of the end and middle sections serves as the farthest point of thermal conduction and exhibits low stress, this area will become the main removal region in the topology optimization process and is therefore designated as the non-design domain. Additionally, the maximum stress on the compressed side wall of the middle section is approximately 5.2 times greater than that of the end section. Therefore, the topology optimization result of the middle section was selected as the overall design reference for the PCM-HC channel.

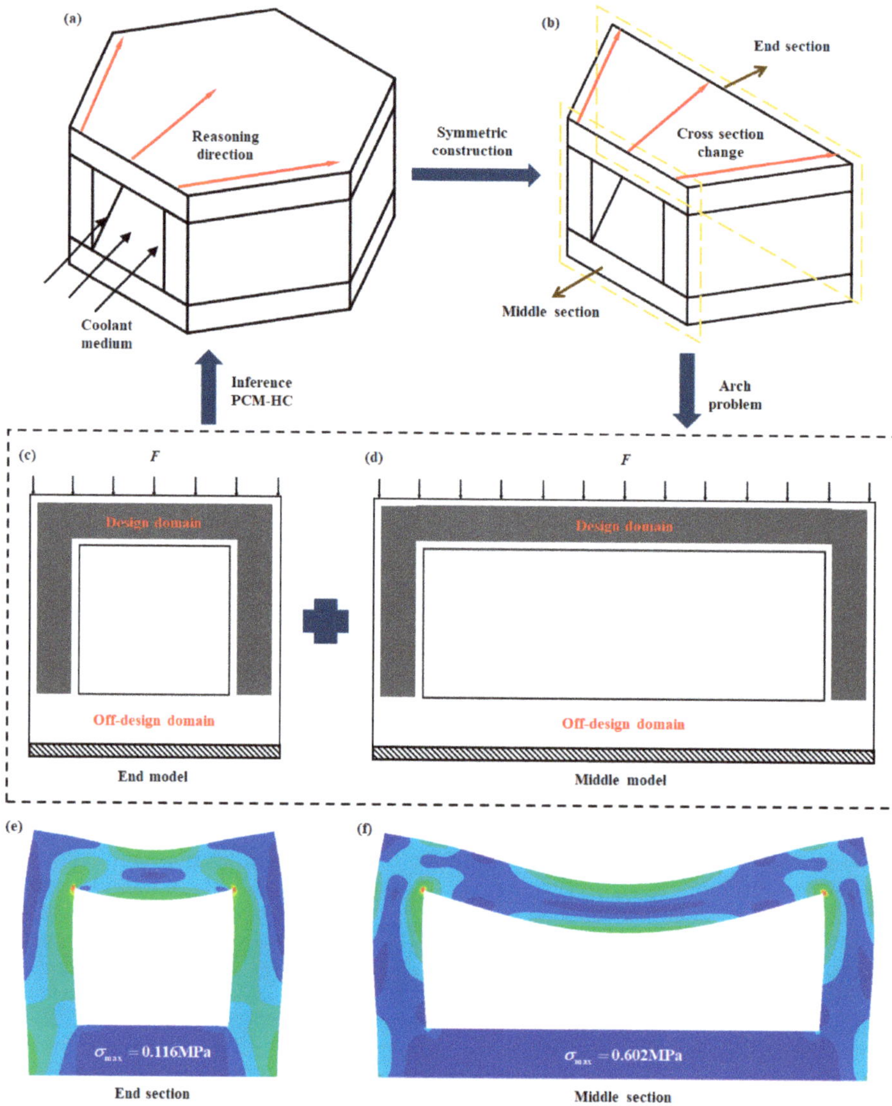

Figure 3. The topology optimization analysis process of HC is transformed from a 3D problem to a 2D problem. (**a**,**b**) HC model simplification process. The HC structure is symmetrical and includes an internal channel for coolant flow, which is simplified into a 2D arch-shaped problem. (**c**,**d**) Topological optimization model. Two cross-sections, one located at the end and another at the middle of the structure, are selected for topology optimization. (**e**,**f**) Stress contour plots for the end section and middle section are presented. The difference of color in different regions of the cross section represents the difference of stress magnitude. The maximum stress values for the end section and middle section are 0.116 MPa and 0.602 MPa, respectively.

In Figure 4a–d, topology optimization is performed with volume fraction gradients set as 100%, 70%, 55%, and 40%, indicating that the optimization mainly focuses on areas of low stress in the structural design region. Therefore, topology optimization primarily optimizes the lowest-performing units in the overall structural performance area by removing inefficient units in order to improve the overall performance and mechanical properties of the structure. As the volume fraction changes, the outline of the main load-bearing structure becomes clearer: for the upper wall surface, the material reduction is mainly concentrated in the middle Section 1, with a trapezoidal shape, and symmetrically distributed along the cross-sectional axis. For the side wall surface, with a volume fraction of 40%, a "Z"-shaped support structure is clearly observed in area 2, indicating that this design method can enhance the lateral load-bearing capacity of the overall structure.

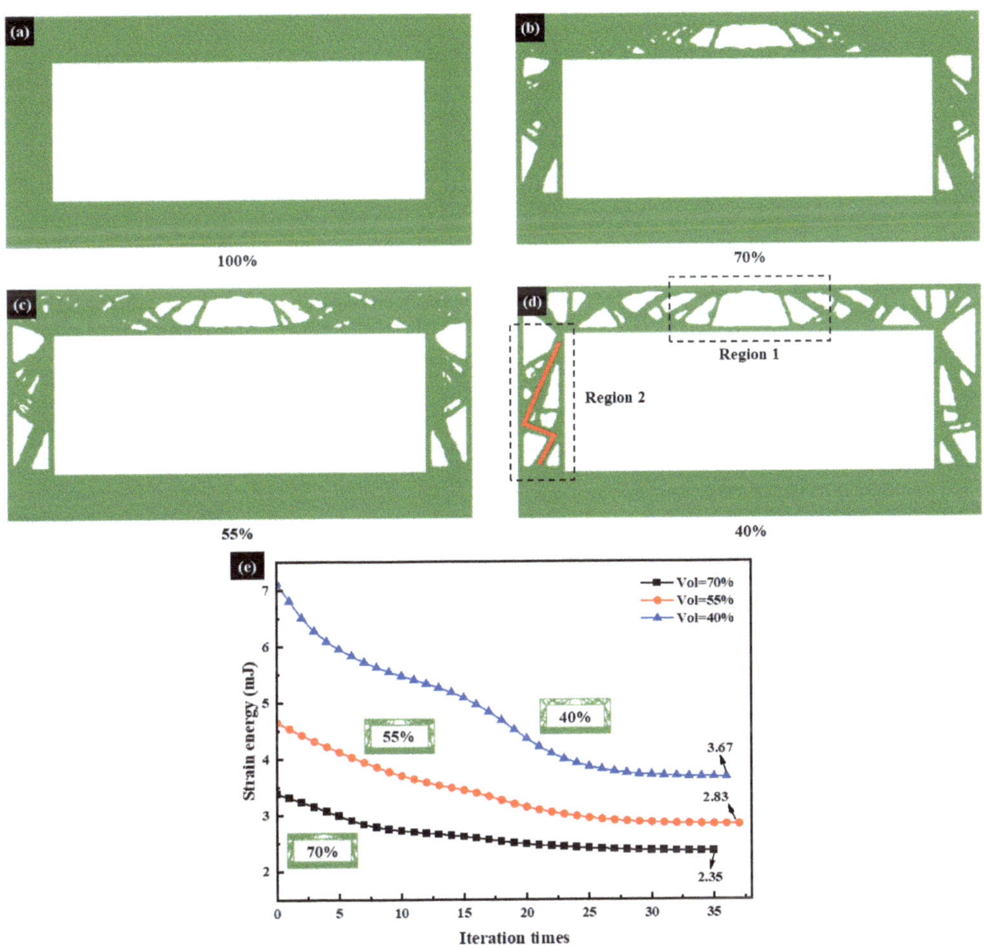

Figure 4. Topology optimization results of the middle section. (**a**) The initial structure diagram of the middle section. (**b–d**) Topology optimization results under volume fraction constraints of 70%, 55%, and 40%. Under a volume fraction constraint of 40%, region 1 exhibits a trapezoidal structure, while region 2 forms a "Z"-shaped pattern. (**e**) Variation of strain energy with iterations at different volume fractions.

In order to compare the mechanical strength of HC sections at different volume fractions, Figure 4e illustrates the variation of strain energy with topology optimization iterations for volume fractions of 70%, 55%, and 40%. Strain energy serves as a measure of structural stiffness, where minimizing strain energy is equivalent to maximizing overall stiffness. As observed in the figure, with an increase in iterations, the optimization effectiveness of HC sections' stiffness improves progressively. At the conclusion of optimization, the strain energy of HC sections with volume fractions of 40% and 55% exceeds that of HC sections with a volume fraction of 70% by 56.2% and 20.4%, respectively. Thus, as the volume fraction decreases, the structural stiffness of HC sections diminishes. Consequently, when the volume fraction is 70%, the mechanical strength is optimal.

3.3. Determination of PCM-HC Model

Based on the topology optimization analysis results of the middle section, a gradual removal of the substrate material SUS304 (Figure 5b) along the cross-sectional direction of HC (Figure 5a) is performed. Subsequently, the cavity is filled with PCM to obtain the PCM-HC model (Figure 5c). In Figure 5d, the belly plate thickness (N) on the side wall varies between 0.15 mm, 0.30 mm, 0.45 mm, and 0.60 mm, while the PCM region on the upper wall has a trapezoidal shape. The characteristic parameters of the trapezoidal PCM region are determined based on the trapezoid area formula $S = (a+b)h/2$. The trapezoidal PCM cross section characteristic vector is defined as follows:

$$\mathbf{W}_i = (\alpha_i, \beta_i, h_i) = (\frac{a_i}{h_i}, \frac{b_i}{h_i}, h_i),\ \ i = 1, 2, \ldots, n \tag{5}$$

where a_i is the length of the upper base of the trapezoid, b_i is the length of the lower base of the trapezoid, and h_i is the height of the trapezoid.

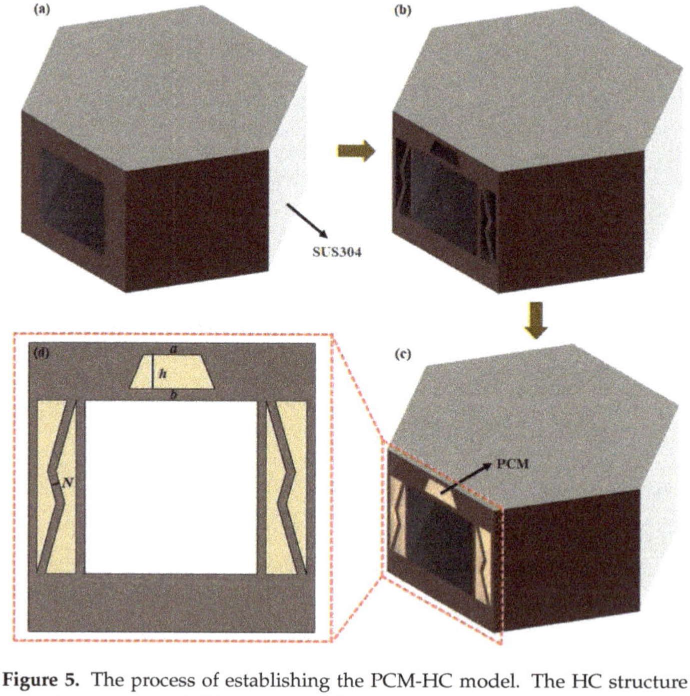

Figure 5. The process of establishing the PCM-HC model. The HC structure (**a**) is modified by removing a portion of SUS304 material (**b**) and filling it with PCM to obtain a PCM-HC structure (**c**). (**d**) The trapezoidal PCM area height is h, the upper bottom length is a, the lower bottom length is b, and the web thickness is N.

At this point, the area of the trapezoidal PCM region can be expressed as follows:

$$S = \frac{1}{2}(\alpha_i + \beta_i)h_i^2 \tag{6}$$

Keeping the area of the trapezoidal PCM region $S = 6\ \text{mm}^2$ constant, let $\mathbf{W_1} = (\frac{10}{3}, 5, \frac{6}{5})$, $\mathbf{W_2} = (\frac{5}{3}, \frac{20}{3}, \frac{6}{5})$, $\mathbf{W_3} = (4, 8, 1)$, $\mathbf{W_4} = (2, 10, 1)$.

When hypersonic vehicles cruise in near-space, they will experience severe aerodynamic loads, imposing significant structural demands on the internal convective cooling channels. The HC and PCM-HC model, inspired by the high mechanical performance of honeycomb structures, are compared with traditional square channels through a simulation of their mechanical properties to verify the structural load-bearing capacity. The square channels maintained the same load-bearing area and internal volume as the HC and PCM-HC models (Figure 6a).

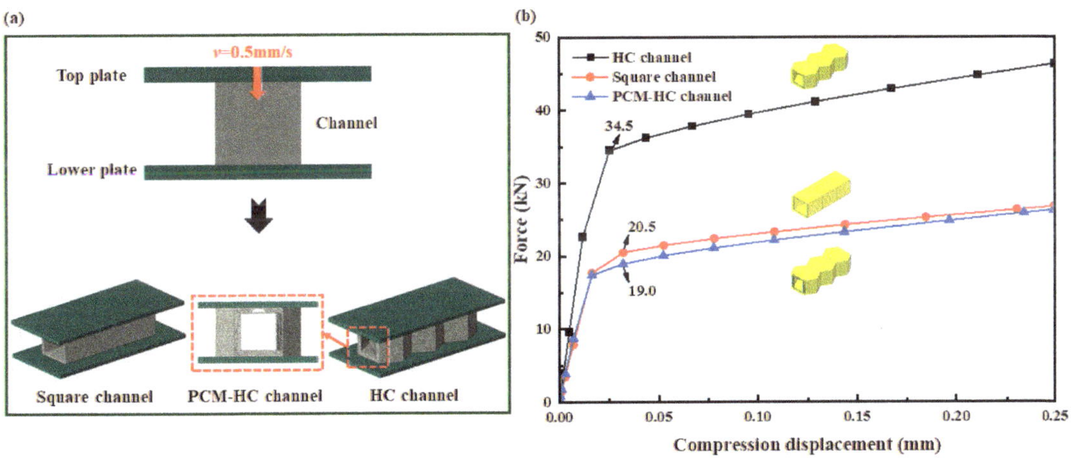

Figure 6. The finite element models and mechanical property analysis of the channel. (**a**) The finite element models of the HC channel, PCM-HC channel, and square channel maintain the same load-bearing area and cavity volume. (**b**) The variation of load capacity with displacement for the HC channel, PCM-HC channel, and square channel under axial compression between rigid plates is investigated.

Mechanical analysis was conducted using the commercial finite element software Abaqus 2021. As depicted in Figure 6a, the finite element model of the channel was placed between upper and lower rigid plates. The lower end of the rigid plate was fixed in all degrees of freedom, while the upper rigid plate compressed downward at a constant speed of 0.5 mm/s. The contact between the upper and lower rigid plates and the channel model was defined as general contact, with tangential penalty contact, a friction coefficient of 0.2, normal hard contact, and allowed separation after contact. The material of channels was defined as SUS304, with material parameters listed in Table 1. HC and square channels are discretized using eight-node linear hexahedral elements (C3D8R), while the PCM-HC channel employs ten-node quadratic tetrahedral elements (C3D10) for discretization, culminating in the completion of the computational solution.

Table 1. Material parameters of SUS304.

Material	ρ/g·cm^{-3}	E/Gpa	σ/Mpa	v
SUS304	7.93	204	800	0.285

The force-displacement curves under axial compression for the channel models depicted in Figure 6b reveal three distinct stages of deformation, namely elastic, yielding, and plastic deformation, for HC, PCM-HC, and square channels. With increasing axial displacement, the mechanical characteristic curves of the models exhibit a gradual upward trend. However, HC demonstrates significantly higher axial load-bearing capacity compared to square and PCM-HC channels, with peak test forces at yield strengths of 34.5 kN, 20.5 kN, and 19.0 kN, respectively. HC exhibits a 68% increase in load-bearing capacity over the square channel under identical bearing area and heat transfer area conditions. Even after removing some material to accommodate PCM, HC maintains comparable load-bearing performance to the traditional square channel.

4. Research on PCM-HC Performance

4.1. Physical Model Description

Due to the uniform heat flux conditions imposed on the upper surface of the convective cooling channel, both the HC and PCM-HC channels, as symmetric three-dimensional structures, can be simplified into a two-dimensional cross-sectional problem, where heat conduction along the length of the channel can offset each other. Under the same heat flux input conditions, if the temperature of the walls surrounding the channel is lower, the heat transfer burden on the cooling medium inside the channel is reduced. Establishing a two-dimensional cross-sectional model of the channel, the thermal distribution on the walls is studied in the absence of coolant flow, aiming to minimize the heat transfer burden on the convective cooling channel as much as possible.

To better simulate the practical application scenario of convective cooling channels, a conventional multi-layer TPS structure is introduced above the HC and PCM-HC structures. A thermal load is applied to the outer surface of the system to simulate the aerothermal load experienced by hypersonic aircraft during flight. The multi-layer TPS consists of an outer high-temperature structural layer and an inner non-load-bearing insulation layer, as shown in Figure 7. The upper layer, depicted in earthy yellow, is made of C/SiC composite high-temperature material to withstand the external thermal load Q_{in}, with a thickness of 2 mm. The middle layer, shown in dark blue, utilizes Saffil alumina fiber insulation material to isolate the significant heat generated by aerothermal effects, with a thickness of 10 mm. The lower layer, represented in deep green, serves as the convective cooling channel with a base material of SUS304. The gray area added in Figure 7b represents the filled PCM. The thermal and physical properties of C/SiC and Saffil refer to relevant literature [50,51], with some parameters varying with temperature.

For the HC model, the heat transfer modes considered include thermal radiation Q_{rad} on the top surface and internal heat conduction Q_{cond} within the structure. For the PCM-HC model, the heat transfer modes include thermal radiation Q_{rad} on the top surface, internal heat conduction, and convective heat transfer Q_{con} which arises only within the PCM due to the phase change process. In addition, the contact thermal resistance between different material interfaces is ignored, and each contact surface is treated as an ideal heat transfer surface. Assuming the liquid phase of the PCM behaves as an incompressible fluid, the volumetric force effects resulting from PCM phase change are neglected [52,53].

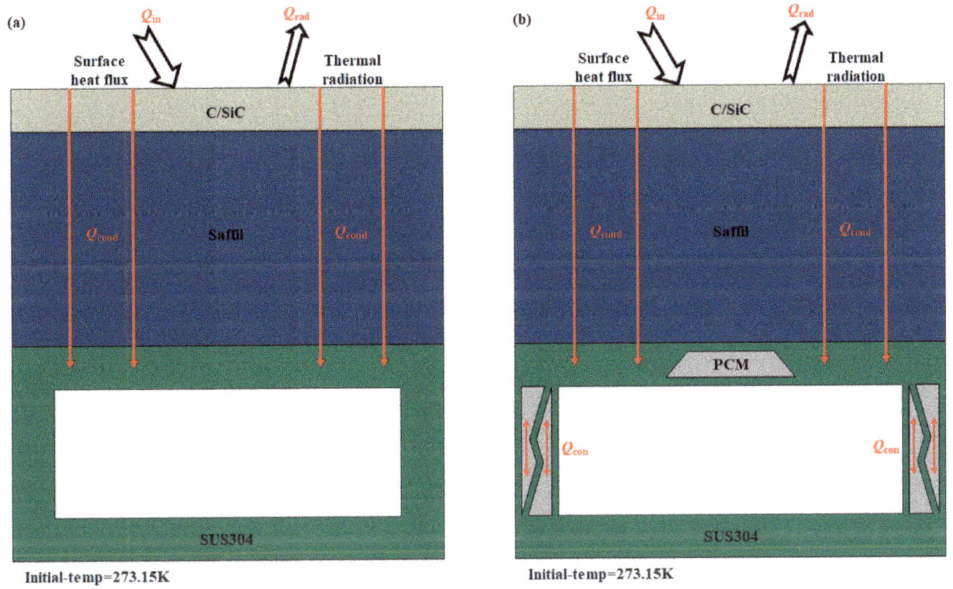

Figure 7. Transient heat transfer analysis models for HC and PCM-HC. In HC (**a**) and PCM-HC (**b**), Q_{in} represents the input heat flux load, Q_{rad} denotes the surface thermal radiation of the system, Q_{cond} signifies the internal heat conduction within the system, and Q_{con} represents the convective heat transfer within the PCM. The arrows represent the direction of heat transfer. The initial temperature of the system is 273.15 K.

4.2. Heat Transfer Mathematical Model

The following equations represent the governing equations for HC and PCM-HC systems, including continuity, energy, and momentum equations, which are described as follows:

(a) Continuity equation [54]:

$$\frac{\partial \rho}{\partial t} + \nabla(\rho \vec{v}) = 0 \tag{7}$$

where ρ is density and $\vec{v}$ is is velocity vector.

(b) Energy equation [55–57]:

$$\frac{\partial(\rho E)}{\partial t} + \nabla(\rho \vec{v} E) = \nabla(k \nabla T) + S \tag{8}$$

where E is the total enthalpy, k is the thermal conductivity coefficient, T is temperature, and S is the energy source term.

Total enthalpy E can also be represented as follows:

$$E = E_0 + \Delta E \tag{9}$$

where E_0 is sensible enthalpy and ΔE is latent enthalpy.

When the PCM is in a solid state, ΔE is 0. When the PCM is in a liquid state, ΔE is L, where L is the latent heat of the PCM. When the PCM is in a semi-solid or mushy state,

ΔE takes on values between 0 and 1. The liquid fraction function δ, which is temperature-dependent, is defined as follows [58]:

$$\delta(T) = \frac{\Delta E}{L} = \begin{cases} 0, T < T_s \\ \frac{(T-T_s)}{(T_l-T_s)}, T_s \leq T \leq \ T_l \\ 1, T > T_l \end{cases} \tag{10}$$

where T_s denotes the lower limit of the phase transition temperature range and T_l denotes the upper limit of the phase transition temperature range.

In Equation (8), S is equal to the sum of the internal heat source and radiative flux, i.e.,

$$S = \nabla(\rho \vec{v} \Delta E) + \nabla \vec{Q_{rad}} \tag{11}$$

When there is no internal heat source, the energy source term S is equal to the radiative flux. $\vec{Q_{rad}}$ represents the presence of outward thermal radiation on the upper surface of the system, which is defined by the following boundary condition equation [56]:

$$\vec{Q_{rad}} = k\frac{\partial T}{\partial y} = Q_{in} - \sigma\varepsilon(T_w^4 - T_e^4) \tag{12}$$

where σ is the Stefan–Boltzmann constant, σ = 5.67 $\times$ 10^{-8} W/(m^2·K^4). Q_{in} is input heat flux density, ε is radiance, and T_w is wall temperature. T_e is ambient temperature, according to the International Standard Atmosphere (ISA), and values are determined for different flight altitudes.

(c) Momentum equation [54]:

$$\frac{\partial \vec{v}}{\partial t} + \nabla(\vec{v}\vec{v}) = -\frac{1}{p}\nabla P + \frac{1}{p}\nabla^2(\mu\nabla\vec{v}) + \vec{F} \tag{13}$$

where μ is dynamic viscosity, P is pressure, and $\vec{F}$ is momentum source term.

When PCM undergoes a liquid phase change, the expression of dynamic viscosity μ is as follows [59,60]:

$$\mu = 0.001e^{(A+\frac{B}{T})} \tag{14}$$

where A is -4.25 and B is 1790 K.

Furthermore, in the analysis of transient heat transfer problems involving PCM, the enthalpy-porosity method is employed to calculate the phase change process [61]. In Equation (13), the momentum source term $\vec{F}$ is defined as follows:

$$\vec{F} = \frac{c_m(1-\delta)^2}{(\delta^3+\eta)}\vec{v}, \tag{15}$$

where c_m is a mushy zone constant that is between 10^5 and 10^7 [62,63]. The introduction of η is to prevent division by zero. The function δ, mentioned earlier, represents the temperature-dependent liquid fraction.

4.3. Numerical Simulation Program

Transient heat transfer analysis of HC and PCM-HC was conducted using the Abaqus finite element software. Structured grids were employed, and a series of 2D models were established to analyze the heat transfer capability of the convective cooling channel when using PCM. The grid type used was DC2D4, with local refinement near the PCM region. The total number of grids was 5059. By solving the two-dimensional equilibrium equations at each node, the temperature distribution at various nodes was calculated, and a temperature contour plot of the model was generated.

As for the boundary conditions, a uniform heat flux load Q_{in} was applied to the top surface of the system, with a loading time of 600 s. Additional time calculations were performed to obtain the maximum temperature on the bottom wall of the channel. The outer surface is subjected to aerodynamic loads, while the two sides and bottom are situated within the aircraft with minimal gap areas where convection effects can be neglected. Therefore, both sides and the bottom of the system are set as adiabatic boundary conditions [56]. Furthermore, there was outward thermal radiation on the top surface with an emissivity (ε) value of 0.85. Additionally, the initial temperature condition for the entire system was set as 273.15 K.

5. Result and Discussion

Firstly, a comparison was made between the heat transfer characteristics of the convective cooling channel with and without the use of PCM filling. Then, in the presence of PCM, the combined effects of two parameters, namely the trapezoidal PCM cross section characteristic vector ($\mathbf{W}$) and the belly plate thickness (N), were evaluated. In this study, five input heat flux densities were considered (30 kW/m^2, 35 kW/m^2, 40 kW/m^2, 45 kW/m^2, and 50 kW/m^2). Lastly, three types of PCM, namely Erythritol, Urea, and Paraffin composite phase change material (CPCM), were selected to investigate their respective impacts on heat transfer.

5.1. Effect of PCM on Heat Transfer Performance

In this section, the heat transfer performance of convective cooling channels with and without PCM filling was compared for five different input heat flux densities (30 kW/m^2, 35 kW/m^2, 40 kW/m^2, 45 kW/m^2, and 50 kW/m^2). Two channel system models were established to compare the effect of PCM. The only difference between them was that the HC channel consisted entirely of SUS304 while the PCM-HC channel was composed of SUS304 and Erythritol. The input heat flux was uniformly applied to the surface of the system, and heat was transferred within the TPS and channel walls primarily through conduction and radiation, as detailed in Section 4.1. It should be noted that in this study, the trapezoidal PCM cross section characteristic vector was $(\frac{10}{3}, 5, \frac{6}{5})$, and the belly plate thickness was 0.30 mm.

Figure 8 illustrates the wall temperature distribution at different time intervals in the PCM-HC channel under a heat flux load of 50 kW/m^2. In general, the heat flow is applied from the upper wall surface, resulting in a gradual decrease in wall temperature from the proximal end of the heat flow input to the distal end. Under the same heat transfer distance, the rate of heat transfer is dependent on the temperature difference between the object and its thermal conductivity, which can be explained by Fourier's law ($q = k\Delta T/L$, where q is the transferred heat flux, k is the material's thermal conductivity, ΔT is the temperature difference, and L is the distance between the heat exchange surfaces). At 510 s, the heat is uniformly diffused from the upper wall surface to both side wall surfaces, with a temperature difference of 14.2 K (Figure 8a). Subsequently, as the PCM reaches its phase change temperature, a significant amount of heat is absorbed during the phase change process. The heat transfer in the trapezoidal PCM region differs from the surrounding area, resulting in lower temperatures and a decreased rate of temperature rise, leading to almost constant temperatures (Figure 8b). At the same time, the PCM on the side walls also undergoes phase change and absorbs heat, resulting in a noticeable delay in temperature rise and a decrease in temperature gradient within the side walls (Figure 8c). At 540 s and 570 s, the temperature difference between the highest and lowest temperatures is reduced to 12.5 K and 13 K, respectively, as the phase change latent heat gradually decreases and the heat dissipation capacity reaches its limit. At 600 s, the PCM completes the phase change process, and the rate of temperature rise gradually recovers, with a reduced temperature difference of 8.8 K between the highest and lowest temperatures (Figure 8d).

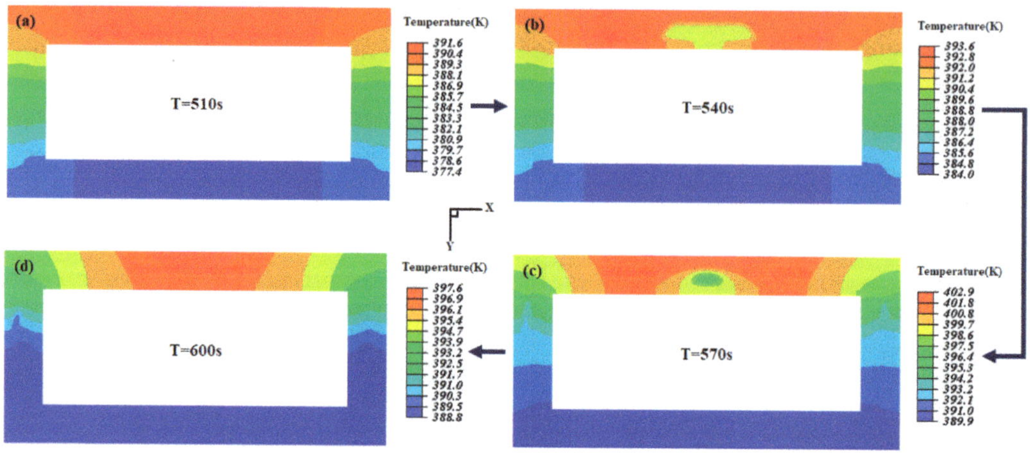

Figure 8. The wall temperature distribution of PCM-HC varies with time under a heat flux load of 50 kW/m^2. (**a–d**) are 510 s, 540 s, 570 s, and 600 s, respectively. The arrows represent the order of time change. $\mathbf{W} = (\frac{10}{3}, 5, \frac{6}{5})$, $N = 0.30$ mn.

As shown in Figure 9, the temperature variations along the AB path at 570 s and the CD path at 590 s under a heat flux load of 50 kW/m^2 are selected to demonstrate the thermal dissipation capability of PCM in the PCM-HC system. In the upper wall surface (Figure 9a), in the non-PCM region, the heat conduction rate is uniform and decreases slowly. However, in the PCM region, the temperature is significantly lower than in the non-PCM region, and the rate of temperature change accelerates. The temperature exhibits a decreasing trend followed by an increasing trend, with the lowest temperature dropping by nearly 5 K compared to the highest temperature in the non-PCM region. Additionally, compared to the HC channel without PCM, the PCM-HC system exhibits an average temperature reduction of 10 K to 15 K. Similarly, the side wall surface (Figure 9b) displays the same pattern of temperature variation. Due to the smaller area covered by PCM, the temperature drop in the PCM region is less pronounced. Nevertheless, it still demonstrates superior thermal dissipation capability compared to HC alone, with an average temperature reduction of 13 K. Thus, in the PCM-HC system, heat conduction near the PCM is slowed down, resulting in lower temperatures compared to the same location in the HC system and impeding heat transfer.

Figure 10a,b illustrate the temperature variations along the EF path (upper wall surface) and CD path (side wall surface), respectively, at the end of the heat flux loading for HC and PCM-HC under a 50 kW/m^2 heat flux input (at 600 s). These paths represent the inner wall surfaces involved in convective heat transfer with the fuel coolant, and a lower temperature distribution is desirable to reduce the burden of heat transfer with the fuel coolant. On the upper wall surface, both HC and PCM-HC exhibit a similar temperature trend, gradually increasing from the sides towards the center, with a temperature difference of approximately 4 K. This is because as the heat flux loading time increases, the phase change process of PCM is completed, leading to a loss in its ability to block heat transfer. The overall temperature of PCM-HC on the upper wall surface is about 12 K lower than that of HC. On the side wall surface, the temperature decrease in PCM-HC is faster than HC in the first half, while in the second half, the temperature decrease in PCM-HC is slower than HC. This is because PCM in the side wall surface is undergoing a phase change, absorbing a significant amount of heat due to latent heat, which attracts the heat flux from the upper wall surface and hinders heat conduction to the lower wall surface. The overall temperature of PCM-HC on the side wall surface is approximately 15 K lower than that of HC.

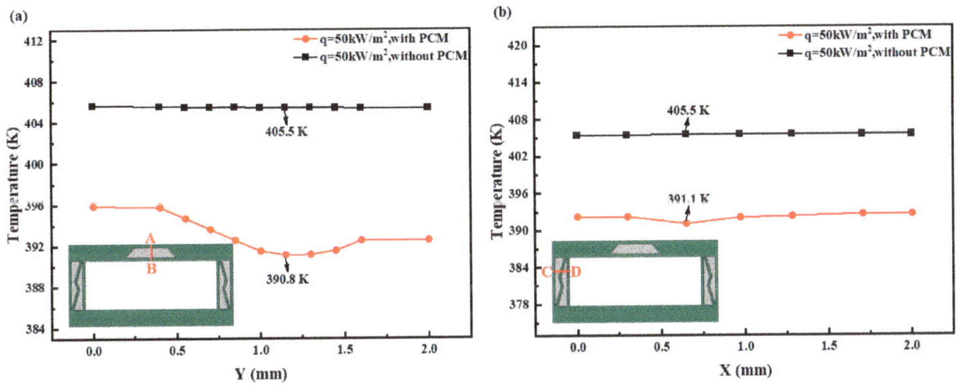

Figure 9. The temperature distribution along a specified path at a specific moment with and without PCM was examined. (**a**) The red line AB indicates the temperature distribution along the vertical trapezoidal PCM at 570 s on the upper wall surface. (**b**) The red line CD indicates the temperature distribution along the horizontal PCM at 590 s on the side wall surface.

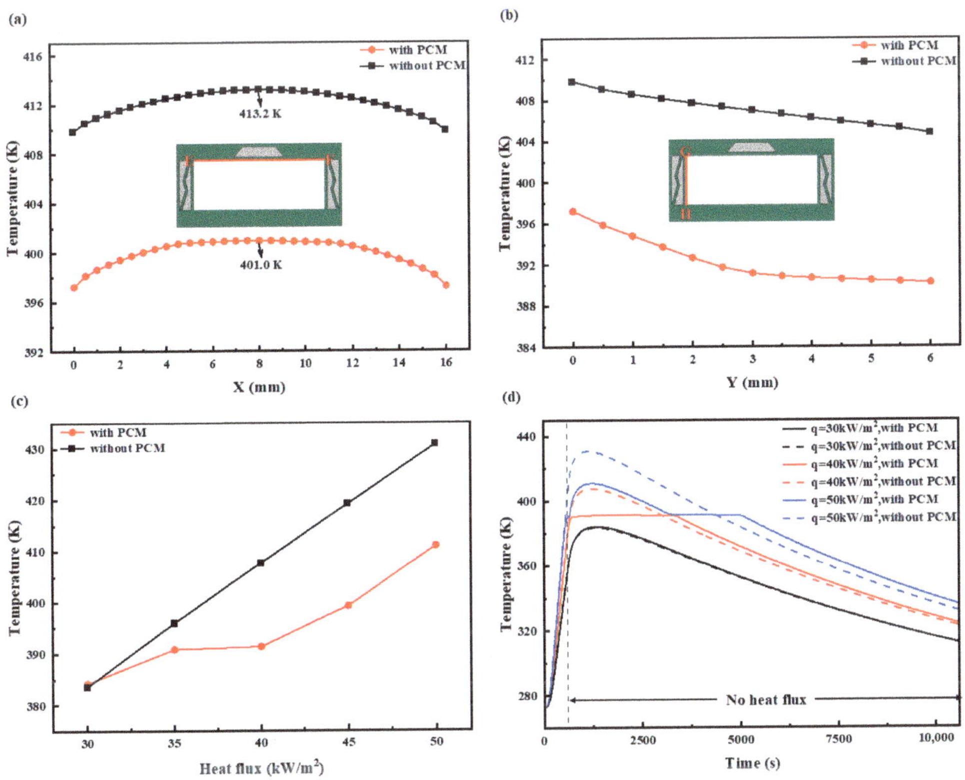

Figure 10. The effect of PCM on the heat transfer performance of HC and PCM-HC. (**a**) The temperature distribution of the upper wall (EF red line). (**b**) The temperature distribution of the side wall (GH red line). (**c**) With the increase in heat flux input (Q_{in} = 30, 35, 40, 45, and 50 kW/m^2), the maximum temperature of the lower wall changes. (**d**) Q_{in} = 30, 40, and 50 kW/m^2; the temperature variation of the lower wall during the entire loading and unloading process after 600 s was recorded.

Figure 10c shows the variation of the maximum temperature on the lower wall surface with and without PCM in the convective cooling channel. As shown in the figure, as the heat flux input increases, the temperature difference between HC and PCM-HC undergoes a change from almost zero to gradually increasing and then maintaining a certain level. This is because when the heat flux input is small, within the 600 s loading time, PCM has not reached its phase change temperature and therefore cannot play a role in heat dissipation. With the increase in heat flux input, after the end of the 600 s heat flux loading, the remaining heat supply enables PCM to slowly reach its phase change temperature, thus exerting a certain heat dissipation effect. Finally, when the heat flux input continues to increase to a level that can quickly meet the latent heat of PCM's phase change, PCM fully exerts its heat dissipation effect. When the heat flux input is 45 kW/m^2, the heat dissipation effect is close to saturation, and the maximum temperature on the lower wall surface with PCM is about 20 K lower than that without PCM. This trend indicates that different scenarios of convective cooling channels correspond to different input heat fluxes and heating durations. Thus, it is necessary to select PCM with a suitable phase change temperature according to the specific requirements to fully utilize its heat dissipation capability.

Figure 10d illustrates the variation trend of the lower wall temperature during the entire loading and unloading process of PCM-HC and HC at different heat flux inputs for 600 s. As the heat flux input increases, both with and without the heat flow input, PCM exhibits good thermal conductivity. When the heat flux input is relatively small, there is no significant difference in temperature between the cases with and without PCM throughout the process, since the PCM has not reached its phase change temperature. When the heat flux input increases to 50 kW/m^2, the highest temperature of the lower wall decreases from 402.7 K to 389.1 K with PCM under heat flow input, and from 430.8 K to 411.0 K without PCM under heat flow input. This reveals that the use of PCM to enhance the heat conduction of the wall in the convective cooling channel is suitable only for high heat flux inputs. Meanwhile, in the section without heat flow input, the temperature begins to drop after reaching its maximum value. As the temperature approaches the phase change temperature of the PCM, the PCM begins to store energy in the form of latent heat, and the temperature remains almost unchanged for a duration of 2000 s, delaying the temperature drop. After the PCM fully solidifies, further heat conduction causes the temperature to continue to decrease, resulting in a higher temperature of the channel with PCM than that without PCM, with a maximum temperature difference of 7.5 K. This implies that the solidification phase change of the PCM during temperature drop is crucial for maintaining the internal temperature stability of the spacecraft when it loses its heat source, consistent with the insulating function of natural honeycomb structures.

Therefore, based on the above analysis, the phase change latent heat of PCM effectively reduces the surface temperature when heat is loaded, improves the heat distribution, promotes heat conduction, and reduces the burden of convective heat transfer between the fuel coolant and channel wall in the convective cooling channel. At the same time, the phase change latent heat of PCM also plays a role in thermal insulation for the spacecraft as the system temperature drops. Therefore, the combination of PCM is an effective approach for designing convective cooling channels with excellent heat conduction capabilities.

5.2. Effect of Geometric Parameter on Heat Transfer Performance

5.2.1. Effect of Trapezoidal PCM Cross Section Characteristic Vector (W)

In this section, the effect of changing the characteristic vector of the trapezoidal PCM ($\mathbf{W_1} = (\frac{10}{3}, 5, \frac{6}{5})$, $\mathbf{W_2} = (\frac{5}{3}, \frac{20}{3}, \frac{6}{5})$, $\mathbf{W_3} = (4, 8, 1)$, $\mathbf{W_4} = (2, 10, 1)$) in PCM-HC was assessed while keeping the PCM area and position unchanged. Other control parameters were kept constant, Q_{in} = 50 kW/m^2.

The graph in Figure 11 illustrates the temperature variation along the EF path (upper wall) at the end of the thermal load (600 s) for different trapezoidal PCM cross section characteristic vectors under a heat flux input of 50 kW/m^2. It can be observed that the temperature distribution is symmetrical, and changes in characteristic vectors have minimal impact on the temperature at the edges of the wall but affect the thermal distribution at the middle position of the wall. As the height h of the trapezoid increases, the temperature at the middle position decreases. Additionally, increasing the difference in length between the upper and lower bases of the trapezoid also leads to a decrease in the middle position temperature. When **W** = (2,10,1), the temperature difference between the highest and lowest temperatures at the middle position is 0.3 K. Although the overall dimensions of the model influence the magnitude of temperature differences, it also indicates that the trapezoidal PCM characteristic vector provides optimal heat dissipation. We also studied the influence of trapezoidal PCM cross section characteristic vectors on the thermal distribution and heat dissipation on the side and bottom walls. However, since the temperature hardly changes, we will not provide detailed explanations here.

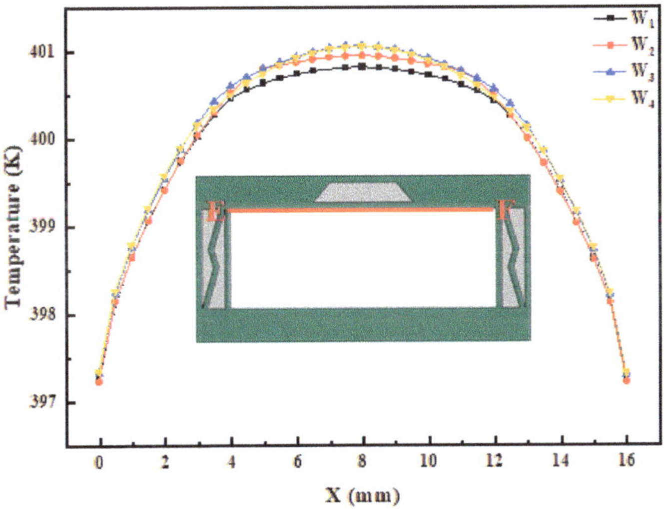

Figure 11. The effect of trapezoidal PCM cross section characteristic vector change ($\mathbf{W_1} = (\frac{10}{3}, 5, \frac{6}{5})$, $\mathbf{W_2} = (\frac{5}{3}, \frac{20}{3}, \frac{6}{5})$, $\mathbf{W_3} = (4, 8, 1)$, $\mathbf{W_4} = (2, 10, 1)$) on the temperature distribution on the upper wall of PCM-HC under a heat flux input of 50 kW/m^2.

5.2.2. Effect of Belly Plate Thickness (N)

In this section, the effect of belly plate thickness (N = 0.15 mm, 0.30 mm, 0.45 mm, 0.60 mm) on heat transfer in PCM-HC will be investigated under the fixed trapezoidal PCM cross section characteristic vector of **W** = (2,10,1), while keeping the other control parameters constant, Q_{in} = 50 kW/m^2.

Figure 12a illustrates that, under the same heat flux input, the temperature on the upper wall decreases with an increase in the baffle thickness. However, this decrease is not uniform; rather, it diminishes less as the thickness increases. This phenomenon arises because the increase in baffle thickness leads to an increase in the volume of the base SUS304 and a decrease in the PCM volume. SUS304 exhibits better thermal conductivity, facilitating faster heat transfer from the upper wall downwards. Simultaneously, the reduction in PCM volume reduces the heat absorption during PCM phase change, weakening its heat dissipation effect and thereby allowing heat to reach the side and lower walls more rapidly, resulting in a decrease in temperature on the upper wall. However, there exists a balance point, due to the limited heat conduction speed in the upper wall region, where the increase in baffle thickness causing accelerated heat conduction reaches equilibrium, hence reducing the decrease in temperature on the upper wall. The maximum temperature difference between baffle thicknesses of 0.60 mm and 0.15 mm is approximately 1.5 K.

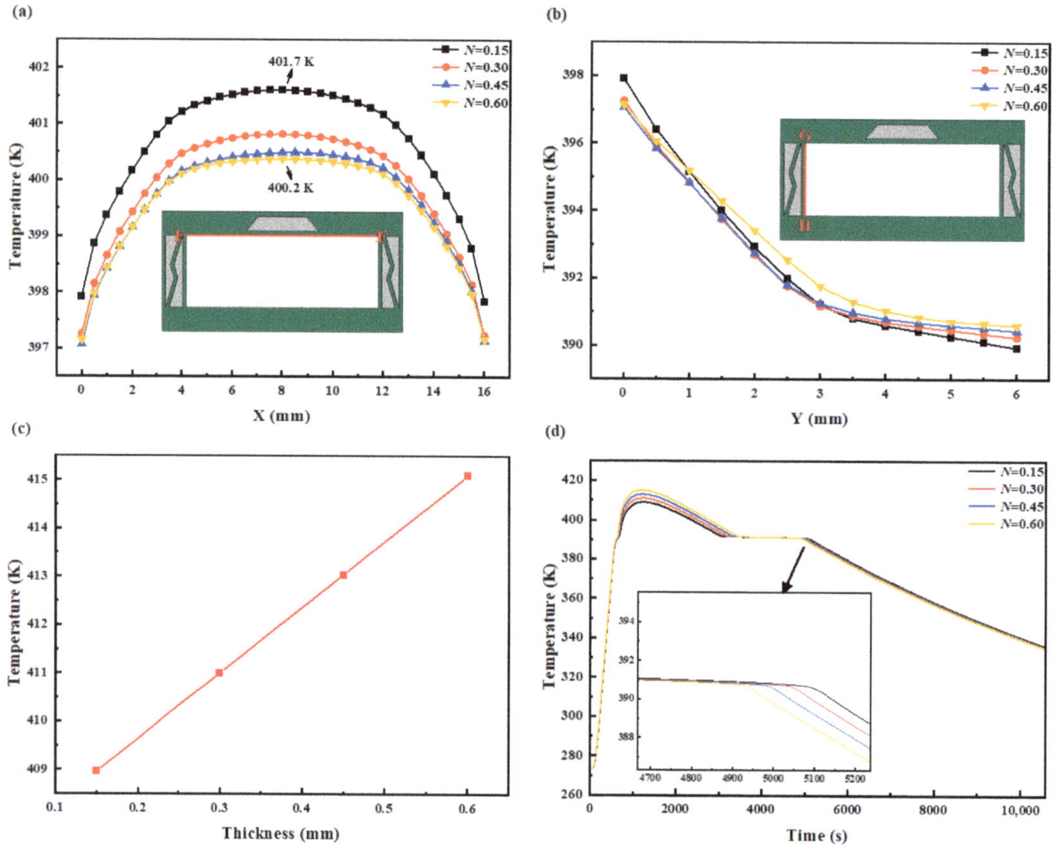

Figure 12. The effect of belly plate thickness changes (N = 0.15, 0.30, 0.45, and 0.60 mm) on the heat transfer performance of PCM-HC (Q_{in} = 50 kW/m^2, **W** = (2,10,1)). (**a**) The temperature distribution variation on the upper wall (EF red line). (**b**) The temperature distribution variation on the side wall (GH red line). (**c**) The variation of the maximum temperature on the lower wall with increasing belly plate thickness N. (**d**) The temperature variation on the lower wall during the entire loading and unloading process after 600 s.

Additionally, in Figure 12b, under the same heat flux input, except for N = 0.60 mm, the temperature on the side wall decreases in the upper half section and increases in the lower half section with an increase in the baffle thickness. The maximum temperature difference between baffle thicknesses of 0.60 mm and 0.15 mm exceeds 0.5 K. The variation inside wall temperature distribution is analogous to that of the upper wall, wherein the combined effects of SUS304 and PCM result in non-uniform changes in heat distribution with changing baffle thickness. Increasing baffle thickness accelerates heat transfer towards the channel's far end, yet weakens PCM's attraction to heat flux on the upper wall and its hindrance to heat transfer on the lower wall simultaneously.

Figure 12c illustrates the trend of the maximum temperature on the lower wall of the PCM-HC channel as the baffle thickness increases, showing a nearly linear relationship. The temperature difference between N = 0.60 mm and N = 0.15 mm is 6 K. This is attributed to the constant thickness of the left and right support pillars on the side wall, resulting in an increase in PCM and a decrease in SUS304 as the baffle thickness decreases. Compared to SUS304, a larger area of PCM enhances the heat dissipation capacity of the channel. PCM possesses latent heat of phase change and a larger specific heat capacity, meaning that when the area of the PCM-HC channel remains constant, PCM can absorb more heat at much lower temperatures. Simultaneously, PCM's latent heat of phase change absorbs more heat, resulting in a lower temperature on the lower wall.

Figure 12d depicts the variation trend of the lower wall temperature throughout the entire process of removing the heat load after 600 s of thermal loading under different baffle thicknesses in the PCM-HC channel. It can be observed that as the baffle thickness decreases, the heat dissipation effect of PCM, both with and without heat flux input, improves. Comparing $N = 0.15$ mm to $N = 0.60$ mm, with heat flux input, the maximum temperature on the lower wall decreases from 390.4 K to 389.5 K, and without heat flux input, it decreases from 415.1 K to 409.0 K. This indicates that adjusting the baffle thickness to enhance heat conduction on the wall of the convective cooling channel yields favorable results.

In summary, reducing the thickness of the baffle significantly decreases the temperature of the lower wall while causing a slight increase in temperature in certain regions of the upper and side walls. However, this increase is minimal compared to the temperature drop on the lower wall and can be considered negligible. Therefore, overall, reducing the baffle thickness is conducive to enhancing the heat dissipation capability of the PCM-HC channel. However, it is imperative to consider the mechanical strength of PCM-HC; decreasing the baffle thickness will inevitably diminish the channel's load-bearing performance. Consequently, the actual baffle thickness should be selected based on the specific load-bearing requirements of the channel. Hence, taking into account both heat transfer and load-bearing performance, a compromise is made in the test design, with a relatively optimal baffle thickness of 0.30 mm chosen.

5.3. *Effect of PCM Species on Heat Transfer Performance*

In this section, the effect of using different PCMs on PCM-HC heat transfer will be studied under the optimal geometric combination with a fixed trapezoidal PCM cross section characteristic vector of $\mathbf{W} = (2,10,1)$ and a fixed belly plate thickness of $N = 0.30$ mm, while keeping other control parameters constant and $Q_{in} = 50$ kW/m^2. To achieve better thermal management performance, PCMs with low thermal conductivity, high latent heat of phase change, and high specific heat capacity are preferred, provided that the PCM phase change temperature is lower than the actual temperature inside the system. Erythritol, Urea, and Paraffin CPCM were selected as the three PCM options, and their thermophysical properties [64–66] are listed in Table 2.

Table 2. PCM thermophysical properties.

Property	Paraffin CPCM	Erythritol	Urea
Density (kg/m^3)	780	1450	1320
Specific heat capacity (J/kg/K)	2350	2610	2100
Thermal conductivity (W/m/K)	1.12	0.73	0.8
Phase transition temperature (K)	343.15	391.15	407.15
Latent heat (kJ/kg)	189.6	340	250

As shown in Figure 13a,b, the filling of different PCMs has produced varying effects on the temperature distribution of the upper and side walls of PCM-HC. Erythritol, with its higher specific heat capacity and phase change heat and lower thermal conductivity compared to the other two PCMs, exhibits excellent thermal conductivity. Compared to the HC without PCM, the average temperature of the upper wall decreases by 12 K, and the average temperature of the side wall decreases by 15 K when using Erythritol. On the other hand, Urea shows poor thermal conductivity due to its higher phase change temperature relative to Erythritol and lower phase change latent heat. Within the limited heat flux loading time, the upper wall completes the phase change latent heat process early and continues to release sensible heat, while some regions of the PCM in the side wall do not reach the phase change temperature, resulting in the incomplete utilization of

PCM's thermal conductivity ability. As a result, the average temperature of the upper wall increases by 1.2 K, and the temperature of the side wall decreases by 0.5 K. The addition of Paraffin CPCM has the opposite effect on the system, leading to an increase in the temperatures of both the upper and side walls. The average temperature of the upper wall rises by nearly 11 K, and the average temperature of the side wall increases by 7.5 K. This is because Paraffin CPCM has a lower phase change temperature. At Q_{in} = 50 kW/m^2 and within a loading time of 600 s, PCM-HC completes the phase change and releases sensible heat much earlier. Additionally, the highest temperature on the lower wall of the PCM-HC channel in Figure 13c reflects the different effects of the three PCMs analyzed above. Compared to the HC without PCM, Erythritol, Urea, and Paraffin CPCM exhibit temperature changes of a decrease by 21.8 K, a decrease by 7.6 K, and an increase by 9 K, respectively.

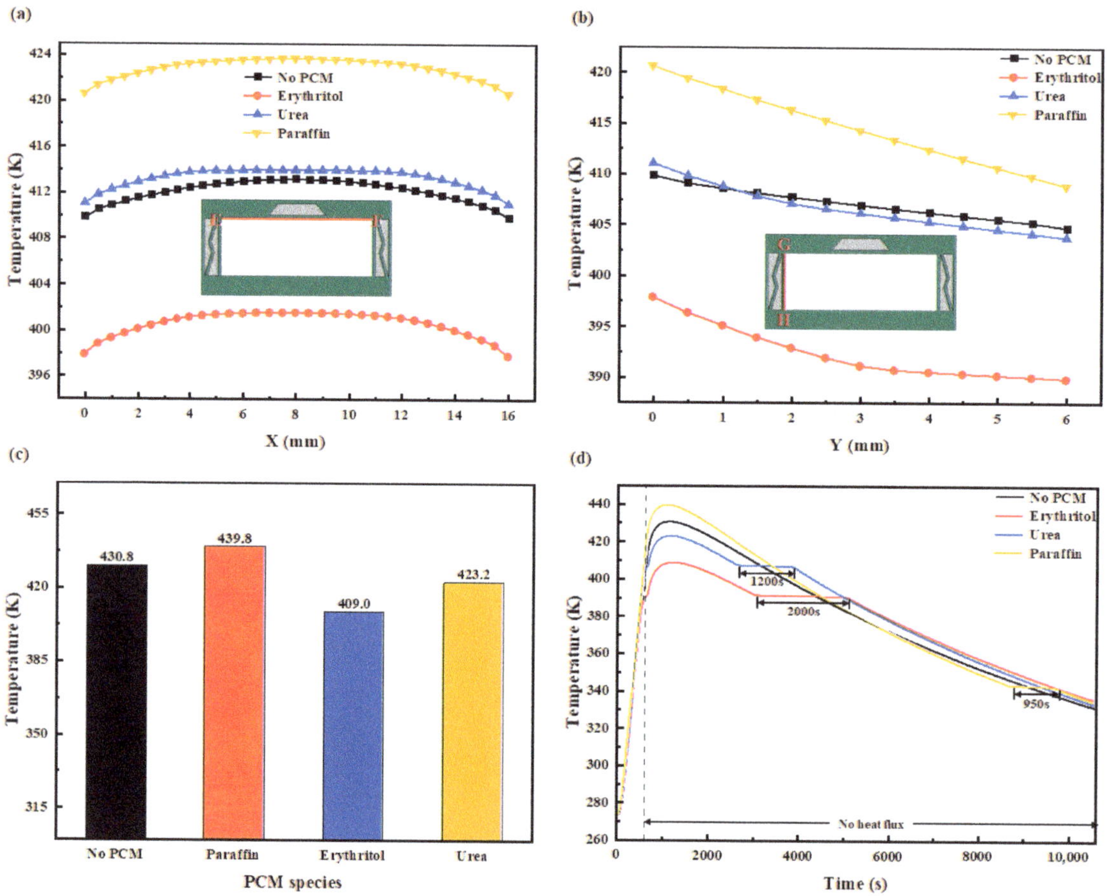

Figure 13. The effect of different PCMs (Erythritol, Urea, and Paraffin CPCM) on the heat transfer properties of PCM-HC (Q_{in} = 50 kW/m^2, **W** = (2,10,1), N = 0.30 mm). (**a**) The variation of temperature distribution on the upper wall (EF red line). (**b**) The variation of temperature distribution on the side wall (GH red line). (**c**) Comparison of the highest temperature on the bottom surface for different PCMs. (**d**) Variation trend of the temperature on the lower wall during the entire process of removing the heat load after 600 s of loading with different PCM fillings.

Figure 13d shows the variation trend of the temperature on the lower wall of PCM-HC during the entire process of removing the heat load after 600 s of loading with different PCM fillings. A comparison reveals that Erythritol exhibits the best thermal conductivity of the PCM-HC structure. During the heat flux loading stage, the highest temperature on the bottom surface decreases by 13 K. In the absence of heat flux loading, the highest temperature on the bottom surface decreases by 20 K. Meanwhile, during the phase change process without heat flux input, the temperature remains almost unchanged for a certain period of time, which is approximately 2000 s for Erythritol, 1200 s for Urea, and 950 s for Paraffin CPCM. This indicates a delayed temperature decrease and demonstrates that Erythritol provides the best insulation effect among the three PCMs.

Therefore, the analysis of heat transfer performance for different PCMs demonstrates that not all PCMs are beneficial for improving the thermal conductivity of PCM-HC. In summary, when using PCM-HC, it is necessary to select a PCM with suitable phase change temperature and latent heat according to the structural parameters and material properties of TPS, as well as the size and duration of the input heat flux load.

6. Conclusions

In this paper, a novel convective cooling channel filled with PCM is proposed. The introduction of PCM aims to improve the heat dissipation capability of the channel, that is, to reduce the wall temperature of the channel. Firstly, a model of a honeycomb-like convective cooling channel (HC) was proposed. Then, two-dimensional topology optimization of HC was conducted using the variable density method based on the SIMP interpolation model, and the PCM-filled PCM-HC structure was constructed. The mechanical load-bearing capacities of HC, PCM-HC, and square channels are then compared. Finally, numerical analysis of heat transfer was carried out from three aspects, i.e., with and without PCM, variation of PCM geometric parameters, and use of different types of PCMs, to determine the optimal geometric combination of the designed convective cooling channel (trapezoidal PCM cross section characteristic vector of (2,10,1), and a thickness of 0.15 mm for the belly plate). The main conclusions are as follows: (1) The convective cooling channel with PCM filling exhibits excellent thermal conductivity. The melting phase change of PCM changes the heat distribution, effectively reducing the wall temperature of the channel and reducing the burden of convective heat transfer between fuel and channel wall. Under the maximum heat flux load input of the tested design, the highest temperature on the inner wall of the channel decreased by 12 K to 20 K. Meanwhile, during the process of PCM solidification latent heat, the temperature remains almost constant when the temperature of the spacecraft system drops, providing insulation for the spacecraft. (2) When the heat flux load remains constant and the area of the trapezoidal PCM cross section is kept unchanged, the temperature distribution on the upper wall of the PCM-HC channel shows variations with changes in the characteristic vector parameters. The optimal thermal conductivity of the channel's upper wall is achieved when the trapezoidal PCM cross section characteristic vector parameters are set to (2,10,1). However, due to the constraints of the overall size of the channel design, this parameter has a less significant impact on the heat transfer performance of the side wall and lower wall. (3) As the thickness of the belly plate decreases from $N = 0.60$ mm to $N = 0.15$ mm, the thermal conductivity of the convective cooling channel exhibits non-uniform variations. Reducing the thickness of the belly plate can significantly lower the temperature of the lower wall of the channel. However, due to the combined effect of the two materials, SUS304 and PCM, in the channel, there is an uncertain impact on the temperature of the side walls. Nevertheless, this impact results in minimal temperature changes. Considering the load requirements of PCM-HC, a belly plate thickness of 0.30 mm is relatively optimal in the tested design, with the highest temperature on the lower wall decreasing by approximately 1 K. (4) Different types of PCMs have different effects on PCM-HC. Erythritol and Urea possess both thermal conductivity and insulation effects. However, Paraffin CPCM did not lower the wall temperature; instead, it increased it, attributed to the thermal physical properties of the PCM. (5) For the convective cooling channel filled with PCM designed in this study for application in the TPS of hypersonic vehicles, it is advantageous for heat dissipation if the difference between the heights of trapezoidal PCM cross-section vectors and the lengths of their upper and lower bases is designed to be greater. Additionally, a smaller thickness of the belly plate is favorable for heat dissipation, provided it meets the channel's load-bearing requirements. Furthermore, the selection of PCM with appropriate phase change temperature and latent heat should be based on the specific operating conditions of the channel, such as the magnitude of the input heat flux load and the duration of loading, to ensure the full utilization of PCM's heat dissipation capabilities.

Author Contributions: Conceptualization, W.L.; methodology, W.L.; investigation, W.L. and X.W.; data curation, W.L., L.L. and X.W.; writing—original draft preparation, W.L.; writing—review and editing, J.Z., X.W., L.L., W.W. and S.Y.; supervision, J.Z.; funding acquisition, J.Z. All authors have read and agreed to the published version of the manuscript.

Funding: This work was supported by the National Key R&D Program of China (2021YFB3400200), the National Natural Science Foundation of China (grant 52075038, 52375282), the National Science and Technology Major Project of China (grant SKS-2022031), and the BIT Teli Young Fellow Recruitment Program (RCPT-20220005).

Data Availability Statement: The research data for this article can be obtained from the corresponding author "Jieliang Zhao" on reasonable request.

Conflicts of Interest: The authors declare no conflict of interest.

References

1. Schmisseur, J.D. Hypersonics into the 21st Century: A Perspective on Afosr-Sponsored Research in Aerothermodynamics. *Prog. Aerosp. Sci.* **2015**, *72*, 3–16. [CrossRef]
2. Sziroczak, D.; Smith, H. A Review of Design Issues Specific to Hypersonic Flight Vehicles. *Prog. Aerosp. Sci.* **2016**, *84*, 1–28. [CrossRef]
3. Glass, D.E. Ceramic Matrix Composite (Cmc) Thermal Protection Systems (Tps) and Hot Structures for Hypersonic Vehicles. In Proceedings of the 15th AIAA International Space Planes and Hypersonic Systems and Technologies Conference, Dayton, OH, USA, 28 April–1 May 2008.
4. Liu, W.; Shu, B.; Guo, L.; He, H. Impacts of Thermal Environment on Modal Frequency of Fgm Shells. *J. Vib. Shock.* **2017**, *36*, 127–131.
5. Jackson, T.A.; Eklund, D.R.; Fink, A.J. High Speed Propulsion: Performance Advantage of Advanced Materials. *J. Mater. Sci.* **2004**, *19*, 39. [CrossRef]
6. Yang, Q.; Xie, W.; Peng, Z.; Meng, S.; Du, S. New Concepts and Trends in Development of Thermal Protection Design and Analysis Technology. *Hangkong Xuebao/Acta Aeronaut. Astronaut. Sinica* **2015**, *36*, 2981–2991.
7. Allouhi, A.; Agrouaz, Y.; Amine, M.B.; Rehman, S.; Buker, M.S.; Kousksou, T.; Jamil, A.; Benbassou, A. Design Optimization of a Multi-Temperature Solar Thermal Heating System for an Industrial Process. *Appl. Energy* **2017**, *206*, 382–392. [CrossRef]
8. Barth, T.; Longo, J.M.A. Advanced Aerothermodynamic Analysis of Shefex I. *Aerosp. Sci. Technol.* **2010**, *14*, 587–593. [CrossRef]
9. Chang, B.; Huang, J.; Yao, W.-X. Thermal Protection Mechanism of a Novel Adjustable Non-Ablative Thermal Protection System for Hypersonic Vehicles. *Aerospace* **2023**, *10*, 1. [CrossRef]
10. Rivers, H.K.; Glass, D.E. Advances in Hot-Structure Development. In Proceedings of the 5th European Workshop on Thermal Protection Systems and Hot Structures, Noordwiji, The Netherlands, 17–19 May 2006.
11. Bao, W.; Li, X.L.; Qin, J.; Zhou, W.X. Modelling and Simulation Methodology of Channel Cooling Using Hydrocarbon Fuel as Coolant under Supercritical Pressures. *Proc. Inst. Mech. Eng. Zhou Part G J. Aerosp. Eng.* **2011**, *225*, 969–984. [CrossRef]
12. Zhang, S.; Qin, J.; Xie, K.; Feng, Y.; Bao, W. Thermal Behavior inside Scramjet Cooling Channels at Different Channel Aspect Ratios. *J. Propuls. Bao Power* **2016**, *32*, 57–70. [CrossRef]
13. Uyanna, O.; Najafi, H. Thermal Protection Systems for Space Vehicles: A Review on Technology Development, Current Challenges and Future Prospects. *Acta Astronaut.* **2020**, *176*, 341–356. [CrossRef]
14. Maruyama, S.; Viskanta, R.; Aihara, T. Active Thermal Protection System against Intense Irradiation. *J. Thermophys. Heat Transf.* **2012**, *3*, 389–394. [CrossRef]
15. Liu, S.; Zhang, B. Effects of Active Cooling on the Metal Thermal Protection Systems. *Aerosp. Sci. Technol.* **2011**, *15*, 526–533. [CrossRef]
16. Zhou, W.; Johnson, B.; Hu, H. Effects of Flow Compressibility and Density Ratio on Film Cooling Performance. *J. Propuls. Power* **2016**, *33*, 964–974. [CrossRef]
17. Lee, Y.J.; Kim, S.J. Thermal Optimization of the Pin-Fin Heat Sink with Variable Fin Density Cooled by Natural Convection. *Appl. Therm. Eng.* **2021**, *190*, 116692. [CrossRef]

18. Vinh Tung, L.; San Ha, N.; Seo Goo, N. Advanced Sandwich Structures for Thermal Protection Systems in Hypersonic Vehicles: A Review. *Compos. Part B Eng.* **2021**, *226*, 109301.
19. Naraghi, M.; Dunn, S.; Coats, D. A Model for Design and Analysis of Regeneratively Cooled Rocket Engines. In Proceedings of the 40th AIAA/ASME/SAE/ASEE Joint Propulsion Conference and Exhibit, Fort Lauderdale, FL, USA, 11–14 July 2004.
20. Paquette, E.T.; Warburton, R.; Eckel, A.; Jaskowiak, M. Cooled Cmc Scramjet Combustor Structure Development. In Proceedings of the 38th AIAA/ASME/SAE/ASEE Joint Propulsion Conference & Exhibit, Indianapolis, IN, USA, 7–10 July 2002.
21. Paquette, E. Cooled Cmc Structures for Scramjet Engine Flowpath Components. In Proceedings of the Aiaa/cira International Space Planes & Hypersonics Systems & Technologies Conference, Capua, Italy, 16–20 May 2006.
22. Eckert, E.R.G.; Livingood, J.N.J. *Comparison of Effectiveness of Convection-, Transpiration-, and Film-Cooling Methods with Air as Coolant*; National Advisory Committee for Aeronautics: Washington, DC, USA, 1953; Volume 1182.
23. Sadique, H.; Murtaza, Q. Samsher Heat Transfer Augmentation in Microchannel Heat Sink Using Secondary Flows: A Review. *Int. J. Heat Mass Transf.* **2022**, *194 Pt 2*, 123063. [CrossRef]
24. Wang, L.-H.; Liu, X.; Wang, H.-Y.; Liu, L.-Z. Study of the Technique of Phase Change Thermal Storage Applied in Aircraft Thermal Control System. *Missiles Space Veh.* **2017**, *6*, 88–92.
25. Olds, J.R. Results of a Rocket-Based Combined-Cycle Ssto Design Using Parametric Mdo Methods. *J. Aerosp.* **1994**, *103*, 154–173.
26. Winston, M.L. *The Biology of the Honey Bee*; Harvard University Press: Cambridge, MA, USA, 1991.
27. Karihaloo, B.L.; Zhang, K.; Wang, J. Honeybee Combs: How the Circular Cells Transform into Rounded Hexagons. *J. R. Soc. Interface* **2013**, *10*, 20130299. [CrossRef] [PubMed]
28. Zhang, K.; Duan, H.; Karihaloo, B.L.; Wang, J. Hierarchical, Multilayered Cell Walls Reinforced by Recycled Silk Cocoons Enhance the Structural Integrity of Honeybee Combs. *Proc. Natl. Acad. Sci. USA* **2010**, *107*, 9502–9506. [CrossRef] [PubMed]
29. Zhang, K.; Si, F.W.; Duan, H.L.; Wang, J. Microstructures and Mechanical Properties of Silks of Silkworm and Honeybee. *Acta Biomater.* **2010**, *6*, 2165–2171. [CrossRef] [PubMed]
30. Duan, H.L.; Wang, J.; Karihaloo, B.L.; Huang, Z.P. Nanoporous Materials Can Be Made Stiffer Than Non-Porous Counterparts by Surface Modification. *Acta Mater.* **2006**, *54*, 2983–2990. [CrossRef]
31. Duan, H.L.; Karihaloo, B.L. Thermo-Elastic Properties of Heterogeneous Materials with Imperfect Interfaces: Generalized Levin's Formula and Hill's Connections. *J. Mech. Phys. Solids* **2007**, *55*, 1036–1052. [CrossRef]
32. Lu, T.J. Heat Transfer Efficiency of Metal Honeycombs. *Int. J. Heat Mass Transf.* **1999**, *42*, 2031–2040. [CrossRef]
33. Wen, T.; Tian, J.; Lu, T.J.; Queheillalt, D.T.; Wadley, H.N.G. Forced Convection in Metallic Honeycomb Structures. *Int. J. Heat Mass Transf.* **2006**, *49*, 3313–3324. [CrossRef]
34. Pei, B.; Chen, Z.; Li, F.; Bai, B. Flow and Heat Transfer of Supercritical Co2 in the Honeycomb Ultra-Compact Plate Heat Exchanger. *J. Supercrit. Fluids* **2019**, *148*, 1–8. [CrossRef]
35. Nkwetta, D.N.; Haghighat, F. Thermal Energy Storage with Phase Change Material—A State-of-the Art Review. *Sustain. Cities Soc.* **2014**, *10*, 87–100. [CrossRef]
36. Kim, T.Y.; Hyun, B.-S.; Lee, J.-J.; Rhee, J. Numerical Study of the Spacecraft Thermal Control Hardware Combining Solid–Liquid Phase Change Material and a Heat Pipe. *Aerosp. Sci. Technol.* **2013**, *27*, 10–16. [CrossRef]
37. Wang, S.; Zhang, D.; Li, C.; Wang, J.; Zhang, J.; Cheng, Y.; Mei, W.; Cheng, S.; Qin, P.; Duan, Q. Numerical Optimization for a Phase Change Material Based Lithium-Ion Battery Thermal Management System. *Appl. Therm. Eng.* **2023**, *222*, 119839. [CrossRef]
38. Tang, L.; Gao, T.; Song, L.; Meng, L.; Zhang, C.; Zhang, W. Topology Optimization of Nonlinear Heat Conduction Problems Involving Large Temperature Gradient. *Comput. Methods Appl. Mech. Eng.* **2019**, *357*, 112600. [CrossRef]
39. Yu, M.; Ruan, S.; Gu, J.; Ren, M.; Li, Z.; Wang, X.; Shen, C. Three-Dimensional Topology Optimization of Thermal-Fluid-Structural Problems for Cooling System Design. *Struct. Multidiscip. Optim.* **2020**, *62*, 3347–3366. [CrossRef]
40. Fawaz, A.; Hua, Y.; Le Corre, S.; Fan, Y.; Luo, L. Topology Optimization of Heat Exchangers: A Review. *Energy* **2022**, *252*, 124053. [CrossRef]
41. Sigmund, O.; Maute, K. Topology Optimization Approaches: A Comparative Review. *Struct. Multidiscip. Optim.* **2013**, *48*, 1031–1055. [CrossRef]
42. Zhuang, C.; Xiong, Z.; Ding, H. Topology Optimization of the Transient Heat Conduction Problem on a Triangular Mesh. *Numer. Heat Transf. Part B Fundam.* **2013**, *64*, 239–262. [CrossRef]
43. Xia, Q.; Shi, T.; Xia, L. Topology Optimization for Heat Conduction by Combining Level Set Method and BESO Method. *Int. J. Heat Mass Transf.* **2018**, *127*, 200–209. [CrossRef]
44. Zhou, M.; Geng, D. Multi-Scale and Multi-Material Topology Optimization of Channel-Cooling Cellular Structures for Thermomechanical Behaviors. *Comput. Methods Appl. Mech. Eng.* **2021**, *383*, 113896. [CrossRef]
45. Li, Q.; Xu, R.; Liu, S.; Liang, G.; Qu, Y. Topology Optimization Design of Multi-Material Quasi-Periodic Cellular Structures for Thermoelastic Responses. *Int. J. Numer. Methods Eng.* **2022**, *123*, 4345–4366. [CrossRef]
46. Huang, M.; Sha, W.; Xiao, M.; Gao, L.; Liu, X.; Wang, Y. Multiscale Topology Optimization of Cellular Structures with High Thermal Conductivity and Large Convective Surface Area. *Int. J. Therm. Sci.* **2024**, *201*, 109053. [CrossRef]
47. Tautz, J.; Heilmann, H.R. *Phänomen Honigbiene*; Springer: Berlin/Heidelberg, Germany, 2007.
48. Sigmund, O. A 99 Line Topology Optimization Code Written in Matlab. *Struct. Multidiscip. Optim.* **2001**, *21*, 120–127. [CrossRef]
49. Rietz, A. Sufficiency of a Finite Exponent in Simp (Power Law) Methods. *Struct. Multidiscip. Optim.* **2001**, *21*, 159–163. [CrossRef]

50. Myers, D.E. *Parametric Weight Comparison of Advanced Metallic, Ceramic Tile, and Ceramic Blanket Thermal Protection Systems;* National Aeronautics and Space Administration, Langley Research Center: Hampton, VA, USA, 2000.
51. Ma, Y.; Xu, B.; Chen, M.; He, R.; Wen, W.; Cheng, T.; Fang, D. Optimization Design of Built-Up Thermal Protection System Based on Validation of Corrugated Core Homogenization. *Appl. Therm. Eng.* **2017**, *115*, 491–500. [CrossRef]
52. Yang, X.-H.; Tan, S.-C.; He, Z.-Z.; Zhou, Y.-X.; Liu, J. Evaluation and Optimization of Low Melting Point Metal Pcm Heat Sink against Ultra-High Thermal Shock. *Appl. Therm. Eng.* **2017**, *119*, 34–41. [CrossRef]
53. Sánchez, P.S.; Ezquerro, J.M.; Fernandez, J.; Rodriguez, J. Thermocapillary Effects During the Melting of Phase Change Materials in Microgravity: Heat Transport Enhancement. *Int. J. Heat Rodriguez Mass Transf.* **2020**, *163*, 120478. [CrossRef]
54. Chen, Y.; Zhuang, Y.; Feng, J.-C.; Huang, S.-M. Numerical Investigation on Flow and Heat Transfer Characteristics in Honeycomb-Like Microchannel Heat Sink Encapsulated with Phase Change Material. *Appl. Therm. Eng.* **2023**, *232*, 121060. [CrossRef]
55. Bhuiya, R.; Shah, N.; Arora, D.; Krishna, N.V.; Manikandan, S.; Selvam, C.; Lamba, R. Thermal Management of Phase Change Material Integrated Thermoelectric Cooler with Different Heat Sink Geometries. *J. Energy Storage* **2022**, *51*, 104304. [CrossRef]
56. Ren, X.M.; Nie, H.; Zhang, M.; Nikitchenko, Y.A.; Tikhonovets, A. Heat Performance Investigation of Phase Change Materials for the Cooling Array System on Vehicles. *Appl. Therm. Eng.* **2020**, *171*, 115115. [CrossRef]
57. Voller, V.R. An Overview of Numerical Methods for Solving Phase Change Problems. *Adv. Numer. Heat Transf.* **1997**, *1*, 341–380.
58. Niezgoda-Żelasko, B. The Enthalpy-porosity Method Applied to the Modelling of the Ice Slurry Melting Process During Tube Flow. *Procedia Eng.* **2016**, *157*, 114–121. [CrossRef]
59. Hasis, F.B.A.; Krishna, P.M.; Aravind, G.P.; Deepu, M.; Shine, S.R. Thermo Hydraulic Performance Analysis of Twisted Sinusoidal Wavy Microchannels. *Int. J. Therm. Sci.* **2018**, *128*, 124–136. [CrossRef]
60. Debich, B.; El Hami, A.; Yaich, A.; Gafsi, W.; Walha, L.; Haddar, M. Design Optimization of Pcm-Based Finned Heat Sinks for Mechatronic Components: A Numerical Investigation and Parametric Study. *J. Energy Storage* **2020**, *32*, 101960. [CrossRef]
61. Yang, Y.-T.; Wang, Y.-H. Numerical Simulation of Three-Dimensional Transient Cooling Application on a Portable Electronic Device Using Phase Change Material. *Int. J. Therm. Sci.* **2012**, *51*, 155–162. [CrossRef]
62. Egolf, P.W.; Manz, H. Theory and Modeling of Phase Change Materials with and without Mushy Regions. *Int. J. Heat Mass Transf.* **1994**, *37*, 2917–2924. [CrossRef]
63. Wang, Y.-H.; Yang, Y.-T. Three-Dimensional Transient Cooling Simulations of a Portable Electronic Device Using PCM (Phase Change Materials) in Multi-Fin Heat Sink. *Energy* **2011**, *36*, 5214–5224. [CrossRef]
64. Sarbu, I.; Dorca, A. Review on Heat Transfer Analysis in Thermal Energy Storage Using Latent Heat Storage Systems and Phase Change Materials. *Int. J. Energy Res.* **2018**, *43*, 29–64. [CrossRef]
65. Zalba, B.; Marín, J.M.; Cabeza, L.F.; Mehling, H. Review on Thermal Energy Storage with Phase Change: Materials, Heat Transfer Analysis and Applications. *Appl. Therm. Eng.* **2003**, *23*, 251–283. [CrossRef]
66. Solé, A.; Neumann, H.; Niedermaier, S.; Martorell, I.; Schossig, P.; Cabeza, L.F. Stability of Sugar Alcohols as PCM for Thermal Energy Storage. *Sol. Energy Mater. Sol. Cells* **2014**, *126*, 125–134. [CrossRef]

micromachines

MDPI

Article

Experimental and Numerical Investigations on Thermal-Hydraulic Performance of Three-Dimensional Overall Jagged Internal Finned Tubes

Shufeng Huang [1,*], Mingjiang Deng [1], Zhixin Chen [1], Dayong Yang [2,*], Yingshuai Xu [1] and Ning Lan [1]

1 School of Mechanical and Electronic Engineering, East China University of Technology, Nanchang 330013, China; xuyingshuai@ecut.edu.cn (Y.X.); ln66645@163.com (N.L.)
2 School of Mechanical and Automotive Engineering, Guangxi University of Science and Technology, Liuzhou 545006, China
* Correspondence: shufeng3030@163.com (S.H.); dyyang@gxust.edu.cn (D.Y.)

Abstract: To satisfy the demand for efficient heat transfer, a novel three-dimensional overall jagged internal finned tube (3D-OJIFT) was fabricated, using the rolling–ploughing/extruding method. The thermal performance of the 3D-OJIFT were studied and compared in experiments and three-dimensional numerical simulations. The RNG k-ε turbulence model is well verified with the experimental results. By analyzing the distributions of velocity, temperature, and turbulence kinetic energy, it was found that the 3D-OJIFT destroyed the development of the velocity and thermal boundary layers, increased the turbulence disturbance, and reduced the temperature gradient, thus improving the heat transfer. The influences of the jagged height and jagged spiral angle of the 3D-OJIFT are discussed. The Nu and f increased as the jagged height of the 3D-OJIFT increased. The Nusselt number of the 3D-OJIFT was 1.67–2.04 times the value for the smooth tube. In addition, the comprehensive heat transfer performance of the 3D-OJIFT improved after increasing the jagged spiral angle. Compared with conventional internal helical-finned tubes and other reinforcement structures reported in the literature, the 3D-OJIFT demonstrated better comprehensive heat transfer performance. Finally, empirical correlations of the 3D-OJIFT were obtained.

Keywords: three-dimensional overall jagged internal finned tube; heat transfer enhancement; experiment; numerical simulation

1. Introduction

Shell-and-tube heat exchangers have been widely applied in many fields, such as air conditioning, power plants, solar heaters, and microreactors [1]. To reduce energy consumption and improve heat transfer, different enhanced heat transfer technologies including passive and active methods have been studied and used in the past few decades [2]. Passive methods typically utilize insertions and/or modified surfaces, which increase heat transfer as the friction factor increases. Finned tubes effectively improve heat transfer with small friction factors and are easy to manufacture and maintain. Owing to these characteristics, finned tubes have been widely used in shell-and-tube heat exchangers.

Many experiments have been conducted to study the heat transfer performance of the tube side of finned tubes. Among these enhanced heat transfer technologies, dimpled tubes, corrugated tubes, internal helical-finned tubes (IHFTs), and three-dimensional internal finned tubes (3D-IFTs) have been investigated experimentally [3]. Dimpled tubes have high heat-transfer efficiency and a relatively small friction factor, owing to the swirls and vortices generated by the dimples [4,5]. The effects of different parameters, such as the concavity, dimple imprint diameter, and depth on the thermal performance of dimpled tubes were investigated [6]. The results of those studies indicated that the heat transfer enhancement could reach 40% with negligible pressure drop [7]. Xie et al. [8] investigated the effects of the number of starts, radius, spiral pitch, depth, and transverse length of a helically dimpled tube on its thermal-hydraulic performance. The correlations for Nu and f were obtained. Corrugated tubes have also been used for improving the thermal performance of the tube side, owing to the swirls induced by corrugations and a larger heat transfer area. Different helical pitches, heights, and characteristics of helical corrugated tubes have been investigated [9,10]. All of the studied corrugated geometries performed better than smooth tubes, but corrugated tubes exhibited worse thermal performance in laminar flows [10,11]. IHFTs are effective at improving the thermal performance of the tube side [12]. Ma et al. [13]. reported that the j-factor for an IFT was approximately 3.5 times that of a smooth tube. The effects of different structure parameters (fin number, diameter, helix angle, width, and fin height) on the heat transfer performance were studied. Lie et al. [14]. found that Nu and f both increased as the values of these parameters increased and that the helical angle had the strongest effect on the heat transfer enhancement. Celen et al. [15]. developed a friction equation for assessing the friction performance. IHFTs have been widely used. However, owing to the low fin height of IHFTs, the spiral flow in such tubes limits the adjustment of the boundary layer [12]. Therefore, IHFTs cannot satisfy the demand for efficient heat transfer.

To address the need for efficient heat transfer, 3D-IFTs have attracted considerable attention in industry, owing to their higher fins. Wang et al. [16] prepared a twisted pin-finned pipe and circumferential pin-finned pipe based on the 3D printing method, and the results showed that the heat transfer coefficients were larger than those of a corresponding plain pipe. However, the productivity and manufacturing costs of these tubes do not meet the requirements of producers. Huang et al. [17] proposed arc-shaped inner-finned tubes and experimentally analyzed their heat transfer performance. However, the heat transfer characteristics and other structural parameters of the heat transfer performance were not addressed. Han et al. [18] studied the heat transfer performance of 3D-IFTs with porous copper fiber inserts. The effects of the diameter, spacing, and porosity of the porous copper fiber inserts on the thermal performance were investigated.

According to the aforementioned literature review, relatively little research has been conducted on 3D-IFTs. The flow inside 3D-IFTs remains poorly understood. In addition, a few empirical correlations have been considered for predicting the thermal performance of 3D-IFTs. In this study, a novel three-dimensional overall jagged internal finned tube (3D-OJIFT) was developed based on the rolling–ploughing/extruding (RPE) method. The characteristics of the 3D-OJIFT were analyzed. Thermal performance experiments were conducted, and the Nu and f experimental values were used for validating the numerical model. By analyzing the distributions of the turbulent kinetic energy, velocity field, temperature field, and local convective heat transfer coefficient, the mechanisms of the 3D-OJIFT heat transfer enhancement were elucidated. Moreover, the effects of the jagged height and jagged spiral angle on the 3D-OJIFT thermal performance were investigated. The 3D-OJIFT was compared with conventional IHFTs. Finally, the empirical correlations for Nu and f were computed and presented.

2. Experimental Studies

2.1. Characterization of the 3D-OJIFT

To satisfy the demand for efficient heat transfer, a 3D-OJIFT was fabricated using the RPE method. The manufacture of a 3D-OJIFT involves two processes. First, the inner thread structure is formed inside the tube by a rolling process, and then, the inner thread structure is ploughed and extruded by using the self-designed tool. Finally, 3D-OJIFTs are obtained. As shown in Figure 1, the novel internal fin has a jagged structure with the characteristics of a three-dimensional integral and rough surface. The 3D-OJIFT comprises a jagged internal fin and staggered grooves, distributed at a certain spiral angle. Compared with conventional IHFTs, the internal fin height of the 3D-OJIFT is significantly higher, which is conducive to increasing the heat transfer area and turbulent flow.

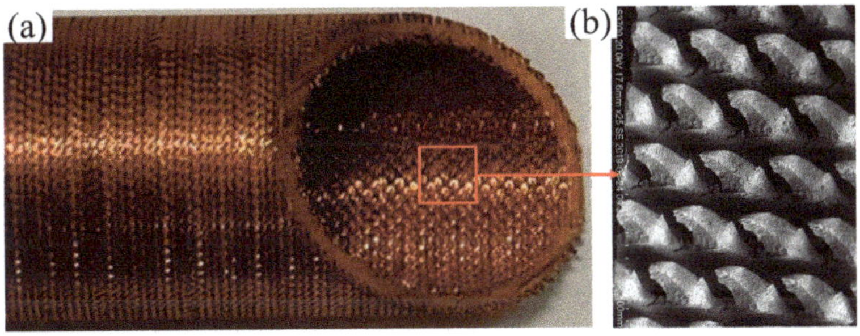

Figure 1. The novel 3D-OJIFT manufactured using the RPE method: (**a**) photographs of 3D-OJIFT, (**b**) SEM of the 3D-OJIF.

2.2. Experimental System

As shown in Figure 2, the experimental device, which was described in detail in our previous study [19], consisted of two coaxial tubes, a hot water circulation loop, a cold water circulation loop, a data-acquisition system, a temperature measuring element, and a pressure measuring element. The two coaxial tubes consist of a plexiglass outer pipe and a copper pipe. The geometric parameters of the plexiglass outer tube are 1800 mm in length, 45 mm in outer diameter, and 5 mm in wall thickness. The geometric parameters of the smooth tube in the test section are as follows: the length is 2000 mm, the inner diameter is 19 mm, and the wall thickness is 1.5 mm. The geometric parameters of the 3D-OJIFT are 2000 mm in length, 22 mm in outer diameter, and 0.8 mm in jagged height. Cold fluid flowed inside the tube, whereas hot fluid on the shell side flowed in the opposite direction. The flow rates of fluid were measured using a rotary flowmeter and regulated using a control valve. The cold water inlet temperature is 25 °C, the hot water inlet temperature is 77 °C, and the inlet temperature was adjusted using compensators and temperature controllers on the water tank. Eight T-type thermocouples were evenly distributed on the outer surface of the internal tube wall for measuring the temperature of the outer tube wall. Four T-type thermocouples were located at the inlet and outlet of the hot and cold water circulation loops to measure water temperature, and the accuracy of these T-type thermocouples was ±0.1 °C. The pressure difference between the pipe inlet and outlet was measured using a pressure transmitter (accuracy, ±0.5%). The experimental data of the temperature and pressure were collected using a data acquisition system. The pipeline and two coaxial tubes were insulated to prevent heat loss.

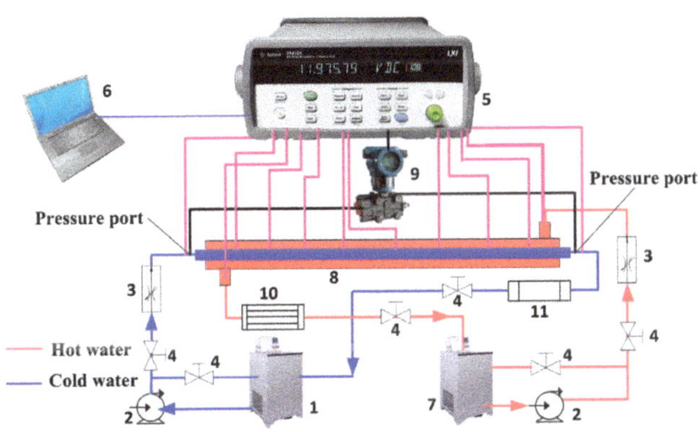

Figure 2. Schematic of the experimental system: (1) cold water tank; (2) pump; (3) rotameter; (4) control valve; (5) data acquisition system; (6) laptop; (7) hot water tank; (8) two coaxial tubes; (9) pressure test element; (10) thermal compensation element; (11) fan coil unit.

2.3. Data Simplification and Uncertainty Analysis

The main objective was to obtain the Nu and f values for the tube side. The method and steps were as follows:

The heat transfer rates (HTRs) of hot and cold water were calculated as

$$Q_c = \dot{m}_c c_p (t_{c,o} - t_{c,i}) \tag{1}$$

$$Q_h = \dot{m}_h c_p (t_{h,i} - t_{h,o}) \tag{2}$$

Considering the heat balance of the HTR, the average HTR was considered to be the HTR of the test section.

$$Q_{\text{ave}} = \frac{Q_c + Q_h}{2} \tag{3}$$

The overall heat transfer coefficient was calculated as

$$U = \frac{Q_{ave}}{A_i \Delta T_m} \tag{4}$$

where ΔT_m denotes the logarithmic mean temperature difference. The overall heat transfer coefficient was expressed as follows:

$$\frac{1}{U} = \frac{1}{h_i} + R_w + \frac{A_i}{h_o A_o} + R_f \tag{5}$$

where R_f is the fouling resistance and R_w is the conduction resistance.

The heat transfer correlation h_o for the annulus side was calculated as follows [20]:

$$Nu_o = 0.023 Re^{0.8} Pr^{0.4} \tag{6}$$

$$h_o = \frac{k}{D_H} 0.023 Re^{0.8} Pr^{0.4} \tag{7}$$

where k is the thermal conductivity of the annulus and D_H represents the hydraulic diameter of the annulus.

The Nu for the tube side was given by

$$Nu_i = \frac{h_i d_i}{k} \tag{8}$$

where d_i is the inside nominal diameter of the tube [21].

The Re and f were calculated as follows:

$$Re = \frac{\rho v d_H}{\mu} \tag{9}$$

$$f = \frac{2\Delta p d_i}{l \rho v^2} \tag{10}$$

The performance evaluation criterion (PEC) was used [22]:

$$PEC = \frac{Nu/Nu_0}{(f/f_0)^{1/3}} \tag{11}$$

where Nu_0 and f_0 are the Nusselt number and friction factor of the smooth tube, respectively.

Based on the method of Schultz and Cole [23], the maximal uncertainties of the $Re, f,$ nd Nu were 1.19%, 2.32%, and 1.66%, respectively.

2.4. Experimental Results

To ensure the accuracy of the experimental data, Nu and f were measured in smooth pipes and compared with the empirical correlations. The Gnielinski correlation [24] was defined as

$$Nu = \frac{(Re - 1000)(f/8)Pr}{1 + 12.7(Pr^{2/3} - 1)(f/8)^{0.5}} \quad (12)$$

The Petukhov correlation [25] was defined as

$$f = (0.79 \ln Re - 1.64)^{-2} \quad (13)$$

As shown in Figure 3, the experimental results agreed well with those empirical results. The maximal error of Nu was approximately 6.7%, while the error of f was approximately 4%. Based on the experimental platform, the thermal performance of the 3D-OJIFT was tested, and Nu and f of the 3D-OJIFT are depicted in Figure 3.

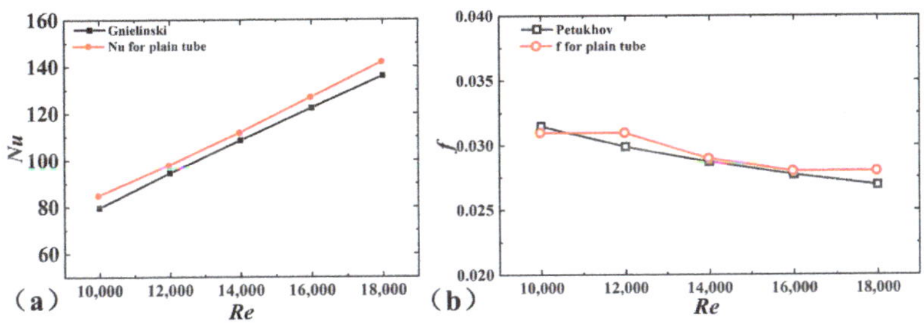

Figure 3. Comparison of the experimental results for a smooth tube and empirical correlations: (**a**) Nu, (**b**) f.

3. Numerical Simulations

3.1. Physical Model

The physical model of the 3D-OJIFT is shown in Figure 4. The wall thickness was 1.5 mm, and the period length (P) of the 3D-OJIFT ranged from 20 mm to 140 mm, which was defined as the length of the calculation domain. The three-dimensional overall jagged internal fin (3D-OJIF) was characterized by the jagged height, jagged spiral angle, and number of 3D-OJIFs. The effects of varying the jagged height (h = 0.4 mm, 0.6 mm, and 0.8 mm) and jagged spiral angle (β = 22°, 42°, and 65°) on the thermal performance were studied.

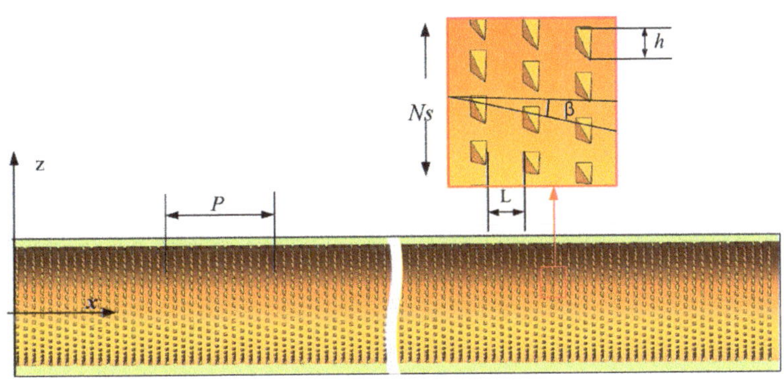

Figure 4. Schematic of the 3D-OJIFT.

3.2. Governing Equations

A three-dimensional computational fluid dynamics (CFD) model of the thermal performance in the 3D-OJIFT was presented. The assumptions, which were commonly made in many studies [26,27], were as follows: (1) the flow is isotropic, continuous, stable, and incompressible and has constant physical properties; (2) the effect of gravity is negligible. The equations were as follows:

Continuity equation:

$$\frac{\partial(\rho u_i)}{\partial x_i} = 0 \tag{14}$$

Momentum equation:

$$\frac{\partial(\rho u_i u_j)}{\partial x_j} = -\frac{\partial p}{\partial x_i} + \frac{\partial}{\partial x_j}\left[u\left(\frac{\partial u_i}{\partial x_j} + \frac{\partial u_j}{\partial x_i}\right)\right] - \frac{\partial\left(\rho \overline{u_i' u_j'}\right)}{\partial x_j} \tag{15}$$

Energy equation:

$$\frac{\partial(u_j(\rho e + p))}{\partial x_j} = \frac{\partial}{\partial x_j}(\lambda\frac{\partial T}{\partial x_j}) \tag{16}$$

where ρ and c_p are the fluid density and fluid specific heat, respectively, p and λ are the pressure and thermal conductivity, respectively, and T and u denote the temperature and fluid viscosity, respectively.

The RNG k-ε turbulence model [28,29] was employed in this paper.

$$\frac{\partial(\rho k u_i))}{\partial x_j} = \frac{\partial}{\partial x_j}\left(\left(\mu + \frac{\mu_t}{\sigma_k}\right)\lambda\frac{\partial k}{\partial x_j}\right) + C_{1\varepsilon}\frac{\varepsilon}{k}G_k - C_{2\varepsilon}\frac{\varepsilon^2}{k}\rho \tag{17}$$

$$\frac{\partial(\rho \varepsilon u_i)}{\partial x_j} = \frac{\partial}{\partial x_j}\left(\lambda\frac{\partial k}{\partial x_j}\left(\mu + \frac{\mu_t}{\sigma_k}\right)\right) + C_{1\varepsilon}\frac{\varepsilon}{k}G_k - C_{2\varepsilon}\rho\frac{\varepsilon^2}{k} \tag{18}$$

3.3. Mesh and Boundary Conditions

As shown in Figure 5, an unstructured mesh was used for discretizing the computational domain. Local mesh refinement was applied to the surface of the 3D-OJIFT, as shown in Figure 5b. The grid independence at a Reynolds number of 14,000 was investigated for the 3D-OJIFT using four sets of meshes (2,493,659, 3,382,197, 4,283,296, and 5,115,893). The relative difference between the Nusselt numbers for the 4,283,296 and 5,115,893 grids was within 3%, and the relative difference between the friction coefficients was within 3%. Thus, the mesh independence analysis indicated that the 4,283,296 grid was sufficient for capturing the characteristics of the flow; thus, this grid was used in subsequent calculations.

Considering the repetitive periodic geometry of the 3D-OJIFT and the computational cost, periodic boundary conditions were used, which have been adopted by much of the literature [30]. Periodic conditions were imposed on the inlet and outlet. The constant temperature condition (350 K) was used for the 3D-OJIFT wall. The upstream temperature was 298 K. No-slip conditions were prescribed for the surfaces of the 3D-OJIFT walls. The working fluid is water. The Reynolds number was in the 10,000–18,000 range. The finite-volume method was implemented for solving the governing equations. The second-order upwind algorithm was used for discretizing the momentum and energy terms. The standard scheme was employed for the pressure term. The convergence criterion was 10^{-8} for energy, while 10^{-6} was used for other equations.

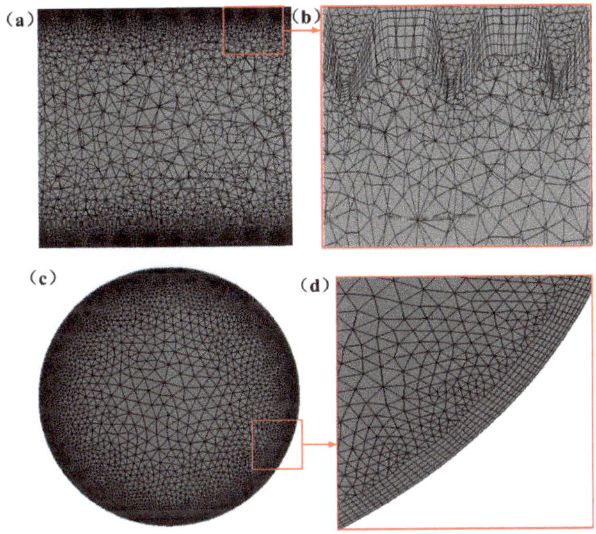

Figure 5. The fluid domain mesh. (**a**,**b**) in the y = 0 plane; (**c**,**d**) in the x = 0 plane.

3.4. Data Reduction

After convergence, the thermo-hydraulic performance was profiled. The convective heat transfer coefficient (h_t) of the tube wall was

$$h_{\mathrm{t}} = \frac{q}{T_{m,w} - T_{m,f}} \tag{19}$$

where q and $T_{m,w}$ denote the heat flux of the pipe wall and the mean temperatures pipe wall, respectively, and $T_{m,f}$ denotes the mean temperatures of the fluid. $T_{m,w}$ and $T_{m,f}$ were defined as follows:

$$T_{m,w} = \frac{1}{A}\int_0^A T_w dS \tag{20}$$

$$T_{m,f} = \frac{\int_0^{D/2} r u_x r T_f dr}{\int_0^{D/2} r u_x r dr} \tag{21}$$

where u_x and T_w are the axial velocity of the cross-section and the temperature of the pipe wall, respectively. T_f denotes the temperature of the water, and A is the surface area of the pipe wall.

3.5. Validation

The RNG k-ε turbulence model was first verified. As depicted in Figure 6, the numerical results for smooth tubes are compared with the experimental results. The maximal errors of Nusselt number and friction factor are below 4.9% and 5%, respectively. In addition, the numerical results for the 3D-OJIFT with β = 22° and h = 0.8 mm were compared with the corresponding experimental results. The data showed that the maximal deviation of Nu was 9.2%, while the maximal deviation of f was 8.9%. Thus, the numerical calculation data agreed well with the experimental results.

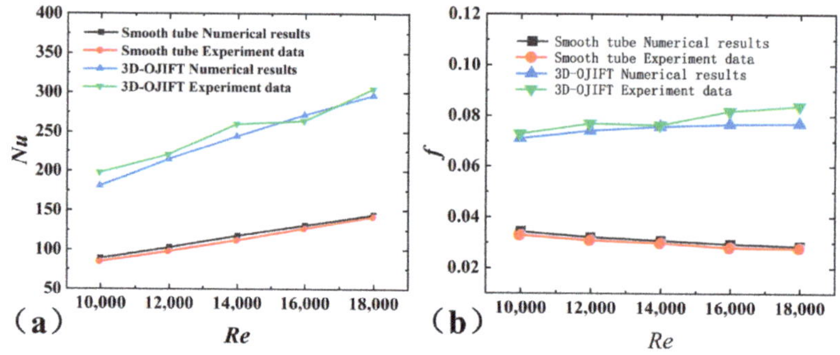

Figure 6. Comparison of numerical results and experimental data: (**a**) *Nu*; (**b**) *f*.

4. Results and Discussion

4.1. Thermo-Hydraulic Characteristics

Considering the distributions of velocity, temperature, turbulent kinetic energy (*TKE*), and the local convective heat transfer coefficient as examples, the thermo-hydraulic characteristics of the 3D-OJIFT were analyzed. The *TKE* quantitatively describes the turbulence fluctuations of the water in a tube and was defined as

$$TKE = \frac{\overline{\mu_z'^2} + \overline{\mu_y'^2} + \overline{\mu_x'^2}}{2} \tag{22}$$

where μ_i' denotes the velocity fluctuation. Compared with smooth tubes, the better thermal performance was mainly owing to the following.

Figure 7 shows the axial flow field for the 3D-OJIFT and the corresponding smooth tube at the y = 0.2 mm section. The axial velocity field of the 3D-OJIFT differs from that of the smooth tube. The axial velocity is low near the 3D-OJIFT tube wall, whereas it is high at the center of the 3D-OJIFT tube wall. Evidently, the 3D-OJIF configuration strongly affects axial velocity, yielding a more mixed flow, secondary flow, and swirl flow in the 3D-OJIFT, thus improving the heat transfer performance.

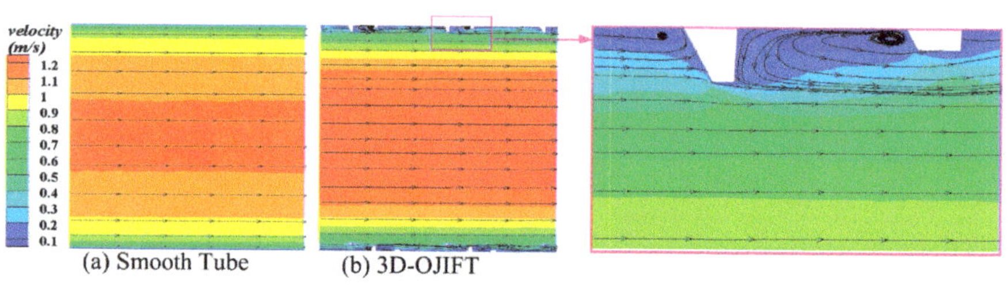

Figure 7. Axial velocity contours of the 3D-OJIFT and smooth pipe at y = 0.2 mm: (**a**) smooth tube; (**b**) 3D-OJIFT.

Figure 8 depicts the radial velocity fields of the 3D-OJIFT and smooth tube at the x = 8.5 mm section. As expected, the 3D-OJIF configuration strongly affects the flow field. As the fluid flows through the 3D-OJIFT, the flow path inside the 3D-OJIFT shrinks owing to the 3D-OJIF surface, causing the velocity to increase. With periodic variation in the velocity field, momentum exchange improves fluid mixing and vortex generation between the mainstream region and the 3D-OJIF region.

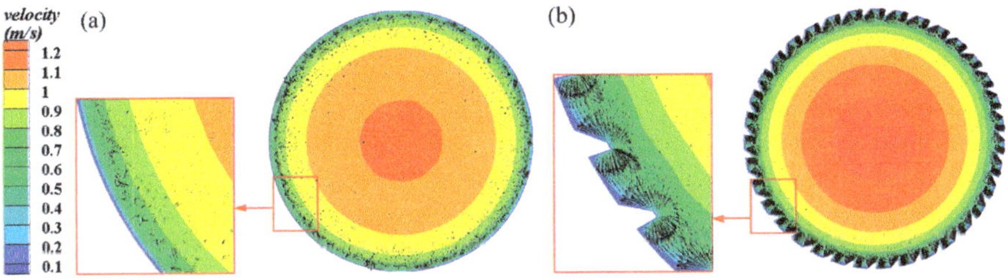

Figure 8. Radial velocity contours of the smooth tube and 3D-OJIFT at x = 8.5 mm: (**a**) smooth tube; (**b**) 3D-OJIFT.

Figure 9 compares the TKE distributions of the 3D-OJIFT and the smooth tube. Evidently, the TKE of the 3D-OJIFT is significantly stronger than that of the smooth tube, owing to the axial flow disturbance near the jagged fin surface. This improves the circulation flow and fluid mixing between the 3D jagged fin surface and the core region, resulting in the destruction of the thermal boundary layer near the jagged fin surfaces. In addition, vortices occur near the jagged fin surface. This may intensify high turbulence, significantly increasing the *Nu* and *f* values of the fluid flowing in the 3D-OJIFT.

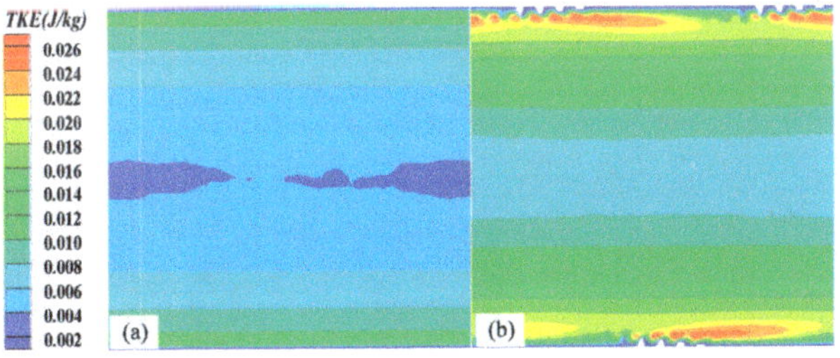

Figure 9. TKE distributions of the smooth tube and 3D-OJIFT at x = 8.5 mm: (**a**) smooth tube; (**b**) 3D-OJIFT.

Figures 10 and 11 show the axial and radial temperature distributions for the smooth tube and 3D-OJIFT, respectively. As shown in Figures 10 and 11, the temperature gradient in the inner wall region of the 3D-OJIFT is steeper than that in the smooth tube. In addition, the 3D-OJIFT precludes the thermal boundary layer from full development. Thus, using a jagged internal fin can improve heat transfer. Figure 12 shows the local heat transfer coefficient for the tube wall along the X direction. Compared with the smooth tube, the local heat transfer coefficient of the 3D-OJIFT is larger and more variable. Therefore, the 3D-OJIFT exhibits better heat transfer performance.

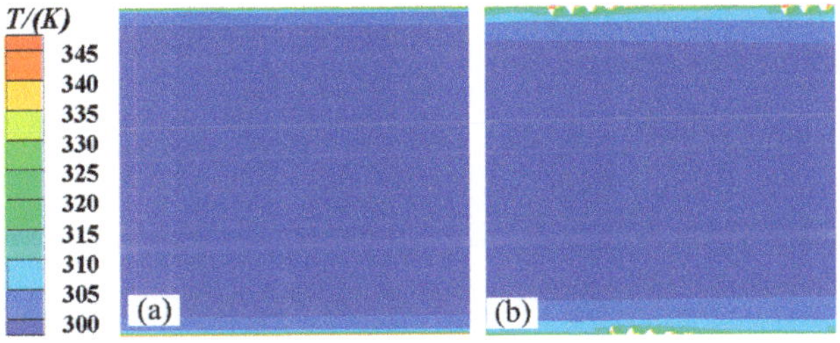

Figure 10. Axial temperature contours of the smooth tube and 3D-OJIFT at x = 8.5 mm: (**a**) smooth tube; (**b**) 3D-OJIFT.

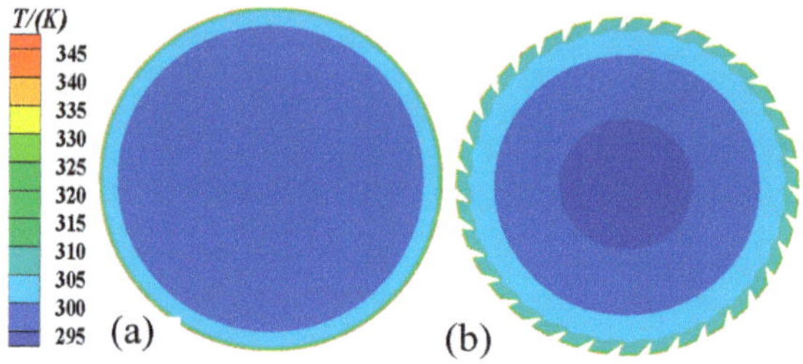

Figure 11. Radial temperature contours of the smooth tube and 3D-OJIFT at y = 0.2 mm: (**a**) smooth tube; (**b**) 3D-OJIFT.

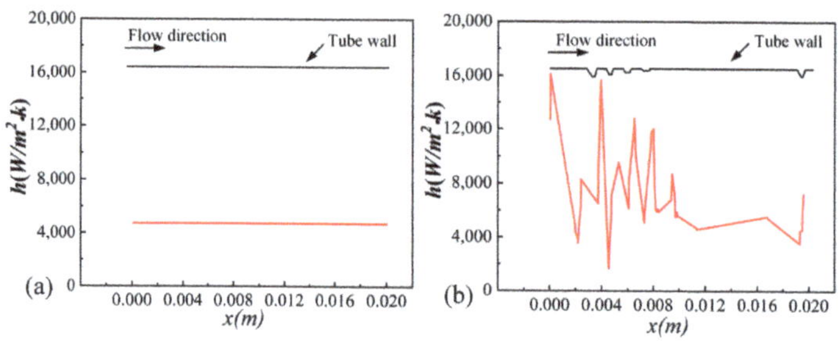

Figure 12. Distributions of the convective heat transfer coefficient: (**a**) smooth tube, (**b**) 3D-OJIFT.

4.2. The Effect of the Jagged Height

Figure 13 shows the Nusselt number versus the Reynolds number for different jagged heights. Evidently, the Nusselt number of the 3D-OJIFT increases with increasing jagged height. The results for the 3D-OJIFT with h = 0.8 mm are better than for the other cases, with approximately a 2.06-fold improvement over the value for the smooth tube at Re = 12,000. The Nu value at h = 0.6 mm is 1.87–2.0 times that for the smooth tube. The results suggest that the 3D-OJIFT increases the thermal boundary layer disruption of turbulence; thus, the 3D-OJIFT exhibits better thermal performance compared with the smooth tube.

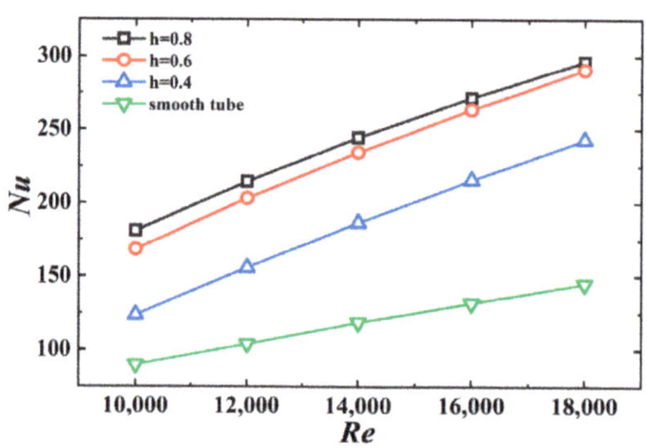

Figure 13. Nu versus Re for 3D-OJIFTs with different h.

The effect of the jagged height on the friction coefficient of the 3D-OJIFT is shown in Figure 14. The results suggest that the friction coefficient increases as the jagged height increases. For h = 0.8 mm, the f value is the largest, 2.06–2.69 times the value for the smooth tube. This is owing to the increased height of the fin, increased contact area between the fin and fluid, and greater flow obstruction of the tube, resulting in a greater f.

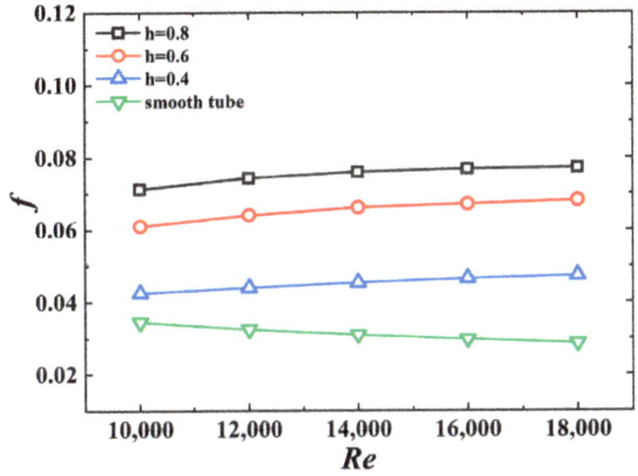

Figure 14. f versus Re for 3D-OJIFTs with different h.

Figure 15 shows the PEC values for different jagged height values. The PEC values are in the 1.28–1.58 range, indicating that the 3D-OJIFT exhibits better comprehensive heat transfer performance. Another conclusion to be drawn from Figure 15 is that the 3D-OJIFTs with h = 0.6 mm and h = 0.8 mm perform better than the 3D-OJIFT with h = 0.4 mm. For Re above 14,000, the comprehensive heat transfer performance of the 3D-OJIFT with h = 0.6 mm is better than that of the 3D-OJIFT with h = 0.8 mm. Therefore, it is important to choose an appropriate height for the inner fin of the 3D-OJIFT.

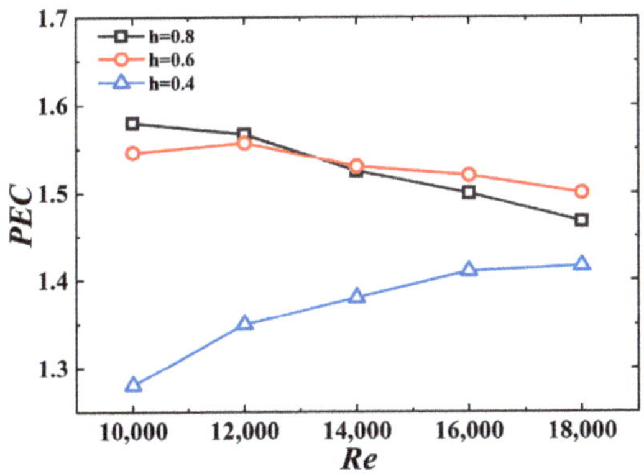

Figure 15. PEC versus Re for 3D-OJIFTs with different h.

4.3. The Effect of the Jagged Spiral Angle

Figure 16 shows the Nusselt number for different jagged spiral angle values. The Nusselt number of the 3D-OJIFT increases as the jagged spiral angle increases. The results for $\beta = 22^\circ$ and $\beta = 65^\circ$ are better than 1.60–1.87-fold and 1.86–2.14-fold, respectively, compared with the corresponding results for the smooth tube. Because a wider jagged spiral angle is conducive to generating a secondary flow of fluid and increasing the turbulence of the fluid, wider jagged spiral angles are associated with better heat transfer performance. However, for the jagged spiral angles exceeding 42°, Nu does not increase significantly. Figure 16 shows that the Nu values for the jagged spiral angles of 42° and 65° are almost the same. This is because when the helix angle exceeds 42°, the disturbance of the two spiral-angle structures to the boundary layer is limited.

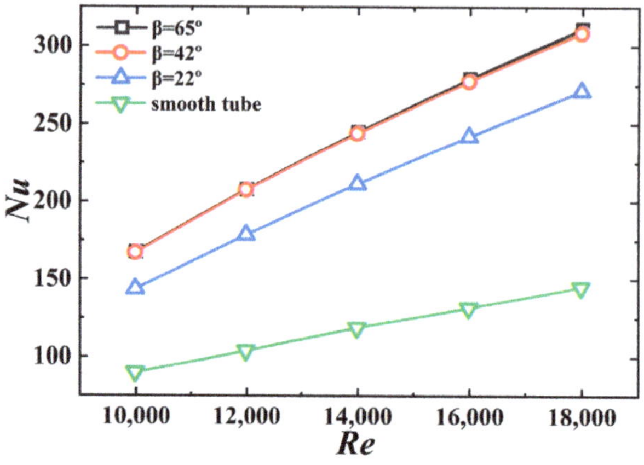

Figure 16. *Nu* versus *Re* for 3D-OJIFTs with different β.

Figure 17 shows the effect of the jagged spiral angle on the friction coefficient of the 3D-OJIFT. Evidently, the wider the jagged spiral angle of the 3D-OJIFT, the larger the friction coefficient. For $\beta = 22°$, the friction coefficient of the 3D-OJIFT is 1.42–1.92 times that for the smooth tube. For $\beta = 65°$, the friction coefficient is 1.67–2.34 times that for the smooth tube. A wide jagged spiral angle increases the area of the 3D-OJIFT exposed to the incoming flow, and the change in the fluid flow direction also increases, implying that the obstruction to the incoming flow increases. Therefore, 3D-OJIFTs with wider jagged spiral angles exhibit more friction.

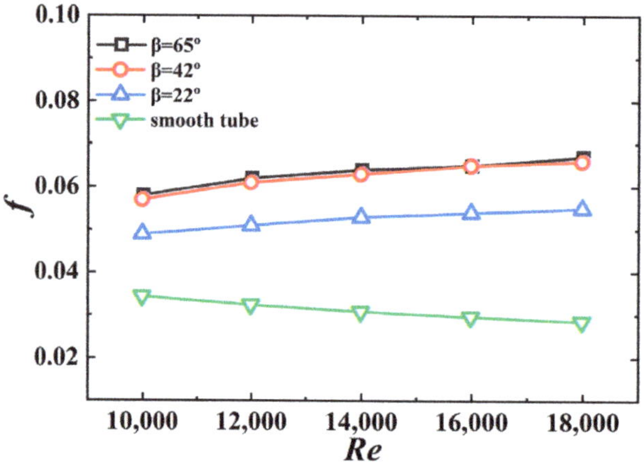

Figure 17. *f* versus *Re* for 3D-OJIFTs with different β.

Figure 18 shows *PEC* versus the *Re* number for different jagged spiral angles. The *PEC* value for the 3D-OJIFT with the above jagged spiral angle exceeds 1 (range, 1.42–1.63), implying that the 3D-OJIFT exhibits good comprehensive heat transfer performance. In addition, 3D-OJIFTs with wider jagged spiral angles should be used for ensuring better comprehensive heat transfer performance at high *Re*.

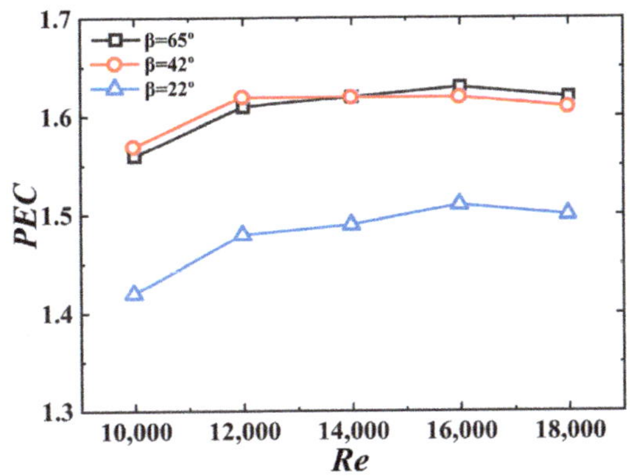

Figure 18. *PEC* versus *Re* for 3D-OJIFTs with different β.

4.4. Comparison with the Conventional Two-Dimensional Spiral Fin Tubes

A conventional IHFT was selected for comparison to evaluate the thermal performance of the designed 3D-OJIFT. The geometric parameters of the IHFT were as follows: the fin height was h = 0.3 mm, and the helix angle was β = 22°. The helix angles of the 3D-OJIFT and IHFT were the same. The height of the 3D-OJIF was 0.6 mm, and fin spacing (L) was 0.6 mm. As shown in Figure 19a, the 3D-OJIFT had a larger *Nu*. The *Nu* of the 3D-OJIFT was approximately 1.87 times that of the smooth tube and 1.1 times that of the IHFT at *Re* = 18,000. As shown in Figure 19b,c, although the friction coefficient of the 3D-OJIFT was larger than that of the conventional IHFT (at 1.39–1.47 times that of the conventional IHFT), the designed 3D-OJIFT exhibited better comprehensive heat transfer performance. The *PEC* value of the 3D-OJIFT was 1.03–1.15 times that of the conventional IHFT.

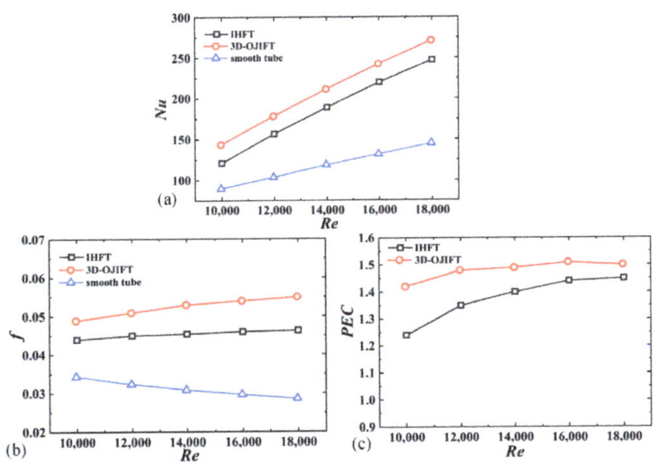

Figure 19. Comparison of the thermal performances of the conventional IHFT and designed 3D-OJIFT for *Re* ranging from 10,000 to 18,000: (**a**) *Nu*; (**b**) *f*; (**c**) *PEC*.

4.5. Comparison with Other Reported Studies

To better evaluate the heat transfer performance of the novel 3D-OJIFT, it was compared with other enhancement techniques reported in the literature, as shown in Table 1. The maximum *Nu*/*Nup* of the 3D-OJIFT is slightly higher than that of other enhancement technologies. However, the *f*/*fp* of the 3D-OJIFT is much smaller than those of an internal helically-finned tube [31], corrugated tube [32], and dimpled tube [33]. The maximum PEC of the 3D-OJIFT is increased by 14.4%, 32.7%, and 16.1%, respectively. Therefore, compared with the other four reinforcement structures reported in the literature, the new 3D-OJIFT presents the best overall heat transfer performance.

Table 1. Comparison between the present study and other reported studies.

Authors	Enhanced Structures	*Nu/Nup*	*f/fp*	*PEC*	*Re*
Present study	3D-OJIFT	2.01–2.04	2.01–2.69	1.46–1.58	10,000–18,000
Li et al. [31]	Internal helically finned tube	1–2.0	1–3.0	1–1.38	2500–15,000
Dizaji et al. [32]	Corrugated tube	1.5–1.75	3.52–3.48	0.98–1.19	7000–17,500
Vicente et al. [33]	Dimpled tube	1.5–2.1	3.2–4.2	1.01–1.36	2000–19,000

4.6. Correlations of the Nusselt Number and Friction Factor

The empirical correlations for the 3D-OJIFT were obtained, as defined by Equations (23) and (24). These empirical Equations (23) and (24) are valid for the following ranges: (10,000 $\leq Re \leq$ 18,000, 0.4 mm $\leq h \leq$ 0.8 mm, 22° $\leq \beta \leq$ 65°). Figure 20 shows that the Nu deviation between the predicted results and simulated data was approximately ±11.1%, and the corresponding deviation for f was approximately ±14.3%. Therefore, the empirical equations are suitable for predicting Nu and f.

$$Nu = 0.012039Re^{1.011559}h^{0.40981}\beta^{0.10465} \tag{23}$$

$$f = 0.011077Re^{0.19686}h^{0.7253}\beta^{0.05752} \tag{24}$$

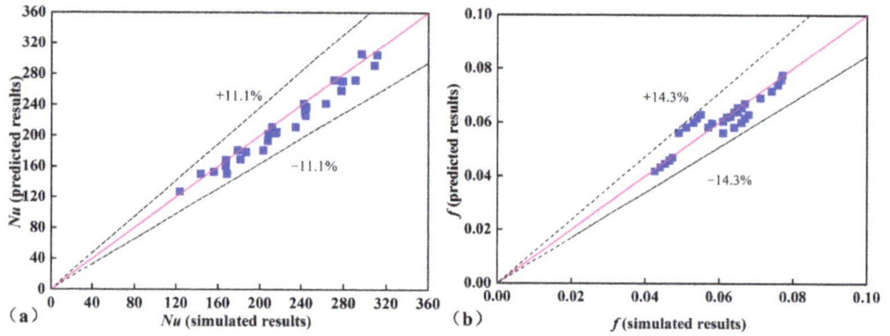

Figure 20. Discrepancies between the numerical and simulated values of Nu and f: (**a**) Nu, (**b**) f.

5. Conclusions

The thermo-hydraulic characteristics of a novel 3D-OJIFT were studied and compared using experimental and numerical methods. The main conclusions are as follows.

1. A novel 3D-OJIFT was fabricated using the RPE method. The 3D-OJIF increased the heat transfer area, prevented the development of the velocity and thermal boundary layers, increased the turbulence disturbance, and reduced the temperature gradient, thus enhancing the heat transfer.
2. The jagged height of the 3D-OJIFT importantly affected the thermal performance. The Nu and f both increased as the jagged height increased. The Nu of the 3D-OJIFT was 1.59–2.14 times that of the smooth tube.
3. The jagged spiral angle of the 3D-OJIFT was very sensitive to the thermal performance. Both Nu and f increased as the jagged spiral angle increased. For better comprehensive heat transfer performance, 3D-OJIFTs with wider jagged spiral angles should be chosen at high Re.
4. Compared with conventional IHFTs and other reported studies, the proposed 3D-OJIFT exhibited better comprehensive heat transfer performance. The PEC value of the proposed 3D-OJIFT was 1.03–1.15 times that of the conventional IHFT. Empirical

correlations were developed to predict the *Nu* and *f* values for 3D-OJIFTs. The discrepancies for *Nu* and *f* were within ±11.1% and ±14.3%, respectively.

Author Contributions: S.H. was responsible for writing the research results and conducting the primary experiments; D.Y. and Y.X. contributed to the article review and provided experimental conditions; M.D., Z.C. and N.L. assisted in completing the necessary experiment and numerical investigations. All authors have read and agreed to the published version of the manuscript.

Funding: This study was supported by the National Natural Science Foundation of China (52065003, 52365054) and Jiangxi Provincial Natural Science Foundation, China (20202BAB204021, 20212BAB204001). The authors acknowledge support from the Postdoctoral Research Funding Project of Jiangxi, China (2021KY50), and Engineering Research Center of Nuclear Technology Application, Ministry of Education, China (HJSJYB2021-2).

Data Availability Statement: All data to support the results of this study are included in this article.

Conflicts of Interest: The authors affirm the absence of any competing interests.

References

1. Khatibi, M.; Kowsari, M.M.; Golparvar, B.; Niazmand, H. Optimum loading of aluminum additive particles in unconsolidated beds of finned flat-tube heat exchangers in an adsorption cooling system. *Appl. Therm. Eng.* **2021**, *196*, 117267. [CrossRef]
2. Tiwari, A.K.; Javed, S.; Oztop, H.F.; Said, Z.; Pandya, N.S. Experimental and numerical investigation on the thermal performance of triple tube heat exchanger equipped with different inserts with WO3/water nanofluid under turbulent condition. *Int. J. Therm. Sci.* **2021**, *164*, 106861. [CrossRef]
3. Ji, W.-T.; Jacobi, A.M.; He, Y.-L.; Tao, W.-Q. Summary and evaluation on single-phase heat transfer enhancement techniques of liquid laminar and turbulent pipe flow. *Int. J. Heat Mass Transf.* **2015**, *88*, 735–754. [CrossRef]
4. Kukulka, D.J.; Smith, R.; Li, W. Comparison of tubeside condensation and evaporation characteristics of smooth and enhanced heat transfer 1EHT tubes. *Appl. Therm. Eng.* **2015**, *89*, 1079–1086. [CrossRef]
5. Li, M.; Khan, T.S.; Al-Hajri, E.; Ayub, Z.H. Single phase heat transfer and pressure drop analysis of a dimpled enhanced tube. *Appl. Therm. Eng.* **2016**, *101*, 38–46. [CrossRef]
6. Kim, J.; Doo, J.; Ha, M.; Yoon, H.; Son, C. Numerical study on characteristics of flow and heat transfer in a cooling passage with protrusion-in-dimple surface. *Int. J. Heat Mass Transf.* **2012**, *55*, 7257–7267. [CrossRef]
7. Afanasyev, V.N.; Chudnovsky, Y.P.; Leontiev, A.I.; Roganov, P.S. Turbulent flow friction and heat transfer characteristics for spherical cavities on a flat plate. *Exp. Therm. Fluid Sci.* **1993**, *7*, 1–8. [CrossRef]
8. Xie, S.; Guo, Z.; Gong, Y.; Dong, C.; Liu, J.; Ren, L. Numerical investigation of thermal-hydraulic performance of a heat exchanger tube with helical dimples. *Int. J. Therm. Sci.* **2022**, *177*, 107530. [CrossRef]
9. Hu, Q.; Qu, X.; Peng, W.; Wang, J. Experimental and numerical investigation of turbulent heat transfer enhancement of an intermediate heat exchanger using corrugated tubes. *Int. J. Heat Mass Transf.* **2022**, *185*, 122385. [CrossRef]
10. Cruz, G.G.; Mendes, M.A.; Pereira, J.M.; Santos, H.; Nikulin, A.; Moita, A.S. Experimental and numerical characterization of single-phase pressure drop and heat transfer enhancement in helical corrugated tubes. *Int. J. Heat Mass Transf.* **2021**, *179*, 121632. [CrossRef]
11. Vicente, P.G.; Garcı, A.; Viedma, A. Mixed convection heat transfer and isothermal pressure drop in corrugated tubes for laminar and transiton flow. *Int. Commun. Heat Mass Transf.* **2004**, *31*, 651–662. [CrossRef]
12. Zdaniuk, G.J.; Chamra, L.M.; Mago, P.J. Experimental determination of heat transfer and friction in helically-finned tubes. *Exp. Therm. Fluid Sci.* **2008**, *32*, 761–775. [CrossRef]
13. Wang, Y.-H.; Zhang, J.-L.; Ma, Z.-X. Experimental determination of single-phase pressure drop and heat transfer in a horizontal internal helically-finned tube. *Int. J. Heat Mass Transf.* **2017**, *104*, 240–246. [CrossRef]
14. Liu, Z.; Yue, Y.; She, L.; Fan, G. Numerical analysis of turbulent flow and heat transfer in internally finned tubes. *Front. Energy Res.* **2019**, *7*, 64. [CrossRef]
15. Celen, A.; Dalkilic, A.S.; Wongwises, S. Experimental analysis of the single phase pressure drop characteristics of smooth and microfin tubes. *Int. Commun. Heat Mass Transf.* **2013**, *46*, 58–66. [CrossRef]
16. Wang, X.; Ho, J.; Leong, K.; Wong, T. Condensation heat transfer and pressure drop characteristics of R-134a in horizontal smooth tubes and enhanced tubes fabricated by selective laser melting. *Int. J. Heat Mass Transf.* **2018**, *126*, 949–962. [CrossRef]
17. Huang, S.; Wan, Z.; Tang, Y. Manufacturing and single-phase thermal performance of an arc-shaped inner finned tube for heat exchanger. *Appl. Therm. Eng.* **2019**, *159*, 113817. [CrossRef]
18. Chen, H.; Wan, Z.; Mo, H.; Huang, S.; Tang, Y. Experimental studies on the compound thermo-hydraulic characteristics in a 3D inner finned tube with porous copper fiber inserts. *Int. Commun. Heat Mass Transf.* **2019**, *100*, 51–59. [CrossRef]
19. Huang, S.; Chen, H.; Zhang, X.; Wan, Z.; Tang, Y. Experimental evaluation of thermal performance in a circular tube with Y-branch insert. *Int. Commun. Heat Mass Transf.* **2019**, *106*, 15–21. [CrossRef]

20. Tan, X.-H.; Zhu, D.-S.; Zhou, G.-Y.; Zeng, L.-D. Experimental and numerical study of convective heat transfer and fluid flow in twisted oval tubes. *Int. J. Heat Mass Transf.* **2012**, *55*, 4701–4710. [CrossRef]
21. Wan, Z.; Tang, Y. Characteristics of uncurled and reversely curled chip during orthogonal cutting. *Int. J. Mach. Tools Manuf.* **2011**, *51*, 831–835. [CrossRef]
22. Webb, R. Performance evaluation criteria for use of enhanced heat transfer surfaces in heat exchanger design. *Int. J. Heat Mass Transf.* **1981**, *24*, 715–726. [CrossRef]
23. Kamboj, K.; Singh, G.; Sharma, R.; Panchal, D.; Hira, J. Heat transfer augmentation in double pipe heat exchanger using mechanical turbulators. *Heat Mass Transf.* **2017**, *53*, 553–567. [CrossRef]
24. Sheikholeslami, M.; Gorji-Bandpy, M.; Ganji, D. Effect of discontinuous helical turbulators on heat transfer characteristics of double pipe water to air heat exchanger. *Energy Convers. Manag.* **2016**, *118*, 75–87. [CrossRef]
25. Tang, X.; Dai, X.; Zhu, D. Experimental and numerical investigation of convective heat transfer and fluid flow in twisted spiral tube. *Int. J. Heat Mass Transf.* **2015**, *90*, 523–541. [CrossRef]
26. Guo, J.; Fan, A.; Zhang, X.; Liu, W. A numerical study on heat transfer and friction factor characteristics of laminar flow in a circular tube fitted with center-cleared twisted tape. *Int. J. Therm. Sci.* **2011**, *50*, 1263–1270. [CrossRef]
27. You, Y.; Fan, A.; Liu, W.; Huang, S. Thermo-hydraulic characteristics of laminar flow in an enhanced tube with conical strip inserts. *Int. J. Therm. Sci.* **2012**, *61*, 28–37. [CrossRef]
28. Launder, B.E.; Spalding, D.B. The numerical computation of turbulent flows. *Comput. Methods Appl. Mech. Eng.* **1974**, *2*, 269–289. [CrossRef]
29. Tu, W.; Wang, Y.; Tang, Y. A numerical study on thermal-hydraulic characteristics of turbulent flow through a circular tube fitted with pipe inserts. *Appl. Therm. Eng.* **2016**, *101*, 413–421. [CrossRef]
30. Huang, S.; Wan, Z.; Wang, Q.; Tang, Y.; Yang, X. Thermo-hydraulic characteristics of laminar flow in a circular tube with porous metal cylinder inserts. *Appl. Therm. Eng.* **2017**, *120*, 49–63. [CrossRef]
31. Li, X.W.; Meng, J.A.; Li, Z.X. Experimental study of single-phase pressure drop and heat transfer in a micro-fin tube. *Experi-Ment. Therm. Fluid Sci.* **2008**, *32*, 641–648. [CrossRef]
32. Dizaji, H.S.; Jafarmadar, S.; Mobadersani, F. Experimental studies on heat transfer and pressure drop characteristics for new arrangements of corrugated tubes in a double pipe heat exchanger. *Int. J. Therm. Sci.* **2015**, *96*, 211–220. [CrossRef]
33. Vicente, P.G.; Garcí, A.A.; Viedma, A. Heat transfer and pressure drop for low Reynolds turbulent flow in helically dimpled tubes. *Int. J. Heat Mass Transf.* **2002**, *45*, 543–553. [CrossRef]

micromachines

MDPI

Article

Gradient-Pattern Micro-Grooved Wicks Fabricated by the Ultraviolet Nanosecond Laser Method and Their Enhanced Capillary Performance

Guanghan Huang [1], Jiawei Liao [1], Chao Fan [1], Shuang Liu [1], Wenjie Miao [1], Yu Zhang [1], Shiwo Ta [2], Guannan Yang [1,*] and Chengqiang Cui [1,*]

[1] State Key Laboratory of Precision Electronic Manufacturing Technology and Equipment, Guangdong University of Technology, Guangzhou 510006, China; ghan.huang@outlook.com (G.H.); liaojiawei0404@163.com (J.L.); fanc_001@163.com (C.F.); lshuang000@163.com (S.L.); 18369147071@163.com (W.M.); zhangyu@gdut.edu.cn (Y.Z.)

[2] State Key Laboratory of Advanced Materials and Electronic Components, Fenghua Advanced Technology Co., Ltd., Zhaoqing 526020, China; tashiwo@china-fenghua.com

* Correspondence: ygn@gdut.edu.cn (G.Y.); cqcui@gdut.edu.cn (C.C.)

Abstract: Capillary-gradient wicks can achieve fast or directional liquid transport, but they face fabrication challenges by traditional methods in terms of precise patterns. Laser processing is a potential solution due to its high pattern accuracy, but there are a few studies on laser-processed capillary-gradient wicks. In this paper, capillary step-gradient micro-grooved wicks (CSMWs) were fabricated by an ultraviolet nanosecond pulsed laser, and their capillary performance was studied experimentally. The CSMWs could be divided into three regions with a decreasing capillary radius. The equilibrium rising height of the CSMWs was enhanced by 124% compared to the non-gradient parallel wick. Different from the classical Lucas–Washburn model describing a uniform non-gradient wick, secondary capillary acceleration was observed in the negative gradient direction of the CSMWs. With the increase in laser power and the decrease in scanning speed, the capillary performance was promoted, and the optimal laser processing parameters were 4 W-10 mm/s. The laser-enhanced capillary performance was attributed to the improved hydrophilicity and reduced capillary radius, which resulted from the increased surface roughness, protrusion morphology, and deep-narrow V-shaped grooves induced by the high energy density of the laser. Our study demonstrates that ultraviolet pulsed laser processing is a highly efficient and low-cost method for fabricating high-performance capillary gradient wicks.

Keywords: enhanced capillary performance; ultraviolet nanosecond laser; capillary gradient wick; micro-grooved wick

1. Introduction

In recent years, with the development of aerospace technology [1], high-speed servers [2], and the miniaturization of electronic devices [3], the heat generated within integrated circuits has significantly increased, posing great challenges to thermal management. Heat pipes play an important role in the thermal management of IC chips due to their high reliability and desirable capacity to cool extremely high heat flux [4]. The wick structure is functional in the transport of working fluid, and it can ensure the rapid backflow of working fluid to the evaporator. Hence, the wick structure is an essential part of heat pipes or other two-phase heat transfer devices. Micro-grooves, owing to their high permeability, low flow resistance, and cost-effectiveness, are one of the most commonly used wick structures [5,6]. Currently, the principal approaches to further enhance the capillary wicking capability of micro-grooves mainly involve the development of advanced manufacturing methods to form special surface morphologies [7–9] and the optimization of capillary structure design [10].

There are various methods for fabricating micro-grooved wick structures. Traditional methods include chemical etching [11], sintering [12], extrusion [13], micro-machining [14], and sintering-cutting composites [15]. Recently, advanced pulse laser processing technology has shown its advantages of high efficiency, high precision, great processing flexibility, and a wide range of applicable materials (including silicon chips [16], sapphire [17], metals [18], etc.), demonstrating the great potential in the preparation of micro-grooved wick structures. Li et al. [19] studied the fabrication process of micro-grooves on nanocarbon foam using laser etching. It was found that the groove morphology was composed of randomly distributed carbon nanotubes, which enhanced the capillary performance of the nanocarbon foam. Jiang et al. [20] employed a two-step nanosecond laser texturing method to manufacture a chemical-free ultrathin aluminum groove wick, demonstrating the enhancement effect of dual-scale micro-grooves on capillary performance. Xie et al. [21] created a supercapillary surface with uniformly arranged micro-grooves on aluminum by femtosecond laser processing, and the best capillary performance was observed at laser energy densities of 18.49 and 52.67 J/cm^2. Jiang et al. [22] investigated the influences of spot overlap rate on micro-groove depth uniformity at lower laser energy densities, and it was revealed that a spot overlap rate of 80% could result in higher ablation efficiency and better depth uniformity. Wang et al. [23] used in-situ sintering and laser ablation to fabricate multi-scale groove-like porous silicon carbide wick structures, which presented enhanced capillary performance. Long et al. [24] prepared supernarrow-deep micro-grooves with a groove width of 50 μm and a depth-to-width ratio of 3.5 with high laser scanning speeds, and the groove wicks demonstrated high permeability and transport capacity. Overall, the above studies mainly focus on the influences of laser parameters on the capillary performance of parallel non-gradient micro-grooves but rarely involve capillary-gradient micro-grooved wick structures.

Capillary-gradient wick structures are a type of wick with increasing or decreasing capillary pressure along the axis, which can be induced by surface wettability gradients or structural capillary radius gradients [25]. As presented in Figure 1b, capillary gradients can be classified into two types: continuous gradients (like a slope) and step gradients (like steps). The directions of capillary gradients include negative direction and positive direction, and the negative one is defined as the direction from the side with a larger effective capillary radius to that with a smaller one. Thus, in the negative gradient direction, the contact angle or the capillary radius decreases, resulting in an increase in capillary force; conversely, in the positive gradient direction, the Laplace pressure becomes negative, and liquid transport is unachievable. Therefore, capillary gradient wick structures offer unique advantages in terms of rapid capillary transport [26] and unidirectional fluid transport capability [10]. Particularly, unidirectional transport wicks enable heat pipes to transfer heat in only one direction, exhibiting their potential applications in challenging heat diodes.

The manufacturing methods and performance research of capillary gradient wick structures have attracted extensive attention. For instance, Zhou et al. [27] used sintered copper wire mesh to create a novel 2 mm thick ultra-thin isothermal plate with a radially capillary-gradient wick structure, significantly enhancing the liquid transport capacity within the capillary structure. Zhang et al. [28] employed a diffusion bonding process to fabricate a copper porous micro-mesh wick structure with gradient porosity, which achieved rapid liquid replenishment during boiling and further enhanced the heat transfer coefficient and critical heat flux. Jiang et al. [29] used hot micro-embossing to process a pyramid-shaped vertical capillary gradient structure, enhancing thin-film evaporation and capillary pumping processes simultaneously. Liu et al. [30] used electroplating on an isothermal plate to manufacture wettability-gradient capillary wicks with a gradient capillary pressure direction from the periphery to the center. It was found that the surface wettability of the capillary wicks changed from hydrophilicity to superhydrophilicity, effectively improving the overall performance of temperature uniformity. Xie et al. [31] employed a pulsed fiber laser to process micro-grooves with varying widths and effective capillary radius, and the adjusted radius gradient enhanced the capillary force of the copper substrate. In general, the above methods have poor pattern controllability and low processing efficiency. Fortunately, precise laser processing offers the advantages of high pattern flexibility, high accuracy, good controllability, and low cost when working with non-uniform, gradient, or patterned wick structures. However, there have been a few reports on laser processing to fabricate capillary gradient wick structures, especially groove-type gradient wick structures.

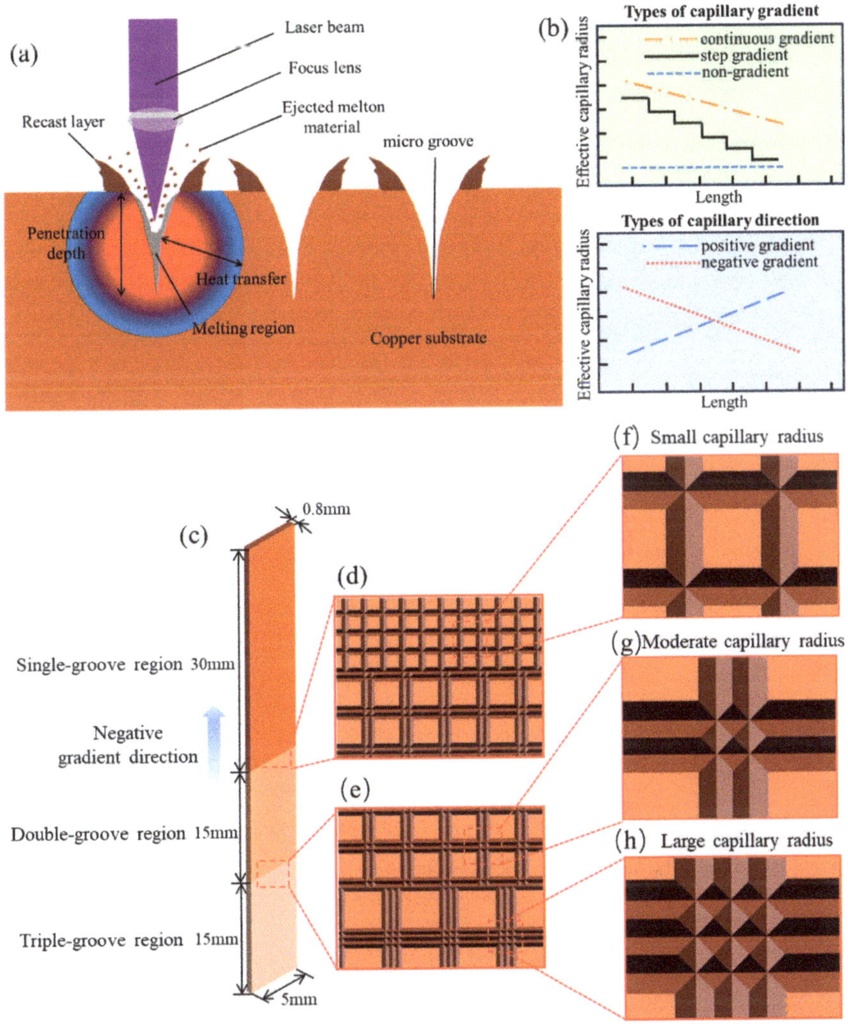

Figure 1. Schematic of CSMWs and their fabrication mechanism by ultraviolet nanosecond laser. (**a**) Fabrication mechanism of high-aspect-ratio V-shaped copper grooves by laser processing. (**b**) Two types of capillary gradient (continuous gradient and step gradient) and two types of gradient direction (positive gradient direction and negative gradient direction). (**c**) Three regions of CSMWs with different capillary radii, including a single-groove region, a double-groove region, and a triple-groove region. (**d**) Junction between the single-groove and double-groove regions. (**e**) Junction between the double-groove and triple-groove regions. Partially enlarged details of (**f**) single-groove region with small capillary radius, (**g**) double-groove region with medium capillary radius, and (**h**) triple-groove region with large capillary radius.

In this study, capillary step-gradient micro-grooved wicks (CSMWs) were fabricated by laser processing. The infrared thermography [32] method was employed to characterize the capillary performance parameter ($\Delta P_{cap} \cdot K$) of the CSMWs, and the influences of laser parameters (laser power and scanning speed) on the capillary performance were investigated. Our study verifies the unidirectionality of capillary transport in gradient micro-grooved wick structures and also demonstrates the enhancement of capillary performance by laser processing on copper surfaces. Furthermore, this study identifies the optimal laser processing parameters for achieving the best capillary performance and reveals the affecting mechanisms of laser parameters on the capillary performance of the CSMWs in terms of shape, size, morphology, and surface wettability. Finally, the study quantitatively establishes the coupling relationship between laser parameters and capillary performance parameters of CSMWs, providing a valuable basis for selecting laser processing parameters to achieve desirable capillary performance.

2. Materials and Methods

2.1. Fabrication of Copper CSMWs by Ultraviolet Nanosecond Laser

In this study, an ultraviolet nanosecond laser was used to fabricate micro-grooves with capillary step-gradients on a copper plate with dimensions of 0.8 mm in thickness, 5 mm in width, and 60 mm in length. The copper plate was made of high-purity copper (99.9% Cu) due to its high thermal conductivity and ease of processing. Its thermophysical properties are shown in Table 1. The laser processing principle is illustrated in Figure 1a. The incident laser beam is focused onto the material surface, and the material is heated by the laser beam to reach the ablation threshold. After rapid heating, melting, evaporation, ionization, evaporation spraying, and fluid ejection, the surface material was removed [33,34].

Table 1. Thermophysical properties of copper [23] and ethanol [35].

Thermophysical Properties	Copper	Ethanol
Density (g/cm^3)	8.93	0.79
Surface tension coefficient (N/m)	-	0.022
Viscosity (10^{-3} Pa·s)	-	1.10
Thermal conductivity (W/(m·K))	386	0.1800
Boiling point(K)	-	351.15
Heat capacity (J/(g·K))	0.386	2.44
Thermal expansion coefficient (10^{-6}/K)	17.5	250
Thermal diffusivity (mm^2/s)	111	8.71679×10^{-8}

The CSMWs are divided into three regions: the triple-groove region, the double-groove region, and the single-groove region, as shown in Figure 1c. In the negative capillary gradient direction, the total groove width decreases from 60 to 20 μm, with a gradually decreasing capillary radius. Taking the CSMW processed at 2.26 W and 90 mm/s as an example, we can calculate that the capillary diameter (the same as the hydraulic diameter) along the axial direction decreases from 25.12 (triple-groove) to 22.16 μm (double-groove), and then further decreases to 19.20 μm (single-groove). Here, the hydraulic diameter is defined as the ratio of four times of the cross-sectional liquid area in saturated grooved wick to the wetting perimeter of the groove. The decrease in hydraulic diameter facilitates the increase in capillary force in the late stage of capillary rise, promoting secondary capillary acceleration.

The laser spot size and processing positions in different regions of the CSMWs are presented in Figures 1c and 2. The lengths of single-groove, double-groove, and triple-groove regions are 30, 15, and 15 mm, respectively. The specific groove dimensions of the three regions are labeled in Figure 2. As can be seen, the single-groove region has a groove spacing S1 of 40 μm (Figure 2a). For the double-groove region (Figure 2b), the spacing between the neighboring double grooves (noted as S2) is 60 μm, and the distance between the centerlines of the adjacent grooves (noted as a) is 20 μm. In the triple-groove region (Figure 2c), the spacing between neighboring sets of triple grooves (noted as S3) is 80 μm. The total grooved widths of the single-groove region, double-groove region, and triple-groove region are 20 μm, 40 μm, and 60 μm, respectively.

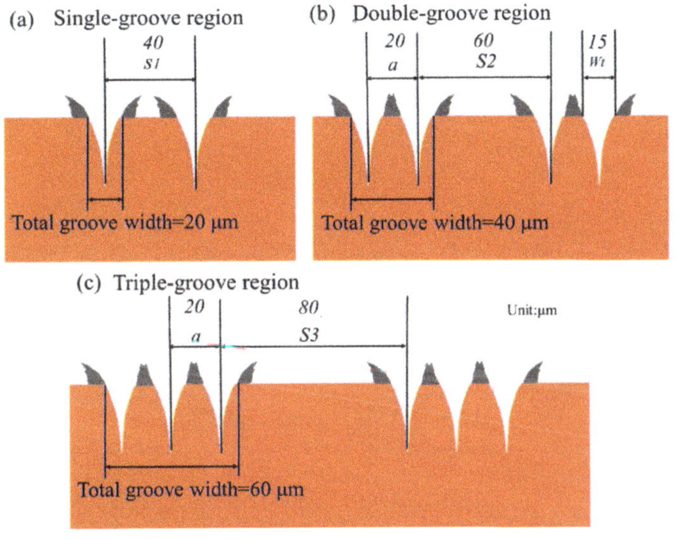

Figure 2. Dimensions of micro-grooves in (**a**) single-groove region, (**b**) double-groove region, and (**c**) triple-groove region.

Table 2 lists the basic parameters of the laser system. A 355 nm ultraviolet nanosecond pulse laser source (FOTIA-355-3-30-W, Inno Laser, Shenzhen, China) was employed, and the scanning speed was controlled using a galvanometric scanner (RAYLASE MS-II-10, Raylase AG, RAYLASE GmbH, Weßling, Germany). The energy distribution of the laser spot followed a Gaussian distribution, resulting in V-shaped grooves. To investigate the influence of laser parameters, a total of 12 samples were prepared by laser processing. The corresponding processing parameters are listed in Table 3. Among these samples, nine were subjected to a capillary rise test in the negative gradient direction, and three were subjected to a capillary rise test in the positive gradient direction (marked as "PG"). The copper plate surfaces for laser processing were fully polished, followed by ultrasonic cleaning with deionized water. Subsequently, an acid treatment process was employed to remove oil stains and insoluble substances from the copper plate surfaces, followed by cleaning and drying using deionized water.

Table 2. Characteristic parameters of an ultraviolet nanosecond pulsed laser system.

Parameters	Value	Units
Wavelength	335	nm
Averaged laser power	2.26~4	W
Pulse width	12	ns
Laser frequency	50	kHz
Spot diameter	20	μm
Single pulse energy	80	μJ

Table 3. Laser parameters of the CSMWs.

Wick Sample Code	Laser Power, *P* (W)	Laser Scan Rate, *V* (mm/s)	Scanning Times, *n*
4 W-90 V	4.00	90	5
4 W-50 V	4.00	50	5
4 W-10 V	4.00	10	5
4 W-90 V-PG	4.00	90	5
4 W-50 V-PG	4.00	50	5
4 W-10 V-PG	4.00	10	5
3 W-90 V	3.12	90	5
3 W-50 V	3.12	50	5
3 W-10 V	3.12	10	5
2 W-90 V	2.26	90	5
2 W-50 V	2.26	50	5
2 W-10 V	2.26	10	5

2.2. *Surface Characterization*

The surface morphology of the CSMWs was characterized using a scanning electron microscope (SEM, HITACHI SU8220, HITACHI, Tokyo, Japan) at an emission voltage of 15 kV. The element distribution of the laser-induced micro-grooves was measured using an energy-dispersive X-ray spectrometer (EDS, Bruker XFlash 6 | 30, Bruker, Karlsruhe, Germany). Contact angle measurements were conducted on double-groove regions. The contact angles were measured by a contact angle measurement instrument (CSCDIC-200S, SINCIN, Dongguan, China) with an accuracy of $\pm 1^{\circ}$. A water droplet of 2 μL was used for contact angle measurements, and the contact angles at five different points on each sample were measured to obtain the average value.

2.3. *Capillary Rise Test by IR Camera*

The capillary rise test of the CSMWs was conducted in both negative and positive gradient directions, and the schematic diagram of the test apparatus is shown in Figure 3. In the test, the infrared thermography method [32] was used to characterize the capillary rise rate of the CSMWs. Ethanol with high wetting capacity was applied as the working fluid, and its thermophysical properties are shown in Table 1. Three samples processed with the same laser power were fixed to a metal bracket for simultaneous testing. The experimental setup was enclosed in a box with black body walls to minimize interference from ethanol evaporation and external temperature fluctuations. A Fotric227s-L23 infrared camera with a resolution of 512×384 was employed, and the meniscus could be clearly captured by the infrared camera due to the emissivity difference between the working fluid and the copper material. The camera was operated at a frame rate of 24 frames per second, providing a time accuracy of 41.6 ms. The test time for each set of samples was 3 min, and the ambient temperature was maintained at 26 ± 0.2 °C throughout the test. The distance between the samples and the camera was 200 mm, with a rise height uncertainty of less than 0.3 mm. The capillary rising height is determined by the average height of the meniscus, which is indicated by the boundary of purple and orange colors. The operating criteria are presented in Supplementary Material (Figure S1).

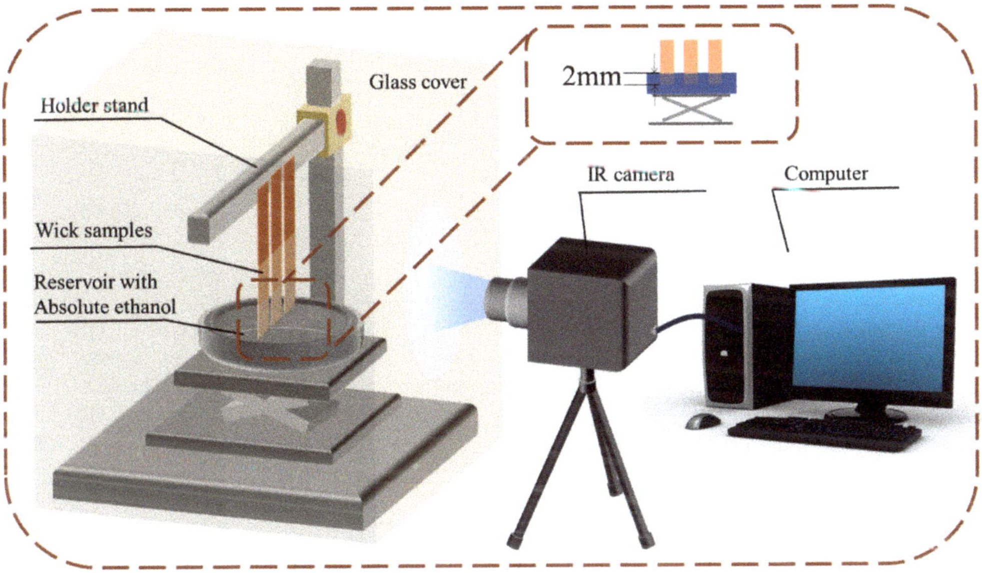

Figure 3. Schematic diagram of the capillary rise test apparatus based on an infrared camera.

3. Capillary Theory and Data Reduction in Capillary Performance Parameter

The capillary pressure induced by the curved liquid–gas interface can be expressed by the Laplace–Young equation [36]:

$$\Delta P_{cap} = \frac{2\sigma cos\theta}{r_p} \tag{1}$$

where r_p, σ, and θ are the capillary radius of the pore, surface tension, and contact angle, respectively. When the effective capillary radius (r_{eff}) is used to represent the ratio of r_p and $cos\theta$, Equation (1) can be rewritten as:

$$\Delta P_{cap} = \frac{2\sigma}{r_{eff}} \tag{2}$$

When a wick comes into contact with the working fluid, the liquid will spontaneously rise along the wick. Since the wick is placed vertically during the test, there is a balance between gravity and capillary pressure. Then, capillary pressure, friction pressure, and gravity can be related as follows [32,37,38]:

$$\Delta P_{cap} = \frac{\mu\varepsilon}{K} h\frac{dh}{dt} + \rho gh \tag{3}$$

where the first term on the right is the viscous friction according to Darcy's Law, and ρgh is the gravitational hydrostatic pressure. In detail, dh/dt is the rate of capillary rise, h is the rising height of the working fluid, μ is the viscosity of the working fluid, ρ is the density of the working fluid, g is the acceleration of gravity, ε is the porosity of the wick, and K is the permeability of the capillary.

In general, the liquid transport capacity can be evaluated by the capillary performance parameter ($\Delta P_{cap}\cdot K$) [32,39], which can be obtained from Equation (3):

$$\Delta P_{cap}\cdot K = h\left(\mu\varepsilon\frac{dh}{dt} + \rho gK\right) \tag{4}$$

The capillary performance parameter ($\Delta P_{cap} \cdot K$) can be calculated by the following method proposed by Deng et al. [32,39]. In Equation (4), the rising velocity, dh/dt, can be obtained from the derivative of the capillary rising height versus time. Rearranging Equation (4) yields:

$$(\Delta P_{cap} \cdot K)\frac{1}{h} - \rho g K = \mu\varepsilon\frac{dh}{dt} \quad (5)$$

When we set $x = 1/h$ and $y = dh/dt$, Equation (5) can be rewritten as:

$$y = \frac{\Delta P_{cap} \cdot K}{\mu\varepsilon}x - \frac{\rho g K}{\mu\varepsilon} \quad (6)$$

Through linear fitting by several data points of x and y from the test, the $\Delta P_{cap} \cdot K$ can be calculated from the slope of the fitting line. The capillary performance parameters of the three regions of the CSMWs are different, so the slopes for different segments of the fitting line are also different. For easy comparison, in this study, the comprehensive capillary performance parameter of the three segments is adopted based on linear fitting for all data.

4. Uncertainties

The uncertainty of the copper substrate dimensions is estimated to be 1.5%. The absolute uncertainty of the contact angle measurements is $\pm 1°$. The uncertainty of the micro-groove dimensions from laser processing is ± 5 μm. The uncertainty of the capillary rise height measured by the infrared camera method is within 1.5%. The uncertainty of dh/dt is estimated to be within 3%. The standard deviations of the capillary performance parameters from linear fitting are presented in Table 4, with a relative standard deviation ranging from 1.51% to 2.21%.

Table 4. Capillary performance parameter ($\Delta P_{cap} \cdot K$) and its deviation from the CSMW samples.

Wick Samples	Capillary Performance Parameter, $\Delta P_{cap} \cdot K$ (10^{-9})	Standard Deviation of $\Delta P_{cap} \cdot K$ (10^{-10})	Relative Standard Deviation of $\Delta P_{cap} \cdot K$
4 W-90 V	7.63	1.21	1.59%
4 W-50 V	7.97	1.23	1.54%
4 W-10 V	9.81	1.48	1.51%
4 W-90 V-PG	6.31	1.13	1.79%
4 W-50 V-PG	6.56	1.15	1.75%
4 W-10 V-PG	8.32	1.46	1.75%
3 W-90 V	5.21	0.97	1.86%
3 W-50 V	6.06	1.10	1.81%
3 W-10 V	7.10	1.24	1.74%
2 W-90 V	2.49	0.53	2.12%
2 W-50 V	2.91	0.62	2.13%

5. Results and Discussion

5.1. *Influences of Laser Parameters on Micro-Groove Morphology*

Figure 4 illustrates the cross-sectional morphology of the double-groove regions processed with different laser parameters. It is evident that laser power and scanning speed have significant influences on the cross-sectional morphology of the micro-grooves. As the laser power increases or the scanning speed decreases, the depth of the micro-grooves gradually increases. For example, at a scanning speed of 10 mm/s, when the power increases from 2.26 to 4 W, the depth of the V-shaped groove increases from 43.4 to 50.1 μm; at a certain laser power of 2.26 W, when the scanning speed decreases from 90 to 10 mm/s, the groove depth increases from 30.2 to 43.4 μm. At higher laser power and lower scanning speeds, more material on the copper substrate reaches the ablation threshold, resulting in deeper grooves due to the increased melting and evaporation [40].

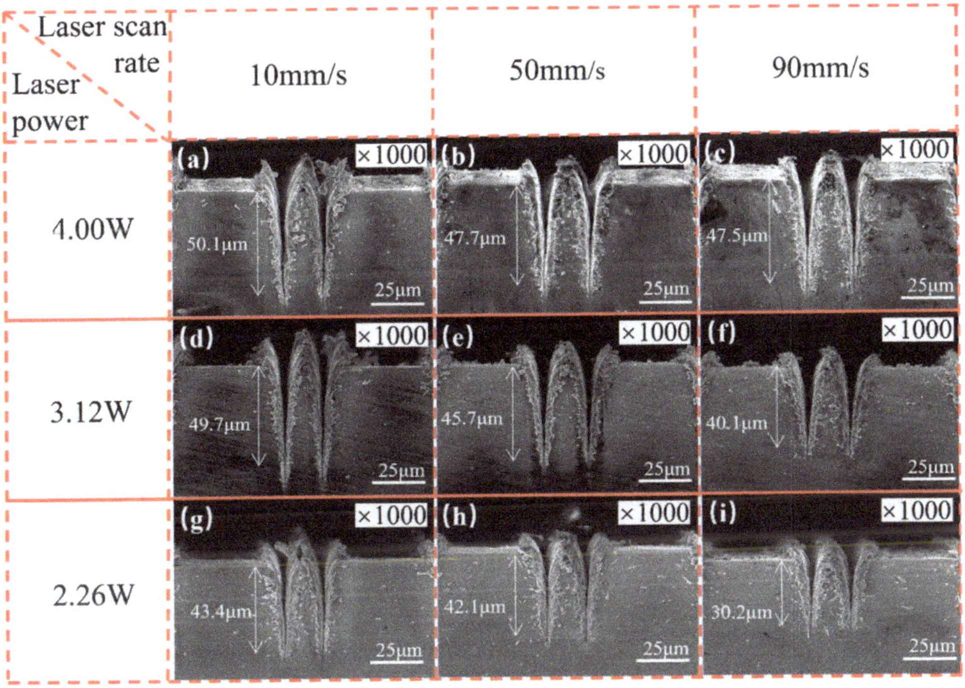

Figure 4. Cross-sectional morphology of double-groove regions fabricated by various laser powers and scanning speeds: (**a**) 4.00 W-10 mm/s, (**b**) 4.00 W-50 mm/s, (**c**) 4.00 W-90 mm/s, (**d**) 3.12 W-10 mm/s, (**e**) 3.12 W-50 mm/s, (**f**) 3.12 W-90 mm/s, (**g**) 2.26 W-10 mm/s, (**h**) 2.26 W-50 mm/s, and (**i**) 2.26 W-90 mm/s.

Laser-processed micro-grooves exhibit unique cross-sectional morphology features distinct from those fabricated by traditional methods. First, laser processing produces a recast layer and raises the top of the micro-grooves above the original copper surface due to the deposition of molten and evaporated copper material on the top of the micro-grooves. The recast layer increases the depth of the grooves, leading to faster capillary transport. Moreover, the spikes and voids within the recast layer also facilitate extensive evaporation of the working fluid on the meniscus surface [24]. Second, laser-processed grooves have a unique V-shaped geometry with a high aspect ratio owing to the Gaussian distribution of laser spot energy. The high aspect ratio is attributed to the extremely high energy density of the laser (25.46 J/cm^2). As a result, the depth-to-width ratio of the grooves fabricated in this study (~3.4 in Figure 4a) is significantly greater than that of the grooves produced through traditional extrusion molding (~1.28) [41] or high-speed spinning methods (~1.25) [42]. From another point of view, the laser-caused material addition method (such as laser-induced electrodeposition) [43] is an effective way to enhance surface capillary performance. For example, the laser-induced maskless electrodeposition method can fabricate complex patterns with high flexibility. Thus, it can be used to fabricate precise capillary-enhanced structures. Besides, the morphology of the fabricated surface is rich (such as stray deposition shape and pyramidal structure), which facilitates surface hydrophilicity.

Figure 5 presents the top view of the SEM images of double-groove regions with various laser parameters. As the laser power increases, the surface roughness of the grooves also shows an increasing trend. For instance, when the laser power increases from 2.26 to 4 W at a fixed scanning speed of 90 mm/s, the surface morphology translates from a smooth cement-like texture to a bark-like texture and then to a rock-like texture, with more and more pronounced surface protrusions. On the other hand, when the laser power is fixed, the smoothness of the groove surfaces gradually diminishes with the increase in scanning speed. This phenomenon is particularly pronounced in the sample group fabricated at a laser power of 2 W. As can be seen, the density of protrusions on the recast layer and groove sidewalls of the 2 W-10 V sample is significantly higher than that of the 2 W-90 V sample. This can be attributed to the increased absorption energy density of the material at higher laser power and lower scanning speed, which intensifies heat transfer and diffusion behaviors, leading to more prominent secondary deposition of laser-ablated copper material [34,40]. As a result, the overlapping phenomenon in the recast layer intensifies and the surface roughness increases [44,45].

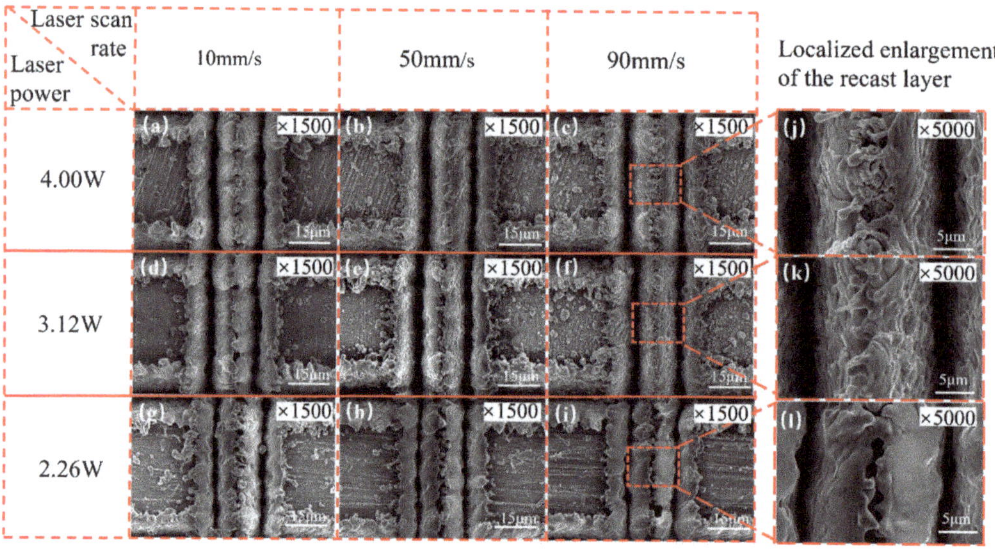

Figure 5. Top-view SEM images of double-groove regions fabricated with different laser powers at different scanning speeds: (**a**) 4.00 W-10 mm/s, (**b**) 4.00 W-50 mm/s, (**c**) 4.00 W-90 mm/s, (**d**) 3.12 W-10 mm/s, (**e**) 3.12 W-50 mm/s, (**f**) 3.12 W-90 mm/s, (**g**) 2.26 W-10 mm/s, (**h**) 2.26 W-50 mm/s, and (**i**) 2.26 W-90 mm/s. Partially enlarged views of the recast layer with a laser power of (**j**) 4.00 W, (**k**) 3.12 W, and (**l**) 2.26 W at a scanning speed of 90 mm/s.

Since the oxidation of the copper surface affects the contact angle, EDS analysis of the CSMWs was conducted to determine the O element mass fraction. Figure 6 shows the EDS images of single grooves, double grooves, and triple grooves with a laser power of 4 W at a scanning speed of 10 mm/s. All samples were exposed to the air for approximately 1 h before testing. The EDS images reveal a pronounced distribution of oxygen element in the recast layer, indicating the oxidation of copper during laser processing. Notably, due to altitude differences, the oxygen element in the central region of the V-shaped groove is not captured, presenting a black color. Figure 7 illustrates the mass fraction change trends of copper and oxygen elements as a function of total groove width. The mass fraction of the oxygen element is strongly related to the total groove width, and the oxidation effect becomes more pronounced as the total groove width increases. This is because the increase in total groove width results in a larger total copper surface exposure to air, leading to a more intense degree of oxidation [20].

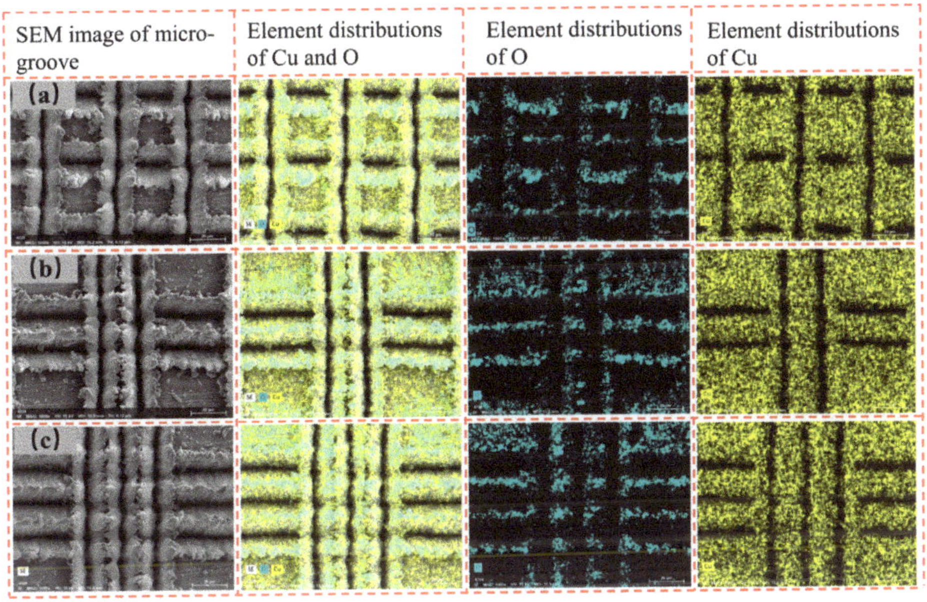

Figure 6. EDS element distribution in micro-groove regions with different total groove widths and capillary radius: (**a**) single-groove regions with a total groove width of 20 μm, (**b**) double-groove regions with a total groove width of 40 μm, and (**c**) triple-groove regions with a total groove width of 60 μm.

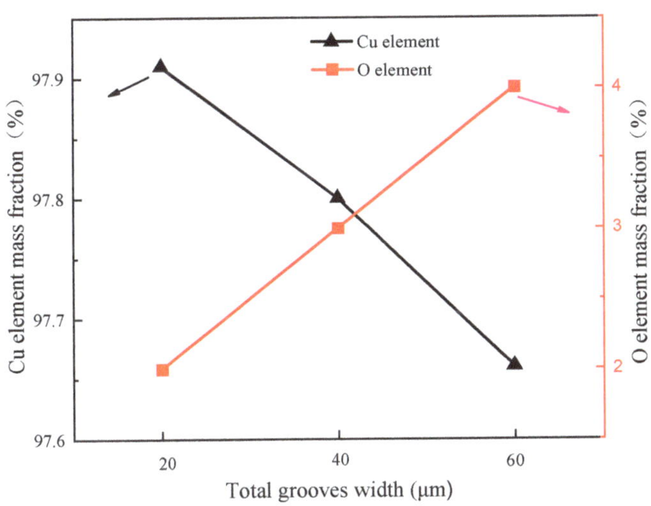

Figure 7. Mass fraction distribution of copper and oxygen elements as a function of total groove width for different micro-groove regions.

5.2. Influences of Laser Parameters on Contact Angle

Figure 8a–i shows the top-view SEM images of double-groove regions fabricated with different laser powers at different scanning speeds. In the images, the axis of the groove is along the left-right direction. It is obvious that the hydrophilicity of laser-induced surfaces is significantly enhanced in comparison to that of the unprocessed surface. As shown in Figure 8j, the original contact angle of the unprocessed surface is 93.1 ± 1°, which is a little hydrophobic due to mild oxidation after short-term storage before testing. The three CSMW samples fabricated with a laser power of 4.00 W present the best hydrophilicity, with the contact angle ranging from 8.9 ± 1° to 17.3 ± 1°. Those fabricated with a laser power of 3.12 W present worse hydrophilicity (with the contact angle ranging from 10.3 ± 1° to 33.3 ± 1°), while the samples fabricated with a laser power of 2.26 W show the worst hydrophilicity (with the contact angle ranging from 41.2 ± 1° to 58.0 ± 1°). As shown in Figure 8a, the contact angle achieves the minimum value of 8.9 ± 1° when the sample is fabricated with a laser power of 4 W at a scanning speed of 10 mm/s. Hence, it can be concluded that the wettability is improved with increasing laser power and decreasing scanning speed, which facilitates the enhancement of capillary performance [23].

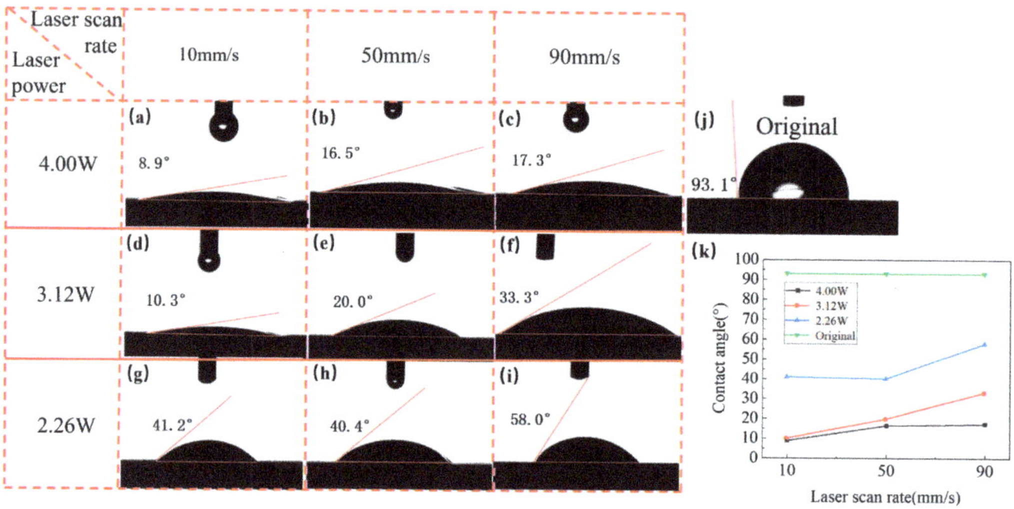

Figure 8. Top-view SEM images of double-groove regions fabricated with different laser powers at different scanning speeds: (**a**) 4.00 W-10 mm/s, (**b**) 4.00 W-50 mm/s, (**c**) 4.00 W-90 mm/s, (**d**) 3.12 W-10 mm/s, (**e**) 3.12 W-50 mm/s, (**f**) 3.12 W-90 mm/s, (**g**) 2.26 W-10 mm/s, (**h**) 2.26 W-50 mm/s, and (**i**) 2.26 W-90 mm/s. (**j**) Original contact angle of an unprocessed copper plate. (**k**) Change the trend of contact angle as a function of scanning speed.

5.3. Enhanced Capillary Performance of the CSMW Compared to the Non-Gradient Parallel Wick

Figure 9 presents the enhanced capillary performance of the CSMW compared to a non-gradient parallel micro-grooved wick. The capillary rising rate and height of the gradient wick substantially outperform the parallel one. As shown in Figure 9a, the rising curve of the parallel wick became stable at a height of 24.0 mm, while the one of CSMW presented two times of secondary accelerations and finally reached about 53.8 mm. The equilibrium rising height of the CSMW was enhanced by 124% compared to the parallel one. The capillary performance parameter was promoted from 9.061×10^{-9} N to 9.81×10^{-9} N, with an improvement percentage of 8.2%. The enhancing mechanism is promoted, balancing between capillary force and permeability in various flowing stages. The CSMW provide a large permeability at the early stage and a large capillary force at the final stage, which promotes capillary flow efficiency.

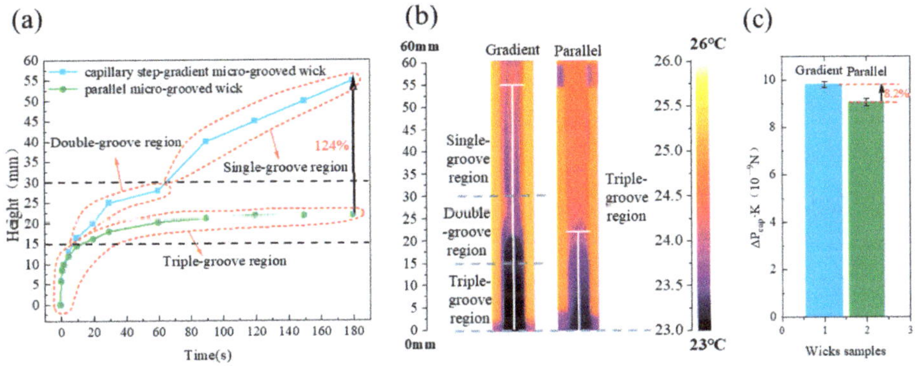

Figure 9. The comparisons between capillary step-gradient micro-grooved wick and parallel micro-grooved wick in terms of (**a**) dynamic capillary rising height, (**b**) equilibrium capillary rising height (in 3 min), and (**c**) capillary performance parameters ($\Delta P_{cap} \cdot K$). The laser parameters are 4.00 W-10 mm/s, and the parallel wick is only composed of a triple-groove region.

5.4. Influences of Capillary-Gradient Direction on Capillary Performance of the CSMWs

To verify the unidirectionality of capillary transport in the CSMWs, the influences of capillary-gradient direction on the capillary rise performance were investigated. Figure 10 presents the SEM images of the CSMWs. As can be seen, the total groove width as well as the capillary radius decrease along the negative gradient direction. Additionally, the grooves are continuous in the vertical direction since the longitudinal grooves are processed in the final step, preventing any recast layer blockage in the longitudinal direction. Figure 11 depicts the capillary rise height of the sample group fabricated with a laser power of 4 W as a function of time in both negative and positive gradient directions. Figure 12 displays the IR images of the equilibrium capillary rise height of the above samples after capillary rise for 3 min.

The CSMWs experimentally exhibit unidirectional liquid transport capacity. In the CSMW with a positive gradient, the working fluid meniscus fails to cross the boundary line (at a height of 30 mm, as shown in Figure 12b) between the single-groove and double-groove regions. This is because liquid transport will be blocked by negative Laplace pressure induced by increasing the capillary radius in the positive gradient direction, which is known as the capillary blocking phenomenon. On the other hand, in the negative gradient direction, the working fluid can cross the boundary lines of neighboring regions, moving from the triple-groove region to the single-groove region, as shown in Figure 12a. These results indicate that the gradient direction significantly affects capillary rise performance. The equilibrium capillary rise heights for the 4 W-10 V, 4 W-50 V, and 4 W-90 V samples under negative gradient conditions are 83%, 55%, and 38%, which are higher than those under positive gradient conditions, respectively, with an average increased percentage of 58.6%.

The CSMWs with a negative gradient exhibit a unique secondary capillary acceleration phenomenon, which contributes to enhanced capillary performance. Taking the 4 W-10 V example as an example, its capillary rise rate does not conform to the classical Lucas–Washburn model [46] for a uniform capillary wick. On its capillary rise curve (Figure 11), there is a secondary acceleration segment while crossing the region boundary, even at the end of the wick. This is due to the sudden reduction in the capillary radius of the CSMW, which produces an increased capillary force and additional driving force. As a result, the equilibrium rising height of the 4 W-10 V sample can reach 55 mm in the negative gradient direction, which is significantly greater than the height (30 mm) in the positive gradient direction for the same sample.

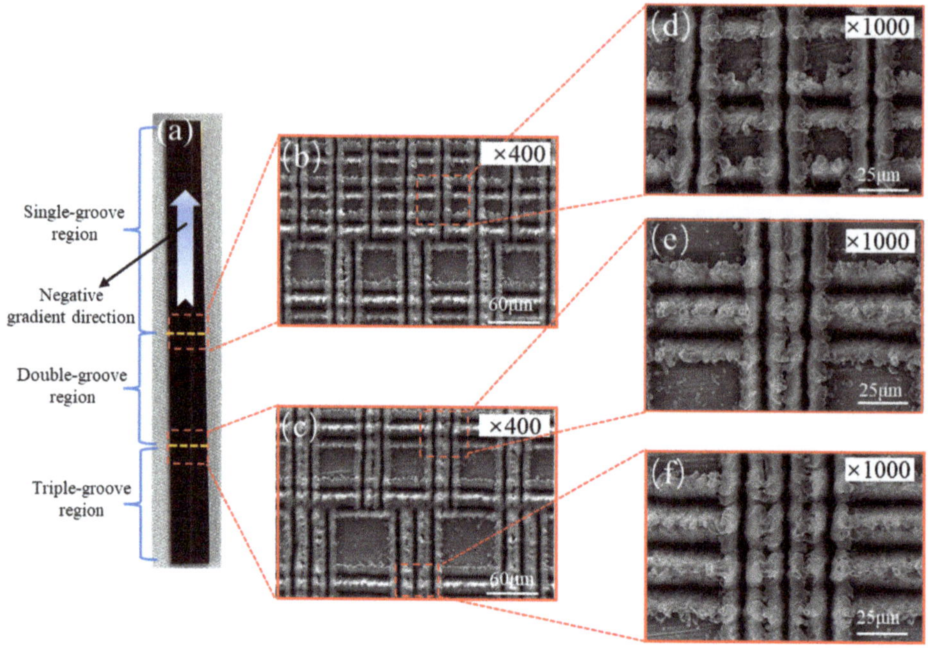

Figure 10. SEM images of the overall CSMW fabricated using an ultraviolet nanosecond laser: (**a**) real image of the CSMW with three regions; SEM images of junctions in (**b**) a single-double groove region and (**c**) a double-triple groove region; and SEM images of (**d**) a single-groove region, (**e**) a double-groove region, and (**f**) a triple-groove region.

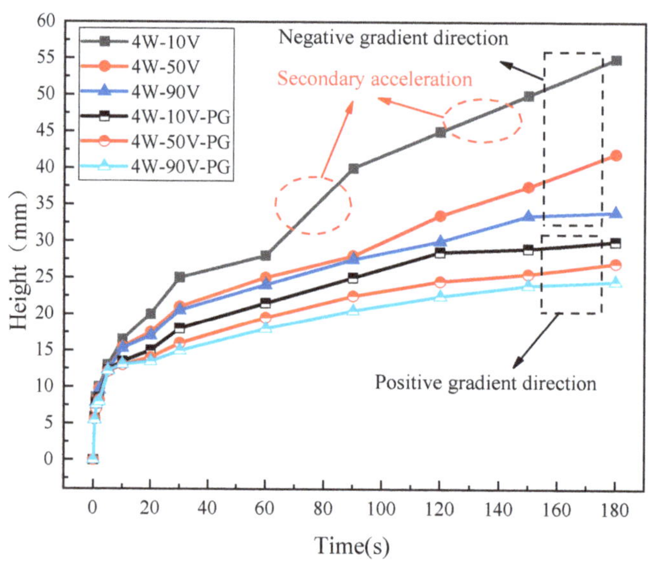

Figure 11. Capillary rise heights of the CSMWs as a function of time in positive and negative gradient directions. The negative gradient direction is from the triple-groove region to the single-groove region with a reduced capillary radius, while the positive gradient direction is the opposite.

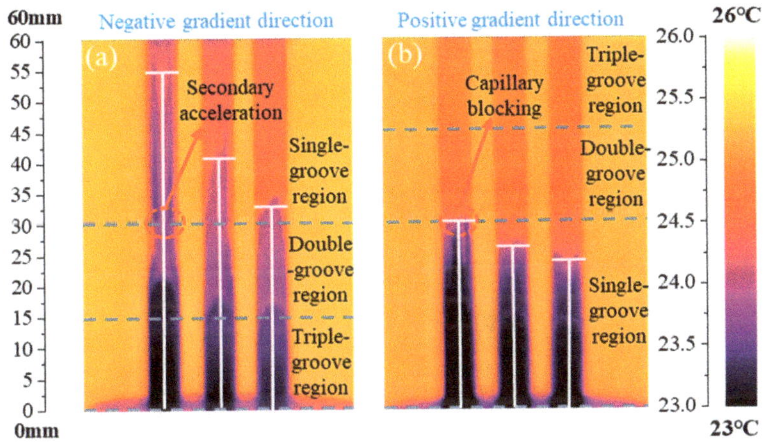

Figure 12. IR images of the equilibrium capillary rise height of the CSMWs in (**a**) negative and (**b**) positive gradient directions. The laser power is 4.00 W, and the scanning speeds are 10, 50, and 90 mm/s from left to right for each subfigure, respectively.

5.5. Influences of Laser Parameters on Capillary Performance of the CSMWs

5.5.1. Influence of Laser Parameters on Capillary Rise Height

Figures 13 and 14 illustrate the capillary rise process and the capillary equilibrium rise heights of the CSMWs fabricated with nine combinations of laser parameters, respectively. The meniscus in all samples increases rapidly in the early stages. The secondary acceleration phenomenon was observed at both the 15 mm height interfaces of the 3 W samples group and the 15 mm and 30 mm height interface of the 4 W samples group. Such a capillary rise height curve of the CSMWs is unique and different from the classical Lucas–Washburn modeling curve for uniform wick with a continuously reducing slope [47,48].

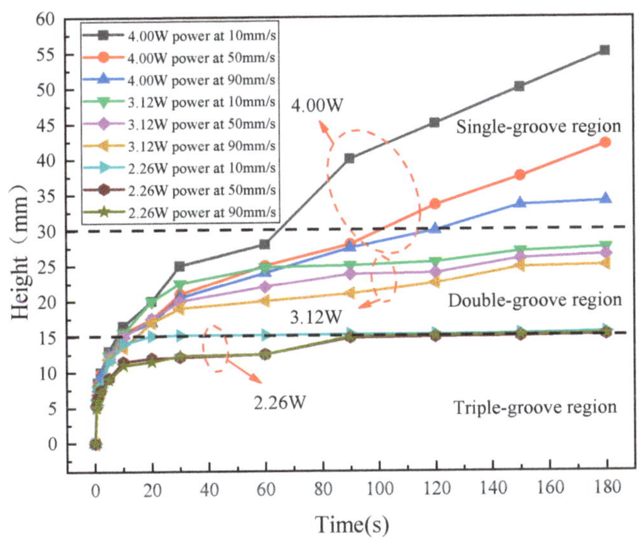

Figure 13. Capillary rise heights of the CSMWs fabricated with different laser powers at different scanning speeds.

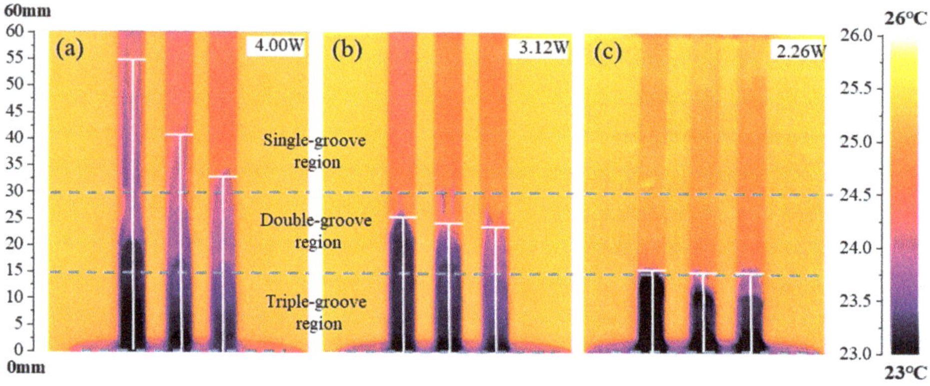

Figure 14. Equilibrium capillary rise heights of the CSMWs fabricated with different laser powers at different scanning speeds: (**a**) 4.00 W, (**b**) 3.12 W, and (**c**) 2.26 W. The laser scanning speeds in each subfigure are 10, 50, and 90 mm/s from left to right, respectively, and the test time is 3 min.

The capillary rise height significantly increases with the increase in laser power. Given a fixed scanning speed of 10 mm/s, when the laser power increases from 2.26 to 3.12 W and then to 4 W, the equilibrium capillary rise height increases from 15.6 to 27.5 mm and then to 55 mm, as shown in Figure 14. As the laser power increases, deeper and narrower V-shaped grooves can be produced, facilitating higher capillary pressure. Additionally, the rough and folded surface morphology resulting from the increased laser power can effectively reduce the contact angle and thus increase the Laplace pressure.

When the laser power is fixed, the capillary rise height significantly increases with the decrease in scanning speed. For any given laser power, the samples fabricated at a scanning speed of 10 mm/s exhibit a higher capillary rise height than those fabricated at the other two higher speeds. Taking the sample group fabricated with a laser power of 4 W as an example, the equilibrium height for the scanning speed of 10 mm/s (55 mm) is higher than that for 50 mm/s (42 mm) or 90 mm/s (34 mm). Furthermore, with the increase in laser power, the sensitivity of capillary rise height to scanning speed becomes more pronounced. For instance, the height difference is relatively small for the samples fabricated with a laser power of 2.26 W at various scanning speeds, but it is significantly larger for those fabricated with a laser power of 4 W. The laser scanning speed affects the laser pulse density on the material surface. When the laser scanning speed is lower, the amount of raw material removed is increased, resulting in a larger structural depth [24]. For ideal V-shaped micro-grooves, the capillary rise rate increases with the increase in structural depth.

5.5.2. Coupled Effect of Laser Parameters on Capillary Performance Parameters of the CSMWs

Figure 15 illustrates the coupled effect of laser power and scanning speed on the capillary performance parameter ($\Delta P_{cap} \cdot K$). With the increase in laser power and the decrease in scanning speed, the capillary performance parameter shows an increasing trend. The 4 W-10 V sample exhibits the highest capillary performance parameter of 9.81×10^{-9} N, which is 293% higher than that of the 2 W-90 V sample (2.49×10^{-9} N). The mechanisms of the enhanced capillary performance of the 4 W-10 V sample can be attributed to the deepest groove depth (50.1 μm), the smallest contact angle (8.9 ± 1°), the roughest surface structure, and the most abundant surface protrusions (as shown in Figure 5). For a new pattern, the optimal parameters in this study are applicable because this parameter produces the best surface hydrophilicity and a smaller capillary radius, which are generally needed by various patterns.

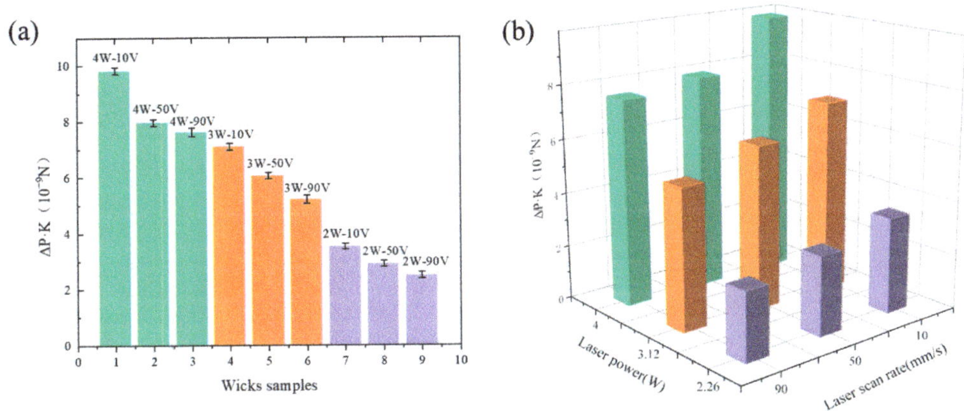

Figure 15. Capillary performance parameters ($\Delta P_{cap} \cdot K$) of the CSMWs fabricated with different laser parameters in the form of (**a**) Two-dimensional bar chart and (**b**) Three-dimensional bar chart.

6. Conclusions

In summary, capillary step-gradient micro-grooved wicks on copper substrate were fabricated using an ultraviolet nanosecond pulse laser, and IR thermography was employed to characterize the capillary rise performance of the CSMWs fabricated with different laser parameters. Some conclusions can be drawn, as follows:

(1) The increase in laser power and the decrease in scanning speed can deepen V-shaped grooves and increase the surface roughness of CSMWs, resulting in more intricate surface morphology and a smaller contact angle. Besides, the equilibrium capillary rising height of the CSMW was greatly enhanced by 124% compared to the non-gradient grooved wick.

(2) The capillary gradient direction of the CSMWs has substantial influences on the capillary performance. In the positive gradient direction, the meniscus experiences capillary blocking at the boundary of neighboring regions. In contrast, for the negative gradient direction, there is a secondary acceleration phenomenon at the region boundary. This can be attributed to the sudden increase in capillary force induced by the decrease in capillary radius, resulting in a unique capillary rise curve that is significantly different from the classical Lucas–Washburn model for uniform capillary wicks.

(3) The laser parameters significantly affect the capillary performance of the CSMWs. The equilibrium capillary rise height increases with the increase in laser power and the decrease in scanning speed. The CSMW sample fabricated with a laser power of 4 W at a scanning speed of 10 mm/s has the highest capillary performance parameter of $9.18 \pm 0.15 \times 10^{-9}$ N, which is 293% higher than that of the 2 W-90 V sample. Therefore, it is very important to select appropriate laser parameters for the enhancement of the capillary performance of the CSMWs.

(4) The enhancement mechanisms of lase processing for the capillary performance of the CSMWs involve improved hydrophilicity and reduced capillary radius. The laser-processed surfaces exhibit increased roughness and abundant protrusion morphology with a recast layer, which can increase the surface area and thereby lower the contact angle, resulting in improved hydrophilicity. The high-energy density and Gaussian distribution of laser energy produce V-shaped grooves, which have elongated the regions at the groove ends, resulting in a large depth-to-width ratio and a reduced capillary radius. To sum up, the increased hydrophilicity and decreased capillary radius result in increased Laplace pressure, thereby enhancing the capillary performance of the CSMWs.

The laser processing technology in this study can provide an efficient and cost-effective method for fabricating patterned wicks with improved capillary performance.

Supplementary Materials: The following supporting information can be downloaded at: https://www.mdpi.com/article/10.3390/mi15010165/s1, Figure S1: The operating criteria for determining the meniscus height.

Author Contributions: Structure design, fabrication, and testing, J.L. and C.F.; data calculation and writing, S.L. and W.M.; writing, G.H., G.Y. and S.T.; funding acquisition, C.C. and Y.Z. All authors have read and agreed to the published version of the manuscript.

Funding: This research was funded by the Science and Technology Program of Guangdong Province (Grant No. 2022A0505050071), the Guangzhou Basic and Applied Basic Research Project (Grant Nos. 2024A04J4375 and 202201010322), the State Key Laboratory of Precision Electronic Manufacturing Technology and Equipment (Grant No. JMDZ202309), and the National Natural Science Foundation of China (Grant No. 62174039).

Data Availability Statement: Upon reasonable request, Guannan Yang, the corresponding author, is willing to provide the data that support the conclusions in this paper.

Conflicts of Interest: Author Shiwo Ta was employed by the company Fenghua Advanced Technology Co., Ltd. The remaining authors declare that the research was conducted in the absence of any commercial or financial relationships that could be construed as a potential conflict of interest.

Nomenclature

a	centerline distance of the adjacent groove (mm)
dh/dt	capillary rise velocity (mm/s)
g	gravitational acceleration (m/s^2)
h	capillary rise height (mm)
K	permeability of wick (mm^2)
P	pressure (Pa)
LP	laser power (W)
$\Delta P_{cap}{\cdot}K$	capillary performance parameter (N)
ΔP_{cap}	capillary pressure (Pa)
r_p	capillary radius of pore (mm)
r_{eff}	effective capillary radius (mm)
$S1$	spacing between single grooves (mm)
$S2$	spacing between double grooves (mm)
$S3$	spacing between triple grooves (mm)
V	laser scanning speed (mm/s)
Greek symbols	
μ	dynamic viscosity (Pa·s)
θ	contact angle (°)
ε	porosity, dimensionless
ρ	density (kg/m^3)
σ	surface tension (N/m)
Subscripts	
cap	capillary
eff	effective
p	pore
Abbreviations	
PG	positive gradient
CSMW	capillary step-gradient micro-grooved wick
IC	integrated circuit

References

1. Wang, J.; Li, Y.; Liu, X.; Shen, C.; Zhang, H.; Xiong, K. Recent Active thermal Management Technologies for the Development of Energy-Optimized Aerospace Vehicles in China. *Chin. J. Aeronaut.* **2021**, *34*, 1–27. [CrossRef]
2. He, W.; Zhang, J.; Li, H.; Liu, S.; Wang, Y.; Lv, B.; Wei, J. Optimal thermal Management of Server Cooling System Based Cooling Tower Under Different Ambient Temperatures. *Appl. Therm. Eng.* **2022**, *207*, 118176. [CrossRef]
3. Sohel Murshed, S.M.; Nieto De Castro, C.A. A Critical Review of Traditional and Emerging Techniques and Fluids for Electronics Cooling. *Renew. Sustain. Energy Rev.* **2017**, *78*, 821–833. [CrossRef]
4. Lu, N.; Li, J.; Sun, Y. Research Progress and Prospect of Heat Pipe Capillary Wicks. *Front. Heat Mass Transf.* **2022**, *18*, 24. [CrossRef]
5. Chen, S.-W.; Hsieh, J.-C.; Chou, C.-T.; Lin, H.-H.; Shen, S.-C.; Tsai, M.-J. Experimental investigation and Visualization on Capillary and Boiling Limits of Micro-Grooves Made by Different Processes. *Sens. Actuators A* **2007**, *139*, 78–87. [CrossRef]
6. Zhong, G.; Tang, Y.; Ding, X.; Chen, G.; Li, Z. Experimental investigation on Wettability and Capillary Performance of Ultrasonic Modified Grooved Aluminum Wicks. *Int. J. Heat Mass Transf.* **2021**, *179*, 121642. [CrossRef]
7. Singh, T.; Dvivedi, A. Developments in Electrochemical Discharge Machining: A Review on Electrochemical Discharge Machining, Process Variants and their Hybrid Methods. *Int. J. Mach. Tools Manuf.* **2016**, *105*, 1–13. [CrossRef]
8. Tang, H.; Tang, Y.; Wan, Z.; Li, J.; Yuan, W.; Lu, L.; Li, Y.; Tang, K. Review of Applications and Developments of Ultra-Thin Micro Heat Pipes for Electronic Cooling. *Appl. Energy* **2018**, *223*, 383–400. [CrossRef]
9. Tang, H.; Xie, Y.; Xia, L.; Tang, Y.; Sun, Y. Review on the Fabrication of Surface Functional Structures for Enhancing Heat Transfer of Heat Pipes. *Appl. Therm. Eng.* **2023**, *226*, 120337. [CrossRef]
10. Huang, G.; Wei, X.; Gu, Y.; Kang, Z.; Lao, L.; Li, L.; Fan, J.; Shou, D. Heterogeneously Engineered Porous Media for Directional and asymmetric Liquid Transport. *Cell Rep. Phys. Sci.* **2022**, *3*, 100710. [CrossRef]
11. Huang, G.; Yuan, W.; Tang, Y.; Zhang, B.; Zhang, S.; Lu, L. Enhanced Capillary Performance in Axially Grooved Aluminium Wicks by Alkaline Corrosion Treatment. *Exp. Therm. Fluid Sci.* **2017**, *82*, 212–221. [CrossRef]
12. Ling, W.; Zhou, W.; Yu, W.; Chu, X. Capillary Pumping Performance of Porous Copper Fiber Sintered Wicks for Loop Heat Pipes. *Appl. Therm. Eng.* **2018**, *129*, 1582–1594. [CrossRef]
13. Zhang, S.; Lin, L.; Chen, G.; Tang, H.; Zeng, J.; Yuan, W.; Tang, Y. Experimental Study on the Capillary Performance of Aluminum Micro-Grooved Wicks with Reentrant Cavity Array. *Int. J. Heat Mass Transf.* **2019**, *139*, 917–927. [CrossRef]
14. Li, P.; Xie, J.; Cheng, J.; Wu, K.K. Anisotropic Wetting Properties on A Precision-Ground Micro-V-Grooved Si Surface Related To their Micro-Characterized Variables. *J. Micromech. Microeng.* **2014**, *24*, 75004–75014. [CrossRef]
15. Huang, S.; Wan, Z.; Zhang, X.; Yang, X.; Tang, Y. Evaluation of Capillary Performance of A Stainless Steel Fiber–Powder Composite Wick for Stainless Steel Heat Pipe. *Appl. Therm. Eng.* **2019**, *148*, 1224–1232. [CrossRef]
16. Lee, S.; Yang, D.; Nikumb, S. Femtosecond Laser Micromilling of Si Wafers. *Appl. Surf. Sci.* **2008**, *254*, 2996–3005. [CrossRef]
17. Katahira, K.; Ogawa, Y.; Morita, S.; Yamazaki, K. Experimental investigation for Optimizing the Fabrication of A Sapphire Capillary Using Femtosecond Laser Machining and Diamond Tool Micromilling. *CIRP Ann.* **2020**, *69*, 229–232. [CrossRef]
18. Ruffino, F.; Grimaldi, M.G. Nanostructuration of Thin Metal Films by Pulsed Laser Irradiations: A Review. *Nanomaterials* **2019**, *9*, 1133. [CrossRef]
19. Li, J.; Zhang, M. Enhanced Capillary Performance of Grooved Nanocarbon Foams as Wicks for Heat Pipes. *Int. Commun. Heat Mass Transf.* **2022**, *130*, 105763. [CrossRef]
20. Jiang, G.; Tian, Z.; Luo, X.; Chen, C.; Hu, X.; Wang, L.; Peng, R.; Zhang, H.; Zhong, M. Ultrathin Aluminum Wick with Dual-Scale Microgrooves for Enhanced Capillary Performance. *Int. J. Heat Mass Transf.* **2022**, *190*, 122762. [CrossRef]
21. Xie, F.; Yang, J.; Ngo, C.-V. The Effect of Femtosecond Laser Fluence and Pitches between V-Shaped Microgrooves on the Dynamics of Capillary Flow. *Results Phys.* **2020**, *19*, 103606. [CrossRef] [PubMed]
22. Jiang, H.; Li, M.; Xie, H. Process Research on the Microgroove Depth Uniformity of Bursting Discs Using Femtosecond Lasers. *Coatings* **2022**, *12*, 567. [CrossRef]
23. Wang, H.; Tang, Y.; Bai, P.; Guo, W.; Luo, Y.; Li, S.; Zhang, X.; Zhou, G. A Multiscale Composite Silicon Carbide Wick with Excellent Capillary Performance. *Int. Commun. Heat Mass Transf.* **2022**, *139*, 106478. [CrossRef]
24. Long, J.; Chu, P.; Li, Y.; Lin, J.; Cao, Z.; Xu, M.; Ren, Q.; Xie, X. Dual-Scale Porous/Grooved Microstructures Prepared by Nanosecond Laser Surface Texturing for High-Performance Vapor Chambers. *J. Manuf. Process.* **2022**, *73*, 914–923. [CrossRef]
25. Lewis, R.; Liew, L.-A.; Xu, S.; Lee, Y.-C.; Yang, R. Microfabricated Ultra-Thin All-Polymer thermal Ground Planes. *Sci. Bull.* **2015**, *60*, 701–706. [CrossRef]
26. Ahmadi, M.; Bigham, S. Gradient Wick Channels for Enhanced Flow Boiling Htc and Delayed Chf. *Int. J. Heat Mass Transf.* **2021**, *167*, 120764. [CrossRef]
27. Zhou, F.; Zhou, G.; Zhou, J.; Huai, X.; Jiang, Y. A Novel Ultra-Thin Vapor Chamber with Radial-Gradient Hierarchical Wick for High-Power Electronics Cooling. *Int. J. Therm. Sci.* **2023**, *183*, 107896. [CrossRef]
28. Zhang, S.; Jiang, X.; Li, Y.; Chen, G.; Sun, Y.; Tang, Y.; Pan, C. Extraordinary Boiling Enhancement Through Micro-Chimney Effects in Gradient Porous Micromeshes for High-Power Applications. *Energy Convers. Manag.* **2020**, *209*, 112665. [CrossRef]
29. Jiang, H.; Wang, X.; Li, X.; Xu, J.; Qi, H.; Shan, D.; Guo, B. Enhanced Evaporation Performance on A Novel Microstructured Surface with Vertical Dimension Gradient. *Int. J. Heat Mass Transf.* **2022**, *199*, 123478. [CrossRef]
30. Liu, C.; Chen, H.; Li, Q. Experimental Study in thermal Expansion Devices of Phase-Change Material Coupled To Vapor Chamber with Gradient Wettability Wick. *J. Energy Storage* **2022**, *55*, 105429. [CrossRef]

31. Xie, X.; Weng, Q.; Luo, Z.; Long, J.; Wei, X. thermal Performance of the Flat Micro-Heat Pipe with the Wettability Gradient Surface by Laser Fabrication. *Int. J. Heat Mass Transf.* **2018**, *125*, 658–669. [CrossRef]
32. Tang, Y.; Deng, D.; Lu, L.; Pan, M.; Wang, Q. Experimental investigation on Capillary force of Composite Wick Structure by Ir thermal Imaging Camera. *Exp. Therm. Fluid Sci.* **2010**, *34*, 190–196. [CrossRef]
33. Leone, C.; Papa, I.; Tagliaferri, F.; Lopresto, V. Investigation of Cfrp Laser Milling Using A 30w Q-Switched Yb:Yag Fiber Laser: Effect of Process Parameters on Removal Mechanisms and Haz formation. *Compos. Part A Appl. Sci. Manuf.* **2013**, *55*, 129–142. [CrossRef]
34. Deng, D.; Wan, W.; Huang, Q.; Huang, X.; Zhou, W. Investigations on Laser Micromilling of Circular Micro Pin Fins for Heat Sink Cooling Systems. *Int. J. Adv. Manuf. Technol.* **2016**, *87*, 151–164. [CrossRef]
35. Tang, J.; Hu, X. Evaluation of Capillary Wetting Performance of Micro-Nano Hybrid Structures for Open Microgrooves Heat Sink. *Exp. Therm. Fluid Sci.* **2020**, *112*, 109948. [CrossRef]
36. Holley, B.; Faghri, A. Permeability and Effective Pore Radius Measurements for Heat Pipe and Fuel Cell Applications. *Appl. Therm. Eng.* **2006**, *26*, 448–462. [CrossRef]
37. Tang, Y.; Deng, D.; Huang, G.; Wan, Z.; Lu, L. Effect of Fabrication Parameters on Capillary Performance of Composite Wicks for Two-Phase Heat Transfer Devices. *Energy Convers. Manag.* **2013**, *66*, 66–76. [CrossRef]
38. Tang, Y.; Tang, H.; Li, J.; Zhang, S.; Zhuang, B.; Sun, Y. Experimental investigation of Capillary force in A Novel Sintered Copper Mesh Wick for Ultra-Thin Heat Pipes. *Appl. Therm. Eng.* **2017**, *115*, 1020–1030. [CrossRef]
39. Deng, D.; Liang, D.; Tang, Y.; Peng, J.; Han, X.; Pan, M. Evaluation of Capillary Performance of Sintered Porous Wicks for Loop Heat Pipe. *Exp. Therm. Fluid Sci.* **2013**, *50*, 1–9. [CrossRef]
40. Hao, X.; Xu, W.; Chen, M.; Wang, C.; Han, J.; Li, L.; He, N. Laser Hybridizing with Micro-Milling for Fabrication of High aspect Ratio Micro-Groove on Oxygen-Free Copper. *Precis. Eng.* **2021**, *70*, 15–25. [CrossRef]
41. Huang, G.; Tang, Y.; Wang, P.; Lu, L.; Yuan, W. thermal Characterisation of Micro Flat Aluminium Heat Pipe Arrays by Varying Working Fluid and inclination Angle. *Appl. Sci.* **2018**, *8*, 1052. [CrossRef]
42. Huang, G.; Li, W.; Zhong, G.; Abdulshaheed, A.A.; Li, C. Optimizing L-Shaped Heat Pipes with Partially-Hybrid Mesh-Groove Wicking Structures. *Int. J. Heat Mass Transf.* **2021**, *170*, 120926. [CrossRef]
43. Zhu, H.; Zhang, M.; Ren, W.; Saetang, V.; Lu, J.; Wu, Y.; Xu, K.; Liu, Y.; Zhang, Z. Laser-induced Localized and Maskless Electrodeposition of Micro-Copper Structure on Silicon Surface: Simulation and Experimental Study. *Opt. Laser Technol.* **2024**, *170*, 110315. [CrossRef]
44. Zhou, W.; Ling, W.; Liu, W.; Peng, Y.; Peng, J. Laser Direct Micromilling of Copper-Based Bioelectrode with Surface Microstructure Array. *Opt. Lasers Eng.* **2015**, *73*, 7–15. [CrossRef]
45. Zhou, W.; Deng, W.; Lu, L.; Zhang, J.; Qin, L.; Ma, S.; Tang, Y. Laser Micro-Milling of Microchannel on Copper Sheet as Catalyst Support Used in Microreactor for Hydrogen Production. *Int. J. Hydrogen Energy* **2014**, *39*, 4884–4894. [CrossRef]
46. Cai, J.; Jin, T.; Kou, J.; Zou, S.; Xiao, J.; Meng, Q. Lucas–Washburn Equation-Based Modeling of Capillary-Driven Flow in Porous Systems. *Langmuir* **2021**, *37*, 1623–1636. [CrossRef]
47. Staples, T.L.; Shaffer, D.G. Wicking Flow in Irregular Capillaries. *Colloids Surf. A Physicochem. Eng. Asp.* **2002**, *204*, 239–250. [CrossRef]
48. Deng, D.; Tang, Y.; Zeng, J.; Yang, S.; Shao, H. Characterization of Capillary Rise Dynamics in Parallel Micro V-Grooves. *Int. J. Heat Mass Transf.* **2014**, *77*, 311–320. [CrossRef]

micromachines

MDPI

Article

Experimental Study of Flow Boiling Regimes Occurring in a Microfluidic T-Junction

Xiangzhong Bao [1], Fei Yang [2] and Xuan Zhang [1,*]

1 Southeast University Architectural Design and Research Institute Co., Ltd., Nanjing 210096, China; baoxiangzhong@outlook.com
2 School of Energy and Environment, Southeast University, Nanjing 210096, China; leo.yang@nexteer.com
* Correspondence: xzhang@mail.usts.edu.cn

Abstract: Microchannel flow boiling is an efficient cooling method for high-heat-flux electronic devices. To understand the flow boiling regime in a T-shaped microchannel, this paper prepared T-shaped microchannels of different sizes and designed an experimental platform for the visualization of flow boiling in microchannels, and aimed to study the evolution characteristics of two-phase flow patterns in T-shaped microchannels. The influences of the flow rate and channel size on the boiling flow pattern inside a T-shaped microchannel were experimentally observed and quantitatively described. The results indicate that the occurrence position of the vaporization core gradually migrates from branch channel to main channel as the wall temperature increases. The flow boiling at the bifurcation of the T-shaped microchannel mainly includes the extrusion fracture flow, bubble flow, plug–annular alternating flow and annular flow, in which the annular flow can be further divided into the intermittent annular flow and the stable annular flow. In addition, a high flow rate and small channel size can lead to the disappearance of the bubble flow, and the presence of the bubble flow delays the appearance of the annular flow.

Keywords: microchannel; boiling; flow pattern; bubble

1. Introduction

In recent years, with the development of science and technology, the development of high-tech, single-phase heat transfer has, in particular, become increasingly unable to meet the needs of production and life. For example, in the field of electronic heat dissipation, the heat dissipation per unit area changed from 1–100 W/cm^2 and quickly soared to its current level of 1–100 W/cm^2. The heat dissipation of electronic equipment used in some special fields even reaches 10^4 W/cm^2, which far exceeded the limit of the single-phase heat transfer. In order to solve these problems, the gas–liquid two-phase heat transfer with large latent heat handling, especially the flow boiling heat transfer, has become a hot topic for scholars to study. Although flow boiling is an effective method to solve the problem of heat dissipation with high heat flux, flow boiling is not a new research direction.

Flow boiling in microchannel (FBM) was introduced as a potential candidate for the heat dissipation of microelectronic and high-power components due to its superior thermal performance and compact structure [1–3]. The bifurcations are key elements in these microchannel devices for the two-phase fluid distribution and collection. In a microfluidic T-junction, the flow boiling involves bubble growth, deformation, breakup and coalescence, as well as instabilities in a confined micro structure, and experiences the complex process of two-phase flow, heat transmission and phase change, as well as interface evolution. During this process, the behavior of the boiling two-phase flow in a bifurcation has a close relationship with the channel dimension, mass and heat flux. In this context, comprehending the flow boiling mechanisms occurring in a microfluidic T-junction is quite important.

In microchannels, the surface tension force becomes the dominant force for the two-phase flow and FBM, while gravitational forces can be considered negligible, causing the flow boiling regime to be significantly different from the macroscale channels [4–6]. The scale effect on flow boiling has been examined experimentally and interpreted by numerical simulations. It is generally accepted that the typical patterns of the flow are slug flow and churn flow, as well as annular flow, which correspond to the region controlled by the surface tension, intermediate region and the region controlled by the inertial force for microscale flow boiling. We can confine the primary regimes of FBM to the nucleate boiling and convective evaporation as well as a combination of these two. In addition, the confined bubble (slug) flow [7,8], vapor back flow [9,10] and microbubble emission boiling [11,12] are observed in microchannels owing to the microscale effect. The above understanding of flow boiling dynamics and heat transfer regimes are related to the straight microchannel. The presence of bifurcation introduces more complex bubble interactions, flow instability and thermo-hydrodynamic coupling for two-phase FBM. However, up to now, how the bifurcation affects the two-phase FBM is less understood [13]. As a consequence, it is necessary to understand the flow boiling regimes in the T-junction microchannel, so as to reveal the flow performance as well as heat transmission regime during FBM with the T-junction.

Considering the potential application of bubble manipulation in micro engineering and technology, a great deal of efforts has been devoted to investigating the interfacial transport phenomena of the adiabatic two-phase flow in microchannels with a branching structure, especially the bubble breakup dynamics at the microfluidic T-shaped junction. Because of the bifurcation, the capillary pressure of the gas–liquid interface experiences a dramatic change during the splitting flow, which results in long-distance interactions among bubbles (slugs) downstream and upstream of the junction. As a result, the two-phase flow at the T-junction is complex and nonlinearly dynamic and possesses a chaotic feature for splitting flow behavior [14]. Studies have demonstrated that two-phase flow behaviors are dependent on a number of parameters, including the channel dimension, gas and liquid flow rate, confinement degree and capillary number. The typical flow regimes for the bubble passing through a T-junction are the breakup, obstructed, asymmetric and non-breakup types of tunnels [15,16]. The flow boiling regimes at a microfluidic T-junction are more complicated due to the multi-interaction among the dynamical evolution of the bubble (bubble formation, growth, collision, coalescence and breakup), the bifurcation effect and the interfacial heat and mass transport in a confined microspace. However, the detailed flow regimes for the boiling in the microchannel with a T-junction are unavailable in the literature. For purpose of understanding the FBM and heat transmission, a unified view of flow boiling regimes in a microfluidic T-junction is required.

Although a large number of studies exists on FBM [17–21], people have placed little attention on heat transfer regimes in microfluidic T-junctions. For the purpose of understanding the underlying features of two-phase FBM with the T-junction, we use a visualization experiment to study the heat transfer behaviors with the T-junction via the technology of high-speed photography. Based on this experiment, the influence of superheat degree and channel dimension on the flow boiling are examined and investigated. In addition, the heat transmission regimes observed in a microfluidic T-shaped junction are analyzed and discussed. Particularly, we present different flow patterns to quantitatively describe the flow boiling regimes, depending on the superheat degree and channel dimension.

2. Description of the Experiment

Aiming to elucidate the flow boiling regimes in microfluidic T-junctions, several microchannels configured with a T-shaped bifurcation are fabricated using glass, and then the flow pattern of boiling in these microchannels is experimentally investigated by a visualization system.

2.1. Experimental Setup and Procedure

The visualization experiment platform of flow boiling in T-shaped microchannels mainly consists of five parts: a liquid flow control and output device, an experimental section, a heating device, a high-speed microscopic imaging system and a data acquisition system. Figure 1 depicts the system diagram of the flow and boiling experiment conducted in T-shaped microchannels. Additionally, Figure 2 illustrates the physical diagram of the system. It can be observed that an injection pump is used to pump the working medium into the experimental section, and the working medium reaches the saturation temperature under the heating device and produces the boiling phenomenon. Meanwhile, the pictures of the FBM are collected by the high-speed microscopic imaging system; then, we save them in the computer, and the data acquisition system collects the temperature signals of each temperature measuring point arranged and saves them. Each component of the experimental system is described below.

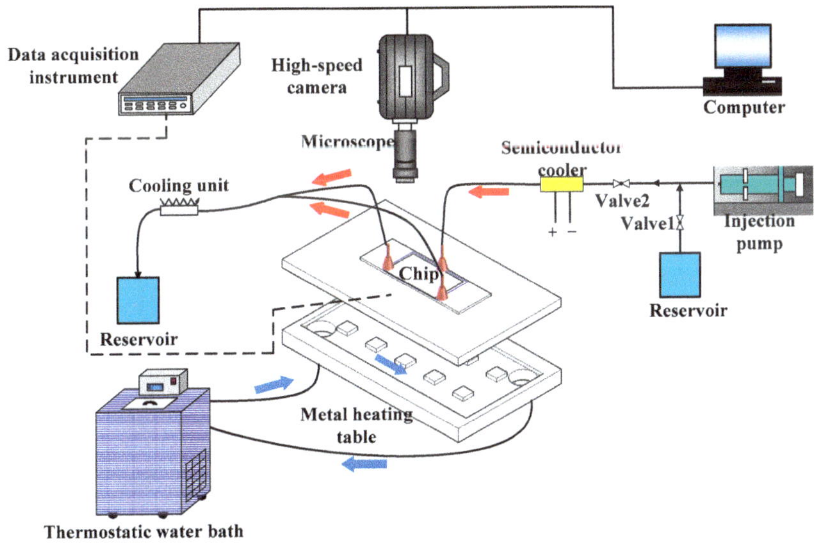

Figure 1. Experimental diagram of flow and boiling in T-shape microchannels.

2.1.1. Liquid Flow Control and Output Device

To achieve a continuous and constant flow of working fluid (refrigerant R141b) for the entire experimental system, two injection pumps, several stainless-steel pipelines, and several one-way ball valves are used in the experiment to form the liquid flow control and output device of the entire system, as can be observed in Figure 3. With the connection method shown in the figure, this device can provide uninterrupted and constant liquid supply during the experiment.

The injection pump used in the experiment is the laboratory-type injection pump LSP01-1BH produced by Baoding Lange Constant Flow Pump Co., Ltd. (Baoding, China), as shown in Figure 4. This injection pump has five working modes: injection, extraction, injection before extraction, extraction before injection, and continuous mode. The injection pump uses a built-in stainless-steel syringe with a maximum capacity of 50 mL and can control the flow error within a range of $\pm 0.5\%$. The ball valve used in the experiment is a 304 stainless steel clamp ball valve, as shown in Figure 5. The pipe connected to this ball valve is a 304 stainless steel pipe (outer diameter: 3 mm; inner diameter: 2 mm).

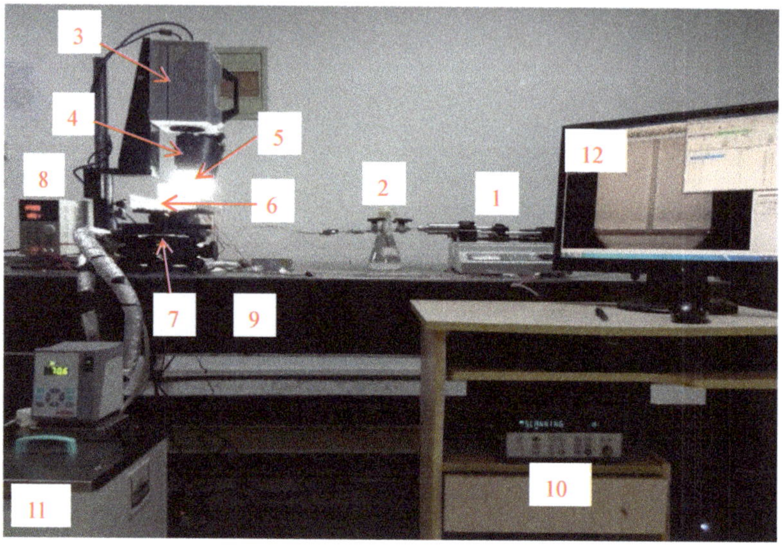

Figure 2. Diagram of flow and boiling in T-shape microchannels (1. Injection pump; 2. Reservoir; 3. High-speed camera; 4. Microscope; 5. High frequency no flashing light; 6. Experimental section; 7. Three-dimensional mobile platform; 8. DC power supply; 9. Optical experimental platform; 10. Agilent data acquisition device; 11. Thermostatic waterbath; 12. Computer).

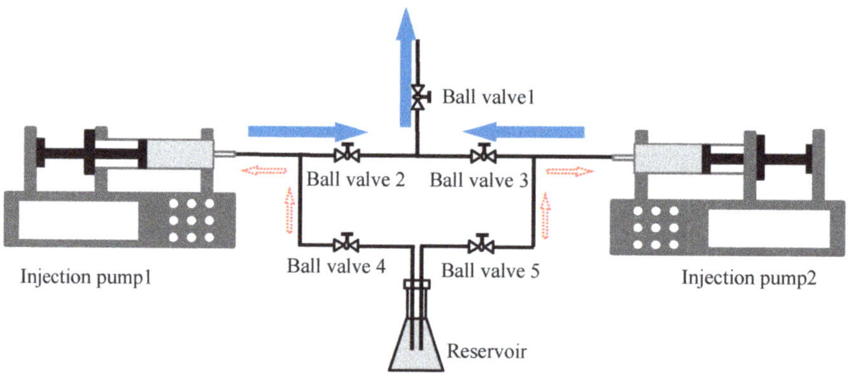

Figure 3. Schematic diagram of liquid flow control and output.

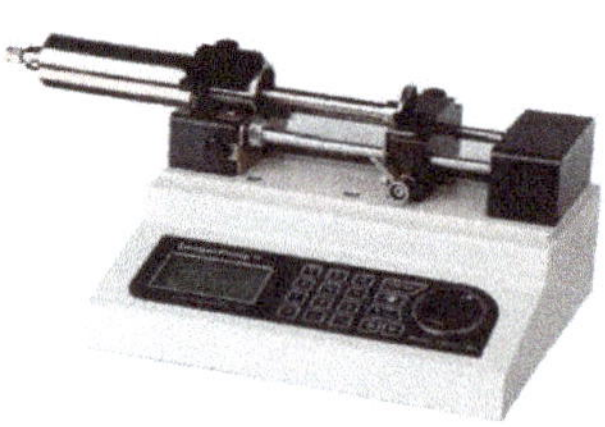

Figure 4. LSP01-1BH injection pump.

Figure 5. A 304 stainless steel ball valve (3 mm).

In the experiment, the continuous and constant flow supply of the entire experimental system is achieved through the following steps:

(1) Before the experiment, close ball valves 1, 2 and 3. Open ball valves 4 and 5, and use injection pump 1 and injection pump 2 to extract the working fluid from the reservoir to full capacity, as shown by the red dashed arrow in Figures 2 and 3.
(2) Before starting the experiment, close ball valves 3, 4 and 5. Open ball valves 1 and 2, and turn on the switch of injection pump 1. Injection pump 1 starts to inject the working fluid into the experimental section. When the injection volume reaches 95% of the full capacity, open ball valve 3 and the switch of injection pump 2, and simultaneously close ball valve 2 and injection pump 1. Injection pump 2 then injects the working fluid into the experimental section, while injection pump 1 stops working.
(3) Open ball valve 4, use injection pump 1 to extract the working fluid from the reservoir to full capacity, and then close ball valve 4. When the injection volume of injection pump 2 reaches 95% of the full capacity, open ball valve 2 and the switch of injection pump 1, and simultaneously close ball valve 3 and injection pump 2. At this time, injection pump 1 injects the working fluid into the experimental section, while injection pump 2 stops working.
(4) Open ball valve 5, use injection pump 2 to extract the working fluid from the reservoir to full capacity, and then close ball valve 5. When the injection volume of the injection pump 1 reaches 95% of the full capacity, repeat the process of steps 2, 3 and 4.

2.1.2. Experimental Section

The experimental section mainly consists of an inlet liquid temperature control device, a T-shaped microchannel chip, and several inlet and outlet pipelines. The inlet liquid temperature control device is mainly composed of a semiconductor refrigeration chip and a DC power supply. Since we use R141b as the refrigerant herein, the saturation temperature at a standard atmospheric pressure is 32.05 °C. In order to maintain the same subcooling of the inlet liquid under all operating conditions, the temperature of the inlet liquid is controlled at 20 °C using the semiconductor refrigeration chip.

The typical technologies for fabricating a microchannel, including silicon wafer etching [22,23], glass etching [24,25] and soft lithography in PDMS [26,27], are able to control the channel size precisely; however, the production processes are very complicated and costly because they fabricate well only with complex manipulations using an expensive experimental apparatus in a clean room. To achieve the convenient laboratory supply, Deng et al. [28] report a straightforward and cost-effective method for fabricating microfluidic devices to produce uniform emulsions using patterned coverslips and microscope glass slides. In this study, this approach is applied to fabricate the microchannel configured with a T-shaped bifurcation (referred as a T-shaped microchannel in the following) for the boiling experiment. As shown in Figure 4, the fabrication process can be listed as follows:

(1) Preparation stage: First, use drafting software such as AutoCAD 2017 to draw the shape of the microchannel. Then, print it as a substrate using a high-resolution printer, as shown in Figure 6(a-1). Cut the prepared cover glass (①, ②, ③ and ④) and carrier

glass (⑤) into the desired dimensions using a glass cutter. Clean the cut glass slides by immersing them in alcohol. This is followed by rinsing with deionized water, and, finally, air drying.

(2) Fixation stage: Place the prepared substrate on the microscope stage and secure it in position. Place the clean carrier glass (⑤) on top of the substrate, aligning it under the microscope, and fix it in position. Then, using the microscope, align and fix the cover glasses (①, ②, ③) along the lines of the substrate to form the shape of the T-shaped microchannel.

(3) Curing stage: With the alignment completed, fill the contact surfaces with a UV glue (Guangzhou Hengda New Materials Technology Co., Ltd. (Guangzhou, China), K3307) using capillary action. Then, expose the contact surfaces to UV light for approximately 2 min to ensure a tight cure.

(4) Cleaning stage: Rinse the cured semi-finished chip with deionized water twice and allow it to air dry.

(5) Sealing stage: Seal the cleaned semi-finished chip with a clean cover glass (④). Fill the contact surfaces with UV glue using capillary action and expose it to UV light for approximately 2 min.

(6) Inlet and outlet fabrication: Finally, place a needle (outer diameter: 9 mm; inner diameter: 0.6 mm) at the entrance and exit of the T-shaped microchannel as the interface for connecting the microchannel and polyethylene tubing. Depending on the requirements, a thermocouple can be placed to monitor the inlet and outlet temperature. Use a modified acrylic acid ester adhesive (GLH-302) to secure the needle to the T-shaped microchannel chip.

In order to conduct comparative experiments, we designed T-shaped microchannels of different sizes to study the flow pattern evolution and temperature instability under different operating conditions. The T-shaped microchannel consists of a main channel and left and right branch channels. The length of which are L_1 and L_2, respectively, and the width of which are w_1 and w_2, respectively. The depth of all channels is h. Four types of T-shaped microchannels were designed for visual experiments on flow boiling. Among them, the first, second, and third types of T-shaped microchannels have the same channel dimensions, and only the depth dimension h is varied. The fourth type of T-shaped microchannel is designed to be closer to conventional channels, with specific dimensions shown in Tables 1 and 2.

Table 1. Dimensions of four T-shaped microchannels.

	Type 1	Type 2	Type 3	Type 4
L_1 (mm)	32	32	32	32
L_2 (mm)	24	24	24	24
w_1 (μm)	800 (857)	800 (877)	800 (898)	1000 (1122)
w_2 (μm)	600 (796)	600 (694)	600 (673)	700 (837)
h (μm)	300 (367)	450 (551)	600 (673)	800 (939)
D_1	436 (514)	576 (676)	686 (769)	888.9 (1022)

Note: The value in brackets is the actual value measured under the microscope, such as 800 (857); the design size is 800 μm; the actual size is 857 μm.

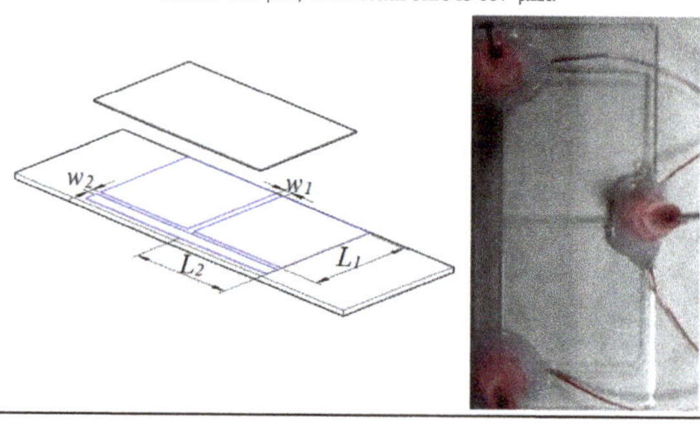

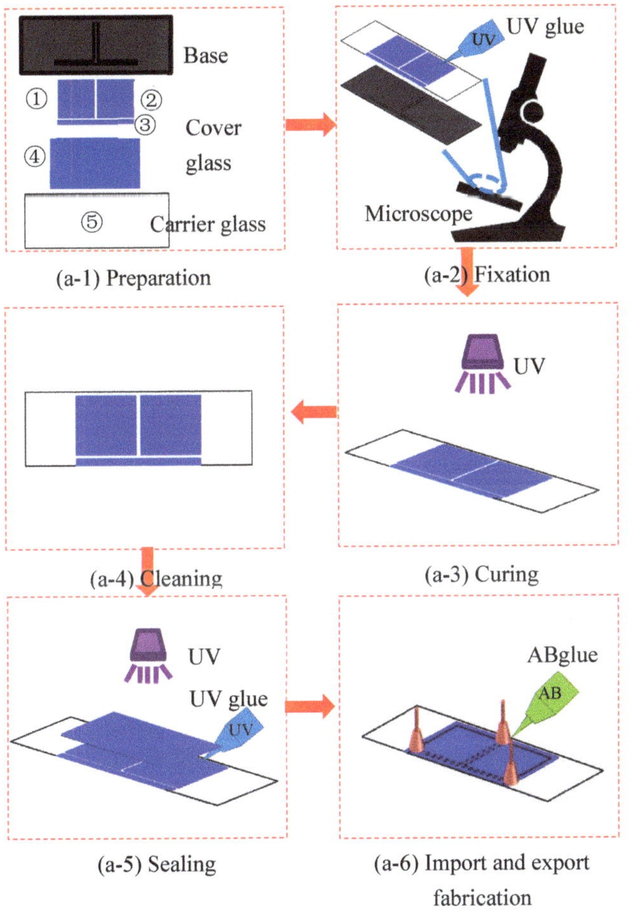

Figure 6. Production process of T-shaped microchannel chip.

Table 2. Information about the instruments and equipment used in the experiment.

No.	Instrument	Type, Equipment Manufacturers and Main Parameters
1	Injection pump	Type: LSP01-1BH, Baoding Lange Constant Flow Pump Co., Ltd. (Baoding, China); comes with 50 mL stainless steel syringe
3	High-speed camera	Type: Photron FASTCAM SA4, Photron Co., Ltd. in Japan (Tokyo, Japan) Maximum frame rate: 500,000 fps; Resolution ratio: 1024 × 1024 (3600 fps)
4	Microscope	Type: AF Micro-Nikkor 60 mm F/2.8D NIKON (Tokyo, Japan)
5	High frequency no flashing light	Type: DAK-C1-0408, Foshan Dongxiong lighting factory (Foshan, China)
7	Three-dimensional mobile platform	Type: LSSP-125WS Translation table, Jiangxi Liansheng Technology Co., Ltd. (Nanchang, China)
8	DC power supply	Type: KORAD KA3005D, Shenzhen Kerui source Technology Co., Ltd. (Shenzhen, China)
9	Optical experimental platform	Type: POT24-12, Jiangxi Liansheng Technology Co., Ltd.

Table 2. *Cont.*

No.	Instrument	Type, Equipment Manufacturers and Main Parameters
10	Agilent data acquisition device	Type: Agilent 34970a, Agilent Technology Co., Ltd. (Santa Clara, CA, USA), equipped with copperConstantan thermocouple produced by American OMEGA Company (Norcross, GA, USA), that is, T-shaped thermocouple
11	Thermostatic waterbath	Type: JULABO MC, Beijing Youlebo Technology Co., Ltd. (Beijing, China)

2.1.3. Heating Unit

During the experiment, a thermostatic water bath was mainly used to heat the T-shaped microchannel chip. The main method is as follows: the inlet and outlet of the constant temperature water bath are connected to a metal heating block, and the constant temperature water bath provides a continuous supply of constant temperature hot water to heat the metal heating block. Through the combined effects of convection and conduction heat transmission, the upper surface of the metal heating block maintained a constant temperature, thereby providing a constant temperature heating condition to the T-shaped microchannel chip.

The constant temperature water bath used in the experiment is the JULABO immersion heating circulator (MC type), as shown in Figure 7. The MC-type JULABO immersion heating circulator adopts an ultra-quiet design and can operate in an ambient temperature range from 5 to 200 °C. The temperature stability of the provided constant temperature water is about ±0.01 °C, and the heating power is 2 kW. At the same time, the pressure pump pressure can be adjusted, providing variable flow rate (11–16 L/min) functionality.

Figure 7. Physical diagram of MC JULABO immersion heating circulator.

The metal heating block is an aluminum block with internal channels for processing, and it is connected to two pipes as the inlet and outlet joints. The physical appearance of the metal heating block is shown in Figure 8, and the 3D sectional view is shown in Figure 9. There are also 10 temperature measurement points arranged on the metal heating block to monitor its temperature and uniformity. The metal heating block is fixed on a PTFE base, and two PTFE chip clamps are used to secure the T-shaped microchannel chip. During the experiment, thermal conductive grease should be applied on the contact surface between the T-shaped microchannel chip and heat source to ensure that the temperature is consistent with that of the T-shaped microchannel chip.

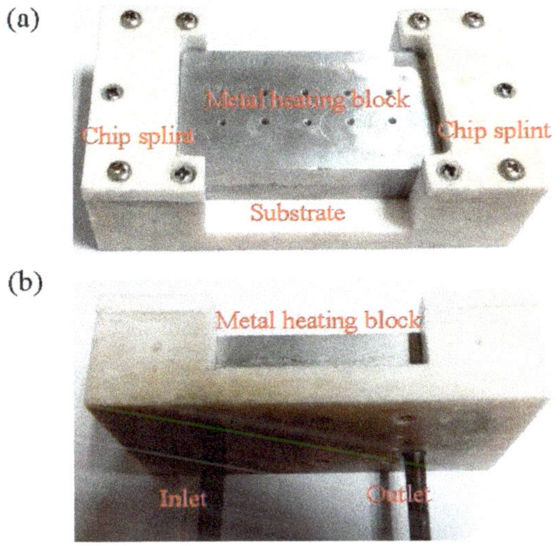

Figure 8. Physical diagram of metal heating block. (**a**) front; (**b**) side.

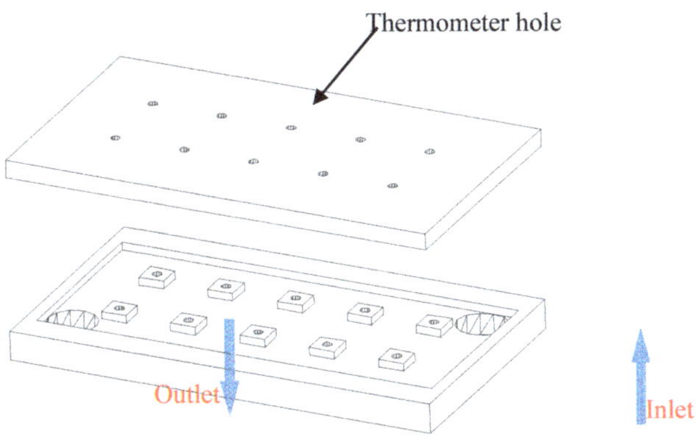

Figure 9. Three-dimensional section view of metal heating block.

2.1.4. High-Speed Microscopic Imaging System

In order to observe the flow boiling phenomenon clearly, a combination of a high-speed camera CCD (presented in Figure 10), microscope lens (as shown in Figure 11), and auxiliary light source is used to capture and store real-time images of the flow boiling phenomenon. The CCD utilized has a resolution of 1024 × 1024 pixels, covering the full frame, and a maximum shooting speed of 3600 fps. The maximum shooting speed in a segmented mode can reach 500,000 fps, and the fastest shutter speed can reach 1 μs. In the experiment, the shooting speed is generally set to 1000 fps~3000 fps. In order to capture the flow boiling phenomenon in the microchannel more clearly, a microscope lens is used for magnification. The maximum magnification of this microscope lens is 32 times. Due to the small aperture of this model of high-speed camera CCD, the lighting effect is poor. Therefore, an auxiliary light source, namely a high-frequency non-flashing lamp, is needed to increase the brightness of the object being captured. The best shooting effect can be achieved by adjusting the incident angle of the auxiliary light source.

Figure 10. High-speed camera: Photron FASTCAM SA4.

Figure 11. Nikon microlens.

2.1.5. Data Acquisition System

The system mainly includes a thermocouple, data acquisition instrument and computer. The thermocouple used is a copper-constantan thermocouple produced by OMEGA, a company in the United States. The thermocouple is a T-type and has a wire diameter of 0.1 mm. The maximum long-term measurement temperature is 150 °C, and the measurement accuracy is ±0.1 °C. The data acquisition instrument used is Agilent 34970a, which is produced by Agilent Technologies, a company in the United States, as can be seen in Figure 12. This model of a data acquisition instrument consists of a data collector and data acquisition board. During the experiment, the positive and negative terminals of the T-type thermocouple are fixed to the data acquisition board through slots. The data acquisition board is inserted into the data acquisition instrument, and the data acquisition instrument converts the potential signal of the T-type thermocouple into a temperature signal., which facilitates the monitoring and recording of the temperatures at various measurement points during the experimental process.

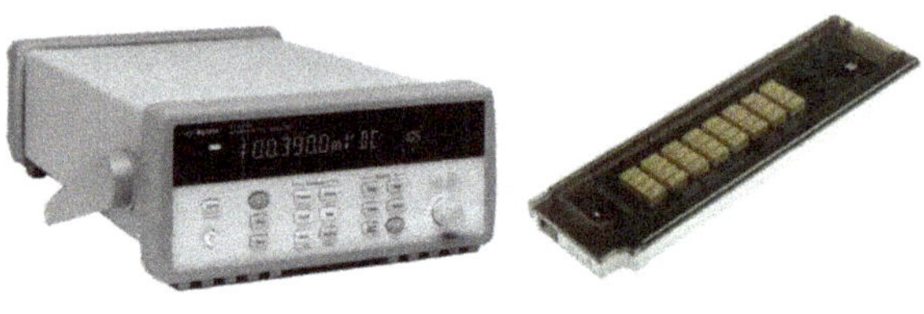

(**a**) (**b**)

Figure 12. Physical diagram of data acquisition instrument. (**a**) Data acquisition unit; (**b**) data acquisition board.

2.1.6. Experimental Method and Procedure

To study the flow and heat transmission phenomena of boiling flow in T-shaped microchannels, we use a high-speed microscopy imaging system to assist in observing the patterns and its evolution. At the same time, a data acquisition system is used to monitor and record the temperature of the entire experimental system. During the experiment, flow boiling experiments are conducted on four different sizes of T-shaped microchannels. The specific experimental steps are as follows:

(1) Develop an experimental plan, clarify the physical quantities to be measured and the flow boiling flow pattern images to be taken during the experiment, determine the various operating conditions that need to be tested, and create tables or other records as required.
(2) Produce the four T-shaped microchannels used in the experiment according to the process presented in Figure 6.
(3) Connect the various components according to the experimental system diagram and set up the high-speed camera microscopy system and the three-dimensional mobile platform on the optical experimental platform. Connect the inlet and outlet of the thermostatic water bath and the entrance and exit of the metal heating table with plastic hoses, and insulate the pipelines. Finally, check the air tightness of the entire experimental system.
(4) Turn on the power of the syringe pump, high-speed camera, constant temperature water bath, data acquisition device and computer, and check whether each component is working properly. Extract the working fluid with both syringe pumps to their full range. Before the experiment starts, check whether all thermocouple readings are normal. Turn on the high-speed camera, and adjust the magnification, shooting frequency, field of view size, and other parameters of the microscope. Then, start the syringe pump to purge the air in the entire experimental system and prepare for the experiment.
(5) Adjust the flow rate of the injection pump for the operating conditions, and change the temperature of the constant temperature water bath to conduct the experiment, monitoring the temperature displayed on the computer. After the temperature of the metal heating table stabilizes, capture the images of the flow pattern using the high-speed microscopy system. Temperature data are collected by the data acquisition system as needed.
(6) After the completion of a set of experiments with a fixed flow rate and different temperatures, wait for the T-shaped microchannel chip to cool to room temperature before adjusting the flow rate of the injection pump for the next set of experiments. Repeat steps (5) and (6).
(7) After the experiment is finished, turn off the CCD, the power of the constant temperature water bath, and the data acquisition device. Close the syringe pump after the T-shaped microchannel chip has cooled down.
(8) Clean up the experimental platform, and disassemble and clean the T-shaped microchannel chip. Replace the chip with a different size and repeat steps (3) to (8).
(9) Organize and analyze the experimental data and the obtained flow pattern images.

2.1.7. Error Analysis

The main parameters involved in the experiment are the inlet mass flow density, temperature of the left and right branches of the T-shaped microchannel, and the heating temperature of the chip. The errors of these parameters are as follows:

(1) Temperature measurement error: In the experiment, temperature is measured using a T-type thermocouple with a single wire diameter of 1 mm. After correction using a constant temperature water bath ice point, the measurement error δT of the temperature measuring instrument is ± 0.1 °C. The lowest measured temperature value T in

this experimental section is 20 °C, and the maximum relative error of the temperature measured by the instrument is $\delta T/T_{\min}$ = 0.5%.

(2) Mass flow density measurement error: Based on the accuracy of the syringe pump used, the error range of the mass flow density G is ±0.5%.

(3) Image quantization error: In this experiment, the resolution of all captured images is 72 ppi, which means there are 72 pixels per inch (1 inch = 25.4 mm). The error in selecting the number of pixels by the pixel software is 1, so the relative error of the position data quantized from the captured images is 38%.

3. Results and Discussions

According to the existing research on FBM, the main focus is on the instability of the flow boiling phenomena and temperature-related parameters on a microscale. In this chapter, experimental research on the flow boiling of refrigerant R141b in T-shaped microchannels is conducted. Experiments are carried out in four different sizes of T-shaped microchannels. The main observed phenomena during the experiments include bubble nucleation, growth, coalescence, sliding, and detachment from the wall. These basic bubble behaviors can evolve into various flow boiling flow patterns. The main flow boiling patterns in conventional channels are bubble flow and slug flow, as well as annular flow. Because of various factors affecting flow boiling, such as wall roughness, fluid velocity, channel shape, and size, flow boiling phenomena exhibit strong randomness and complexity. Moreover, compared to conventional channels, the flow range in microchannels is more narrow, making the variation of flow boiling flow patterns more complex. The features of flow boiling and heat transmission in T-shaped microchannels will be discussed from three aspects: bubble dynamics, flow boiling flow patterns and their evolution at the bifurcation, and the instability of outlet temperature.

3.1. *Bubble Behavior in T-Shaped Microchannel Flow Boiling*

3.1.1. Bubble Generation

During the experiment, the subcooled fluid (liquid) enters the chip through the pipeline, and the chip is in close contact with a temperature-controlled metal heating platform, with thermal grease filling the contact surface. When the flow in the entire experimental system reaches a steady state, it can be considered as a continuous flow of subcooled fluid passing through the temperature-controlled heating surface, being continuously heated, and the temperature of the microchannel and the fluid will continuously increase. When the temperature reaches a certain level, nucleation bubbles will appear, indicating the occurrence of nucleate boiling. Since the fluid is continuously heated, there is a certain temperature gradient along the flow direction. Therefore, nucleation bubbles will first appear in the branch channel during the experiment, and then in the main channel. We can find in Figure 13 that, in the type 2 chip, under the condition that the mass flow rate of the main channel is 113.6 kg/m^2·s, nucleation bubbles first appear in the branch channel. As the channel wall temperature increases (Figure 13b–e), the location of nucleation bubbles moves from the branch channel to the main channel, getting closer to the inlet of it. This phenomenon of nucleation core movement is particularly evident under a high mass flow rate density. Under low mass flow rate density, the temperature gradient decreases due to the slower flow velocity of fluid, making it easier to reach thermal stability. Therefore, the initial position of nucleation cores appears in the main channel.

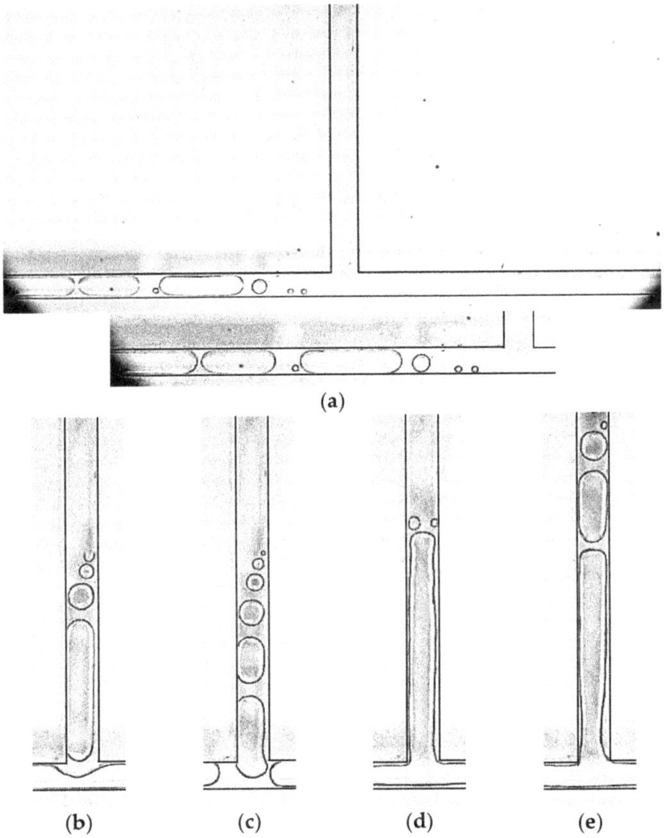

Figure 13. Nucleated bubbles appear at different locations in the channel, in working conditions: Type 2, G = 113.6 kg/m^2·s, (**a**) T_{heat} = 37.5 °C, (**b**) T_{heat} = 40°C, (**c**) T_{heat} = 42 °C, (**d**) T_{heat} = 46 °C and (**e**) T_{heat} = 54 °C.

3.1.2. Combination of Bubbles

Bubble coalescence is the most typical behavior of bubbles during the FBM. After the appearance of nucleation bubbles in the boiling nucleus, the vapor–liquid interface of the bubbles absorbs heat and evaporates, increasing the gas volume inside the bubbles, which is the process of bubble growth. Nucleation bubbles grow in the direction of the flow, and in the normal direction, as they flow with the mainstream. If there are two or more adjacent nucleation nuclei on the same side wall, the nucleation bubbles generated by these nuclei will grow in the axial direction. When the volume of the bubbles reaches a certain level, these two nucleation bubbles come into contact and collide, merging into a larger bubble under the influence of surface tension. This phenomenon is called an axial coalescence of bubbles, as shown in Figure 14. If there are nucleation nuclei on both side walls, the nucleation bubbles they generate will also collide and merge in the normal direction. This merging phenomenon is called the normal coalescence of bubbles, as shown in Figure 15. Before the nucleation bubbles detach from the boiling nucleus, the bubble interface absorbs heat from the wall, which also proves that bubbles enhance the heat transmission in nucleate boiling, and the coalescence of bubbles helps the nucleation bubbles detach from the boiling nucleus, thus generating new nucleation bubbles faster and further enhancing the heat transfer.

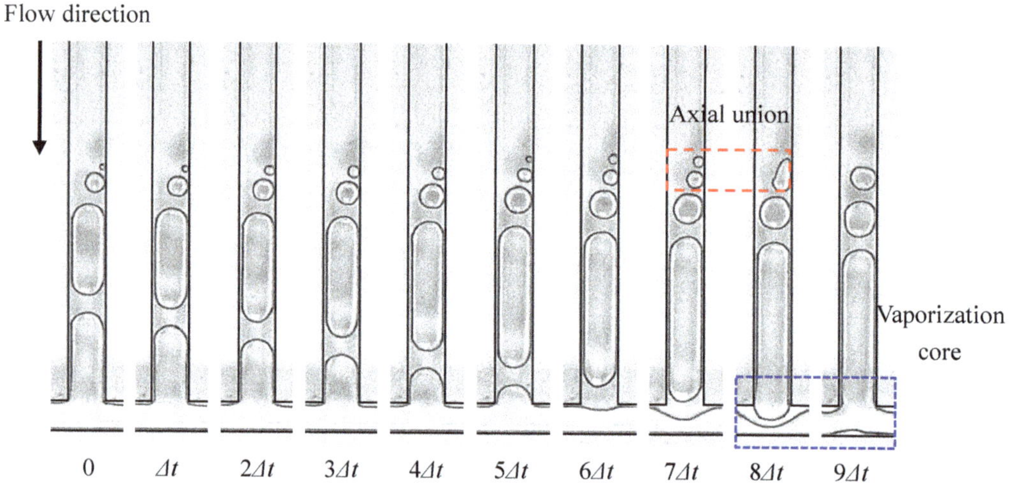

Figure 14. Axial merging of bubbles, Δt = 2 ms, and working condition: Type 2, G = 113.6 kg/m^2·s and T_{heat} = 37.5 °C.

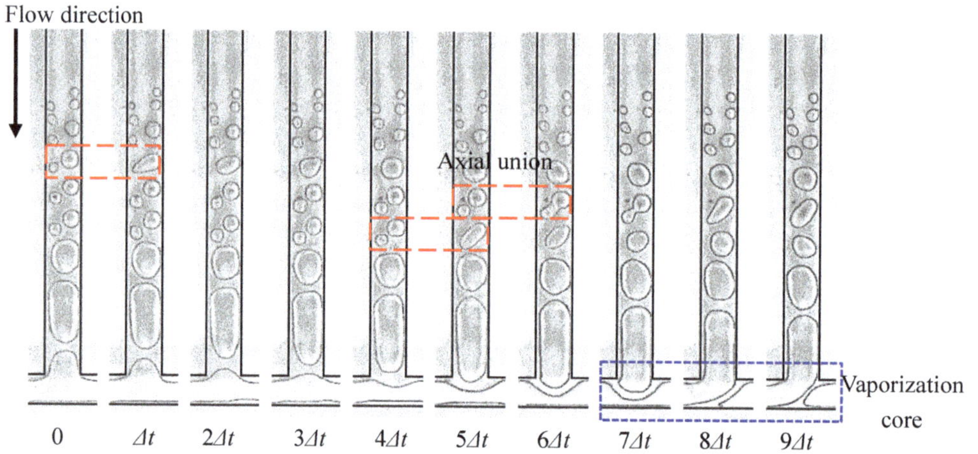

Figure 15. Normal merging of bubbles, Δt = 0.33 ms, and working condition: Type 2, G = 113.6 kg/m^2·s and T_{heat} = 45.8 °C.

In addition, the coalescence between bubbles and plugs, between plugs and between plugs and annular flows were observed. These coalescence phenomena are the basis for the evolution of flow patterns of FBM. For example, the coalescence of bubbles will cause a further growth of the bubbles. The vapor–liquid interface formed after the coalescence continues to absorb heat and evaporate, increasing the gas volume. When the normal direction of bubble growth is restricted by the width of microchannels, the bubbles can only grow along the axial direction, which change from initially spherical to cylindrical with spherical ends; this is one of the formation modes of plugs. Similarly, the coalescence of multiple plugs in microchannels is also one of the ways in which the slug flow transitions to annular flow.

3.1.3. The Rupture of the Bubble

The rupture of bubbles in T-shaped microchannels is a unique bubble dynamics phenomenon caused by the presence of bifurcation structures. Generally speaking, during FBM or parallel microchannels, bubble coalescence is the main phenomenon observed, and bubble rupture is not observed. However, in T-shaped microchannels, when bubbles or plugs in the main channel pass through the bifurcation, they exert a squeezing effect on the plugs in the branch channel. The plugs in the branch channel may either merge with the bubbles or plugs in the main channel or rupture under the squeezing effect. As shown in Figure 14 from $t = 8\Delta t$ to $t = 9\Delta t$, a T-shaped plug is formed by the merging of a plug in the main channel and a plug in the branch channel at the bifurcation. In Figure 15, during the time period from $t = 7\Delta t$ to $t = 9\Delta t$, the plug in the branch channel is squeezed and ruptures into two sub-plugs. The sub-plug in the left branch then merges with the plug in the main channel to form an L-shaped plug. This situation of rupture and merging forms the characteristic flow pattern of FBM. In studies of two-phase flows without a phase change, it has been found that the merging and rupture of bubbles at the bifurcation are mainly related to the critical pressure and the length of the plug. However, in flow boiling processes, the merging and rupture of bubbles at the bifurcation are more complex than without the phase change, because the plug length and the pressure inside the channel change in real time.

3.2. *The Flow Boiling Flow Pattern and Its Evolution at the Bifurcation of T-Shaped Microchannel*

From existing studies, it can be found that in straight microchannels, the subcooled liquid enters the channel and is continuously heated during the flow process. When the liquid temperature approaches the saturation temperature, axial temperature gradients contribute to the formation of nucleation bubbles on the channel wall. This is known as subcooled boiling. As the liquid continues to flow, the bulk temperature reaches the saturation temperature, resulting in saturated boiling. During these two stages, only small bubbles are generated on the channel wall, so the flow boiling flow pattern is mainly in the form of a bubble flow. These small bubbles continue to flow and are heated, resulting in an increase in the bubble volume and occupation of the central part of the channel. As the slug flows continue to flow and are heated, they merge with each other to form a slug flow pattern. Finally, an annular flow pattern is formed, where only a ring-shaped liquid film flows around the channel wall. Because of the ongoing evaporation of the liquid film, it becomes thinner until it completely disappears, resulting in a dryout phenomenon. Therefore, flow patterns of FBM mainly include bubble flow, slug flow, annular flow, dryout flow, and so on. The main focus is on the effect of the bifurcation structure on the flow patterns of FBM with T-junction; thus, we mainly focus on the flow patterns at the bifurcation. Through visual experimental research, it can be found that the flow patterns at the T-junction of the microchannel mainly include the bubble flow, squeeze and rupture flow, slug–annular alternating flow and annular flow. During the experiment, a dryout flow was also observed. However, due to the rapid increase in wall temperature after the dryout, considering the heat resistance of the experimental glass chip and the safety of the entire experiment, the dryout flow was not allowed to occur for a long time during the experiment, and the dryout phenomenon is not discussed in this paper. Below, we will specifically discuss the flow boiling flow patterns at the bifurcation.

3.2.1. Squeeze and Rupture Flow

The squeeze and rupture flow is a special flow pattern that occurs during FBM with the T-junction. Because of the presence of bifurcation, bubbles (plugs) will inevitably merge or rupture when passing through the bifurcation, forming the unique squeeze and rupture flow pattern in T-shaped microchannels.

During the experiment, we mainly observed two situations of the squeeze and rupture flow. One situation is that at low mass flow density, as can be seen in Figure 16, the primary channel is single-phase liquid flow, while nucleation bubbles appear in the branch channel.

The nucleation bubbles in the branch channel quickly grow into plugs, which occupy the branch channel. Due to the fluctuation of the outlet pressures, the plug pulsates back and forth in the branch channel, causing a blockage to the liquid flow in the primary channel at the junction. The accumulation of liquid in the main channel exerts squeezing pressure on the front end of the plug, causing the plug to neck down at the bifurcation, and eventually the plug ruptures into two subplugs. This is the most common form of the squeeze and rupture flow at the bifurcation of T-shaped microchannels, and it lasts for a relatively long time.

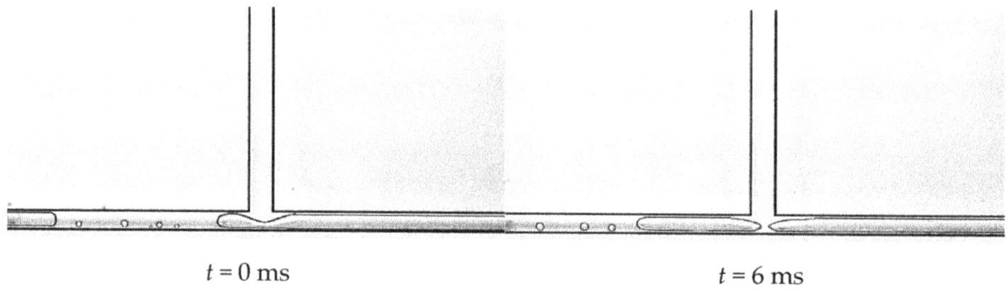

Figure 16. Squeeze and rupture flow, working condition: Type 2, G = 14.2 kg/m^2·s and T_{heat} = 39.6 °C.

The other situation is that there are only a few circular nucleation bubbles that gradually grow with the main flow in the main channel, and plugs and circular small bubbles are distributed alternately in the branch channel, as shown in Figure 17. The bifurcation is occupied by alternating plugs in the branch channel and the liquid containing a small number of circular bubbles. When the plug blocks the bifurcation, the front end of the plug is pressurized and ruptures into a circular or slug bubble due to the pressure from the liquid containing circular bubbles in the main channel. At the bifurcation, there is also merging between the circular small bubbles in the main flow and the slug or circular bubbles in the branch channel. When the liquid containing circular bubbles is present at the bifurcation, the circular bubbles in the main channel directly enter the branch channel with the main flow. This situation mainly occurs when the mass flow density is relatively high and the nucleation bubbles in the main channel enter the bifurcation in the form of small bubbles. If the mass flow density is too low at this time, the nucleation bubbles in the main channel will enter the bifurcation via plugs, forming a slug flow pattern. This flow pattern can also be treated as a common flow pattern from the squeeze and rupture flow to the bubble flow.

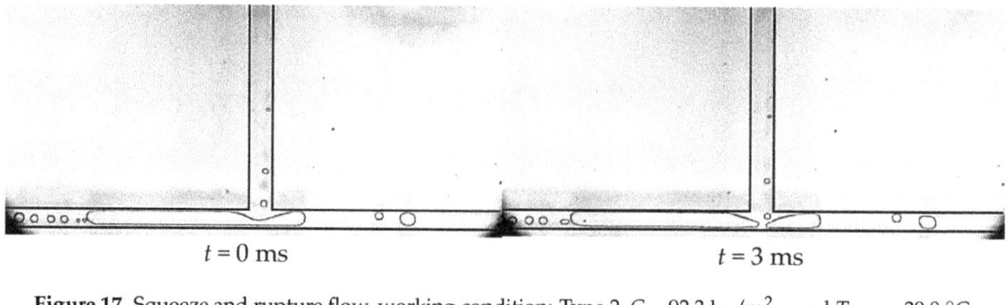

Figure 17. Squeeze and rupture flow, working condition: Type 2, G = 92.3 kg/m^2·s and T_{heat} = 39.8 °C.

3.2.2. Slug Flow

Slug flow is a typical flow pattern of FBM, and the main characteristics of slug flow at the bifurcation in T-shaped microchannel flow boiling are that there are only a few circular nucleation bubbles that grow gradually with the main flow in the main channel, while there are dispersed bubbles flowing downstream in the left and right branch channels, as shown in Figure 18. A few nucleation bubbles form in the main channel and grow gradually with the main flow, directly entering the branch channels at the bifurcation without a bubble rupture or coalescence. Most nucleation bubbles are generated in the branch channels, and their growth rate is higher than that in the main channel. They grow into slug bubbles in the downstream of the branch channels, changing from circular to elongated shapes. The slug flow mainly occurs in T-shaped microchannel flow boiling under high mass flow density and low heat load conditions. Under the condition of low mass flow density, the growth rate of nucleation bubbles in the branch channels is extremely fast, quickly occupying the entire bifurcation, and suppressing the vaporization core at the bifurcation. Therefore, the occurrence of the slug flow was not observed in the experimental process under low mass flow density.

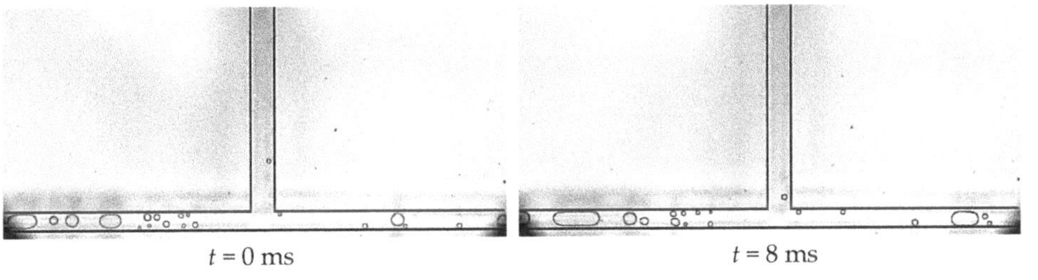

Figure 18. Slug flow, working condition: Type 2, G = 113.6 kg/m^2·s and T_{heat} = 46.7 °C.

3.2.3. Plug–Annular Alternated Flow

The plug–annular alternating flow is a special gas–liquid two-phase flow pattern that occurs during FBM with the T-junction. Due to the bifurcations, the annular flow in the branch channel is difficult to sustain when the slug flow or plug flow exists in the main channel. The bubbles or plugs in the main channel will inevitably disrupt the annular flow liquid film in the branch channel when passing through the bifurcation, resulting in the alternate presence of plug and annular flows at the bifurcation, which is the plug–annular alternated flow pattern.

The main characteristics of the plug–annular alternated flow pattern are as follows: circular small bubbles and slug bubbles are randomly distributed in the main channel, while annular flow exists in the branch channel, and slug bubbles may also appear at the end. There are two reasons for the appearance of slug bubbles at the end of the branch channel: one reason is that the circular small bubbles in the main channel exist independently until flowing into the branch channel; another reason is that when the branch channel is in an annular flow, the vaporization core on the branch channel wall is suppressed. When the annular flow in the branch channel is disrupted, the inhibitory influence of the liquid film on the vaporization core disappears, and some nucleation bubbles will reappear in the branch channel. These nucleation bubbles quickly grow, and some of them merge with the plugs in the branch channel, while others grow independently into slug bubbles. This is the reason why slug bubbles sometimes appear at the end of the branch channel.

In the main channel, slug bubbles and circular small bubbles are irregularly distributed along the main flow direction, and the bubble size does not increase sequentially. We can observe from Figure 19 that the channel end has circular bubbles, but slug bubbles also exist in the middle, which is caused by the random merging of bubbles during the flow process. When the bubbles in the main channel pass through the bifurcation in the form of slug bubbles, there will be a merging situation with the annular flow in the branch channel. At this time, the annular flow in the branch channel is not disrupted, but a necking section appears at the bifurcation. The front end of the slug bubble merges with it before the necking section breaks, and the branch channel eventually returns to the annular flow. When the bubbles in the main channel pass through the bifurcation in the form of circular small bubbles, similar to the squeeze rupture flow pattern, the annular flow liquid film is blocked at the bifurcation. Under the compression of circular small bubbles and slug bubbles in the main channel, a necking section is formed in the liquid film until the film finally ruptures, and the annular flow gas in the branch channel is divided into two parts. The circular small bubbles and slug bubbles in the main channel quickly grow and merge after passing through the bifurcation, and most of the bubbles merge with the broken gas band in the branch channel to form a new annular flow, while a small part of the bubbles form a plug flow and flow directly to the outlet. We call this flow pattern, with an alternate presence of plug and annular flows at the bifurcation, the plug–annular alternated flow.

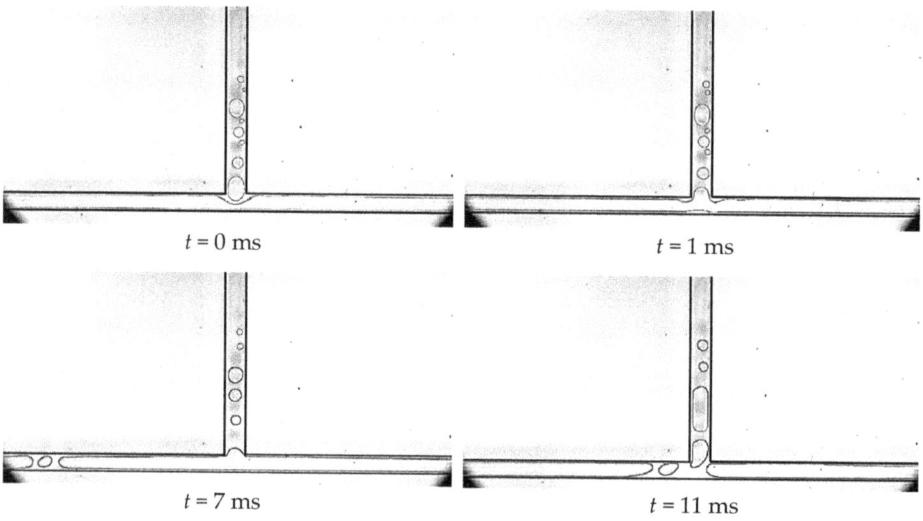

Figure 19. Plug–annular alternated flow, working condition: Type 2, $G = 113.6$ kg/m^2·s and $T_{heat} = 55.7$ °C.

3.2.4. Annular Flow

The annular flow is a typical two-phase flow pattern that occurs during FBM, and it is also one of the most common flow patterns in T-shaped microchannels during flow boiling. In our experiments, we observed two types of annular flow patterns: An intermittent annular flow and stable annular flow. The intermittent annular flow refers to the continuous presence of annular flow at the bifurcation point, which is always covered by a liquid film. A stable annular flow refers to the continuous presence of annular flow not only at the bifurcation point but also throughout the entire T-shaped microchannel.

The main flow characteristics of an intermittent annular flow are as follows: slug bubbles and annular flow alternate in the main channel, while the branch channel maintains a continuous annular flow that extends to the end of the main channel, as can be observed in Figure 20. In the experiments, nucleate bubbles are generated at the upper-middle position of the main channel, and these nucleate bubbles quickly grow into slug bubbles after leaving the nucleation point. The length of slug bubbles increases sequentially in the flow direction, and during the flow process, the slug bubbles merge with each other to form longer plugs. The plugs periodically merge with the annular flow in the branch channel before reaching the bifurcation point. Throughout the experiments, no annular flow rupture phenomenon similar to the slug–annular flow pattern occurs at the bifurcation point. The merging of plugs in the main channel and the annular flow in the branch results in the annular flow throughout the entire T-shaped microchannel. However, the nucleate bubbles in the upper-middle position disrupt the annular flow in the main channel, resulting in an unstable annular flow throughout the entire T-shaped microchannel. Despite this, the bifurcation point maintains an annular flow state throughout the process, so we refer to this flow pattern as the intermittent annular flow.

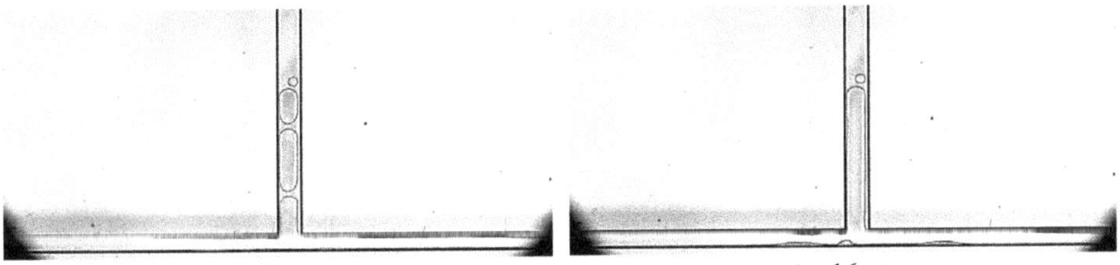

$t = 0$ ms $t = 16$ ms

Figure 20. Intermittent annular flow, working condition: Type 2, $G = 14.2$ kg/m^2·s and $T_{heat} = 42.5$ °C.

In contrast to the intermittent annular flow, the stable annular flow maintains a stable annular flow pattern throughout the entire T-shaped microchannel during the flow process. Its flow characteristics are as follows: the main channel and branch channel continuously maintain an annular flow pattern. Most of the liquid quickly vaporizes after entering the entrance of the main channel (as presented in Figure 6) and flows towards the outlet through the annular gas film, while a small amount of unvaporized liquid flows towards the outlet along the annular liquid film. Because of the bifurcation, the liquid flowing accumulates at the bifurcation, resulting in an increase in film thickness at the end of the main channel, as can be seen in Figure 21. Throughout the flow process, the bifurcation point maintains an annular flow state without rupture, coalescence, or other phenomena, and the entire T-shaped microchannel maintains a stable annular flow. We refer to this flow pattern as the stable annular flow. As the heating temperature continues to rise, the annular liquid film evaporates continuously, causing it to become thinner and eventually disappear completely, leading to a drying-out phenomenon.

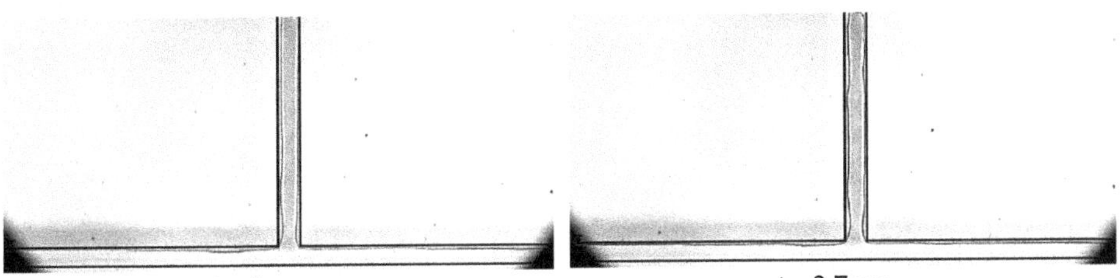

$t = 0$ ms $t = 3.7$ ms

Figure 21. Steady annular flow, working condition: Type 2, $G = 14.2$ kg/m^2·s and $T_{heat} = 53.7$ °C.

3.3. *Flow Boiling Flow Pattern at the Bifurcation of T-Shaped Microchannel and Its Influencing Factors*

The experiment was conducted on T-shaped microchannels of four different sizes to study flow boiling. During the experiment, the operating conditions were changed by adjusting the mass flow rate at the channel inlet and the heating temperature of the chip. Figure 22 shows the flow regime map at the bifurcation point under different operating conditions. The y-axis represents the mass flow rate at the entrance of microchannel (G), and the x-axis represents the degree of the superheat of the chip (ΔT). As shown, the factors that primarily affect the flow mechanism at the bifurcation point of the microchannel are the mass flow rate at the inlet, the degree of superheat (ΔT), and the size of it. The specific effects of these three factors on the flow regime at the bifurcation point of the T-shaped microchannel will be discussed below.

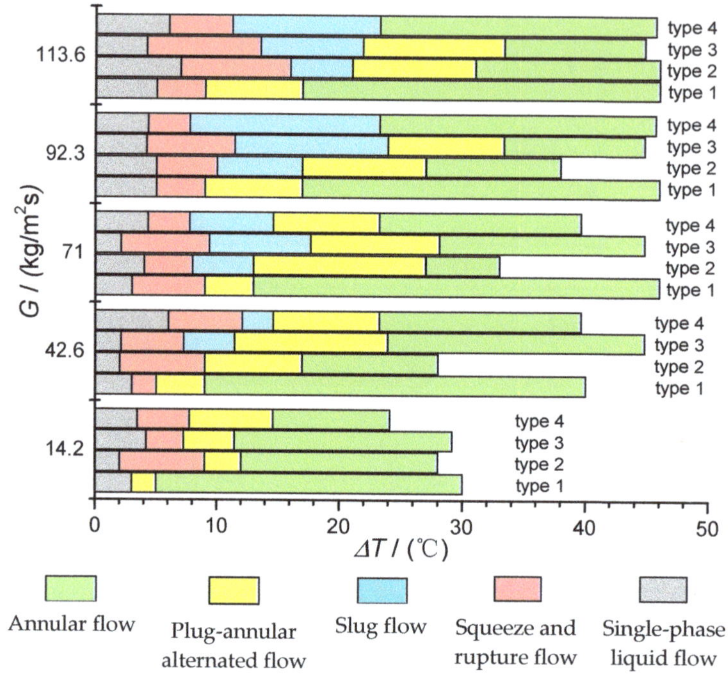

Figure 22. Flow boiling flow pattern at the bifurcation of a T-shaped microchannel.

3.3.1. Effect of Fluid Mass Flow Density on the Flow Pattern of Flow Boiling at the Bifurcation of T-Shaped Microchannel

This experiment studied the flow boiling behavior of chips of the same size with five different mass flux densities. As shown in Figure 23(b1,b2,e), in type 2 chips, no bubble flow was observed at the bifurcation under the condition that the mass flux density was less than 42.6 $kg/m^2{\cdot}s$, while the bubble flow appeared when the mass flux density was bigger than 71 $kg/m^2{\cdot}s$. This is because, at the same heating temperature, when the mass flux density is low, the nucleate bubbles generated in the main channel rapidly grow into slug bubbles after leaving the vaporization nucleus, and even merge and grow into slug plugs during the downstream flow; thus, the two-phase always exists in the form of slug plugs or gas bands at the bifurcation. In addition, at the minimum mass flux density of 14.2 $kg/m^2{\cdot}s$, the bubble flow was absent in all four chip sizes, while when the mass flux density increases, the bubble flow appeared successively in type 3, type 4 and type 2 chips, and only the type 1 chip did not show a bubble flow in this range of mass flux. This reflects the influence of the microchannel size effect.

Furthermore, in the four different chip sizes, when the mass flux density increases, the appearance time of the annular flow gradually delays, as can be observed in Figure 22. Under the condition that the mass flux density is low, the time for the slug–annular alternating flow pattern is short, and it quickly evolves into an annular flow pattern; thus, the appearance of the annular flow pattern is relatively early. The increase in mass flux density provides more subcooled liquid; thus, the previously missing bubble flow appears during the flow process, and the appearance time of the annular flow is delayed. However, at high mass flux densities, the time difference for the appearance of the annular flow gradually decreases, indicating that the influence of mass flux density on the appearance time of the annular flow gradually weakens.

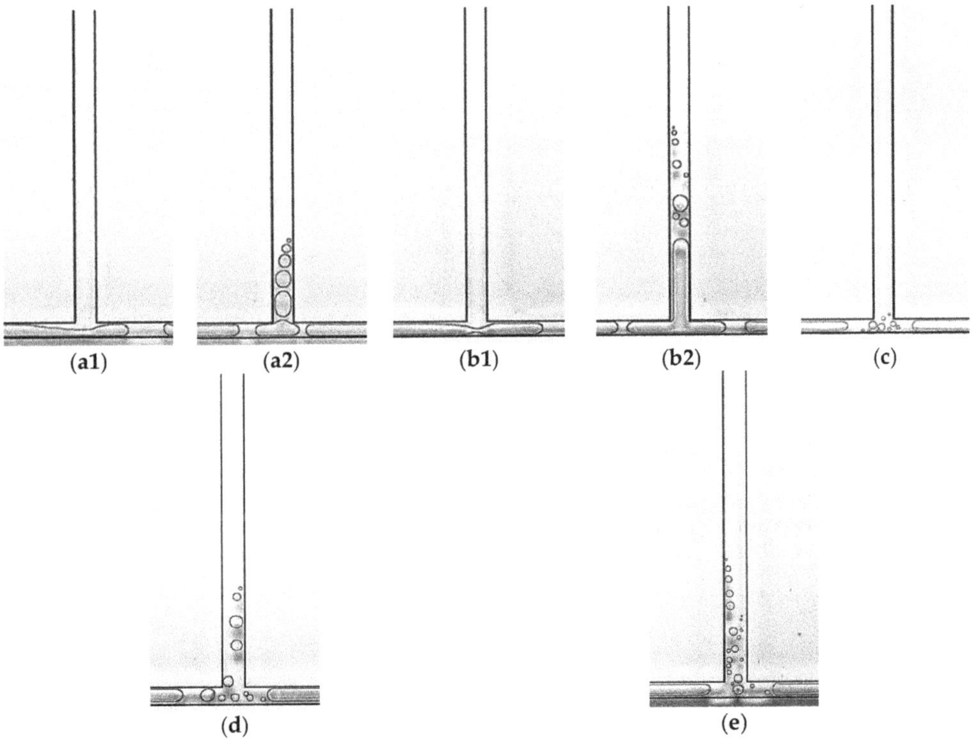

Figure 23. Effect of microchannel size on slug flow in T-shaped microchannel, working conditions: (**a1**) Type 1, $G = 42.6$ kg/m^2·s and $T_{heat} = 35.8$ °C; (**a2**) type 1, $G = 42.6$ kg/m^2·s and $T_{heat} = 37.9$ °C; (**b1**) type 2, $G = 42.6$ kg/m^2·s and $T_{heat} = 39.8$ °C; (**b2**) type 2, $G = 42.6$ kg/m^2·s and $T_{heat} =41.9$ °C; (**c**) type 3, $G = 42.6$ kg/m^2·s and $T_{heat} = 39.9$ °C; (**d**) type 1, $G = 42.6$ kg/m^2·s and $T_{heat} = 46.7$ °C; (**e**) type2, $G = 71$ kg/m^2·s and $T_{heat} = 41.8$ °C.

3.3.2. Influence of Microchannel Size on the Flow Pattern of FBM with T-Junction

The flow boiling phenomena were investigated in four different-sized chips through visualization experiments. The main channel width of the type 1, type 2, and type 3 chips is 800 μm, and the width of the branch channels is 600 μm. The depths of the three chips are 300 μm, 450 μm and 600 μm. The main channel width of the type 4 chip is 1000 μm, and the width of the branch channels is 700 μm. All of them have a depth of 800 μm. Based on the division of microchannel sizes, type 1, type 2, and type 3 chips can be regarded as microchannels, while the type 4 chip can be considered as a conventional small channel. As shown in Figure 23, under the operating condition of a mass flow rate density of 42.6 kg/m^2·s, the flow pattern at the bifurcation of the type 1 chip is the squeeze-induced rupture flow (Figure 23(a1)), with liquid flow in the main channel and slug flow in the branch channel. With a slight increase in the heating temperature to 37.9 °C, vapor bubbles produced by nucleation appeared in the main channel, but they had grown into slugs by the time they reached the bifurcation, and no bubble flow was observed at the bifurcation (Figure 23(a2)). Similarly, at a heating temperature of 37.9 °C, the bifurcation of the type 2 chip exhibited squeeze-induced rupture flow (Figure 23(b1)). As the heating temperature increased to 39.8 °C, the bifurcation displayed a slug–annular alternating flow (Figure 23(b2)). The main reason for this phenomenon is that, under a mass flow rate density of 42.6 kg/m^2·s, the nucleation bubbles generated in the main channel of type 1 and type 2 chips quickly grow into slugs because of the channel size limitation, and the two-phase flow at the bifurcation directly transforms from a single-phase liquid to slugs. Therefore, the bubble flow pattern is absent at the bifurcation. On the contrary, due to the greater depth of the type 3 chip, the growth of nucleation bubbles in the main channel is limited in the depth direction compared to type 1 and type 2 chips. Hence, small bubbles can be clearly observed at the bifurcation at a heating temperature of 39.9 °C (Figure 23c). Similarly, the bubble flow at the bifurcation is also more pronounced in the type 4 chip (Figure 23d). At mass flow rate densities of 71, 92.3 and 113.6 kg/m^2·s, bubble flow was only absent at the bifurcation of the type 1 chip, while it was observed at the bifurcation of the other three chips. The absence of bubble flow under relatively high mass flow rate densities further demonstrates the effect of microchannel size on the flow pattern.

The limitation of microchannel size contributes to the absence of bubble flow in shallow-depth chips, because the two-phase flow in the restricted channel exists as small bubbles for a very short time or even directly transforms from a single-phase liquid to slugs. At the same time, due to the limitation of the channel size, the appearance time of the annular flow will also change. As shown in the flow pattern diagram in Figure 22, under a constant mass flow rate density, such as 92.3 kg/m^2·s, the annular flow first appears in the type 1 chip and last appears in the type 3 chip. Therefore, reducing the size of microchannels will cause the annular flow to appear earlier.

3.4. *Unstable Outlet Temperature in T-Shaped Microchannels*

Flow instability is a key issue in microscale flow and heat transfer research, and temperature instability is the most common research content. During the experiments, the inlet temperature of the supercooled liquid in the main channel was kept constant, and the instability of the outlet temperature was studied by changing the heating temperature and liquid flow rate. Experimental results show that both the heating temperature as well as the mass flow rate density of the inlet liquid have an influence on the fluctuation of the outlet temperature.

3.4.1. Influence of Heating Temperature on Outlet Temperature Instability

As shown in Figure 22, under different heating temperatures, the bifurcation of the T-shaped microchannel exhibits different flow patterns. Therefore, the effect of the heating temperature on the instability of the outlet temperature reflects, to some extent, the influence of different flow patterns on the instability of the outlet temperature. For a qualitative and quantitative analysis, we selected the type 1 chip and analyzed the operating condition with a main channel inlet flow rate of 113.6 kg/m^2·s. When the heating temperatures were 35.8 °C, 41.9 °C, 50 °C and 62 °C, as shown in Figure 22, the corresponding flow patterns at the bifurcation of the T-shaped microchannel were the squeeze-induced rupture flow, slug–annular alternating flow, unstable annular flow and stable annular flow.

When the heating temperature is 35.8 °C, the bifurcation of the T-shaped microchannel exhibits a squeeze-induced rupture flow, with a liquid flow in the main channel and a pulsating slug flow in the branch channel. The slug flow will be squeezed and ruptured into two sub-slugs at the bifurcation. Therefore, at the left and right outlets of the microchannel, sometimes there is a single-phase supercooled liquid flow and sometimes there is a slug flow, with a small void fraction. This leads to the fluctuation of the temperatures at the left and right outlets, as shown in Figure 24a. The fluctuation of the outlet temperature is more pronounced due to the fluctuation of the outlet pressure (caused by the rupture of slugs at the outlet, etc.), which leads to the pulsating slug flow in the branch channel having a significantly increased residence time, allowing more sufficient heat exchange between the gas phase in the slug wall. The pulsating slug flow in the branch channel is the main reason for the more intense fluctuation of the outlet temperature in the squeeze-induced rupture flow. Figure 25 shows the fluctuation of the outlet temperature of the type 1 chip under a different heat load. We can conclude that the variations of the outlet temperature reduce when the heating temperature increases.

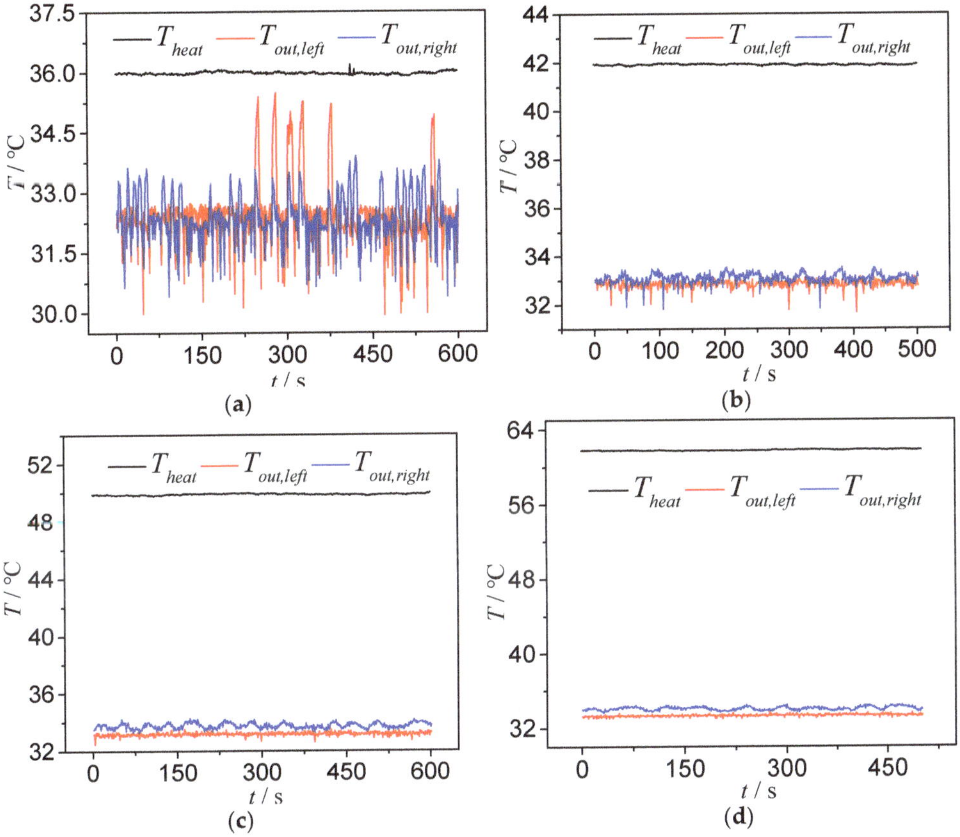

Figure 24. Outlet temperature fluctuations at different heating temperatures: type 1, G = 113.6 kg/m^2·s. (**a**) Type 1, G = 113.6 kg/m^2·s and T_{heat} = 35.8 °C; (**b**) type 1, G = 113.6 kg/m^2·s, T_{heat} = 41.9 °C; (**c**) type 1, G = 113.6 kg/m^2·s and T_{heat} = 50 °C; (**d**) type1, G = 113.6 kg/m^2·s and T_{heat} = 62 °C.

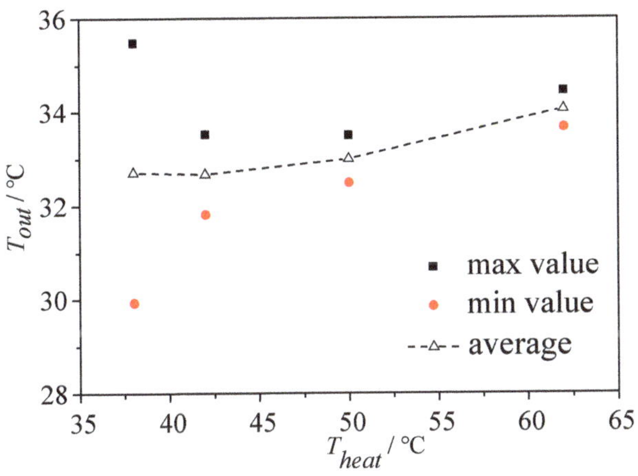

Figure 25. Outlet temperature at different heating temperatures: Type 1, G = 113.6 kg/m^2·s.

As the heating temperature increases, as presented in Figure 24b and with the decrease in heating temperature by 41.9 °C, the bifurcation of the T-shaped microchannel exhibits a plug-ring alternating flow. The inlet section is still a subcooled single-phase liquid flow, while the end of the main channel is mainly a slug or plug flow, with the plug and annular flow alternating in the branch channel. The bubbles (plugs) in the main channel split into two sub-bubbles (plugs) at the bifurcation, which drives the two-phase flow in the left and right branch channels, eliminating the pulsation phenomenon of the plug flow and squeezing in the broken flow. All the bubbles (plugs) flow directly to the outlet, reducing the relative heat exchange time between the gas and the microchannel. Therefore, the peak change of the outlet temperature decreases, and at the same time, due to the increase in the proportion of the outlet void, the fluctuation amplitude of the outlet temperature decreases.

When the heating temperature further increases, as presented in Figure 24c,d, when the heating temperature reaches 50 °C and 62 °C, the bifurcation of the T-shaped microchannel shows the annular flow. The bifurcation and the branch channel maintain the annular flow, and the void fraction at the outlet approaches one. At this time, the variations of the outlet temperature are further weakened. The difference is that at a heating temperature of 50 °C, it is unstable annular flow, and the annular flow at the outlet fluctuates. At a heating temperature of 62 °C, it is stable as the outlet becomes thinner and relatively stable; thus, the fluctuation of the outlet temperature is relatively reduced.

3.4.2. Effect of Mass Flow Density on Outlet Temperature Instability

From Section 3.3.1, we can find that that the mass flow rate density of the inlet exerts a vital impact on the flow regime of FBM with the T-junction, which is consistent with the flow boiling regime diagram at the bifurcation in Figure 22. As shown in Figure 26, in the type 3 chip, when the heating temperature is 39.6 °C, the temperature at the left and right outlets fluctuates when the inlet mass flow rate densities are around 42.6 kg/m^2·s, 71 kg/m^2·s, 92.3 kg/m^2·s, and 113.6 kg/m^2·s. Figure 27 presents the temperature fluctuation range at the left and right outlets under different mass flow rate densities. The results show that around the heating temperature of 39.6 °C, with the increase in the mass flow rate density, the temperature fluctuation range at the left and right branches gradually increases, but the average temperature at the outlets decreases.

As shown in Figure 22, in the type 3 chip, when the heating temperature is 39.6 °C, the flow regimes at the bifurcation correspond to a slug–annular alternating flow, bubble flow, squeeze and rupture flow, and squeeze and rupture flow for inlet mass flow rate densities of 42.6 kg/m^2·s, 71 kg/m^2·s, 92.3 kg/m^2·s, and 113.6 kg/m^2·s, respectively. Similar to the impact of heating temperature on outlet temperature instability, the influence of the mass flow rate density on the outlet temperature instability is ultimately reflected in the influence of flow regimes on the outlet temperature instability. In the type 3 chip, when the heating temperature is 39.6 °C and the inlet mass flow rate density is 42.6 kg/m^2·s, the flow regime at the bifurcation is the slug–annular alternating flow. The left and right outlets are occupied alternately by slugs and gas regions. At this moment, the void fraction at the outlet cross-section is relatively large and mainly occupied by the gas phase. Therefore, the temperature fluctuation range at the left and right outlets is relatively small, as can be observed in Figure 26a.

We can find in Figures 26b and 26 that, when the mass flow rate density increases to 71 kg/m^2·s, at a heating temperature of 39.6 °C, the flow regime at the bifurcation in the type 3 chip becomes the bubble flow. At this time, the flow through the left and right outlets is mainly composed of bubbles, slugs, and subcooled liquid. Since a small amount of subcooled liquid passes through the left and right outlets, the minimum value of the temperature fluctuation at the outlets is significantly lower than that at a mass flow rate density of 42.6 kg/m^2·s and lower than the saturation temperature of 32.05 °C.

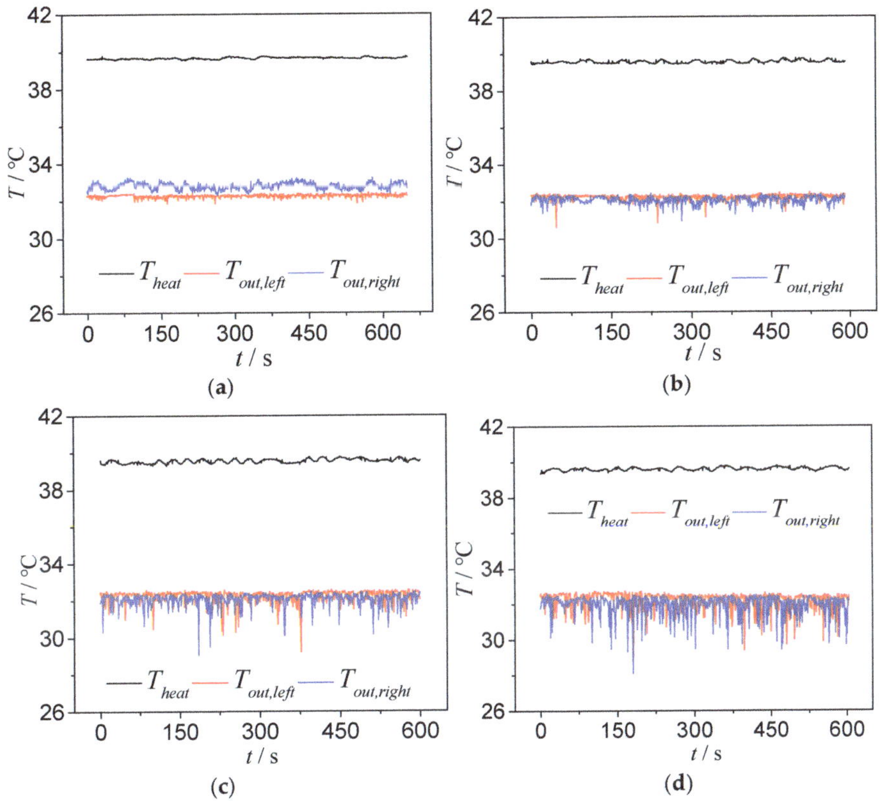

Figure 26. Outlet temperature fluctuations under different mass flow density: type 3, T_{heat} = 39.6 °C. (**a**) Type 3, G = 42.6 kg/m^2·s and T_{heat} = 39.6 °C; (**b**) type 3, G = 71 kg/m^2·s and T_{heat} = 39.6 °C; (**c**) Type 3, G = 92.3 kg/m^2·s and T_{heat} = 39.5 °C; (**d**) type 3, G = 113.6 kg/m^2·s and T_{heat} = 39.6 °C.

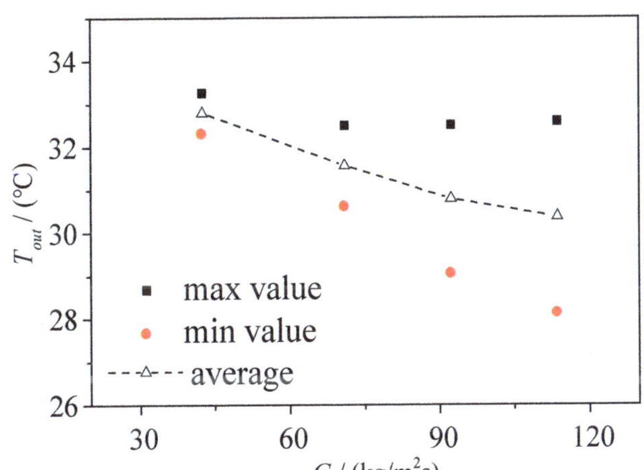

Figure 27. Outlet temperature under different mass flow density: type 3, T_{heat} = 39.6 °C.

When the mass flow rate density further increases to above 92.3 kg/m^2·s, as presented in Figure 26c,d, the temperature fluctuation range at the inlet and outlet further increases. This is because the flow regime at the bifurcation in the type 3 chip becomes a squeeze and rupture flow. At this time, the void fraction at the cross-section is smaller, and the flow through the left and right outlets is mainly composed of slugs and subcooled liquid. Moreover, for the same period of time, the proportion of the subcooled liquid flow is larger compared to that in the bubble flow regime. This is determined by the characteristics of the squeeze and rupture flow regime, where the slug oscillates back and forth in the branch channel, and most of the flow through the outlet is subcooled liquid after the oscillating slug is squeezed out at the bifurcation. This alternating flow of slugs and subcooled liquid through the outlet is the main reason for the largest amplitude of temperature fluctuation at the left and right outlets.

4. Conclusions

To study the flow boiling phenomenon and mechanism in bifurcated microchannels, this chapter designs and fabricates four different-sized T-shaped microchannel chips, and establishes an experimental platform for flow boiling in T-shaped microchannels. The flow patterns of FLM with the T-junction can be observed through visualization experiments, and the variation of flow patterns and influencing factors in four different-sized T-shaped microchannels under different operating conditions are analyzed. Moreover, the temperature instability of FBM with the T-junction is also investigated, and the experimental results show that:

(1) In the flow boiling of T-shaped microchannels, the vaporization nuclei first appear in the branch channels, and as the heating temperature increases, the vaporization nuclei gradually extend into the main channel.

(2) The flow patterns of FBM with the T-junction mainly include the squeeze and rupture flow, slug flow, plug–annular alternating flow, and annular flow.

(3) Apart from the heating temperature, the inlet mass flow rate also affects flow patterns at the bifurcation of microchannels. When the mass flow rate at the entrance decreases, the bubble flow at the bifurcation disappears, and when the inlet mass flow rate increases, the appearance of the annular flow at the bifurcation is delayed. Furthermore, the size of the T-shaped microchannel also has an influence on the flow boiling flow patterns at the bifurcation. The bubble flow disappears in smaller microchannels, and with the increase in microchannel size, the timing of the appearance of the annular flow at the bifurcation is also delayed.

(4) Flow boiling in T-shaped microchannels results in temperature instability at the outlet, which mainly manifested as temperature fluctuations. The heating temperature and outlet mass flow rate affect the variations of the outlet temperature. As the heating temperature increases, the variations of the outlet temperature gradually weaken. Conversely, as the mass flow rate at the entrance increases, the variations of the outlet temperature gradually weaken. These two influencing factors ultimately stem from the effect of flow patterns on temperature fluctuations at the outlet.

(5) Future studies can be carried out from the following aspects: (1) Flow boiling phenomenon of different working fluids in bifurcated microchannels, and a more applicable flow pattern diagram, are established. (2) Thickness change of the bubble necking section in the nitrogen–water two-phase flow at a high mass flow density. (3) Effects of bifurcation angle and shape on gas–liquid two-phase flow in a bifurcated microchannel. The research work in this paper will help to understand the flow boiling two-phase flow characteristics in complex microchannel networks. With the increase in electronic device layout density, the T-type microchannel structure will be a method to solve the high heat flux of electronic devices.

Author Contributions: Conceptualization, X.B. and F.Y.; methodology, X.B.; software, X.B.; validation, X.B., F.Y. and X.Z.; formal analysis, X.B.; investigation, X.B.; resources, X.B.; data curation, X.B.; writing—original draft preparation, X.B.; writing—review and editing, X.B.; visualization, F.Y.; supervision, X.Z.; project administration, X.Z.; funding acquisition, X.Z. All authors have read and agreed to the published version of the manuscript.

Funding: This research was funded by National Natural Science Foundation of China (Grant Nos. 52022020).

Data Availability Statement: Data are contained within the article.

Conflicts of Interest: The authors declare no conflict of interest. Xiangzhong Bao and Xuan Zhang are employees of Southeast University Architectural Design and Research Institute Co., Ltd. The paper reflects the views of the scientists, and not the company.

References

1. Liu, D.; Garimella, S.V. Flow Boiling Heat Transfer in Microchannels. *J. Heat Transf.* **2007**, *129*, 1321–1332. [CrossRef]
2. Zhu, Y.; Antao, D.S.; Chu, K.-H.; Chen, S.; Hendricks, T.J.; Zhang, T.; Wang, E.N. Surface Structure Enhanced Microchannel Flow Boiling. *J. Heat Transf.* **2016**, *138*, 091501. [CrossRef]
3. Harirchian, T.; Garimella, S.V. A comprehensive flow regime map for microchannel flow boiling with quantitative transition criteria. *Int. J. Heat Mass Transf.* **2010**, *53*, 2694–2702. [CrossRef]
4. Wu, J.F.; Chen, Y.P.; Shi, M.H.; Fu, P.P.; Peterson, G.P. Three-Dimensional Numerical Simulation for Annular Condensa-tion in Rectangular Microchannels. *Nanoscale Microscale Thermophys. Eng.* **2009**, *13*, 13–29. [CrossRef]
5. Chen, Y.P.; Yao, F.; Huang, X.Y. Mass transfer and reaction in methanol steam reforming reactor with fractal tree-like mi-crochannel network. *Int. J. Heat Mass Transf.* **2015**, *87*, 279–283. [CrossRef]
6. Wu, J.F.; Shi, M.H.; Chen, Y.P.; Li, X. Visualization study of steam condensation in wide rectangular silicon microchannels. *Int. J. Therm. Sci.* **2010**, *49*, 922–930. [CrossRef]
7. Chen, L.; Tian, Y.; Karayiannis, T. The effect of tube diameter on vertical two-phase flow regimes in small tubes. *Int. J. Heat Mass Transf.* **2006**, *49*, 4220–4230. [CrossRef]
8. Tsoligkas, A.; Simmons, M.; Wood, J. Two phase gas–liquid reaction studies in a circular capillary. *Chem. Eng. Sci.* **2007**, *62*, 5397–5401. [CrossRef]
9. Fabbri, G. Analysis of vapor back flow in single-pass air-cooled condensers. *Int. J. Heat Mass Transf.* **1997**, *40*, 3969–3979. [CrossRef]
10. Fabbri, G. Effect of disuniformities in vapor saturation pressure and coolant velocity on vapor back flow phenomena in sin-gle-pass air-cooled condensers. *Int. J. Heat Mass Transf.* **2000**, *43*, 147–159. [CrossRef]
11. Suzuki, K.; Kokubu, T.; Nakano, M.; Kawamura, H.; Ueno, I.; Shida, H.; Ogawa, O. Enhancement of heat transfer in subcooled flow boiling with microbubble emission. *Exp. Therm. Fluid Sci.* **2005**, *29*, 827–832. [CrossRef]
12. Suzuki, K.; Saitoh, H.; Matsumoto, K. High heat flux cooling by microbubble emission boiling. *Ann. N. Y. Acad. Sci.* **2002**, *974*, 364–377. [CrossRef]
13. Chen, Y.P.; Shi, M.H.; Cheng, P.; Peterson, G.P. Condensation in microchannels. *Nanoscale Microscale Thermophys. Eng.* **2008**, *12*, 117–143. [CrossRef]
14. Jin, N.-D.; Dong, F.; Zhao, S. The complexity measure analysis of conductance fluctuation signals of gas-liquid two-phase flow and its flow pattern characterization. *Acta Phys. Sin.* **2007**, *56*, 720–729.
15. Yan, Z.; Huang, H.; Pan, W.; Deng, B.; Pan, Z. Influences of wall superheat and channel width ratio on bubble behaviors and heat transfer within a heated microchannel T-junction. *Int. Commun. Heat Mass Transf.* **2021**, *126*, 105481. [CrossRef]
16. Chen, Y.P.; Wu, R.; Shi, M.H.; Peterson, G.P. Visualization study of steam condensation in triangular microchannels. *Inter-Natl. J. Heat Mass Transf.* **2009**, *52*, 5122–5129. [CrossRef]
17. Law, M.; Lee, P.-S. A comparative study of experimental flow boiling heat transfer and pressure characteristics in straight- and oblique-finned microchannels. *Int. J. Heat Mass Transf.* **2015**, *85*, 797–810. [CrossRef]
18. Tiwari, N.; Moharana, M.K. Conjugate effect on flow boiling instability in wavy microchannel. *Int. J. Heat Mass Transf.* **2021**, *166*, 120791. [CrossRef]
19. Xia, G.; Cheng, Y.; Cheng, L.; Li, Y. Heat Transfer Characteristics and Flow Visualization during Flow Boiling of Acetone in Semi-Open Multi-Microchannels. *Heat Transf. Eng.* **2019**, *40*, 1349–1362. [CrossRef]
20. Mao, Z.; Yoshida, K.; Kim, J.-W. Active sorting of droplets by using an ECF (Electro-conjugate Fluid) micropump. *Sensors Actuators A Phys.* **2020**, *303*, 111702. [CrossRef]
21. Mao, Z.; Yoshida, K.; Kim, J.-W. Developing O/O (oil-in-oil) droplet generators on a chip by using ECF (electro-conjugate fluid) micropumps. *Sensors Actuators B Chem.* **2019**, *296*, 126669. [CrossRef]
22. Bai, F.; Li, M.; Song, D.; Yu, H.; Jiang, B.; Li, Y. Metal-assisted homogeneous etching of single crystal silicon: A novel approach to obtain an ultra-thin silicon wafer. *Appl. Surf. Sci.* **2013**, *273*, 107–110. [CrossRef]

23. Xu, D.; Xiong, B.; Wu, G.; Wang, Y.; Sun, X. Wang, Isotropic Silicon Etching with XeF2 Gas for Wafer-Level Micromachining Applications. *J. Microelectromech. Syst.* **2012**, *21*, 1436–1444. [CrossRef]
24. Piret, N.; Santoro, R.; Dogot, L.; Barthélemy, B.; Peyroux, E.; Proost, J. Influence of glass composition on the kinetics of glass etching and frosting in concentrated HF solutions. *J. Non-Crystalline Solids* **2018**, *499*, 208–216. [CrossRef]
25. Weigel, C.; Phi, H.B.; Denissel, F.A.; Hoffmann, M.; Sinzinger, S.; Strehle, S. Highly Anisotropic Fluorine-Based Plasma Etching of Ultralow Expansion Glass. *Adv. Eng. Mater.* **2021**, *23*, 2001336. [CrossRef]
26. Bernardeschi, I.; Tricinci, O.; Mattoli, V.; Filippeschi, C.; Mazzolai, B.; Beccai, L. Three-Dimensional Soft Material Micropatterning via Direct Laser Lithography of Flexible Molds. *ACS Appl. Mater. Interfaces* **2016**, *8*, 25019–25023. [CrossRef]
27. Hamouda, F.; Barbillon, G.; Gaucher, F.; Bartenlian, B. Sub-200nm gap electrodes by soft UV nanoimprint lithography using polydimethylsiloxane mold without external pressure. *J. Vac. Sci. Technol. B* **2010**, *28*, 82–85. [CrossRef]
28. Deng, N.-N.; Meng, Z.-J.; Xie, R.; Ju, X.-J.; Mou, C.-L.; Wang, W.; Chu, L.-Y. Simple and cheap microfluidic devices for the preparation of monodisperse emulsions. *Lab Chip* **2011**, *11*, 3963–3969. [CrossRef]

micromachines

MDPI

Article

A Study on the Simulation and Experiment of Evaporative Condensers in an R744 Air Conditioning System

Thanhtrung Dang * and Hoangtuan Nguyen

Department of Thermal Engineering, HCMC University of Technology and Education, Ho Chi Minh City 71307, Vietnam; tuannh@hcmute.edu.vn
* Correspondence: trungdang@hcmute.edu.vn; Tel.: +84-913606261

Abstract: The heat transfer characteristics of evaporative condensers in an R744 air conditioning system were evaluated using the numerical and the experimental methods. Two configurations of condensers were studied: Case 1 with five layers of tubes and Case 2 with eight layers of tubes. In order to evaluate the heat transfer characteristics, the temperature field, the phase change, the pressure distribution, and thermodynamic parameters were considered. For Case 2, it indicated the capability of R744 condensation from the superheated status to the liquified status by analyzing the outlet temperature of the condenser changed from 28.7 °C to 30.3 °C with a change in condensation pressure from 72.6 bar to 68.5 bar. In this study, R744 mass flow rate increases from 14.34 kg/h to 46.08 kg/h, and the pressure drop also increases from 0.23 bar to 0.47 bar for the simulation and 0.4 bar to 0.5 bar for the experiment, respectively. The results indicate that the five-layer configuration causes a higher pressure drop and lower COP than those obtained from the eight-layer one (splitting into two sets for smaller pressure drop). Furthermore, the evaporative condensers using mini tubes that are flooded in the cooling water tank are suitable for the subcritical R744 air conditioning system. In addition, the results obtained from the experimental data are in good agreement with those obtained from the numerical results, with a deviation of less than 5%.

Keywords: evaporative condenser; heat transfer; R744; subcritical; air conditioning

1. Introduction

Currently, scientists are concerned about greenhouse effect problems. One of these problems is the exhaust gas that is generated by refrigerant leakage from air conditioning systems as well as industrial and commercial refrigeration systems. Therefore, research on environmentally friendly refrigerants is an interesting goal for experts. In natural refrigerants, R744 (known as CO_2) has the potential to replace traditional refrigerants with many advantages such as having low Global Warming Potential (GWP = 1) and Ozone Depletion Potential (ODP = 0) and not being able to cause fire or explosion. Based on the above advantages, many previous studies have been conducted to experimentally evaluate subcritical and transcritical CO_2 refrigeration systems.

The R744 transcritical cycle has been widely researched to find optimal working points as well as optimal designs to improve compressor energy efficiency. Related to this field, Zhang et al. [1] conducted simulations based on the R744 transcritical cycle involving a few performance parameters and CFD (Computational Fluid Mechanics) models for tubular and wing air cooling condensers. The models were compared and confirmed with experimental measurements and literature reviews. The simulation and experiment results in [2–4] were analyzed with an R744 transcritical cycle operating in heat pump mode with auxiliary thermoelectric cooling equipment. These studies show the effects of coolant flow rate and temperature, auxiliary cooling capacity, and compressor discharge pressure on the COP (Coefficient of Performance) and the heating efficiency factor. Deng et al. [5] theoretically analyzed an ejector refrigeration cycle operating on a transcritical cycle. Using an expansion device and controlling the suction rate, the ejector ensures the amount of vapor returning to the compressor and the amount of liquid to the evaporator. The COP was studied in a variety of injection ratios while simultaneously comparing it with conventional vapor compression refrigeration cycles. Zhang et al. [6,7] experimentally compared the R744/R290 subcritical cycle in the cascade system to the R744 transcritical cycle. The analysis results of the discharge pressure, the outlet temperature of the gas cooler, and the mass ratio between R744 and R290 show that using an IHX (Internal Heat Exchanger) for the subcritical cycle is less effective than the transcritical cycle. Gullo et al. [8] evaluated aspects related to the R744 transcritical cycle in a refrigeration plant in supermarkets. The study concluded that R744 can be favorably utilized at supermarkets, especially suitable for areas with high ambient temperatures. Hazarika et al. [9] conducted an experimental study and numerical simulation of an R744 transcritical air conditioning system. Not only were the variations in the heat transfer area of the evaporator proposed, to study the system COP further, but also an evaluation of the effect of air velocity and pressure of the air cooler was presented. Khedher et al. [10] predicted the heat loss of buildings by applying artificial neural networks to study the walls of buildings and coating materials with respect to the effect of the heat transfer coefficient as well as indoor and outdoor surface temperatures. However, the studies in [1–10] did not mention CO_2 air conditioning systems with throttle inlet temperatures below the critical point.

To reduce the condensation pressure and increase the overall refrigeration performance, Sanz-Kock et al. [11] conducted an experimental evaluation of the energy efficiency of a cascade refrigeration system using R134a/CO_2, where they used semi-hermetic compressors in subcritical cycles. Parameters such as compressor efficiency, temperature difference at the cascade heat exchanger, and compressor discharge temperature were evaluated under conditions of various evaporation and condensation temperatures. Zhang et al. [12] proposed a cascade refrigeration system with R1270/CO_2. Particularly, R1270 was used in the high temperature cycle, while CO_2 was used as a refrigerant in the low temperature cycle. The results showed that R1270 is a better replacement than R290 and R717 in the high-stage cycle. Llopis et al. [13] implemented an experimental evaluation of welded plate heat exchangers in subcritical cycles using R744 with water cooling evaporative condenser and semi-hermetic compressors. The influence of the internal heat exchanger in the cascade cycle was analyzed theoretically. The results indicate that welded plate heat exchangers for subcritical cycles had less impact on energy efficiency but were suitable for energy improvement in the cascade cycles. Nicola et al. [14] analyzed the thermal efficiency of the cascade refrigeration cycle, in which the low-stage R744 substance was mixed with other refrigerants such as R170, R290, R1150, and R717. Based on the Carnahane–Starlinge–DeSantis equation of state, the COP results indicated that R744 should only be used in its mono-component form, in combination with other refrigerants. The experimental results in [15,16] discussed the operating parameters and energy efficiency in a cascade cycle using the R717/R744. Dopazo et al. [15] performed a comparison with a two-stage cycle using R717 under the same operating conditions with evaporation temperatures and condensation temperatures corresponding to cascade cycles ranging from −17.5 °C to −7.5 °C. Lee et al. [16] conducted a further exploration of the evaporation temperature, the condensation temperature, and the temperature difference in the cascade heat exchanger. From refs. [11–16], most of the research conducted on subcritical cycles focused on experimental research to find suitable refrigerants in the high-temperature cycle, find out the relationship between COP and the operation parameters, and determine the optimal operation regime. They used the cascade refrigeration system and did not use the single-stage subcritical refrigeration system.

From the literatures above, the studies have focused on exploring the influence of parameters such as evaporation temperature, condensation temperature, condensation pressure, different ways of configuring equipment, design parameters, and refrigerants to find the most suitable case in each specific application. While there are fewer R744 subcritical cycle applications that use evaporative condensers using water and air, especially with the condensing mini tubes that are flooded in the cooling water tank. Therefore, the numerical simulation and experimental investigation for this application are potential directions that should be developed, especially in tropical countries. To sum up, the numerical simulation on and experimental verification of the heat transfer characteristics of flooded-type condenser applied in residential air conditioning systems is one of the breakthrough studies that will be presented in more detail in this paper.

2. Methodology

2.1. Calculation Design

With the initial design data for an R744 air conditioning system, the testing cycle has a cooling capacity of 1000 W, an evaporation temperature of 10 °C, and a condensation temperature of 30.5 °C. Design parameters were calculated and expressed according to Figure 1 and Table 1.

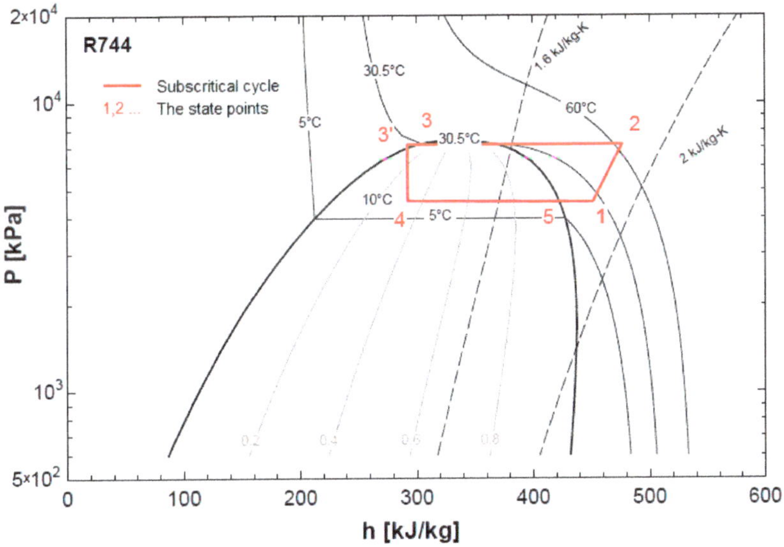

Figure 1. The P–h diagram of testing cycle.

Table 1. Thermodynamic parameters at state points according to the initial design.

State Points	Temperature (°C)	Pressure (bar)	Enthalpy (kJ.kg)	Entropy (kJ.kg)	Volume (m³/kg)
1	22	45	444	1.89	0.0086
2	64	73	474	1.85	-
3	30.5	73	309	1.28	0.0016
3′	28.5	73	289	1.28	-
4	10	45	289	1.32	-
5	10	45	422	1.78	0.007

In this study, the key components of a testing cycle were designed based on the computational theory with the following equations:

The cooling capacity was calculated as follows:

$$Q_0 = G(h_5 - h_4) \quad (1)$$

The condensing capacity for G kg was calculated as follows:

$$Q_k = G(h_2 - h_{3'}) \quad (2)$$

The work of adiabatic compression was determined as follows:

$$N = G(h_2 - h_1) \tag{3}$$

The Coefficient of Performance (COP) of the cycle was calculated as follows:

$$COP = \frac{Q_0}{N} \tag{4}$$

The overall heat transfer rate was calculated as follows:

$$A = \frac{Q_k}{U\Delta t_{lm}} \tag{5}$$

where Q_o is cooling capacity (W); G is mass flow rate (kg^{-1}); h is enthalpy of fluid ($kJ \cdot kg^{-1}$); Q_k is condensing capacity (W); N is adiabatic compression power (kW); A is the heat transfer area (m^2); U is the overall heat transfer coefficient ($W \cdot m^{-2} \cdot K^{-1}$); and Δt_{lm} is log mean temperature difference (°C).

Based on the same overall heat transfer coefficient, this study offers two design configurations for the condenser, consisting of Case 1 (five layers of copper tubes) and Case 2 (eight layers of copper tubes), with dimensions and parameters shown in Figure 2 and Table 2. These condensers use mini-size tubes. For Case 2, there are two tube sets (banks) arranged in parallel; each set has four tube layers (or four passes). Each layer has 25 mini tubes with the outer and inner diameters of 4 mm and 3 mm, respectively. The actual image of the condenser with eight layers using mini tubes is shown in Figure 3. The condensing mini tubes are flooded in the cooling water tank.

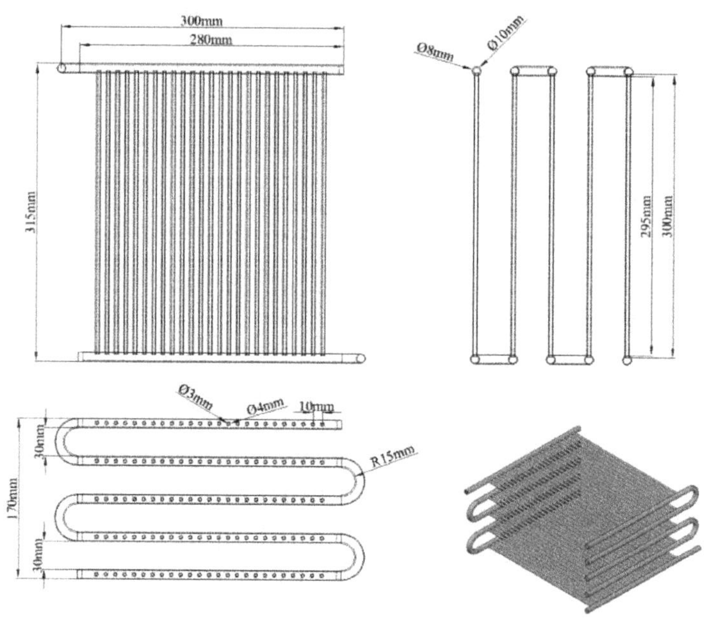

(**a**) Case 1—the condenser with five layers.

Figure 2. *Cont.*

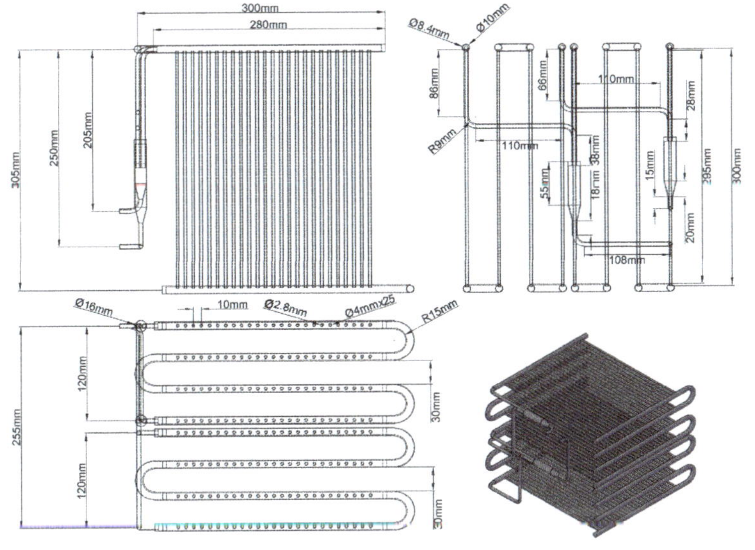

(**b**) Case 2—the condenser with eight layers.

Figure 2. The design parameters of two condensers using mini tubes.

Table 2. Design parameters for two configurations.

Parameters	Case 1 Five Layers	Case 2 Eight Layers
Overall heat transfer rate ($W{\cdot}m^{-2}{\cdot}K^{-1}$)	120.4	
Log mean temperature difference (°C)	17.6	11.07
Heat transfer area (m^2)	0.47	0.75
Reynolds number range	1000–3000	
The number of pipes per layer	25	25
Overall dimension L × W × H (mm)	315 × 300 × 170	305 × 300 × 255
Number of passes	5	4
Number of sets	1	2

Figure 3. The actual photo of the condenser with eight layers using mini tubes.

2.2. Numerical Simulation

This model was solved based on Conjugate Heat Transfer model [17,18], which was formed of two models: (1) Heat Transfer in Solids and Fluids and (2) Turbulent Flow.

For Heat Transfer in Solids and Fluids model, the assuming equations are expressed as follows:

$$\rho C_p u \nabla T + \nabla q = Q + Q_{ted} \tag{6}$$

$$q = -k\nabla T \tag{7}$$

For the Turbulent Flow model, the assuming equations are expressed as follows:

$$\rho(u\nabla)u = \nabla[-pI + K] + F \tag{8}$$

$$\nabla(\rho u) = 0 \tag{9}$$

The transport equation for k reads is as follows:

$$\begin{gathered} \rho(u\nabla)k = \left[\left(\mu + \tfrac{\mu_T}{\sigma_k}\right)\right]\nabla k + P_k - \rho\epsilon \text{ where} \\ K = (\mu + \mu_T)\left(\nabla u + (\nabla u)^T\right) - \tfrac{2}{3}(\mu + \mu_T)(\nabla u)I - \tfrac{2}{3}\rho k I \end{gathered} \tag{10}$$

The transport equation for ϵ is as follows:

$$\rho(u\nabla)\epsilon = \left[\left(\mu + \frac{\mu_T}{\sigma_\epsilon}\right)\right] + C_{\epsilon 1}\frac{\epsilon}{k}P_k - C_{\epsilon 2}\rho\frac{\epsilon^2}{k}, \tag{11}$$

$$\text{Where } \epsilon = \text{ep and } \mu_T = C_\mu \rho \frac{k^2}{\epsilon} \tag{12}$$

The momentum formulation is as follows:

$$P_k = \mu_T\left[\nabla u \div \left(\nabla u + (\nabla u)^T\right) - \frac{2}{3}(\nabla u)^2\right] - \frac{2}{3}\rho k \nabla u \tag{13}$$

The phase change equation is as follows:

$$\rho = \theta_1\rho_1 + \theta_2\rho_2 \tag{14}$$

The specific heat capacity at constant pressure is expressed as follows:

$$C_p = \frac{1}{\rho}\left(\theta_1\rho_1 C_{p,1} + \theta_2\rho_2 C_{p,2}\right) + L_{1-2}\frac{\partial \alpha_m}{\partial T} \tag{15}$$

The mass fraction of two-phase flow is as follows:

$$\alpha_m = \frac{1}{2}\frac{\theta_2\rho_2 - \theta_1\rho_1}{\theta_1\rho_1 + \theta_2\rho_2} \tag{16}$$

The relationship between the vapor quality and the phase indicator is as follows:

$$x = \theta_2 \text{ and } 1 - x = \theta_1 \tag{17}$$

where μ_T is the turbulence viscosity (Pa·s); u is velocity of fluid ($m \cdot s^{-1}$); ρ is density of fluid ($kg \cdot m^{-3}$); k is thermal conductivity ($W \cdot m^{-1} \cdot K^{-1}$); T is the absolute temperature (K), C_p is the specific heat capacity at constant pressure ($J \cdot kg^{-1} \cdot K^{-1}$)), Q contains additional heat sources ($W \cdot m^{-3}$), q is the external heat flux ($W \cdot m^{-2}$); θ is phase indicator fraction; α is thermal diffusion ($W \cdot m^{-2}$); ∇u is the gradient of velocity ($m \cdot s^{-2}$); and x is vapor quality.

The numerical simulation for water-cooled condensers was conducted using COMSOL Multiphysics 6.1 software with the Conjugate Heat Transfer model [17]. This model consists of two models: the Turbulent Flow model and Low-Reynolds k-ε combined with the Heat Transfer in Solids and Fluids model.

To simplify the simulation model for Case 2, there is an even division of rows of pipes between two sets; the R744 entering the condenser is evenly divided. Therefore, the simulation model uses a symmetrical method by reducing the number of elements, saving time, improving convergence, and increasing accuracy.

Complete mesh predefined as normal level consists of 692,399 domain elements, 134,042 boundary elements, and 24,441 edge elements as shown in Figure 4. With the quality of the simulation mesh supporting a faster simulation process, the average Skewness acceptable indexes in the rows of pipes at the condenser are from 0.5–0.7 for the tubes and from 0.75–0.8 for the manifolds, as shown in Figure 5. A grid independence analysis was implemented to verify the simulation, as shown in Table 3. Regardless of mesh quality, the R744 outlet temperature converges at 28.7 °C.

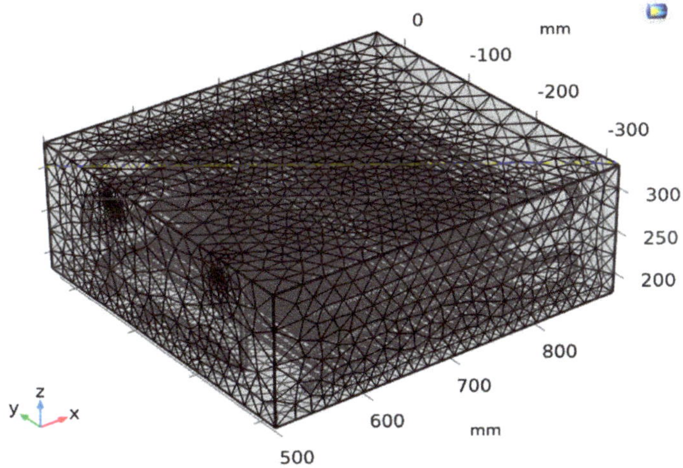

Figure 4. Meshing for the test sample.

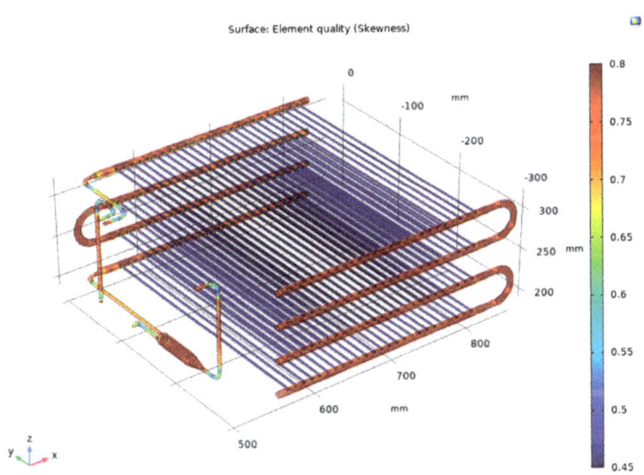

Figure 5. Skewness index for simulation model.

Table 3. Grid independence verification.

Mesh Type Predefined	Grid Elements	Simulation
Coarse	232,519	28.62
Normal	692,399	28.70
Fine	1,184,533	28.71
Finer	7,634,022	28.81

For heat transfer variables, the GMRES (Generalized Minimal Residual Method) iterative solver was used. This method produces an approximate solution to the linear system after a finite number of steps. These methods are useful for large systems of equations where it is reasonable to trade-off precision for a shorter run time. Iterative methods use the coefficient matrix indirectly, through a matrix–vector product or an abstract linear operator. Iterative methods can be used with any matrix, but they are typically applied to large sparse matrices for which direct solves are slow. Furthermore, for turbulence variables, the PARDISO (Parallel Sparse Direct Solver) was used to improve sequential and parallel sparse numerical factorization performance by pipelining parallelism with a combination of left-looking and right-looking supernode techniques. PARDISO is a multithreaded on platform to save the solving time. In each case study, the boundaries conditions include the inlet temperature of R744 (°C), the outlet pressure of R744 (bar), the mass flow rate of R744 (g.s^{-1}), the average water temperature (°C), and velocity of water (m/s).

The computer for solving this model has the following configuration: CPU-Intel Core i7 6820HQ (2.7 GHz Turbo 3.6 GHz, 4 cores, 8 threads), Ram of 16 GB DDR4 with bus 2400 MHz, and VGA-Nvidia Quadro M3000M. The average computation time fluctuates from 30 min to 2 h.

2.3. Experiment Set Up

The diagram for the testing system and the data acquisition establishment are shown in Figure 6. The experimental system is designed with a complete R744 air conditioning cycle (which is cooled by the evaporative cooling method) and operated in Vietnamese climatic conditions. For this experimental system (in Figure 6a), the low-pressure superheated vapor coming from the evaporator enters the compressor to perform the process of isotropy adiabatic compression that converts external work into high-pressure superheated vapor. The superheated vapor continues to enter the condenser, releasing heat to water, performing complete liquefied isobaric condensation. The liquid refrigerant proceeds to the hand throttle valve to become a wet saturated vapor with low pressure (low temperature). Moreover, R744 enters the evaporator-receiving heat from the ambient that performs isobaric evaporation to become the dry saturated vapor. Leaving the evaporator, R744 returns to the compressor to complete a closed cycle. A real photo of the system is shown in Figure 6b.

Furthermore, measuring devices such as flowmeters, temperature, and pressure at the cycle's state points were installed at specified locations to collect the experimental parameters. The types of testing apparatuses and their accuracies are shown in Table 4. The experimental uncertainties were estimated following the method described by Holman [19]. The uncertainty values of several parameters such as the outlet temperature of condenser, the mass flow rate, and the pressure drop are 0.5%, 0.5%, and 0.7%, respectively.

Table 4. Measuring devices and their accuracies.

Devices	Accuracy	Range	Manufacturers
Thermometer	±0.1 °C	−270–400 °C	Extech
Thermocouples–T type	±0.1 °C	0–100 °C	Omega-USA
Clamp meter	±2%FS	0–600 A	Hioki
Humidity meter	±3%FS	1.0–99.9%	Tenmars
Pressure sensor	0.5%FS	0–100 bars	Sensys
Turbine flow rate sensor	0.5%	400 to 5000 kg/s	DGT

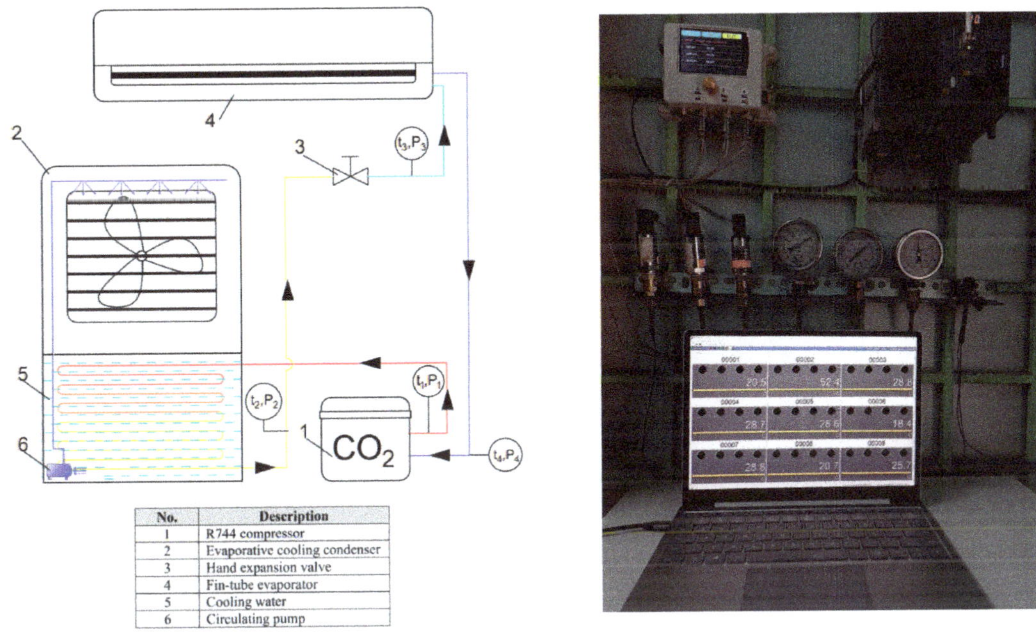

(**a**) The test loop

(**b**) A real photo

Figure 6. The diagram and a real photo of the test loop.

The test loop operated for 20 min in order to have the system reach its steady state. Then, the experimental data (temperature, pressure, flow rate, etc.) were recorded for 10 min. The values for the system as a whole were calculated from the average values of all temperatures being recorded.

3. Results and Discussion

3.1. Temperature Field at R744 Side in Numerical Simulation

The simulation model is solved with the boundary conditions, as shown in Table 5. For example, the inlet temperature of the condenser on the R744 side is 63.3 °C, and the mass flow rate of R744 is 3.28 g/s. These values were verified with the experimental data.

Table 5. Boundary conditions for the numerical simulation.

Input Parameters	Setting Value
Inlet temperature of R744 (°C)	63.3
Outlet pressure of R744 (bar)	72.6
Mass flow rate of R744 ($g \cdot s^{-1}$)	3.28
Average water temperature (°C)	26.4
Velocity of water (m/s)	0.003

Figure 7 shows the temperature field of R744 in the condenser for Case 2. Based on 3D images and temperature scales (Celsius), under current simulation conditions, the system operates under a pressure of 72.6 bar, and the superheated vapor enters the condenser at a temperature of 63.3 °C. It is observed that the temperature distribution through the unit has changed with the inlet R744 temperature of 63.3 °C after exchanging heat with water at the temperature of 26.4 °C (the average velocity of water is 0.003 m/s). The temperature field of the R744 at the inlet decreases gradually through each layer until the end of the condenser with the outlet temperature of 28.7 °C. Furthermore, Figure 8 indicates the phase change of R744, and the vapor phase is indicated as transparent, while the liquid phase is indicated on a blue scale. From the numerical simulation, it is determined that R744 is completely liquefied in the last two layers. This is the peculiarity that numerical simulation can show while the experiment is difficult to observe.

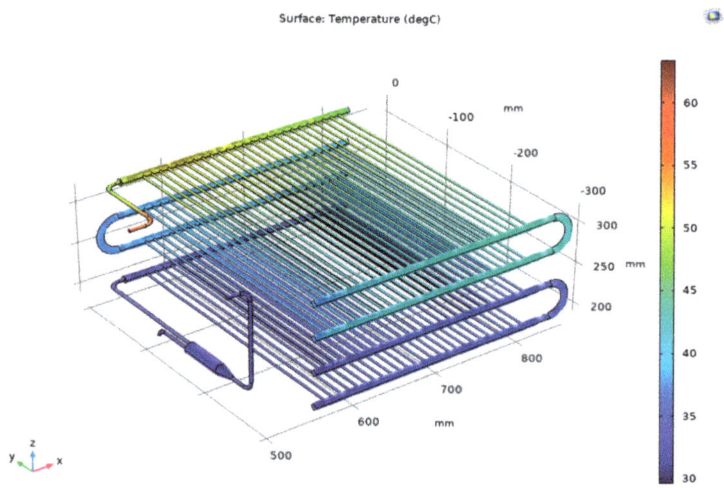

Figure 7. Temperature field at R744 side for Case 2.

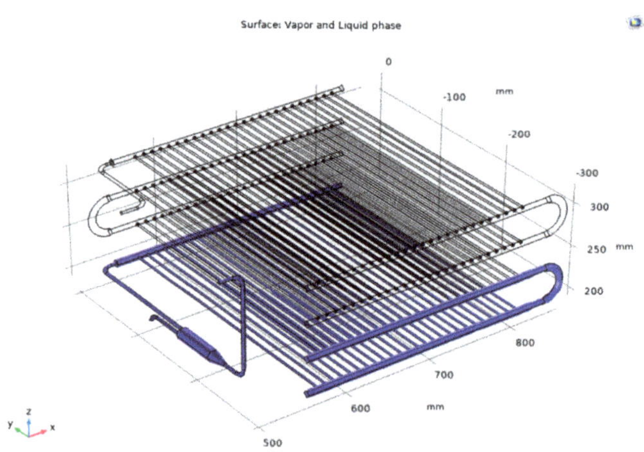

Figure 8. Phase indication at R744 side for Case 2.

3.2. Validation

The outlet R744 temperatures of the condensers from the simulation results were compared experimentally to verify the reliability under the same conditions. Table 6 presents comparative data between simulation and experiment under typical operating conditions for R744 side: the evaporation temperature of 9.2 °C, the mass flow rate of 14.34 kg/h, the inlet pressure of 72.6 bar, and the inlet temperature of 63.3 °C. It is observed that the deviation number is quite small (1.9% with temperature and 0.3% with pressure), which demonstrates the predictive simulation is very good and reliable when compared to the experiment for studying the evaporative condenser.

Table 6. Comparing between experiment and simulation of the condensers for Case 2.

Parameters	Unit	Experiment	Simulation	Error
Outlet R744 temperature	°C	30.3	29.7	1.9%
Outlet R744 pressure	bar	72.6	72.83	0.3%

3.3. The Effect of the Inlet and Outlet R744 Temperatures of Condenser

In Case 2, the evaporation temperature was varied by adjusting the opening of the throttle valve, which gradually increased the evaporation pressure in the range from 43.38 bar to 50 bar as well as the evaporation temperature in the range from 9.2 °C to 16.2 °C. In this case, the mass flow rates of R744 through the condenser are increasing from 14.34 kg/h to 46.08 kg/h. The boundary conditions for the simulation are shown in Table 7.

Table 7. Boundary conditions for the numerical simulation.

Input Parameters	Setting Range
Inlet of R744 (°C)	48.5–63.3
Outlet pressure of R744 (bar)	68.5–72.6
Mass flow rate of R744 ($g \cdot s^{-1}$)	3.28–6.89
Average water temperature (°C)	24.9–26.4
Velocity of water (m/s)	0.003

The data obtained from the experiments and the predictions from the simulation are shown in Figure 9 for Case 2. In the study cases, the outlet R744 temperatures of the condenser gradually decrease from 29.7 °C to 28.03 °C in the simulation, while the outlet temperatures decrease from 30.3 °C to 27.8 °C in the experiment. Here, the similarity is quite high between the simulation and the experiment results, with the maximum error of 1.9%.

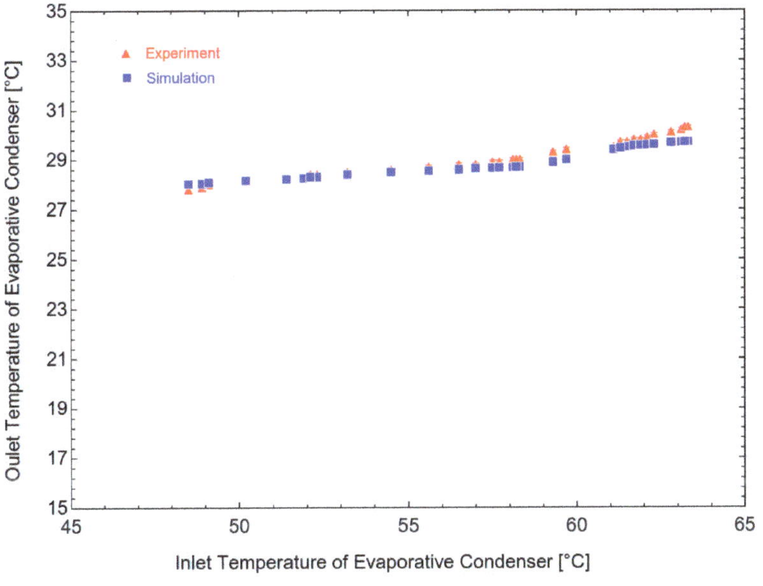

Figure 9. The inlet and outlet R744 temperatures of condenser from simulation and experiment.

3.4. Pressure Distribution at R744 Side in Numerical Simulation

Figure 10 shows the pressure distribution on the R744 side for Case 2. The pressure profile result shows that the pressure value has an uneven decrease. However, the result is quite evenly distributed between the layers, and the pressure drop is uniform for the mini tube inside. When passing through the manifolds, R744 creates vortex zones that increase pressure loss at the manifolds. The pressure at the inlet of the unit is 72.83 bar and falls at the outlet with a value of 72.6 bar. The above results show that the pressure drop between the inlet and outlet of the unit is 0.23 bar.

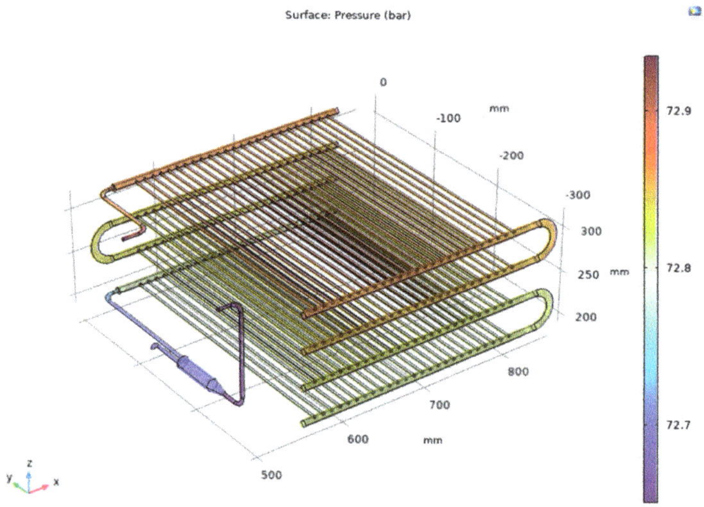

Figure 10. Pressure distribution on the R744 side.

3.5. The Effect of the Mass Flow Rate on the Pressure Drop

Figure 11 shows the relationship between the inlet pressure of the condenser and the mass flow rate of R744, both numerically and experimentally. The study included 30 steady states to figure out the pressure drop when R744 passes through the condenser while the outlet pressure is kept constant (green color point), and the inlet pressure is collected and verified with the experimental data. It is observed that the mass flow rates increase from 14.34 to 21.6 kg/h, the pressure drops are relatively stable at 0.4 bar (the inlet of 73 bar and the outlet of 72.6 bar) in the experiment; the pressure drops increase negligibly from 0.23 bar to 0.25 bar in the simulation. However, when increasing the mass flow rate from 25.08 to 46.08 kg/h, the pressure drop from the experiment is relatively stable at 0.5 bar while it changes from 0.3 bar to 0.47 bar in the simulation. This phenomenon is caused by the fact that the flowmeter in the experiment exhibits a high error at low flow values. When the flow increases, the accuracy of the measuring device is improved. To sum up, the numerical simulation can predict the pressure drop with greater accuracy at high values of the mass flow rate.

Figure 12 shows a comparison of the pressure drop between Case 1 (5 layers) and Case 2 (8 layers) in the experiment. The mass flow rates increase from 14.9 to 44.9 kg/h, and the pressure drops increase from 0.86 bar to 1.07 bar for Case 1; the pressure drops increase from 0.4 bar to 0.5 bar for Case 2. The pressure drops of Case 1 are much higher than those obtained from Case 2. This phenomenon is caused by the fact that the flowing path of R744 being longer with more passes than in the case of eight layers, which are divided into two sets (only four passes). The local pressure drop and the friction pressure drop generated are higher, leading to an increase in the overall pressure drop.

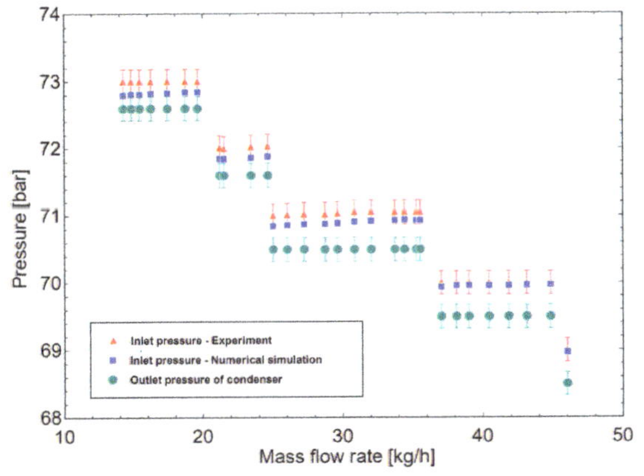

Figure 11. A comparison between experiment and numerical simulation on the inlet pressure with an increase in R744 mass flow rate.

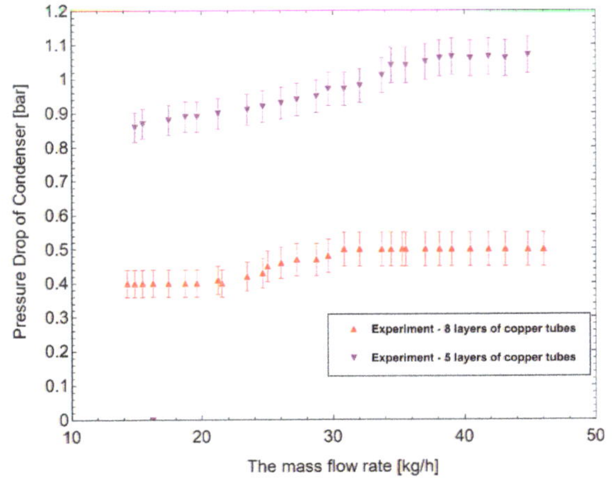

Figure 12. A pressure drop comparison between two cases with increasing of R744 mass flow rate.

3.6. The Comparison of the Optimal Design between Two Case Studies

In a typical case, the thermodynamic parameters when operating at the same cooling capacity of 960 W are given in Table 8 for comparing the two case studies.

Table 8. A comparison of thermodynamic parameters in two case studies.

Parameters	Case 1	Case 2
Log mean temperature difference (°C)	8.56	14.7
Pressure drop (bar)	1.07	0.4
Coefficienct of Performance	2.72	4.34
Adiabatic compression energy (W)	353	214

Based on the data obtained for Case 1 (five layers), it is observed that the pressure drop is 1.07 bar, which is quite high compared to the Case 2 (eight layers) pressure drop with a value of 0.4 bar; so, this causes an increase in the adiabatic compression. At the same time, the log mean temperature difference is 8.56 °C for Case 1 compared to 14.7 °C for Case 2, demonstrating the heat transfer efficiency. In addition, the efficiency of the system is expressed with a COP value of 4.34 in Case 2, and it is quite high when compared with

Case 1 (only 2.72). Thus, the configuration of eight layers divided into two sets is more optimal in the distribution of R744 to ensure a low-pressure drop, increasing the efficiency factor of the system.

From the results above, the evaporative condensers using mini tubes that are flooded in the cooling water tank are suitable for the subcritical R744 air conditioning system. This method could decrease the inlet temperature of the throttle valve and increase system efficiency.

4. Conclusions

Two evaporative condensers using mini tubes were simulated with the numerical method using the COMSOL Multiphysics 6.1 software. The numerical results were verified with the experimental method. The study has achieved the following results.

The temperature field and the phase change were simulated under specific operating conditions; the outlet R744 temperature of the condenser changed from 28.7 °C to 30.3 °C, combined with the operating pressure that changed from 72.6 bar to 68.5 bar. The numerical results showed the completely liquefied state of R744.

Within the study limits, the mass flow rate increases from 14.34 kg/h to 46.08 kg/h, and the pressure drop also increases from 0.23 bar to 0.47 bar for the simulation and from 0.4 bar to 0.5 bar for the experiment, respectively. A turbulent flow increases the local pressure drop at the manifold, which in turn affects the pressure drop throughout the evaporative condenser.

For the design of condensers, it is essential to arrange the number of tube layers so as to achieve higher heat transfer efficiency while reducing the pressure drop. The pressure drop of the five-layer tube configuration is higher than that obtained from the eight-layer one (splitting into two sets).

The evaporative condensers using mini tubes that are flooded in the cooling water tank are suitable for the subcritical R744 air conditioning system. The method could decrease the inlet temperature of the throttle valve and increase system efficiency.

The numerical results are in good agreement with those obtained from the experimental results, with the maximum percentage of errors being less than 5%. These results will make additional contributions to numerical simulation studies of evaporative condensers, especially with CO_2.

In addition, further research will focus on exploring tubes with smaller diameters, such as microscale, tube numbers, water velocity, etc., in harsher climatic conditions using the numerical and experimental methods.

Author Contributions: Investigation, H.N.; Writing—original draft, T.D. All authors have read and agreed to the published version of the manuscript.

Funding: This work was supported by project no. T2022-107 (sponsored by the HCMC University of Technology and Education, Vietnam) and is deeply appreciated.

Conflicts of Interest: The authors declare no conflict of interest.

Nomenclature

T	the absolute temperature (K)	Q_0	cooling capacity (W)
u	the fluid velocity (m/s)	Q_k	condensing capacity (W)
p	pressure (Pa)	h	enthalpy of fluid (kJ/kg)
μ_T	the turbulence viscosity (Pa·s)	N	work of adiabatic compression (W)
F	external forces (N/m^3)	G	mass flow rate of fluid (kg/s)
ρ	the density of fluid (kg/m^3),	A	heat transfer area (m^2)
C_P	the specific heat (J/(kg·K))	U	overall heat transfer area (m^2)
Q	additional heat sources (W/m^3)	Δt_{lm}	log mean temperature difference (°C)
q	external heat flux (W/m^2)	∇u	the gradient of velocity ($m.s^{-2}$)
k_T	thermal conductivity (W/m·K)		

References

1. Zhang, X.; Ge, Y.; Sun, J.; Li, L.; Tassou, S.A. CFD Modelling of Finned-tube CO_2 Gas Cooler for Refrigeration Systems. *Energy Procedia* **2019**, *161*, 275–282. [CrossRef]
2. Sánchez, D.; Catalán-Gil, J.; Cabello, R.; Calleja-Anta, D.; Llopis, R.; Nebot-Andrés, L. Experimental Analysis and Optimization of an R744 Transcritical Cycle Working with a Mechanical Subcooling System. *Energies* **2020**, *13*, 3204. [CrossRef]
3. Yang, J.; Wang, L.; Han, Y.; Zhang, X.; Du, Y. Simulation and Experimental Study of CO_2 Transcritical Heat Pump System with Thermoelectric Subcooling. *Designs* **2022**, *6*, 115. [CrossRef]
4. Zhu, Y.; Huang, Y.; Li, C.; Zhang, F.; Jiang, P.X. Experimental investigation on the performance of transcritical CO_2 ejector–expansion heat pump water heater system. *Energy Convers. Manag.* **2018**, *167*, 147–155. [CrossRef]
5. Deng, J.Q.; Jiang, P.X.; Lu, T.; Lu, W. Particular characteristics of transcritical CO_2 refrigeration cycle with an ejector. *Appl. Therm. Eng.* **2007**, *27*, 381–388. [CrossRef]
6. Zhang, F.Z.; Jiang, P.X.; Lin, Y.S.; Zhang, Y.W. Efficiencies of subcritical and transcritical CO_2 inverse cycles with and without an internal heat exchanger. *Appl. Therm. Eng.* **2011**, *31*, 432–438. [CrossRef]
7. Zhang, X.P.; Wang, F.; Fan, X.W.; Wei, X.L.; Wang, F.K. Determination of the optimum heat rejection pressure in transcritical cycles working with R744/R290 mixture. *Appl. Therm. Eng.* **2013**, *54*, 176–184. [CrossRef]
8. Gullo, P.; Hafner, A.; Banasiak, K. Transcritical R744 refrigeration systems for supermarket applications: Current status and future perspectives. *Int. J. Refrig.* **2018**, *93*, 269–310. [CrossRef]
9. Hazarika, M.M.; Ramgopal, M.; Bhattacharyya, S. Studies on a transcritical R744 based summer air-conditioning unit: Impact of refrigerant charge on system performance. *Int. J. Refrig.* **2018**, *89*, 22–39. [CrossRef]
10. Khedher, N.B.; Mukhtar, A.; Md Yasir, A.S.H.; Khalilpoor, N.; Foong, L.K.; Nguyen Le, B.; Yildizhan, H. Approximating heat loss in smart buildings through large scale experimental and computational intelligence solutions. *Eng. Appl. Comput. Fluid Mech.* **2023**, *17*, 2226725. [CrossRef]
11. Sanz-Kock, C.; Llopis, R.; Sánchez, D.; Cabello, R.; Torrella, E. Experimental evaluation of a R134a/CO_2 cascade refrigeration plant. *Appl. Therm. Eng.* **2014**, *73*, 41–50. [CrossRef]
12. Zhang, Y.; He, Y.; Wang, Y.; Wu, X.; Jia, M.; Gong, Y. Experimental investigation of the performance of an R1270/CO_2 cascade refrigerant system. *Int. J. Refrig.* **2020**, *114*, 175–180. [CrossRef]
13. Llopis, R.; Sanz-Kock, C.; Cabello, R.; Sánchez, D.; Torrella, E. Experimental evaluation of an internal heat exchanger in a CO_2 subcritical refrigeration cycle with gas-cooler. *Appl. Therm. Eng.* **2015**, *80*, 31–41. [CrossRef]
14. Di Nicola, G.; Polonara, F.; Stryjek, R.; Arteconi, A. Performance of cascade cycles working with blends of CO_2 + natural refrigerants. *Int. J. Refrig.* **2011**, *34*, 1436–1445. [CrossRef]
15. Dopazo, J.A.; Fernandez-Seara, J. Experimental evaluation of a cascade refrigeration system prototype with CO_2 and NH_3 for freezing process applications. *Int. J. Refrig.* **2011**, *34*, 257–267. [CrossRef]
16. Lee, T.S.; Liu, C.H.; Chen, T.W. Thermodynamic analysis of optimal condensing temperature of cascade-condenser in CO_2/NH_3 cascade refrigeration systems. *Int. J. Refrig.* **2006**, *29*, 1100–1108. [CrossRef]
17. COMSOL Multiphysics Sofware. *Heat Transfer Module, Version 6.1*; COMSOL Multiphysics Sofware: Stockholm, Sweden, 2022.
18. .Nguyen, T.; Dang, T. The Effect of Fin Shape on the Heat Transfer and the Solution Time of a Microchannel Evaporator in a CO_2 Air Conditioning System—A Numerical Investigation. *Micromachines* **2022**, *13*, 1648. [CrossRef]
19. Holman, J.P. *Experimental Methods for Engineers*; McGraw-Hill: New York, NY, USA, 1984.

micromachines

MDPI

Article

Pseudo-Desublimation of AdBlue Microdroplets through Selective Catalytic Reduction System Microchannels and Surfaces

Claudiu Marian Picus, Ioan Mihai * and Cornel Suciu *

Faculty of Mechanical Engineering, Automotive and Robotics, Stefan cel Mare University, 720229 Suceava, Romania; claudiu.picus@usm.ro

* Correspondence: mihai.i@usm.ro (I.M.); suciu@usm.ro (C.S.); Tel.: +40-735173288 (I.M.); +40-742007820 (C.S.)

Abstract: In the present paper, the occurrence and development of the pseudo-desublimation process of AdBlue microdroplets in the microchannels and surfaces of catalytic reduction systems (SCR) are reported. In order to understand how the pseudo-desublimation process develops, the influence of heat flux values on the heat transfer of AdBlue injection was analysed, taking into account the structure of the microchannels inside the SCR and the overall configuration of the installation. The evolution of the AdBlue vapour flow in the SCR system was simulated, as well as the temperature variation along an SCR microchannel through which the mixture flows. An experimental set-up was designed in order to visualise and interpret the processes at the onset of pseudo-desublimation. The results described in this paper confirm the existence of a pseudo-desublimation process that occurs only under certain temperature conditions when AdBlue is injected into SCR systems. The characteristics of the crystals formed and their growth rate depend on the working temperature, which could be controlled by efficient preheating methods immediately after engine start. A better understanding of the process will allow the development of methods of avoiding solid depositions on SCR system components, which has a direct impact on SCR catalyst performance and durability.

Keywords: microchannels; pseudo-desublimation; SCR; AdBlue; NOx

1. Introduction

The Euro 7 standards for cars, proposed to come into effect in 2025, include a number of stricter emission regulations. Although the new values of NOx (nitrogen oxides) emission limits are not much lower than in the Euro 6d standard, the cold emissions test will mean big increases in costs, mainly for diesel engines. At present, coding to Euro 6d regulations, petrol engines have a limit of 60 mg/km for NOx emissions, while diesel engines have a limit of 80 mg/km. With the Euro 7 rules, the NOx emission limit will be 60 mg/km for all types of internal combustion engines. There is a one-off derogation from 60 mg/km to 75 mg/km NOx for small commercial vehicles up to 3.5 tonnes with a specific power output of up to 35 kW (48 hp). Emissions will be measured using the same method as in Euro 6d, called RDE (Real Driving Emissions), which has been found to be viable and conclusive. The European Commission's communication does not explicitly mention the introduction of a short-distance cold emissions test for small commercial vehicles with a whole-trip specific power of 10 km. Basically, the kilometre limit will be identical to the normal long-distance test, but will be very difficult to achieve [1,2]. Other changes brought about by the Euro 7 rules include extending the period of compliance of vehicles with emission standards from 100,000 km and 5 years, as was the case under Euro 6d rules, to 200,000 km and 10 years, as the average age of cars in Europe approaches 12 years [1,2].

The amount of pollutants according to European requirements for engines up to 35 Kw diesel and petrol are shown in Table 1, comparatively for the Euro 6d and Euro 7 standards, where PM represents the accepted particulates or atmospheric particulate matter and PN is the particle number (symbol N), which is the number of constituent particles in the analysed thermodynamic system. The carbon monoxide (CO) emissions are currently limited to 1000 mg/km for diesel and 500 mg/km for petrol engines, while the Euro 7 standard will limit CO emissions to 500 mg/km for both types of engines. The total hydrocarbon (THC) emissions are also limited at 100 mg/km. Non-methane hydrocarbons (NMHCs) are also limited for all engines at 69 mg/km. To meet the European requirements shown in Table 1, a selective catalytic reduction (SCR) system is implemented on the exhaust manifold. Selective catalytic reduction (SCR) is a widely used technology to reduce NOx (nitrogen oxide) emissions from engines in both light- and heavy-duty vehicles. It involves the use of a catalyst, typically made of various ceramic materials, with an active catalytic component such as vanadium. In the SCR process, a urea solution is injected upstream of the catalyst.

Table 1. The amount of pollutant according to European requirements for engines up to 35 Kw diesel and petrol.

Pollutant	Unit	Euro 7	Euro 6d
NO_x	mg/km	60	60/80
PM	mg/km	4.5	4.5/4.5
PN	-	6×10^{11}	$6 \times 10^{11}/6 \times 10^{11}$
CO	mg/km	500	1000/500
THC	mg/km	100	100/-
NMHC	mg/km	68	68/-

In the European Union, this solution is known as AdBlue, while in the United States it is called Diesel Exhaust Fluid (DEF). The urea solution plays a crucial role in initiating the catalytic reaction within the catalyst. For catalytic conversion to be effective, the water content of the AdBlue solution must be evaporated and the NH_3 (ammonia) molecules must be mixed with the exhaust gas. It is important to achieve an even distribution of NH_3 on the catalyst surface to maximise the efficiency of catalytic conversion. If there is a high local concentration of ammonia, an excessive ammonia emission may result after the process taking place in the SCR. On the other hand, a low local concentration of ammonia may lead to poor NOx conversion. In the Euro 6 emission standard, the permissible limit for excess ammonia in the catalyst has been regulated to ten particles per million (ppm) [3]. This stricter limit has increased the complexity of dosing and mixing urea. To achieve both a high conversion efficiency and lower ammonia emissions, a more advanced and controlled mixing process is required. Urea dosing and mixing technologies aim to ensure an efficient distribution of NH_3 molecules on the catalyst surface, minimising excess ammonia and achieving optimal catalytic conversion of NOx emissions. These developments contribute to reducing the environmental impact of vehicles and improving air quality by reducing harmful NOx emissions [4,5].

The present study reports how the evaporation and vaporisation of AdBlue droplets in the microchannels of an SCR system takes place in a pseudo-desublimation process. The pseudo-desublimation of AdBlue microdroplets through microchannels of selective catalytic reduction systems is a process that starts by converting the liquid AdBlue solution into microdroplets by injection. Once atomised, due to the temperature in the SCR system, these microdroplets are transformed into aerosols which enter the SCR microchannels as vapour. In the microchannel, depending on temperature, the AdBlue vapour undergoes a desublimation process, which consists of a quasi-instantaneous transformation of the microdroplets into vapour or, at high temperatures, directly into solids. This process occurs due to the small size of the microchannels, their various shapes, as shown in Figure 1, and the velocity of the fluid flowing through it [6–8].

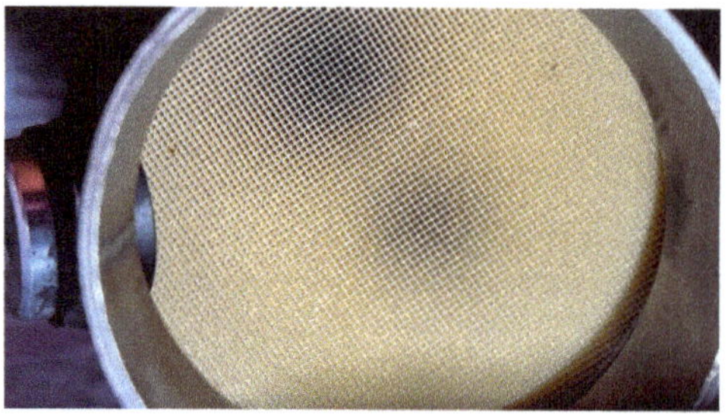

Figure 1. Example of microchannels of a SCR system.

The role of AdBlue injection is to disperse the solution in micrometre-sized aerosols in order to achieve a more uniform distribution in the exhaust gas, which contributes to improving the efficiency of the catalytic reduction process [9].

Various studies, mostly theoretical, regarding micro- and nanochannel shapes and their influence on ion transfer phenomena can be found in literature [10–12]. The selective catalytic reduction of NOx with ammonia was modelled in previous studies found in the literature [13]. As can be seen from Figure 1, the shape of the microchannels used in the investigated SCR systems is that of a honeycomb, but in the practice of making SCR systems other shapes of microchannels can be used, such as cylindrical, triangular, square, rectangular or complex shapes. Previous theoretical models [13] showed that a reduction in the number of microchannel cross-section sides leads to an increase in the reaction rate due to an increase in the area to volume ratio. However, the results indicated that the rate of reaction is strongly affected by diffusion at the corners of noncircular channels.

Microchannels are used in SCR practice in order to increase the contact area between the exhaust gas of diesel engines and the catalyst surfaces. SCR systems are designed to create an environment conducive to chemical reactions that contribute to the reduction in pollutant emissions, with nitrogen oxides (NOx) being targeted. At the same time, microchannels can also be used to distribute the AdBlue solution evenly in the exhaust gas, thus contributing to the efficient distribution of the reactant throughout the SCR system [14–17]. Inside these microchannels, the occurrence of a pseudo-desublimation process is undesirable, as it results in solid deposits that ultimately clog the system (see Figure 2).

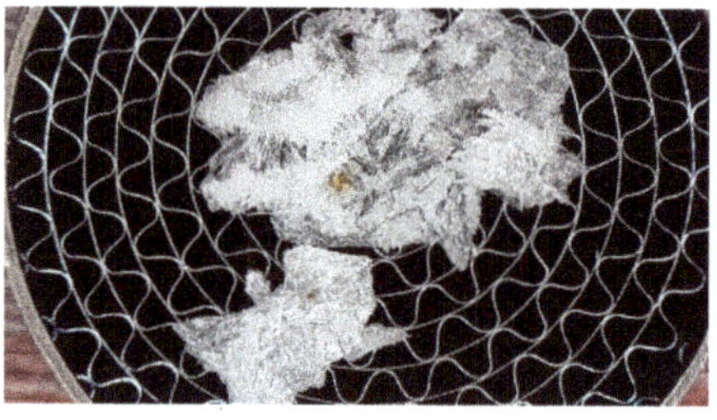

Figure 2. Pseudo-desublimation effect in SCR systems.

2. Heat Exchange in Microchannels and Other SCR Components

In a first step, the conditions under which the desublimation process occurs as a function of the change in exhaust gas temperature and temperature evolution in certain SCR components were investigated. Literature such as [11,13,18] investigated different geometries and heat transfer aspects in such systems. In the study of heat transfer of an SCR catalyst, a circular microchannel geometry was adopted. In the case of an SCR catalyst, heat transfer takes place between the walls of the microchannels and the AdBlue vapours that are found diffused in the exhaust gas through several mechanisms [18], as shown in Figure 3.

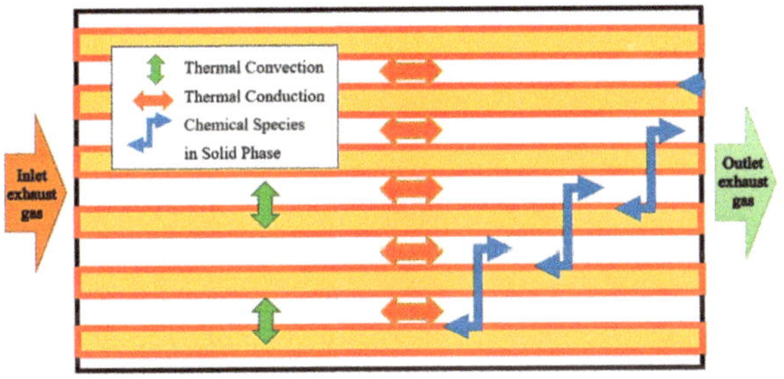

Figure 3. Thermal phenomena occurring in microchannels with circular geometry.

The abovementioned mechanisms include the following:

- Convection from/to the AdBlue gas–vapour mixture flowing between the microchannel walls: heat is transferred by convection between the mixture flowing inside the channels and the microchannel walls. Typically, heated exhaust gases transfer heat to the walls, and cooled exhaust gases take heat from the walls in the case of the pseudo-desublimation of AdBlue.
- Radiation from/to the AdBlue gas–vapour mixture flowing inside the channels: thermal radiation can also contribute to heat transfer between the monolith walls and the gas flowing inside the channels. This means that thermal energy is transferred by radiation between the catalyst surface and the AdBlue gas–vapour mixture from the exhaust.
- Convection from/to the environment: heat can be transferred by convection between the monolith walls and the environment. For example, in a vehicle-mounted SCR catalyst system, the surrounding environment (air) can pick up heat from the catalyst by convection.
- Radiation from/to the environment: radiative heat transfer between the monolith walls and the environment. This means that heat energy is transferred by radiation between the surface of the catalyst and the surrounding environment.

The abovementioned heat transfer mechanisms are important in evaluating and optimising the performance of SCR catalysts, as the distribution and efficiency of heat transfer can influence chemical reactions and the performance of the SCR system in reducing NOx emissions.

To study the processes in SCR systems, the following variables were considered for the analysis of pseudo-desublimation of AdBlue droplets in a cylindrical microchannel: microchannel diameter D (in m), microchannel length L (in m), AdBlue solution velocity u (in m/s), exhaust gas velocity v (in m/s) and inlet AdBlue solution concentration C_{in}.

By applying the mass and momentum conservation equations, partial differential equations describing the pseudo-desublimation process in the cylindrical microchannel are obtained. These equations were adopted according to the models developed by [19–21] and can be expressed as follows:

$$\frac{\partial C}{\partial t} + u\frac{\partial C}{\partial x} = 0. \tag{1}$$

Equation (1) represents the mass conservation law in a microchannel, where C is the concentration of the AdBlue solution, t is the time, x is the axial position in the microchannel and u is the velocity of the AdBlue solution. The equation indicates that the concentration variation as a function of time and position is compensated by the mass transport given by the velocity of the AdBlue solution. The momentum conservation equation in a microchannel is given by

$$\rho\frac{\partial v}{\partial t} + \rho \cdot u\frac{\partial v}{\partial x} = -\frac{\partial p}{\partial x}, \tag{2}$$

where v is the exhaust gas velocity, p is the pressure and ρ is the density. The equation indicates that the variation in momentum as a function of time and position is compensated by the momentum transfer given by the pressure difference in the axial direction and the fluid density [22].

To simplify the system of equations, we will appreciate the following assumptions that the exhaust gas velocity is much higher than the velocity of the AdBlue solution, so we can neglect the terms containing u in the momentum conservation equation.

With this assumption, the system of equations can be reduced to:

$$\frac{\partial C}{\partial t} + u\frac{\partial C}{\partial x} = 0; \frac{\partial p}{\partial x} = -\rho\frac{v^2}{D}. \tag{3}$$

The solution of this system of equations depends on the initial and boundary conditions. To determine the concentration distribution of the AdBlue solution in the microchannel, the initial concentration C_{in} and the boundary conditions (e.g., the concentration at the microchannel outlet) must be known. The convection–diffusion heat transfer equation or convection–conduction equation is as follows:

$$\frac{\partial T}{\partial t} + v\frac{\partial T}{\partial x} = \alpha\frac{\partial^2 T}{\partial x^2}. \tag{4}$$

Equation (4) describes the temperature variation T as a function of time t and axial position x, given the fluid velocity v and thermal diffusion coefficient α [23].

The calculation of the convective transfer coefficient of the exhaust gas h_i upon impact with the SCR catalyst [24] can be performed with the following equation:

$$h_i = \alpha_p \cdot d \cdot L\left(\frac{\mu}{Pr}\right)^{0.33} \tag{5}$$

In Equation (5), α_p is a proportionality or correction factor that takes into account the heat transfer at the catalyst surface, d is the catalyst honeycomb diameter, L is the catalyst length, μ is the dynamic viscosity of the exhaust gas and Pr represents the Prandtl number of the exhaust gas [22,23,25].

$$Pr = \mu\frac{C_p}{k}. \tag{6}$$

In Equation (6), C_p represents the specific heat capacity of the exhaust gas, k represents the thermal conductivity of the catalyst and μ is the dynamic viscosity of the exhaust gas.

The proportionality factor α_p will be

$$\alpha_p = \frac{h_{gas} \cdot k}{h_{gas} + k}. \tag{7}$$

In Equation (7), h_{gas} is the heat transfer coefficient of the gas at the catalyst surface and k represents the thermal conductivity of the catalyst [26,27].

$$\frac{1}{U} = \frac{1}{h_{gas}} + \frac{\Delta x}{k} + \frac{1}{h_{cat}}, \tag{8}$$

The notations in the overall heat transfer coefficient, U, Equation (8), have the following meanings: h_{gas} is the exhaust gas heat transfer coefficient, h_{cat} is the catalyst heat transfer coefficient and Δx is the catalyst thickness.

In order to determine the heat transfer coefficient between the combustion gas and the catalyst, we can use the general formula [28]

$$dq_{conv} = h \cdot dA_s\left(T_{gas} - T_{cat}\right), \tag{9}$$

where dq_{conv} is the convective heat flux (in W/m^2), h is the heat transfer coefficient (in W/m^2K), A_s is the heat transfer area (in m^2), T_{gas} is the exhaust gas temperature (in K) and T_{cat} is the catalyst temperature (in K) [29]:

The SCR parameters taken into account in the calculations are shown in Table 2.

Table 2. SCR parameters.

Parameter	Value
Catalyst honeycomb total size	100 mm × 50 mm × 200 mm
Catalyst honeycomb microchannel diameter	0.8 mm
Catalyst honeycomb microchannel length	200 mm
Composition of exhaust gases	NO_x = 50 ppm, NH_3 = 100 ppm, CO_2 = 10%
Exhaust gas temperature	400 °C
Ambient temperature	25 °C
Exhaust gas pressure	1.32×10^5 Pa
Exhaust gas velocity	10 m/s
Catalyst thermal conductivity	0.5 W/mK
Heat capacity of catalyst	900 J/kgK
Catalyst density	800 kg/m^3
Initial catalyst temperature	200 °C

By replacing the above-described equations in the energy balance, the following equality is obtained:

$$\frac{d}{dx}\left(A_s\frac{dT}{dx}\right) - \frac{h}{k}\frac{dA}{dx}\left(T_{gas} - T_{cat}\right) = 0 \quad (10)$$

The global heat transfer rate q_g is given by the relation

$$q_g = mC_{pcat}\left(T_{gas} - T_{cat}\right), \quad (11)$$

where C_{pcat} represents the thermal capacity of the catalyst.

The average temperature over the entire length of the microchannel at different temperatures and exhaust gas mass m is determined with the relation

$$T_b = T_{cat} - \left(T_{cat} - T_{gas}\right)\exp\left(\frac{-h\pi DL}{mC_{pcat}}\right), \quad (12)$$

where D represents the catalyst microchannel diameter and L is the total considered length.

The temperature variation various gas temperature values (marked in the graph legend) determined by aid of Equation (11) are graphically represented for various temperature levels, as shown in Figure 4.

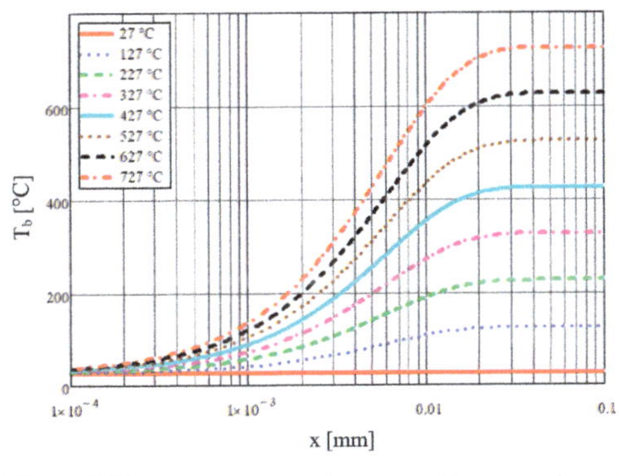

Figure 4. Temperature variations over the microchannel length for various gas temperature values (marked in the graph legend).

The global thermal phenomena for the heat transfer to catalysts with circular microchannels are governed in the case of convection, radiation and conduction by the following equations [23,30].

Heat loss (q) from the SCR catalyst occurs by convection to the ambient temperature (q_{conv}) (Equation (13)) and by radiation exchange with its walls (q_{rad}) (Equation (14)).

$$q_{conv} = h_{gas}\pi DL(T_{cat} - T_{amb}) \tag{13}$$

$$q_{rad} = \varepsilon\pi DL\sigma\left(T_{cat}^4 - T_{amb}^4\right), \tag{14}$$

Therefore, as the total heat loss is $q = q_{conv} + q_{rad}$ and with $A = \pi D$, where A represents the surface area of the pipe, it results that

$$q = h_{gas}AL(T_{cat} - T_{amb}) + \varepsilon AL\sigma\left(T_{cat}^4 - T_{amb}^4\right), \tag{15}$$

where ε is the emissivity of the SCR catalyst and T_{amb} is the ambient temperature around the SCR catalyst.

Figure 5 shows the variation in the heat transfer, depending on the temperature change of the AdBlue gas–vapour mixture, considering the different heat transfer modes inside the SCR.

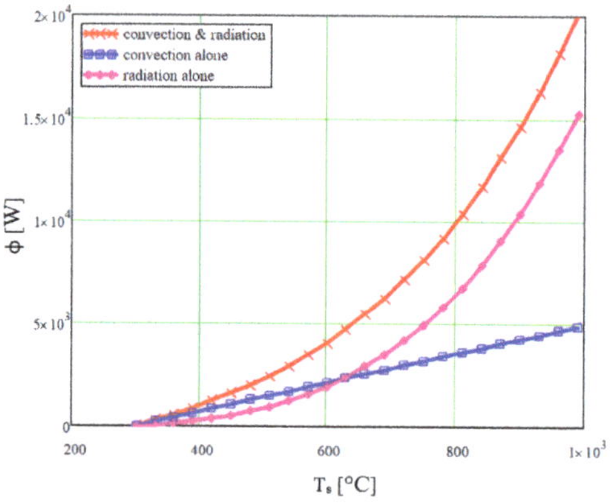

Figure 5. Variation in heat transfer with exhaust gas temperature for convection, radiation and both in a SCR.

We assume that pseudo-desublimation takes place in a cylindrical microchannel of diameter D and length L, through which AdBlue droplets flow at a constant velocity u. After heating, the AdBlue droplets are dispersed as aerosols in the exhaust gas flowing through the microchannel at a constant velocity v in the axial direction of the microchannel. The Nusselt number (Nu) as a function of the friction factor (f), the Reynolds number (Re) and the Prandtl number (Pr) are used to calculate the convective heat exchange in microchannels, taking into account the roughness of the microchannel. If the Nusselt number allows the appreciation of the convective phenomenon, the friction factor takes into account the pressure losses caused by friction in the fluid flow [21,31–34]:

$$Nu(f) = \frac{\left(\frac{f}{8}\right) \times (Re - 100) \times Pr}{1 + 12.7 \times \left(\frac{f}{8}\right)^{\frac{1}{2}} \times \left(Pr^{\frac{2}{3}} - 1\right)}, \tag{16}$$

Using a calculation code developed under Mathcad, the variation in the Nusselt number with the friction coefficient f was calculated for the flow of the AdBlue exhaust gas–vapour mixture through the SCR microchannels, as shown in Figure 6.

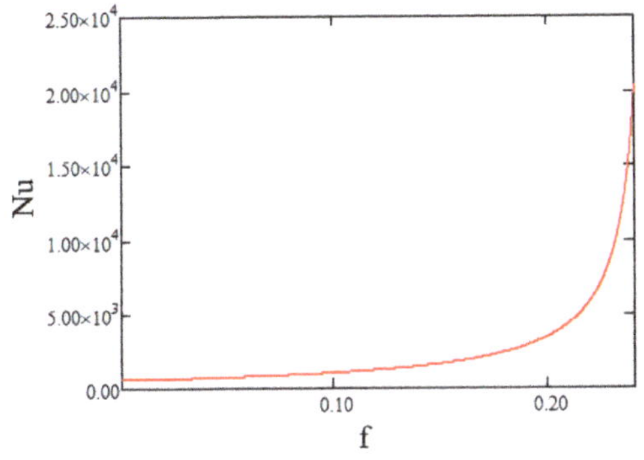

Figure 6. Variation of Nusselt number as a function of friction coefficient.

By analysing the obtained results, it can be seen that the Nusselt number reaches a maximum value of 2×10^4 for the maximum value of the friction coefficient.

Figure 7 shows the results of the temperature variation in the catalyst microchannels in the transient regime considering that they have a diameter of 0.8 mm and a length of 200 mm. In order to accurately study the temperature variation, the x-distance of the mixture flowing through the microchannels was also taken into account.

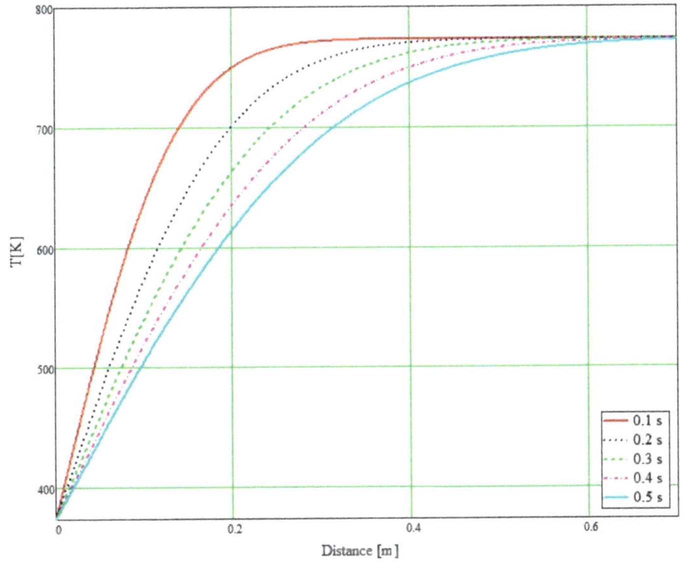

Figure 7. Transient temperature variation along the SCR.

For the transient calculation of the temperature variation along the SCR catalyst traversed by a high-temperature mixture, the relation given in Equation (17) was used.

$$T(x,t) = T_{in} + (T_{out} - T_{in})\,\mathrm{erf}\left(\frac{x}{2\mathrm{sqrt}(\alpha t)}\right). \tag{17}$$

In Equation (17), the parameters T_{in} and T_{out} represent the temperatures (in K) at the inlet and outlet of the catalyst, respectively. The thermal diffusivity coefficient is α (in m^2/s), and the "erf" function is the Gaussian error function.

Figure 7 shows the results of the calculations obtained by implementing Equation (17) in Mathcad V14.0. The abscissa of the graph represents the distance travelled by the mixture of hot exhaust gas and AdBlue vapour. The multiple plots in Figure 7 represent the temperature evolution along the travelled distance, with increasing of time. Time was considered between 0 and 0.5 s in 0.1 s increments, showing that the SCR temperature increases with the distance travelled but also depends on time.

In a previous study [35], the temperature variation in the wall where pseudo-desublimation occurs was analysed as a function of time and microchannel wall thickness, given the concentration of AdBlue when the rate of heat flow passes through it.

Similar research on the vaporisation of liquid droplets in a high-temperature gaseous environment revealed that the variation in wall temperature is directly dependent on the wall thickness and AdBlue concentration. It was also observed that the temperature increased with time, but the rate of this increase was influenced by the microchannel wall thickness and AdBlue concentration.

In conclusion, the study conducted by Mihai et al. [35] highlighted the importance of controlling variables such as wall thickness and heat flow rate in order to optimise the pseudo-desublimation process and obtain the most accurate results. To determine the microchannel wall temperature T_p, dependent on time (t) and a specific depth or position z_1 and taking into account the space travelled by the mixture, the following relation was used:

$$\mathrm{Tp}(t, z_1) = \dot{q}\frac{\delta_T(t)}{2k}\left(1 - \frac{z_1}{\delta_T(t)}\right)^2, \tag{18}$$

where $\dot{q}$ represents the density of thermal flow (internal source of generating heat), $\delta_T(t)$ is the distance travelled along the z_1 direction, through the microchannel, k is the specific constant of the chemical reaction, dependent on temperature, as shown in Equation (19), and C represents the AdBlue vapour concentration.

The results obtained in Mathcad for a specific density of the thermal flow value of 6×10^{-3} W are shown in Figure 8, and it can be seen that the temperature increases with time depending on the thickness of the microchannels, whose values are shown in the figure legend (in millimetres).

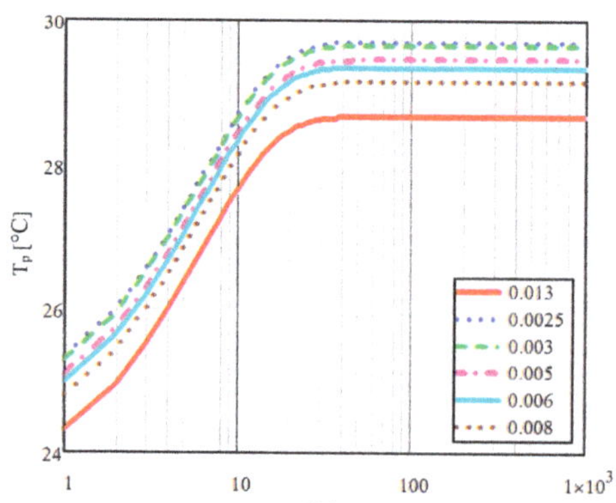

Figure 8. Transient heat flow at different microchannel wall thicknesses.

Another concern was the analysis of the exhaust gas conversion rate as a function of temperature and microchannel diameter. Since temperature and microchannel diameter influence mass and heat transfer due to the increased contact area between the gas and the catalyst walls, a significant impact on the conversion process of gaseous pollutants occurs.

To determine the conversion rate, a Matlab code was used using the Arrhenius equations. Using these equations, the specific constant of the chemical reaction k_a as a function of temperature was calculated:

$$k_a = k_0 \exp\left(\frac{-E_a}{RT(j)}\right), \tag{19}$$

where k_0 is a pre-exponential constant, E_a is the activation energy, R is the ideal gas constant and T(j) is the temperature.

The conversion rate C(j) was calculated using the formula given in Equation (20).

$$C(j) = \frac{k_a\, A\, L\, P}{R\, T(j)} \tag{20}$$

where A is the microchannel area, L is the microchannel length, P is the pressure and j is the index of the allocation of different temperatures of the mixture. In the calculation, different microchannel diameters were considered: values between 0.1 mm and 0.6 mm for a microchannel length of 200 mm. This process allowed obtaining the conversion rate depending on the specified parameters according to models given in the literature [24,36–38].

The graph in Figure 9 represents the conversion rate as a function of the gas inlet temperature through the microchannel for different diameter sizes. Diameter values can be found in the figure legend.

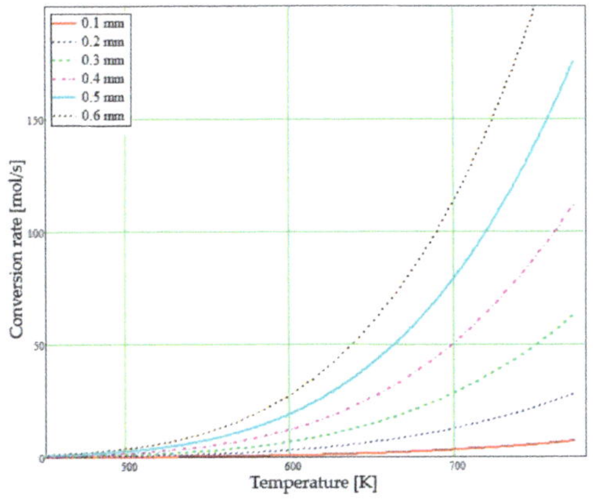

Figure 9. Conversion rate of exhaust gas through microchannels with different diameter sizes.

The analysis of this graph (Figure 9) allows observing that the conversion rate evolves exponentially with increasing temperature of the mixture through the microchannel diameter.

A study was carried out taking into account different values of the activation energy (E_a). For example, at an activation energy of 50,000 J/mol and a microchannel diameter of 0.10 mm, a lower conversion rate was obtained than for an activation energy of 120,000 J/mol and a diameter of 0.30 mm. This result indicates the existence of a dependence between the two variables, showing that microchannel size and activation energy significantly influence the conversion rate.

The concentrations of AdBlue aerosols and their evolution along the microchannel as a function of time were considered, and it was found that their concentration decreases over time due to the pseudo-desublimation phenomenon. This decrease in concentration can be explained by the rapid transformation of the AdBlue solution into a fine aerosol in the microchannel upstream and their dispersion in the exhaust gas. Therefore, as the aerosols move along the microchannel, their concentration gradually decreases.

These results demonstrate the importance of considering the influence of variables such as microchannel diameter, activation energy and aerosol concentration in the catalytic reduction process and in the mathematical modelling of this process. Understanding and controlling these variables can help to optimise the performance of catalytic reduction systems and improve the efficiency of the gaseous pollutant conversion process.

An increase in the activation energy value will lead to a decrease in the NOx conversion rate at lower temperatures. In other words, the higher the activation energy E_a, the more the chemical reaction requires important temperatures in order to take place to a significant extent. This shows that E_a is a key factor in influencing the rate of NOx conversion.

Figure 10 shows how the AdBlue vapour concentration (C_{vap}) varies with microchannel length (L) and time (t), obtained in Matlab based on Equation (21).

$$C_{vap}(t, L) = C_0 \cdot \exp\left(-\frac{K \cdot t}{L}\right), \tag{21}$$

where C_0 is the initial AdBlue vapour concentration and K is the evaporation rate. This formula describes how the concentration of AdBlue vapour decreases over time as a function of microchannel length and evaporation rate. The higher the evaporation rate K, the faster the AdBlue evaporates, and the faster its concentration decreases over time.

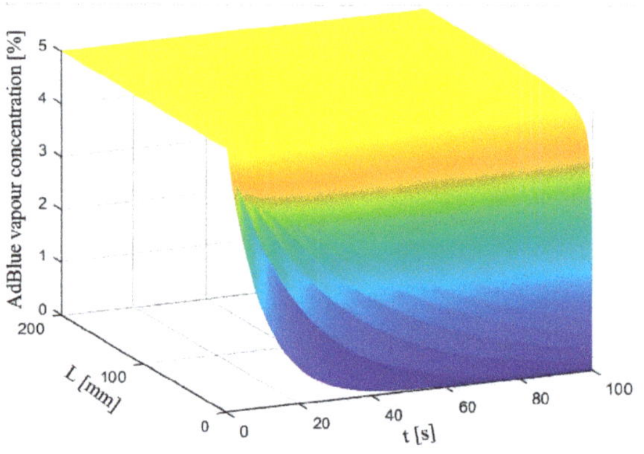

Figure 10. Aerosol concentration evolution as a function of microchannel length and time.

The change in aerosol concentration shows a sharp drop in the initial value in the first 50 s. This indicates that the pseudo-desublimation process has a rapid evolution in the initial stages and generates a significant amount of aerosols in a relatively short time. Studying and modelling this variation in aerosol concentration is important to understand the dynamics and behaviour of aerosols in the catalytic reduction system. By obtaining detailed information about the aerosol concentration variation, the process can be optimised and measures can be taken to ensure a uniform and efficient aerosol distribution in the exhaust gas.

3. Modelling of AdBlue Vapour Mass Change in the Pseudo-Desublimation Process

Simulations carried out in Ansys Release 2023 R1 for the mass exchange analysis and study of the volume fractions of liquid AdBlue coming into contact with the exhaust gas considered an initial temperature of 25 °C for the liquid solution and a temperature of 500 °C for the exhaust gas.

Figure 11 shows the simulation results for the turbulent kinetic energy, obtained by aid of Ansys, for an SCR system subjected to a process of AdBlue injection in volume (given by the exhaust gas) and on surfaces (walls of the SCR upstream tubing).

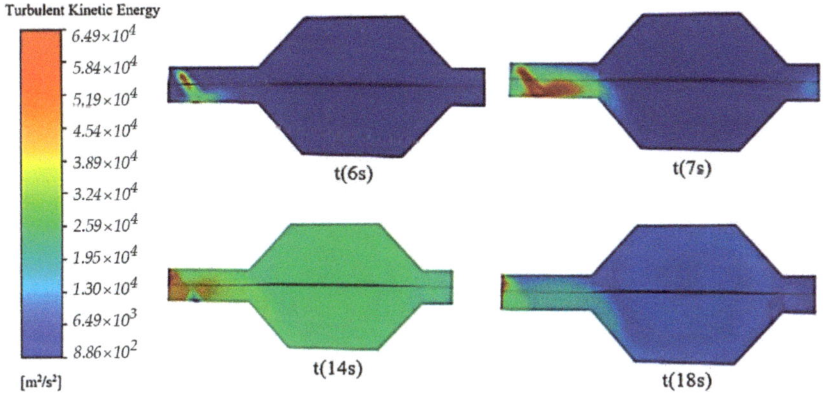

Figure 11. AdBlue vapour flow simulation in SCR system.

From Figure 11, it can be seen that at time t(6 s), the injection of the AdBlue solution takes place, which causes a flow of the droplet jet forming a ring that evolves partially in volume (first part) and then on the surfaces (reaches the pipe wall). At time t(7 s), a rapid diffusion and evaporation of the aerosols can be observed; in fact, the pseudo-desublimation process takes place. The pseudo-desublimation process consists of the direct transformation of a liquid directly into a solid, but unlike pure desublimation, in this case there is an intermediate phase. The intermediate phase consists of the appearance of a vapour phase (between the liquid and solid state) which proceeds so rapidly that it does not remain entirely between the liquid and solid state. Particularly important for the occurrence of pseudo-desublimation is the working temperature. It can be seen from the figure that part of the AdBlue vapour is integrated and entrained in the combustion gases. At time t(14 s), it can be seen from the simulation how the AdBlue vapour–combustion gas mixture propagates through the entire volume of the SCR catalyst.

At time t(18 s), diffusion is observed along the entire length of the catalyst, and most likely under certain conditions even within the microchannels a pseudo-desublimation phenomenon occurs. The statement is based on the experimental finding that solid deposits of the resulting AdBlue salts occur as a result of the pseudo-desublimation process taking place at well-specified high temperatures. The simulations allow us to study the behaviour of the mixture inside an SCR system with the visualisation of the vapour mass exchange and its flow behaviour. The temperature variations in the SCR system contribute fundamentally to the occurrence or non-occurrence of the pseudo-desublimation process (a totally undesirable phenomenon) so that in operation it can be avoided by using the appropriate programming environment specific to the gas cleaning system during the use of AdBlue.

Figure 12 shows the temperature variation in a microchannel of a honeycomb in an SCR catalyst with a diameter of 0.8 mm, obtained by simulation in Ansys.

In this representation, a time interval up to t(17 s) is considered, in which the operating regime is transient. AdBlue injection takes place at t(8 s), as illustrated in Figure 12.

The first image corresponds to the initial time t(0 s), which is associated with the period when the exhaust gas mixture encounters the microchannels of the SCR system (SCR cartridge). It can be seen that the velocity of the mixture oscillates when it enters the catalyst (indicated by the detail). In the first two sections, only combustion gases are introduced into the SCR, and between t(4 s) and t(6 s) a gradual temperature increase occurs. After this point, the injection of AdBlue takes place at time t(8 s), when the exhaust gas undergoes cooling. The rest of the sections in the figure show how the temperature changes over time, which can be analysed on the basis of the gradient in the legend.

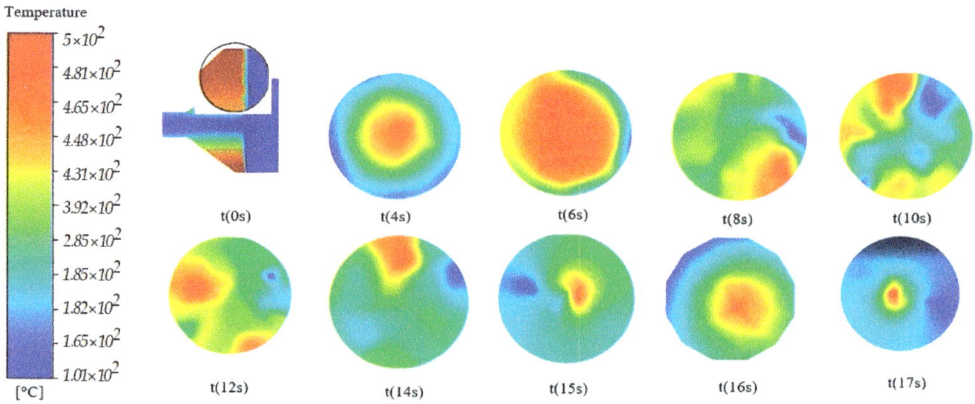

Figure 12. Temperature variation along an SCR microchannel traversed by the mixture.

An overview of the temperature change obtained in Ansys for the SCR assembly is shown in Figure 13, where it can be seen that we have a different temperature range depending on each area of the catalyst.

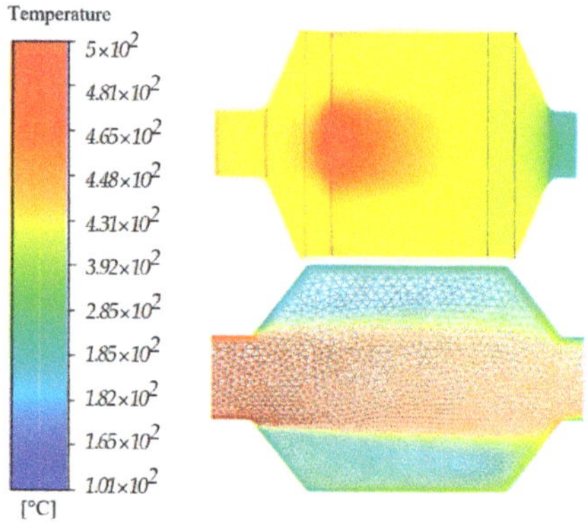

Figure 13. Temperature field in an SCR system.

The microchannel temperature analysis is shown at the top, with maximum values obtained centrally on the SCR cartridge and mixture velocity vectors at the bottom.

4. Experimental Set-Up

In the study of AdBlue injection processes in an SCR system, an experimental set-up was designed and built to allow laboratory tests with the aim of demonstrating that under certain conditions the phenomenon of pseudo-desublimation can occur in the case of contact with a strongly heated surface. The main components of the set-up are shown in Figure 14. The experimental set-up allows spraying a controlled jet of AdBlue droplets onto a heated surface in order to study the pseudo-desublimation process.

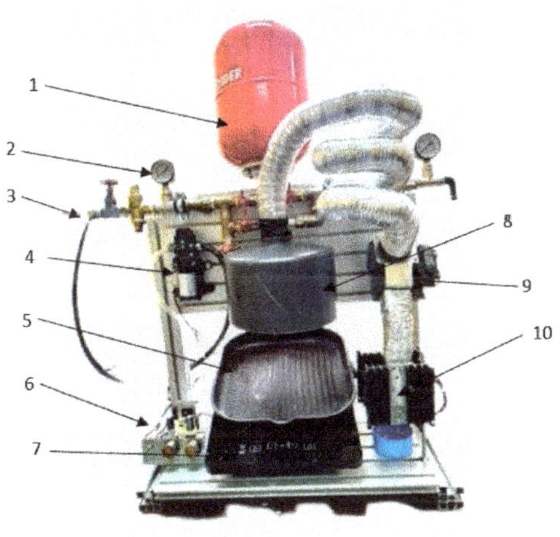

Figure 14. Experimental set-up for the study of AdBlue pseudo-desublimation.

The meanings of the numbered items in Figure 14 are as follows: 1, AdBlue tank; 2, pressure gauge; 3, tap; 4, AdBlue pump; 5, test surface; 6, electronic module; 7, inductive heating system; 8, hood; 9, fan; 10, Peltier vapour cooling module.

The AdBlue solution is taken from tank 1 after opening tap 3 and sent via pump 4 to an injector (not shown). The AdBlue injector is controlled by control module 6, which can adjust the time between two injections and the amount of solution injected. Pressure gauge 2 measures the pressure in the system, generated by pump 4. The jet of AdBlue microdroplets reaches test surface 5, which can reach temperatures ranging from ambient to a maximum of 600 °C. Heating to any temperature of the test surface can be achieved by means of induction hob 7. Fume extractor 8 with vent 9 captures the vapours generated as a result of evaporation and the pseudo-desublimation of AdBlue (pseudo-desublimation does not occur permanently). The vapours formed are sucked into a tube, which is cooled with Peltier element 10, so that their condensation takes place.

Experiments were carried out at different temperatures in order to know exactly whether and under what conditions pseudo-desublimation occurs for an AdBlue mixture. The results of these tests can be used to better understand the pseudo-desublimation phenomenon and to optimise the efficiency of the catalytic emission reduction process. Experiments were performed at different temperature thresholds using various injection times and varying the flow rate of AdBlue. For the analysis of the results, fast 240 fps movies were taken, both from top and side views of the test surface, using Trouble Shooter equipment produced by the Fastec Imaging Corporation Model TSHRCS.

After performing the experiments, the microtopography of a 1.5 mm by 2 mm region of the test surface, containing both a clean surface and solidified deposits, was mapped by aid of a Marsurf CWM 100 confocal microscope and interferometer. By evaluating the obtained microtopography parameters of the crystal deposits created on the test surface, the pseudo-desublimation process can be analysed.

5. Experimental Results

The images shown in Figures 15–19 show extracts of the most relevant frames (selected frame numbers are indicated in figures in red) obtained for different combinations of temperature, injection time and injection flow rate. In these images, the darker features represent the test surface and the lighter features illustrate the solidified AdBlue droplets. In each figure, for constant experimental parameters regarding temperature, injection time and injection flow rate, the shown images present the time evolution of the phenomena.

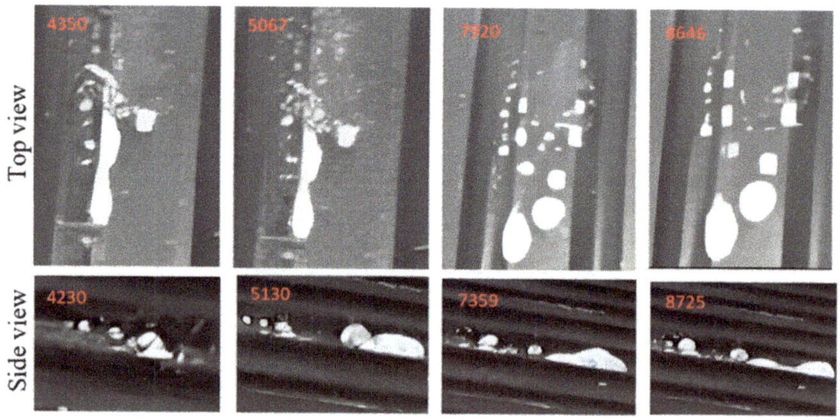

Figure 15. Injection of AdBlue without vaporisation.

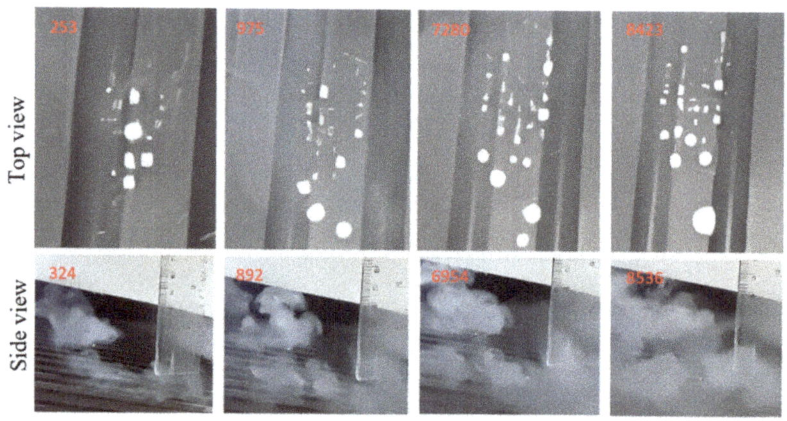

Figure 16. AdBlue injection when vaporisation starts.

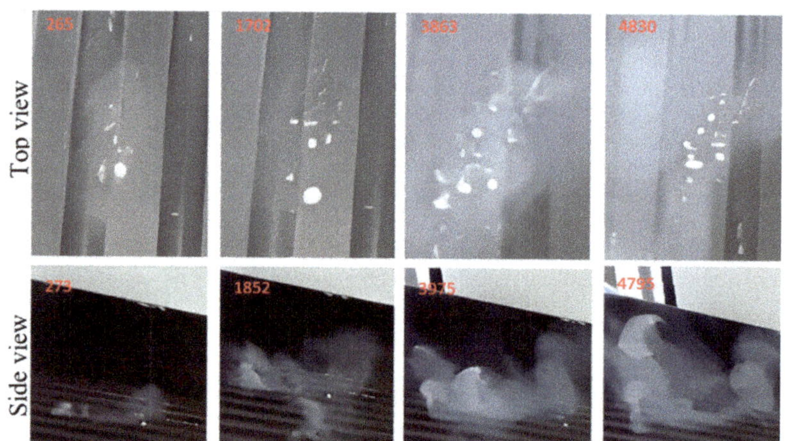

Figure 17. AdBlue injection in case of pseudo-desublimation.

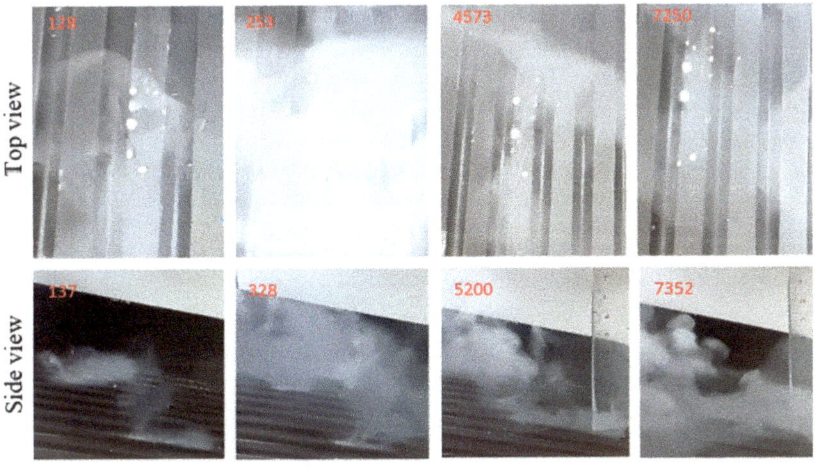

Figure 18. Development of the pseudo-desublimation process.

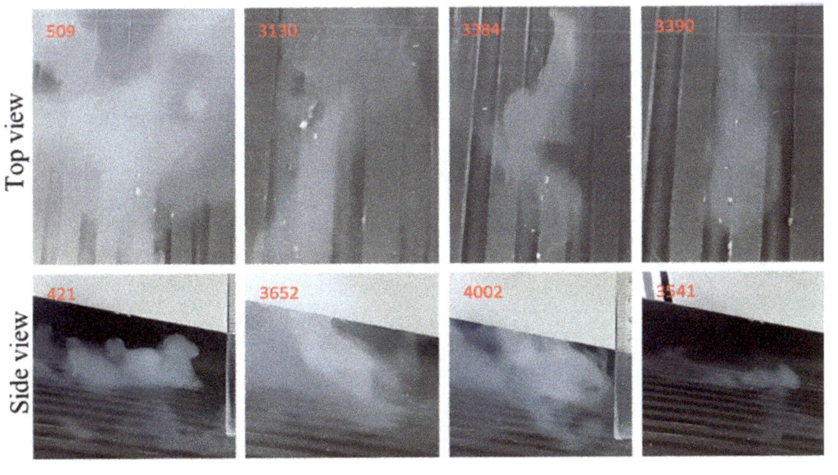

Figure 19. Occurrence of deflagration phenomenon of solid formations obtained after pseudo-desublimation.

The results of the experiments performed with the same flow rate of AdBlue injected on the heated surface showed the following cases:

(a) Injection of AdBlue without vaporisation, Figure 15.

This situation occurs if the temperature of the combustion gases is not very high, for example, at start-up or idling when the maximum threshold does not exceed the vaporisation temperature. In this situation, the droplet jet encounters the walls of the tube and the SCR catalyst without instant vaporisation. Since the temperature is low in the SCR assembly, we have the worst case when the system does not catalytically convert the AdBlue solution, so we will have the highest level of nitrogen-based NOx.

(b) Injection of AdBlue in the case that the vaporisation process starts, Figure 16.

The AdBlue solution contains 32.5% urea solution and 67.5% demineralised water. Injected into the exhaust gas, AdBlue® reduces harmful nitrogen oxide (NOx) emissions and complies with Euro 4, 5 and 6 pollution standards. According to ISO 22241 [39], Specifications for AdBlue, the boiling point is 100 °C. In other words, any temperature value at which the AdBlue solution is injected that exceeds this threshold will lead to the onset of vaporisation. The phenomenon is favoured by the lower droplets and flow rates, as well as the increased velocity.

(c) AdBlue injection in the case of pseudo-desublimation, Figure 17.

At temperatures between 130 and 140 °C for the tested surface in the case of low flow rates of the injected AdBlue, pseudo-desublimation is observed. This consists of the formation of a solid part adhering to the surface and of vapours which are entrained and incorporated into the microchannels of the SCR catalyst. From the experimental images in Figure 17, it can be seen that at the beginning of the process, the volume of the solid part Isocyanate (HCNO) is very small, but it adheres to the tested surface.

(d) Development of the pseudo-desublimation process.

At temperatures above 140 °C, a pronounced deposition on the heated surface with a lower evaporation rate was observed from the images in Figure 18. This may indicate that at this temperature, the evaporation process was accelerated and the AdBlue droplets have a higher tendency to deposit and crystallise on the heated surface.

(e) The appearance of the deflagration phenomenon of solid formations obtained after pseudo-desublimation.

In the temperature range between 200 and 300 °C, as shown in Figure 19, an instantaneous increase in the vaporisation rate of AdBlue was observed. Vaporisation occurs instantaneously from the first fraction of a second and manifests itself as a deflagration phenomenon. This result suggests that, at these particularly high temperatures, AdBlue vaporises so rapidly that the solid part no longer settles on the surface tested because it is separated into micrometric pieces by deflagration. As such, this phenomenon is favourable, as controlling the temperature inside the SCR can avoid the formation of solid deposits in the microchannels. We can conclude that lower temperatures favour the formation of crystals and the deposition of AdBlue, while higher temperatures lead to complete vaporisation and a reduction in residues.

These findings highlight the importance of temperature in the evaporation, reaction and deposition process of AdBlue, and temperature control and optimisation can play a crucial role in ensuring a uniform and efficient distribution of AdBlue in the SCR catalytic reduction system.

A detailed microscopic analysis of the deposits generated at a temperature of 230 °C, using the Marsurf CWM 100 confocal microscope, allows a quantitative evaluation of the crystals deposited on the test surface, as shown in Figure 20. The surface microtopography, shown in a 3D representation in Figure 20, indicates the distribution of solid deposits over the analysed sample. In the analysis, crystal height values, both surface and profile, were also taken into account for the studied sample.

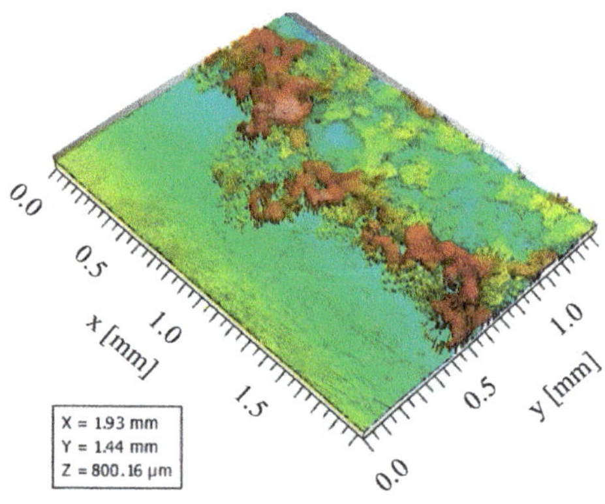

Figure 20. Surface microtopography of pseudo-desublimation deposits.

In Figure 21, the profile of urea deposition on the same tested surface as Figure 20 is shown in 2D. A single profile taken from the surface microtopography and placed on the right side of Figure 21 shows the surface peaks and valleys that allow for further analysis, as described further.

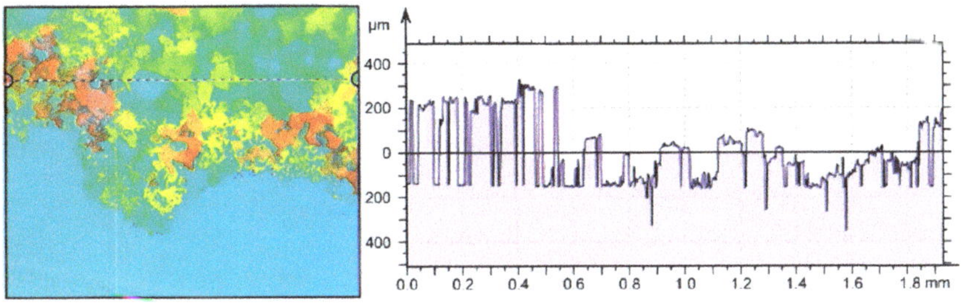

Figure 21. Area of interest on the test plate and profiles of crystals deposited as a result of the pseudo-desublimation process.

Based on the obtained surface values found in Table 3, the profile roughness parameters Ra and Rz were calculated. The values obtained indicate that Ra = 31.122 µm (arithmetic mean of the absolute ordinate values within a sampling length of a profile) and Rz = 150.20 µm (arithmetic mean of the added distances between the five highest profile peaks and lowest profile valleys). It was found that the height of the crystalline deposit profile is influenced by the temperature at which urea decomposition took place. Thus, the temperature factor plays an important role in crystal formation and deposition on the surface as well as on the profile.

Table 3. Surface amplitude parameters (ISO 4287 [40]).

Amplitude Parameter	Unit	Values
Rz	µm	150.20
Ra	µm	31.122

Sq surface roughness values were obtained using the Mahr MarSurf CWM 100 confocal microscope and interferometer (Figure 22). Sq is a surface texture parameter (ISO 25178 [41]) equivalent to the standard deviation of heights and is calculated as the root mean square value of the ordinate values within the defined area. Measurements were performed in the central region of the samples in 2×1.5 mm^2 areas. Surface roughness was taken into account, so both directions were considered. The parameters were determined according to ISO 25178 using the default settings of the MountainsLab 8.1 software, which uses a predefined robust L-Gaussian filter of 0.8 mm.

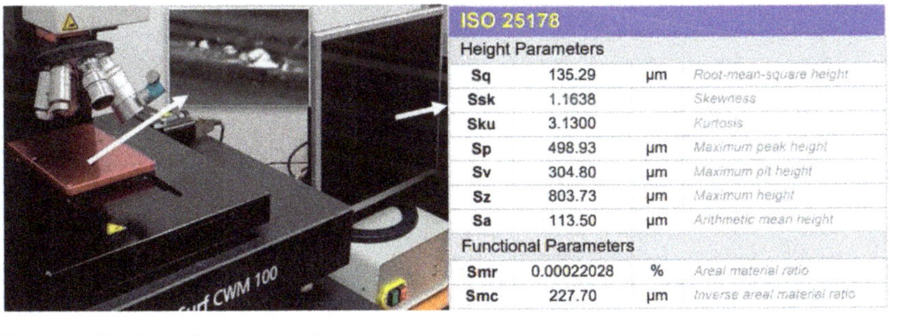

Figure 22. Surface microtopography measurement.

These measurements allowed the surface roughness of the samples to be accurately assessed and the surface texture to be characterised in detail. The Sq values obtained provide important information about the variability in surface height and can be used to compare roughness levels between samples and to assess the effects of different methods or treatments on surface texture. By using appropriate tools and software, we can gain a deeper understanding of surface properties and their impact on the performance and functionality of components and systems. A sample of the profile of crystals deposited on the test plate can be observed in Figure 23. The highlighted regions of the profile indicate a clean surface region (right side of the image, between 0.86 mm and 1.21 mm) as well as a region with solidified deposits (left side of the image, between 0.07 mm and 0.81 mm). The measured parameters indicate for the 0.74 mm wide region with solidified deposits a maximum profile height of 138 μm and a mean height of 53.6 μm.

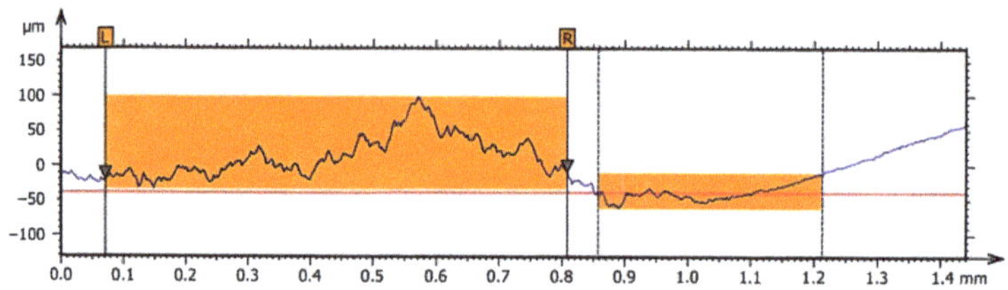

Parameters	Step 1	Unit
Width	0.73859	mm
Maximum height	138.02	μm
Mean height	53.617	μm

Figure 23. Profile of crystals deposited on the test plate observed using the CWM 100 confocal microscope.

6. Conclusions

The present paper proves experimentally that in SCR systems the exhaust gas temperature is essential as a value for the occurrence of the pseudo-desublimation process. Knowing and accurately controlling the temperature of the exhaust gas–AdBlue mixture allows the SCR system software to ensure the injection process at optimal times to avoid the formation of solid deposits on the SCR manifold and inside the microchannels of this system.

The pseudo-desublimation of AdBlue microdroplets inside the microchannels of SCR systems is a process started by the conversion of the liquid AdBlue solution into microdroplets by injection. Once atomised, the high temperatures present in the SCR system transform the microdroplets into vapour that enters the SCR microchannels. Inside these microchannels, depending on temperature, the AdBlue vapour undergoes a desublimation process, which consists of a quasi-instantaneous transformation of the microdroplets into vapour or, at high temperatures, directly into solids. This process occurs due to the small size of the microchannels, their various shapes and the velocity of the fluid flowing through it.

The characteristics of the crystals formed and the rate of their growth depend on the working temperature of the gaseous medium, which can be controlled by efficient preheating methods immediately after engine start-up to the operating temperature. The use of such methods can help to avoid solid depositions on SCR system components, which have a direct impact on SCR catalyst performance and durability.

Author Contributions: Conceptualization, C.M.P. and I.M.; methodology, C.M.P. and I.M.; software, I.M.; validation, I.M., C.S. and C.M.P.; formal analysis, I.M.; investigation, I.M., C.S. and C.M.P.; resources, I.M. and C.S.; data curation, I.M.; writing—original draft preparation, C.M.P.; writing—review and editing, C.S.; visualization, I.M., C.S. and C.M.P.; supervision, I.M.; project administration, I.M.; funding acquisition, I.M. and C.M.P. All authors have read and agreed to the published version of the manuscript.

Funding: This work was funded by the Ministry of Public Works, Development and Administration within REGIONAL OPERATIONAL PROGRAM 2014–2020, PRIORITY AXIS 1: PROMOTING TECHNOLOGICAL TRANSFER, PRIORITY OF INVESTMENTS 1.1, OPERATION B: Supporting science and technology parks, under contract no. 6592/24.12.2020 (SMIS: 128730), project title: ESTABLISHMENT OF THE EAST EUROPEAN BORDER SCIENTIFIC AND TECHNOLOGICAL PARK IN SIRET.

Data Availability Statement: Some or all data, models or code generated or used during the study are available from the corresponding author by request.

Conflicts of Interest: The authors declare no conflict of interest.

References

1. Magureanu, R. Aerul mai curat costa scump. *Auto Motor si Sport*, December 2022; 14.
2. Wikipedia. European Emission Standards. 2023. Available online: https://en.wikipedia.org/wiki/European_emission_standards#cite_note-eurlex715-18 (accessed on 30 July 2023).
3. European Comission. Commission Staff Working Document—Impact Assessment Report, Annex 5–8, Accompanying the Document, Proposal for a Regulation of the European Parliament and of the Council on Type-Approval of Motor Vehicles and of Engines and of Systems, Components and Separate Technical Units Intended for Such Vehicles, with Respect to Their Emissions and Battery Durability (Euro 7) and Repealing Regulations (EC) No 715/2007 and (EC) No 595/2009. 2022. Available online: https://single-market-economy.ec.europa.eu/system/files/2022-11/SWD_2022_359_1_EN_impact_assessment_part3_v6.pdf (accessed on 11 May 2023).
4. DIESELNET. Emission Standards EURO VI. Available online: https://www.dieselnet.com/standards/eu (accessed on 29 May 2023).
5. Larsson, P.; Lennard, W.; Dahlstrom, J.; Andersson, O.; Tunestal, P. NOx-Conversion and Activation Temperature of a SCR Catalyst Whilst Using a Novel Biomimetic Flash-Boiling AdBlue Injector on a LD Engine. In Proceedings of the SAE 2016 International Powertrains, Fuels & Lubricants Meeting, Baltimore, MD, USA, 24–26 October 2016.
6. Šimčík, M.; Vala, R.; Vavřík, D. Modeling of AdBlue spray evaporation and urea deposition in an SCR system. *Energy Procedia* **2017**, *128*, 59–66.
7. Sadeghi, F.; Rajendran, A.; Peters, B. Experimental and numerical study of AdBlue evaporation and decomposition in a urea-SCR system. *Int. J. Therm. Sci.* **2017**, *117*, 330–343.
8. Matsumoto, Y.; Oikawa, T.; Makino, K. Numerical simulation of urea-water solution spray evaporation and mixing process in an automotive SCR system. *Energy Procedia* **2014**, *61*, 1656–1659.
9. Maricq, M.M.; Chase, R.E.; Xu, N.; LaVoie, G.A.; Kim, J.; Lin, H.C. Particle size distributions and nitrogen oxide emissions from modern diesel engines. *Environ. Sci. Technol.* **2017**, *106*, 4522–4529.
10. Zhenping, Z.; Li-Hsien, Y.; Mingkan, Z.; Shizhi, Q. Ion Transport and Selectivity in Biomimetic Nanopores with pH-Tunable Zwitterionic Polyelectrolyte Brushes. *Nanoscale* **2015**, *7*, 17020–17029. [CrossRef]
11. Hossein, D.; Mahdi, K.; Seyed, N.A. Importance of nanochannels shape on blue energy generation in soft nanochannels. *Electrochim. Acta* **2022**, *431*, 141175. [CrossRef]
12. Tseng, S.; Lin, S.-C.; Lin, C.-Y.; Hsu, J.-P. Influences of Cone Angle and Surface Charge Density on the Ion Current Rectification Behavior of a Conical Nanopore. *J. Phys. Chem. C* **2016**, *120*, 25620–25627. [CrossRef]
13. Sadeghi, F.; Tirandazi, B.; Khalili-Garakani, A.; Nasseri, S. Investigating the effect of channel geometry on selective catalytic reduction of NOx in monolith reactors. *Chem. Eng. Res. Des.* **2017**, *118*, 21–30. [CrossRef]
14. Wang, Y.; Li, Z.; Fan, R.; Guo, X.; Zhang, C.; Wang, Y.; Ding, Z.; Wang, R.; Liu, W. Deactivation and Regeneration for the SO_2-Poisoning of a Cu-SSZ-13 Catalyst in the NH_3-SCR Reaction. *Catalysts* **2019**, *9*, 797. [CrossRef]
15. Hu, J.; Wang, B.; Xu, Y.; Shi, X.; Liu, F.; Hao, J. NH_3-SCR of NOx over zeolite catalysts: A review on reaction mechanism. *Appl. Sci.* **2017**, *7*, 727.
16. Chen, D.; Muthusamy, S.; Ray, A. Heterogeneous Photocatalysis in Environmental Remediation. *Dev. Chem. Eng. Miner. Process.* **2008**, *8*, 505–550. [CrossRef]
17. Zuo, X.; Jia, Y.; Xu, J.; Zhang, X. Numerical simulation of AdBlue distribution in the selective catalytic reduction system of a heavy-duty diesel engine. *Energy Procedia* **2019**, *158*, 1840–1845.
18. Petkovic, S.; Pesic, R.; PesicJovanka, R.; Lukić, L. Heat transfer in exhaust system of a cold start engine at low environmental temperature. *Therm. Sci.* **2010**, *14*, 219–232. [CrossRef]
19. Zhu, Z.; Midlam-Mohler, S.; Canova, M. Development of physics-based three-way catalytic converter model for real-time distributed temperature prediction using proper orthogonal decomposition and collocation. *Int. J. Engine Res.* **2019**, *22*, 146808741987612. [CrossRef]
20. Bird, R.B.; Stewart, W.E.; Lightfoot, E.N. *Transport Phenomena*; John Wiley & Sons: Hoboken, NJ, USA, 2007.
21. Incropera, F.P.; DeWitt, D.P.; Bergman, T.L.; Lavine, A.S. *Fundamentals of Heat and Mass Transfer*, 7th ed.; John Wiley & Sons: Hoboken, NJ, USA, 2013.

22. Çengel, Y.A.; Cimbala, J.M. Momentum Analysis of Flow Systems. In *Fluid Mechanics: Fundamentals and Applications*, 2nd ed.; McGraw-Hill: New York, NY, USA, 2019.
23. Incropera, F.P.; DeWitt, D.P.; Bergman, T.L.; Lavine, A.S. *Fundamentals of Heat and Mass Transfer*, 8th ed.; John Wiley & Sons: Hoboken, NJ, USA, 2019.
24. Froment, G.F.; Bischoff, K.B.; De Wilde, J. *Chemical Reactor Analysis and Design*, 3rd ed.; John Wiley & Sons: Hoboken, NJ, USA, 2011; ISBN 978-0-470-56541-4.
25. Perry, R.H.; Green, D.W. *Chemical Engineering Handbook*; McGraw-Hill: New York, NY, USA, 2018.
26. Holman, J.P. Steady-State Conduction. In *Heat Transfer*, 10th ed; McGraw-Hill Education: New York, NY, USA, 2009; ISBN 978-0-07-352936-3.
27. Incropera, F.P.; DeWitt, D.P.; Bergman, T.L.; Lavine, A.S. One-Dimensional Steady-State Conduction. In *Fundamentals of Heat and Mass Transfer*, 7th ed.; John Wiley & Sons: Hoboken, NJ, USA, 2011.
28. Holman, J.P. External Forced Convection. In *Heat Transfer (Mcgraw-hill Series in Mechanical Engineering)*, 10th ed.; McGraw-Hill Education: New York, NY, USA, 2009; ISBN 978-0-07-352936-3.
29. Pletcher, R.H.; Tannehill, J.C.; Anderson, D.A. *Computational Fluid Mechanics and Heat Transfer*; CRC Press: Boca Raton, FL, USA, 2012.
30. Modest, M.F. *Radiative Heat Transfer*; Academic Press: Cambridge, MA, USA, 2013.
31. Colebrook, C.F.; White, C.M. Experiments with fluid friction in roughened pipes. *Math. Phys. Sci.* **1937**, *161*, 367–381.
32. Moody, L.F. Friction factors for pipe flow. *Trans. ASME* **1944**, *66*, 671–684. [CrossRef]
33. Dittus, F.W.; Boelter, L.M.K. *Heat Transfer in Automobile Radiators of the Tubular Type*; University of California Press: Berkeley, CA, USA, 1930; Volume 2, pp. 443–461.
34. Zheng, W.; Sun, J.; Ma, C.; Yu, Q.; Zhang, Y.; Niu, T. Numerical Study of Fluid Flow and Heat Transfer Characteristics in a Cone-Column Combined Heat Sink. *Energies* **2021**, *14*, 1605. [CrossRef]
35. Mihai, I.; Suciu, C.; Picus, C.M. Considerations for the Maximum Heat Load and Its Influence on Temperature Variation of the Evaporator in Flat MHPs in Transient Regimes. *Micromachines* **2022**, *13*, 979. [CrossRef] [PubMed]
36. Molina, R.; Orcajo, G.; Martinez, F. KBR (Kinetics in Batch Reactors): A MATLAB-based application with a friendly Graphical User Interface for chemical kinetic model simulation and parameter estimation. *Educ. Chem. Eng.* **2019**, *28*, 80–89. [CrossRef]
37. Smith, J.M.; Van Ness, H.C.; Abbott, M.M. *Introduction to Chemical Engineering Thermodynamics*; McGraw-Hill: New York, NY, USA, 2017.
38. Karagiannidis, S. Catalytic Microreactors for Portable Power Generation. 2011. Available online: https://beckassets.blob.core.windows.net/product/readingsample/7514267/9783642176678_excerpt_001.pdf (accessed on 18 September 2023). [CrossRef]
39. *ISO 22241*; Diesel Engines—NOx Reduction Agent AUS 32—Part 1: Quality Requirements. ISO: Geneva, Switzerland, 2009.
40. *ISO 4287*; Geometrical Product Specifications (GPS)—Surface Texture: Profile Method—Terms, Definitions and Surface Texture Parameters. ISO: Geneva, Switzerland, 1997.
41. *ISO 25178*; Surface Texture Characterization. ISO: Geneva, Switzerland, 2012.

micromachines

MDPI

Article

Flow Boiling of Liquid n-Heptane in Microtube with Various Fuel Flow Rate: Experimental and Numerical Study

Muhammad Tahir Rashid [1], Naseem Ahmad [2,*], Raees Fida Swati [3,*] and Muhammad Bilal Khan [4]

1 School of Aerospace Engineering, Beijing Institute of Technology, Beijing 100081, China
2 Department of Mechanical Engineering, Institute of Space Technology, Islamabad 44000, Pakistan
3 Department of Aeronautics and Astronautics, Institute of Space Technology, Islamabad 44000, Pakistan
4 Faculty of Mechanical Engineering, Ghulam Ishaq Khan Institute of Engineering Sciences and Technology, Topi, Swabi 23640, Pakistan
* Correspondence: naseem_saddiqui@yahoo.com (N.A.); raees.swati@ist.edu.pk (R.F.S.)

Abstract: The evaporation of liquid hydrocarbon n-heptane is discussed in detail with experimentation and numerical techniques. A maximum wall temperature of 1050 K was reported during an experimental process with a two-phase flow that was stable and had a prominent meniscus at a small fuel flow rate (FFR) $\leq$ 10 μL/min. At medium to high FFR (30–70 μL/min), the flow field was unstable, with nucleating bubbles and liquid droplets inside the microtube and the maximum temperature recorded was 850 K for 70 μL/min. For the numerical model, the temperature of the wall was used as a boundary condition. Using the numerical model, the evaporative flux at the meniscus, pressure drop, pressure oscillation, and heat transfer coefficient (HTC) were investigated. A single peak in HTC was obtained at a low fuel flow rate, while multiple peaks were obtained for high FFR. At low FFR, the pressure peak was observed to be 102.4 KPa, whereas at high FFR, the pressure peak increased to 105.5 KPa. This shows a 2% increase in pressure peak with an increase in FFR. Similarly, when the FFR increased from 5 μL/min to 70 μL/min, the pressure drop increased from 500 Pa to 2800 Pa. The high amplitude of pressure drops and a high peak of HTC were found, which depend on the mass flow rate. The coefficient of variation for pressure drop depends mainly on the fuel flow rate.

Keywords: volume of fluid; microtube; evaporating meniscus; flow boiling; pressure drop fluctuation; heat transfer

1. Introduction

Numerous research efforts are directed towards the manufacturing and downsizing of micro-scaled devices, including micro gas turbines [1–3], micro swing engines [4–7], micro Swiss-roll combustors [8,9], and micro photovoltaic devices [10,11], and have attracted considerable research for further development. The typical power source for such systems is batteries, but due to weight restrictions on micro devices and the low energy density of batteries, researchers are seeking alternate methods of producing electricity on a small scale while utilizing the micro combustor to benefit from the high energy density (40 KJ/g) of hydrocarbon fuel. This energy density is approximately 100 times greater than that of the majority of sophisticated batteries [12–15]. However, it is noted that flow boiling and flame stability are critical concerns in the development of micro combustors [16,17].

Several experimental studies [18–20] have focused on microchannel boiling heat transfer and flow pattern. Leong et al. [18] gave a critical review of the boiling model and bubble dynamics in the microchannel. Parjapati et al. [19,21] reported the instabilities in micro channels and also described their promising features, whereas Cebi et al. [20] worked on flow boiling in mini and microchannels for enhanced geometries. Both are time-consuming and expensive due to their design and operating techniques. Furthermore, the properties inside the microtube cannot be measured correctly due to its small size. Due to the above reasons, numerical simulation is a widely used technique for such purposes.

A few theoretical models for microchannel flow patterns were also proposed with empirical correlation and simplifications in the past. Thome [22,23] proposed a theoretical model that accounts for surface tension, gravity, the near-wall effect, wall conduction, axial shear stress, and the unstable term to calculate the heat transfer caused by the presence of a liquid layer in the mini and microchannel. Zhao and Liao's [24] research was based on a theoretical model for vapor flowing in a vertical mini-triangular tube to condense into a film. Due to the thin liquid covering on the sidewall of the triangular channel and the surface tension effect, they revealed that the HTC at the cross-section is higher than the circular channel with similar characteristics. Wang and Rose [25,26] developed a mathematical model for a non-circular microchannel that took into account gravity, interfacial shear stress, and surface tension. They elucidated that the surface tension effect is what causes the increase in heat transfer.

Several numerical simulations of flow boiling have been also reported in recent years. Numerical simulation in square microchannels was conducted by Ferrari et al. [27], who demonstrated that a bubble moves more quickly in a square microchannel than in a tube. Additionally, they enhanced flow mixing, which affected the efficacy of the heat transfer. The annular flow boiling in microchannels was studied by Guo et al. [28] using a numerical model and it was revealed that a modest input perturbation has very little impact on boiling heat transfer. Moreover, the authors showed that uncertainties at the interface caused by phase change have a substantial influence on the rate of heat transfer via interfacial wave advection. A three-dimensional numerical simulation on a 0.6 mm microtube was conducted by Tiwari et al. [29]. They discovered that the relationship between the thermal conductivity of the microtube wall and the fluid is crucial to the behavior of bubbles during their formation and that the thickness and material of the microtube wall change during bubble confinement. The influence of hydrogen and carbon dioxide on biogas was studied by Ahmed et al. [30,31]. They revealed that the utilization of biogas with diesel is attractive to cut down discharges and enhance the performance of the engine.

Liquid hydrocarbon flow boiling has drawn a lot of interest because of its high energy density, low storage requirements, ease of transportation, and small space requirements for microenergy systems. Muller et al. [32,33] carried out an experimental study in an annulus with heptane fluid to evaluate the sub-cooled boiling and HTC in a convective regime. They noticed several convective and nucleate boiling heat transfer regimes. The heat transfer coefficients for nucleate boiling were discovered to be independent of the mass velocity. The boiling hysteresis phenomenon was only observed at small mass velocities, and the flow direction's influence was not detected. Hapke et al. [34] employed a thermographic approach to examine how n-heptane flows and boils in a microchannel. They observed temperature profiles over the width of the channel and noticed different boiling regimes. Moreover, they concluded that both heat and mass flux disturb pressure drop in the microchannel. However, they argued that numerical simulation should be implemented to solve these types of problems. A numerical simulation of the influence of injection time was conducted [35,36] and it was found that near the contact line, large diffusion fluxes exist due to the intensity of thin film evaporation, and local interfacial temperature and evaporative flux are slightly increased. Very little literature is available for the numerical investigation of liquid n-heptane in a microtube.

The main aim of the current research is to develop a numerical model for understanding flow boiling in the microtube, the interfacial evaporation rate, and the oscillation of liquid pressure. Non-uniform temperature supplied to the inner wall causes the flow to boil. Additionally, the non-uniform temperature is measured experimentally with an infrared camera.

2. Experimental Setup

The experimental setup contains a straight microtube, a delivery system for fuel, a temperature measurement system, and an observation system, and is shown in Figure 1. The microtube is mounted horizontally on a syringe during experimentation. In this experiment, quartz (QT) with an outer diameter of 0.5 mm and an interior diameter of 0.2 mm is employed. Liquid fuel is injected using a syringe pump (LSP01-1A) and a plastic syringe with an inner diameter of 10 mm. The steel tube is first connected to the syringe and then the QT tube is glued with a steel tube. There is a small diffusion flame anchored at the tube's exit. Every experiment was carried out in a single environment and at room temperature.

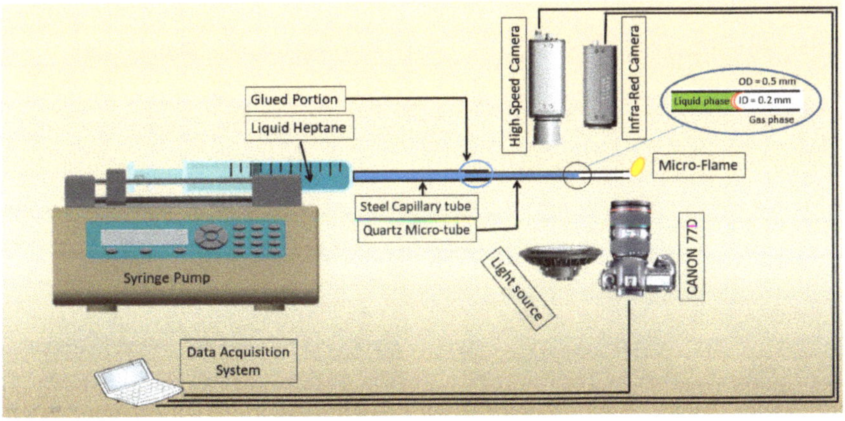

Figure 1. Sketch of the experimental setup.

The syringe pump precisely regulates the fuel flow rate with an inaccuracy of less than 1.0%. The camera captures the tiny flame at 128 fps. An infrared radiation (IR) camera (FLIR, A40) is used to record the surface temperature of the microtube. The IR camera has a 150 mm macro lens and a 64 mm by 48 mm field of vision. The spectral range that was measured is 7.5 to 13 μm. The temperature variations on the tube surface are shown in the IR images. The calibration of the IR camera is discussed in Tahir et al. [17].

3. Observation

N-heptane is employed as a liquid fuel in the experiment, and the flow rate varies from 5 μL/min to 70 μL/min. Figure 2a depicts that the flame shape is spherical with a flame length of ~1 mm at Qf $\leq$ 10 μL/min. Figure 2a also depicts the equivalent liquid–vapor interface with segregated liquid and vapor phases to demonstrate the phenomenon of laminar evaporation. The two phenomena are captured by an ordinary camera at 128 fps. The instantaneous times of 0.032 s and 0.007 s show the capture time of the flame oscillation and flow pattern oscillation, respectively. A linear trend can be seen in the succeeding flame length and height depicted in Figure 2b along with the interface distance. This suggests that laminar evaporation in the microtube causes a continuous flame to form at low fuel flow rates. Due to a low fuel flow rate, the liquid n-heptane is entirely absorbed in the liquid–vapor interface. The interface distance was approximately 2.30–2.35 mm and 2.14–2.25 mm, respectively, at FFR of 5 μL/min and 10 μL/min. Nucleate bubbles and liquid droplets were not observed in the subcooled liquid phase or vapor phase, respectively.

The flame is stretched out and no longer resembles a ball when Qf is less than 30 μL/min. Because of the incomplete combustion of the hydrocarbons, the flame is more yellow and is referred to as unstable. The blue light-containing lower portion of the flame (Figure 3a) extends at 0.302 s, which may be due to the burning of liquid droplets. Figure 3a illustrates the flame at Qf 70 μL/min, which has an extensive reaction zone and a bright yellow color from the burning of slugs or liquid droplets. Additionally, the flame is referred to as an explosive flame even though it has no particular configuration and burns violently.

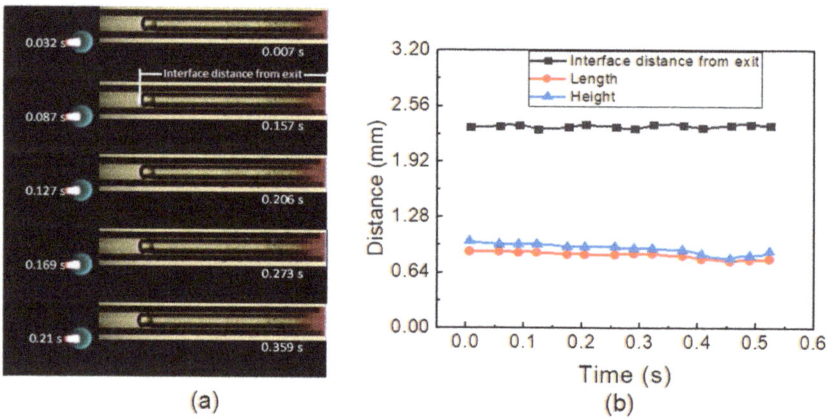

Figure 2. (**a**) Flame picture, flow pattern at 5 micro. (**b**) Interface distance and flame characteristics at low FFR.

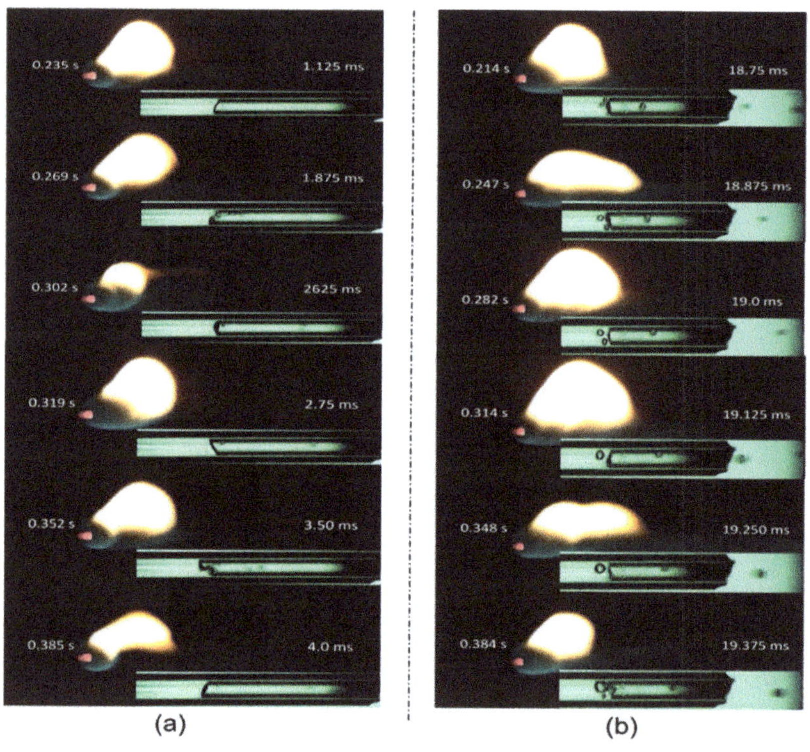

Figure 3. Flame characteristics and evaporation phenomena: (**a**) 30 μL/min, (**b**) 70 μL/min.

A high-speed camera at 8000 fps is used to record the evaporation phenomena with FFR > 30 μL/min. The two-phase flow phenomena with FFR > 30 μL/min are depicted in Figure 3. Convective boiling predominates with an FFR of 30 μL/min, and we did not observe the commencement of nucleation boiling (ONB), but when the FFR is increased to 70 μL/min, nucleate boiling is observed. Figure 3b depicts the progression of the nucleate bubble's rupture cycle from 18.75 ms to 19.37 ms. Figure 3 shows the liquid droplet at FFR > 30 μL/min. As shown in Figure 2a, a stable flame forms at the tube exit when the FFR is small because the fuel is completely evaporated at the interface. However, as the FFR increases, more liquid fuel must be vaporized, but the liquid–vapor interface becomes too narrow and small, making it impossible to vaporize excess liquid. As shown in Figure 3, the extra liquid fuel may vaporize as a continuous liquid film along the tube's interior wall or as a liquid droplet in the vapor domain.

While the evaporation changes from laminar to nucleate boiling with an increase in FFR, the flame shape changes from a steady flame to an explosive flame. The temperature distribution along the micro tube's exterior wall is depicted in Figure 4. At a low fuel flow rate, the maximum temperature gradient is 350 K/m. As the FFR increases to medium flow conditions, the temperature gradients increase up to 14%. A further increase in flow conditions to a higher FFR results in a 65% increase compared to the gradient of low FFR. The percentage increase in the temperature gradient from low to medium FFR is small, i.e., 14%, and from medium to high FFR the percentage increase is 65%. The maximum temperature of the wall drops as the FFR rises, and this is due to two factors. The first explanation is that a high FFR results in greater heat absorption for gasification, which lowers the temperature of the wall. The second reason is that the presence of liquid slugs and droplets in the vapor domain lowers the wall temperature by requiring more heat for gasification.

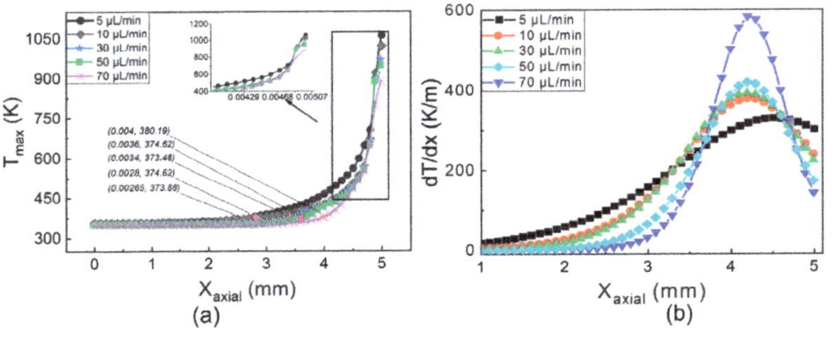

Figure 4. Temperature distribution. (**a**) Temperature profile. (**b**) Thermal gradient.

The aforementioned observation leads to the hypothesis that FFR has impacted flame, which, in turn, has impacted wall temperature, which then has impacted interface movements. In order to study the impact of fuel flow rate and wall temperature on two-phase flow, pressure drop, pressure oscillation, and heat transfer, numerical simulation is conducted in Ansys 16.

4. Mathematical Modeling

The following assumptions are made in the current research:

1. In the microtube, the gravity effect is dominated by the surface tension force.
2. The physical properties like surface tension, latent heat, and density are constant for each phase.
3. The liquid n-heptane boiling point is assumed to be 373.15 K.

4.1. VOF Model

The Volume of Fluid (VOF) method in the commercial software Fluent 16.0 with a user-defined function (UDF) was used to describe two-phase flow boiling. A fixed grid method called the VOF model was created for two or more immiscible fluids. The VOF model resolves a single set of fluid-specific momentum equations. The VOF method records the volume fraction of each phase in the computing cell, and the total volume fraction of all phases equals unity.

$$\alpha_l + \alpha_v = 1 \tag{1}$$

The equation for the vapor and liquid phase is given as:

$$\frac{\partial \alpha_l}{\partial t} + \nabla.\left(\vec{v}\alpha_l\right) = -\frac{S}{\rho_l} \tag{2}$$

$$\frac{\partial \alpha_v}{\partial t} + \nabla.\left(\vec{v}\alpha_v\right) = -\frac{S}{\rho_v} \tag{3}$$

The continuity, energy, and momentum equations are:

$$\frac{\partial(\rho v)}{\partial t} + \nabla.\left(\rho\vec{v}\vec{v}\right) = -\nabla p + \nabla.\left[\mu\left(\nabla\vec{v} + \nabla\vec{v}^{T}\right) - \frac{2}{3}\mu\nabla.\vec{v}I\right] + \rho\overline{g} + F_{vol} \tag{4}$$

$$\frac{\partial(\rho E)}{\partial t} + \nabla.\left[\vec{v}(\rho E + p)\right] = \nabla.(K\nabla T) + Q \tag{5}$$

where

$$\rho = \alpha_l\rho_l + \alpha_v\rho_v \tag{6}$$

$$k = \alpha_l k_l + \alpha_v k_v \tag{7}$$

$$\mu = \alpha_l\mu_l + \alpha_v\mu_v \tag{8}$$

$$E = \frac{\alpha_l\rho_l E_l + \alpha_v\rho_v E_v}{\alpha_l\rho_l + \alpha_v\rho_v} \tag{9}$$

The Continuum Surface Force (CFS) model is used for the surface tension effect, and surface tension can be described as a pressure jump over the surface. The divergence theorem can be used to express the force at the surface into a volume force, F_{vol}. The momentum equation is represented as:

$$F_{vol} = \sigma_{lv}\frac{\alpha_l\rho_l k_v\nabla\alpha_v + \alpha_v\rho_v k_l\nabla\alpha_l}{0.5(\rho_l + \rho_v)} \tag{10}$$

where σ_{lv} is the surface tension at the interface. The definition of curvatures of the liquid and vapor phase is defined as:

$$k_l = \frac{\Delta\alpha_l}{|\Delta\alpha_l|} \qquad k_v = \frac{\Delta\alpha_v}{|\Delta\alpha_v|} \tag{11}$$

4.2. Heat and Mass Transfer Due to Evaporation at the Interface

The Lee model [37] is used for mass transfer and predicts phase transition at constant pressure. In a quasi-thermo-equilibrium condition, the key determinant of mass transfer is saturation temperature. The Lee model is defined as:

$$S = \frac{r_l\alpha_l\rho_l(T_l - T_{sat})}{T_{sat}} \qquad T_l \geq T_{sat} \tag{12}$$

$$= \frac{r_v\alpha_v\rho_v(T_l - T_{sat})}{T_{sat}} \qquad T_l < T_{sat} \tag{13}$$

The parameters r_l and r_v are equal to 100 s^{-1} in order to maintain the interface temperature in the range of 1 K [28,38,39]. Small values of these coefficients produce a significant difference between saturation and interfacial temperature, while excessively large values cause numerical divergence. The direct measurements of the heat transmission are:

$$Q = -h_{lh}S \tag{14}$$

The following equation is used to determine the interface evaporative flux [40]

$$\ddot{m}_{eva} = \left[\frac{2\,.\,\overline{\sigma}}{2-\overline{\sigma}}\right] \cdot \left(\frac{MB_{vap}}{2\,.\,\pi\,.\,\overline{R}}\right)^{1/2} \cdot \left(\frac{P_{sat,liq}}{T_{liq}^{1/2}} - \frac{P_{sat,vap}}{T_{vap}^{1/2}}\right) \tag{15}$$

4.3. Simulation Condition and Grid System

The computational domain consists of a 2D axisymmetric with both the solid and liquid regions having the same specification as the quartz tube used in the experiments, as depicted in Figure 5. The length of the tube is 5 mm as the major temperature variation is in this range. This small length also gives benefits to fewer computation resources. The surface tension force's dominance over gravity at a small tube diameter is confirmed using the Bond number criterion [41]. The gravitational impact was therefore neglected, and the 2D calculation region is validated. Table 1 provides the heptane characteristics.

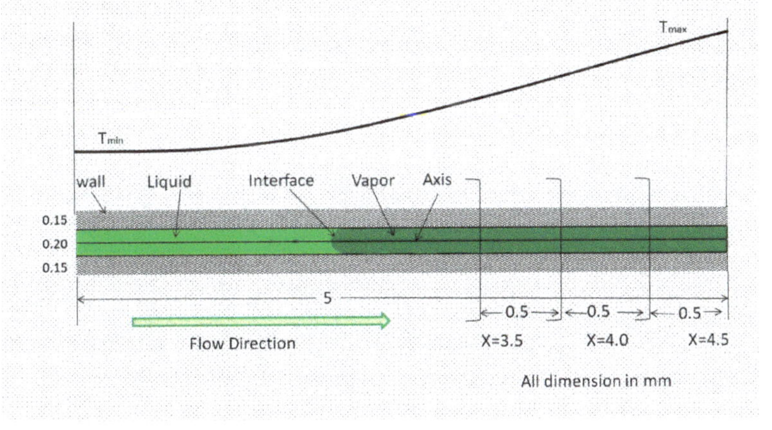

Figure 5. Schematic of the capillary tube.

Table 1. Heptane properties [35].

Property	Value
Density (kg/m^3)	666.95 at 295.45 K
Thermal conductivity (W/m-K)	0.12746
Thermal capacity (J/kg)	2298.4
Viscosity (kg/m-s)	0.000331
Vapor molecular weight	100
Evaporation latent heat (J/kg)	3.506×10^5
Saturated vapor pressure (Pa)	5316 at 295.45 K
Accommodation coefficient	1

For an FFR of 5 µL/min, the grid independence analysis was carried out. In the current research work, a quadrilateral mapped mesh is employed and it is shown in Figure 6. The same mesh size is used for both the solid and liquid parts. Table 2 represent a different grid for both solid and liquid domain at FFR of 5µL/min. To calculate the relative error (e%) the following equation is used:

$$e\% = \left|\frac{X_1 - X_2}{X_1}\right| \times 100 \tag{16}$$

where X_1 is the parameter value obtained from the finest grid; X_2 is the value obtained from other grids; and X_i (i = 1, 2, 3...) shows the variable of interest, such as the rate of evaporation at the interface, the average heat transfer coefficient, or the pressure drop.

The grid independence analysis was carried out at a mass flow rate of 5 μL/min. Three different mesh types are used in the current research. Table 2 provides the mesh detail information.

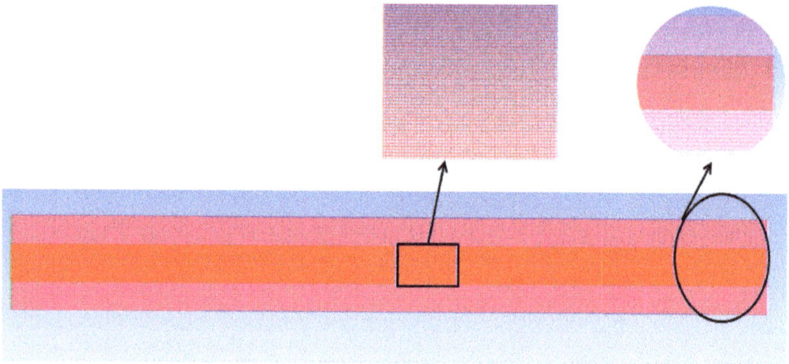

Figure 6. Computational grid: liquid and solid region.

Table 2. Mesh dependency result.

Grid No.	Grid Size	Average HTC (W/m^2-K)	e%	Evaporative Flux at Meniscus (Kg/m^2-s)	e%	ΔP (Pa)	e%
1	64,500	967	-----	8.63×10^{-9}	----	1324	----
2	34,000	960	0.72	8.56×10^{-9}	0.81	1315	0.67
3	22,000	945	2.2	8.5×10^{-9}	1.51	1290	2.5

Based on the above test results, mesh 2 will be employed in all simulations as adapted because the deviation from mesh 1 is minimal, and also in this mesh, the computational cost is low. The edge length along the x-axis is 4.5 μm, while the edge length in the y-axis is 4 μm. The courant number, which changes depending on the fuel flow rate, determines the time step. Table 3 provides the volumetric flow rate of fuel and the corresponding mass flow rate.

Table 3. Volumetric flow rate of fuel and corresponding mass flow rate.

FFR (μL/min)	5	10	30	50	70
MFR (mg/s)	0.056	0.113	0.343	0.58	0.797

The geometric reconstruction scheme [38] is employed for a volumetric fraction. Pressure-implicit with the splitting of operators, or PISO, is used to couple pressure and velocity because it provides a more approximate relationship between the two corrections. The momentum and energy equations were discretized using an upwind approach of the first order.

4.4. Data Reduction

The microtube's outer surface receives heat addition and is given by:

$$Q = \dot{m}c_p(T_{out} - T_{in}) \tag{17}$$

The heat flux at the microtube's outer surface is calculated using energy balance and is stated as:

$$q''_{applied} = \frac{Q}{A_s} = \frac{\dot{m}c_p(T_{out} - T_{in})}{2\pi r_o L} \tag{18}$$

The definition of the local Nusselt number is:

$$Nu_l = \frac{2\delta_f h_l}{\lambda_f} \tag{19}$$

where λ_f is n-heptane's thermal conductivity and h_l is the local heat transfer coefficient. To calculate the local heat transfer coefficient as:

$$h_l = \frac{q_z''}{T_w - T_f} \tag{20}$$

T_w is the local wall temperature at the microtube's inner wall, and T_f for the local bulk temperature at the tube's axis. Additional average local *HTC* over the microtube length information is provided by:

$$HTC_{ave} = \frac{1}{L}\int_0^L h_l dl \tag{21}$$

5. Results and Discussion

Table 4 lists the simulation conditions. The numerically simulated cases are solved for a time period of 1 s.

Table 4. Flow conditions.

Case. No	Case Name	Investigation	Remarks
1	FL5TB5	Experiment and Simulation	Boundary condition retrieved from experiment with 5 μL/min of mass flow rate
2	FL10TB10	Experiment and Simulation	Boundary condition retrieved from experiment with 10 μL/min of mass flow rate
3	FL30TB30	Experiment and Simulation	Boundary condition retrieved from experiment with 30 μL/min of mass flow rate
4	FL50TB50	Experiment and Simulation	Boundary condition retrieved from experiment with 50 μL/min of mass flow rate
5	FL70TB70	Experiment and Simulation	Boundary condition retrieved from experiment with 70 μL/min of mass flow rate
6	FL10TB5	Simulation	Flow rate is 10 μL/min while boundary condition is from case 1
7	FL30TB5	Simulation	Flow rate is 30 μL/min while boundary condition is from case 1
8	FL50TB5	Simulation	Flow rate is 50 μL/min while boundary condition is from case 1
9	FL70TB5	Simulation	Flow rate is 70 μL/min while boundary condition is from case 1
10	FL5TB10	Simulation	Flow rate is 5 μL/min while boundary condition is from case 2
11	FL5TB30	Simulation	Flow rate is 5 μL/min while boundary condition is from case 3
12	FL5TB50	Simulation	Flow rate is 5 μL/min while boundary condition is from case 4
13	FL5TB70	Simulation	Flow rate is 5 μL/min while boundary condition is from case 5

5.1. Validation

A separate validation of heat transfer, pressure drop, and evaporative flux is carried out because the two-phase flow in a microtube is a complicated process involving the heat transfer, pressure drop, and evaporation rate. Generalized correlations for 0.2 mm hydraulic diameter micro channels employing water fluid were proposed by Lee and Garimella [42]. Figure 6a depicts the results of a steady-state single-phase simulation with a chosen mesh at a heat flux of 50 W/cm^2 and water entering the microchannel at a temperature of 300 K. The findings are in good agreement with the correlation. The second validation for the Nusselt number is conducted utilizing data from Tiwari and Mohrana [29]. A separate single flow with water fluid is investigated in a hydraulic tube of 0.6 mm at a heat flux of 45 KW/m^2; the results are depicted in Figure 7a, which are consistent with the available literature data [29].

To build a correlation for heat transferal, a collection of 3899 data points covering 12 different wetting and non-wetting fluids with hydraulic diameter ranges of 0.16 to 2.92 mm, mass fluxes of 20 to 3000 Kg/m^2 s, and heat fluxes of 0.4 to 115 W/cm^2 was used by Bertsch et al. [43]. A separate steady-state simulation with n-heptane fluid at a mass flux of 140 Kg/m^2 s with varying heat fluxes of 30–75 KW/m^2 was calculated and compared with the HTC calculated by Bertsch correlation. The results are shown in Figure 7b and match with the literature [43].

An unsteady simulation was carried out with R141B at a fuel flow rate of 15 L/h and heat flux of 24 KW/m^2, similar to the experimental condition of Yang et al. [44], and the results are shown in Figure 6c. The properties of R141B were taken from REFPROP [45]. The results show that till 0.04 s, the oscillation of the pressure drop with the present model and published data is similar; however, at 0.05 s and 0.11 s, the pressure drop peaks are opposite. This difference is due to two reasons. First, the fluid properties used by Yang et al. [44] may vary with temperature, and in the present research, the fluid properties are taken as constant. The second reason is due to mesh size. The liquid patches inside the vapor region are almost similar in qualitative comparison, as shown in Figure 7c.

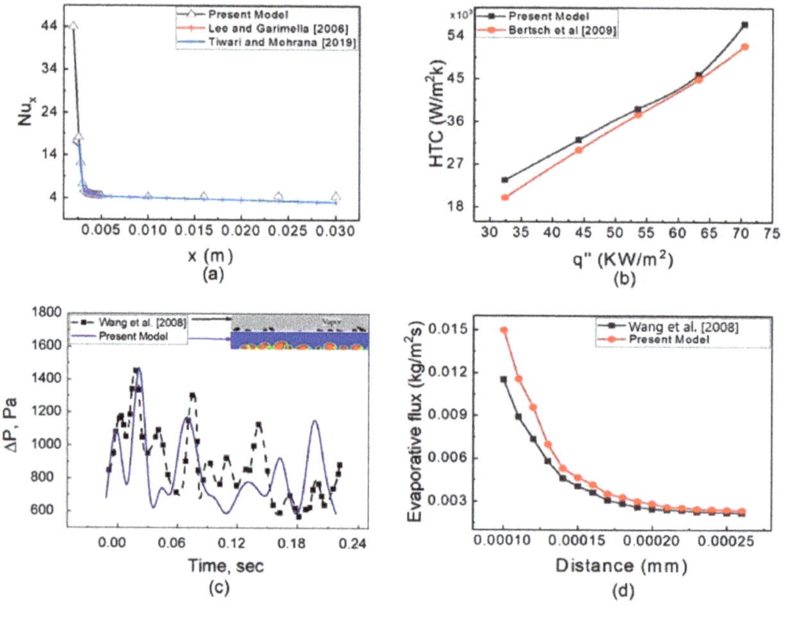

Figure 7. Validation of the present numerical model. (**a**) Validation of Nusselt number, Lee and Garimella [42], Tiwari and Mohrana [29]. (**b**) Validation of HTC, Bertsch et al. [43]. (**c**) Validation of pressure drop, Wang et al. [46]. (**d**) Validation of evaporative flux, Wang et al. [46].

Wang et al. [46] suggested a generalized model to calculate the evaporative flux at the interface with methanol fluid. Methanol's characteristics were acquired from REFPROP [45]. Using the present model, we calculate the evaporative flux in a 0.6 mm ID open microtube filled with methanol. The results are shown in Figure 7d. The evaporative flux at the corner is high with the present model and this is may be due to mesh size across the radial direction, whereas the evaporative flux at the middle of the meniscus is similar in both the present model and the previous literature [46].

5.2. Mass Flow Rate Effect on Two-Phase Flow and Evaporation Rate

At low fuel flow, the stable flow with dynamic meniscus inside the microtube is obtained, whereas a circular blue flame is anchored at the tube exit, as shown in Figure 2a. This blue flame indicates complete combustion due to the complete evaporation of liquid fuel at the interface [17,23]. The dynamic behavior of the evaporating meniscus is due to surface tension, vapor recoil, thermos-capillarity, viscous force, geometric constraints, and associated thin-film geometry [47]. At a high FFR, the nucleate bubble downstream of the evaporating meniscus was obtained. Similarly, at a low fuel flow rate, liquid droplets were not obtained; however, at a high fuel flow rate, liquid droplets with a random diameter were observed.

Figure 8a,b illustrates distinctive interface experimental and simulation results, respectively, with different time t_0. This time is different due to the involvement of several factors in experimentation and simulation. However, the sub-time step is the same. The liquid film is observed in the simulated results, whereas in the experiments, it was not observed. Figure 8c shows the dynamics of evaporating meniscus in both experimental and simulated conditions with a maximum difference of 2%. The courant number for a low fuel flow rate was kept under 2.

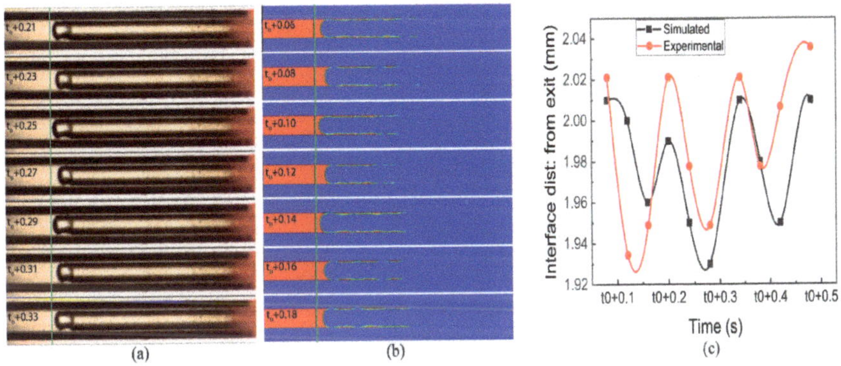

Figure 8. (**a**) Experimental results of interface dynamics. (**b**) Simulated results of interface dynamics. (**c**) Comparison of interface distance in experimental and simulated results (time in sec).

At FFR > 30 μL/min, an unstable flow with the nucleate bubble in the liquid region and liquid droplets at the vapor region were formed. Figure 9 shows the growth cycle of the nucleating bubbles in both the experimental and simulation work. Figure 9a indicates that the nucleating bubble first drops from the wall, increases in size, sticks to the wall again, and finally smears the interface with an asymmetric geometrical configuration. However, Figure 9b indicates the nucleating bubble at the axis due to the axisymmetric solver used in the calculation. Figure 9c shows a good comparison of the nucleating bubble diameter observed in the experimental and simulated results against the normalized time. Figure 9d shows the detachment distance between the nucleating bubble and the liquid–vapor interface. The detachment distance found in the simulated results was not in accordance with the experimental results, which may be due to the use of an axisymmetric solver; however, the profile of the detachment distance is nonlinear, as in the experimental results. The courant number for a high fuel flow rate was kept under 0.1.

Figure 10a shows the two-phase flow configuration with a nucleating bubble in the downstream and a liquid droplet in the upstream both in an experimental and simulated analysis at an instantaneous time. Figure 10b shows the liquid droplet diameter at different FFRs with an error of 10%, which is acceptable for the simulation method.

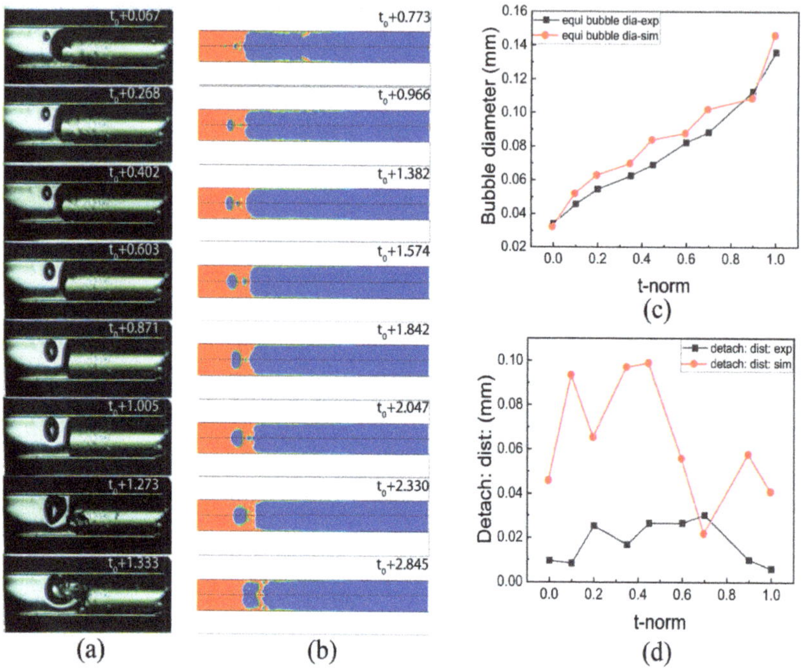

Figure 9. Nucleating bubble: (**a**) experimental result, (**b**) simulation result, (**c**) equivalent bubble diameter vs. normalize time, and (**d**) detachment distance vs. normalize time (time in m/s).

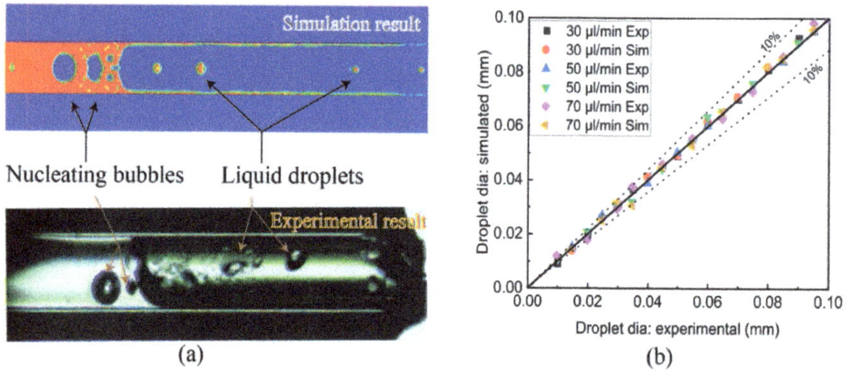

Figure 10. (**a**) Two-phase flow with nucleate bubble and liquid droplet at an instantaneous time. (**b**) Droplet diameter experimental vs. simulation at different fuel flow rates.

Figure 11 shows the pressure drop across the interface, evaporation rate at the interface, and interface distance with flow time. The distinctive interface moves 0.1 mm in the microtube due to the vapor pressure difference across the interface, and this movement has a significant effect on the pressure drop and evaporation rate. This oscillatory behavior of the evaporating meniscus is due to many factors like vapor recoil, surface tension, and thermos-capillarity, as discussed by [47]. The pressure drop and evaporation rate have the same trend in flow time. If we look at Figure 7d, we can see that the maximum evaporative flux is found near the wall and this is due to high diffusion flux. The non-uniformity of diffusion flux at the meniscus results in strong evaporation near the wall.

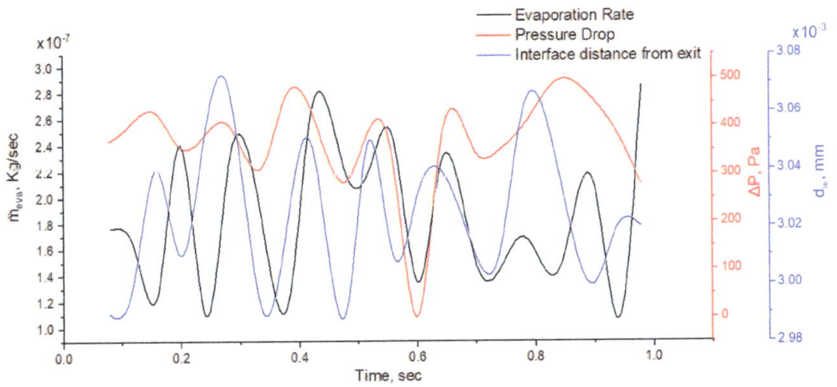

Figure 11. Plot of interfacial distance, pressure drop, and evaporation rate vs. time.

5.3. Mass Flow Rate Effect on Heat Transfer

The impact of FFR and TB on HTC is discussed in this section. Figure 12 depicts the axial fluctuation of HTC in a microtube at 0.20, 0.60, and 0.90 s and Equation (20) is also used to calculate the average HTC. Figure 11 illustrates that the HTC increases with an increase in FFR for both subcooled and bulk boiling, and this relationship is also endorsed by Lie et al. [48]. However, Bertsch et al. [49] oppose the rise of HTC with an increased mass flow rate. Moreover, Basu et al. [50] revealed that HTC mainly depends on surface tension. At FFR $\leq$ 10 μL/min, i.e., for the condition of FL5TB5 and FL10TB10, there is a single peak observed at all time steps, as shown in Figure 12a–c, which is possibly due to a very small temperature gradient across the interface. The HTC for FL10TB10 is higher from FL5TB5 because higher temperature gradients cause the temperature to increase more quickly, which induces stronger evaporation and results in the movement of the interface to the outlet of the tube.

At increased FFR $\geq$ 30 μL/min, there are multiple peaks of HTC, as shown in Figure 12. The precipitous peak of HTC value 16,000 W/m^2-K is found for FL50TB50 at 0.20 s, as shown in Figure 12a, while the remaining peaks are of the order of 12,000 W/m^2-K and this is due to nucleating bubbles, as seen in Figures 9 and 10. The same rapid peak is also revealed by Huang et al. [51]. The HTC for FL70TB70 first increases and then suddenly decreases due to two reasons. The first is due to the existence of a vapor bubble before the interface, which results in low heat transfer due to the low thermal conductivity of the quartz tube. The second is due to the presence of nucleating bubbles with low specific heat. The average HTC coefficient for five simulated settings during a time period of one second is shown in Figure 12d. The average HTC for FL5TB5 and FL10TB10 is less than 4000 W/m^2-K, whereas it ranges from 5000 W/m^2-K to 16,000 W/m^2-K for FL50TB50 and FL70TB70. The nucleating bubble is the reason why the average HTC for FL30TB30 is between 2950 and 6550 with only two peak values of 12,500 W/m^2-K and 14,500 W/m^2-K at 0.64 s and 0.89 s, respectively.

In Figure 12, the combined effect of FFR and TB is presented, which makes it difficult to recognize the dominating factor on HTC peaks. We further segregated the simulation into two types: the first type has the same FFR and the TB was changed, whereas in the second type, the TB was kept constant and FFR was changed. Figure 13 shows the axial HTC with the same FFR and different TB at 0.20, 0.60, and 0.90 s. The single peak is observed for both FL5TB5 and FL5TB10 at all time steps. Multiple peaks of HTC are formed for FL5TB30 at 0.90 s; similarly, multiple peaks with a maximum value of 7500 W/m^2-K and 7000 W/m^2-K are found for FL5TB50 and FL5TB70, respectively. The HTC in the subcooled regions remains the same; however, at 0.60 s, the HTC is higher, which is due to the nucleate bubble in the liquid region. Figure 12 demonstrates that as the interface location approaches the tube's exit, the values of HTC increase and this is due to an increase in the thermal gradient, which induces the nucleating bubble phenomenon that creates the liquid film, which forms high heat flux during liquid–solid thermal interaction. Figure 12d shows the average HTC at different intervals of time. The maximum average HTC is found for FL5TB70 at 0.16 s, whereas the other HTCs for the different conditions are superimposed at all intervals of time.

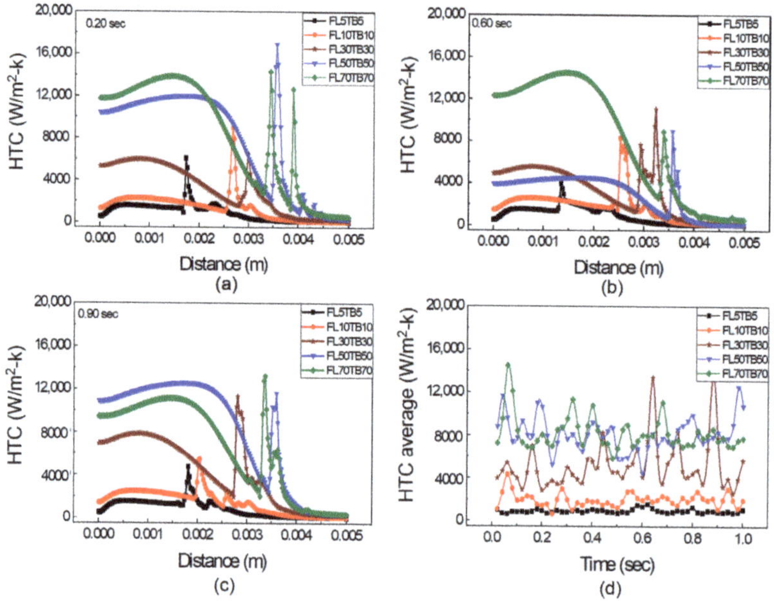

Figure 12. HTC at axial distance with different FFR and TB: (**a**) 0.20 s, (**b**) 0.60 s, and (**c**) 0.90 s. (**d**) Average HTC vs. flow time.

Figure 14 shows the axial HTC with different FFR and the same TB. The single peaks for both FL5TB5 and FL10TB5 are shown and multiple peaks for higher FFR are also shown. At 0.20 s, the maximum peak for 50 and 70 μL/min is 8000 and 11,200 W/m^2-K, respectively, which shows that an increase in FFR increases the HTC. At 0.60 s, the maximum peak for 50 and 70 μL/min is 11,400 and 11,600 W/m^2-K, respectively; however, this behavior of HTC is contradicted at 0.90 with 50 μL/min having a maximum HTC value of 11,500 W/m^2-K, while at 70 μL/min, the maximum value of HTC is 8000 W/m^2-K. Due to the existence of liquid droplets in the vapor area and nucleate bubbles in the liquid zone, there is a direct and inverse link between FFR and maximal HTC. These bubbles and droplets form the liquid film in the respective region, which increases and deteriorates the HTC, respectively. Figure 14d shows the average HTC at different intervals of time. After this time step, the average HTC for 50 μL/min is higher than 70 μL/min, which may be caused by flow phenomena inside the microtube. The average HTC and FFR have a direct relationship up to 0.6 s. The average value of HTC for 30 μL/min is greater than 50 μL/min and 70 μL/min at 0.60 s.

The discussion in this section leads to the conclusion that conjugate heat transfer and annular flow are the dominant processes in the microtube. This type of flow is also reported by Yen et al. [52]; however, they did not find a peak in the HTC calculation while investigating the convective boiling in a 0.51 mm ID tube with FC72 fluid. Furthermore, it is concluded that TB controls the peak position, whereas FFR controls the peak value of HTC.

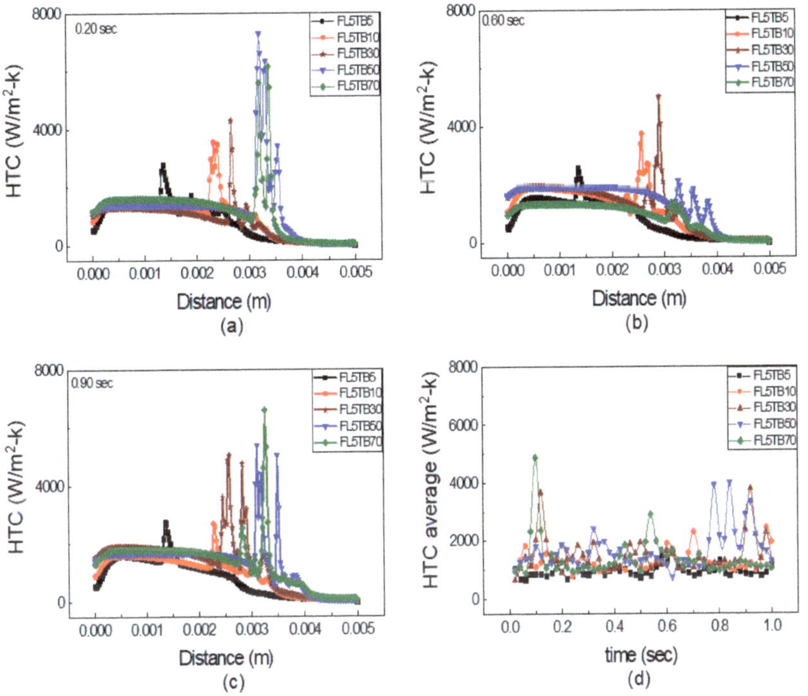

Figure 13. Axial HTC for same FFR and different TB: (**a**) 0.20 s, (**b**) 0.60 s, and (**c**) 0.90 s. (**d**) Average HTC vs. flow time.

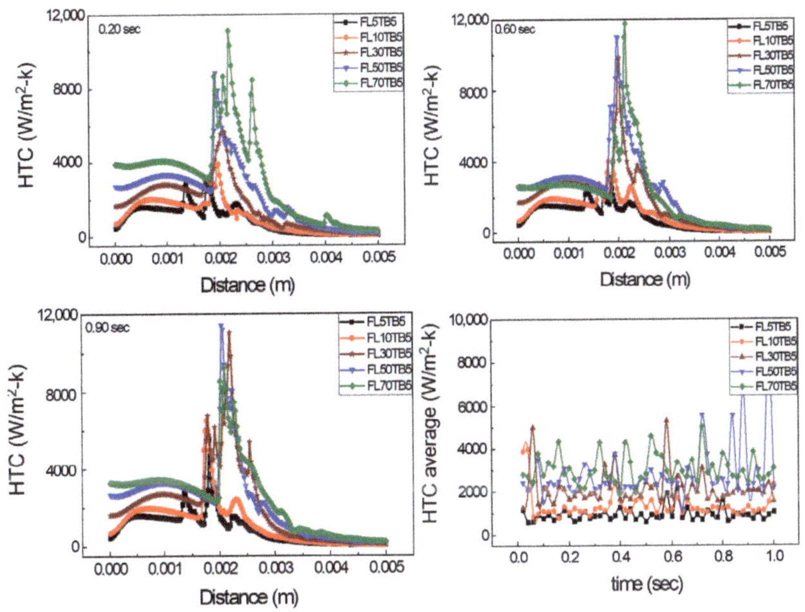

Figure 14. Axial HTC for different FFR and same TB: (**a**) 0.20 s, (**b**) 0.60 s, and (**c**) 0.90 s. (**d**) Average HTC vs. flow time.

5.4. Mass Flow Rate Effect on Pressure Oscillation and Volume Fraction

The influence of FFR is investigated on pressure oscillations and volume fraction. The results of three flow conditions, i.e., FL5TB5, FL30TB30, and FL70TB70, will be discussed. Figure 15 represents the pressure oscillation and spatial discretization of VF for FL5TB5. The pressure oscillations in the microtube are due to the rise in the vapor quality. Figure 14 shows that the increase in vapor quality decreases the two-phase flow density. This decrease in density further increases the velocity, and as a result, the pressure of the vapor phase increases. Wang et al. [53] estimated the capillary pressure in the stable flow using the following formula:

$$P_c = P_v - P_l \tag{22}$$

where P_l, P_c, and P_v are liquid pressure, capillary pressure, and vapor pressure, respectively. The interfacial tension present at the contact causes capillary pressure. The instability in the molecules' molecular force of attraction actually causes the interfacial tension. Figure 14 demonstrates that the capillary pressure varies over the course of the time step, which may be caused by conjugate heat transfer.

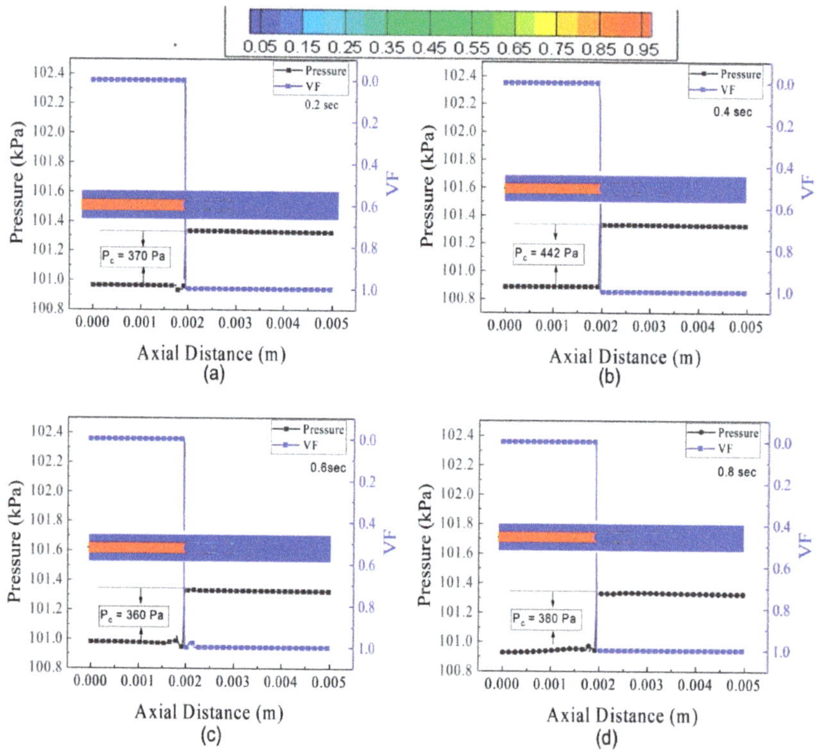

Figure 15. Pressure and volume fraction in the microtube at different intervals of time for FL5TB5: (**a**) 0.20 s, (**b**) 0.40 s, (**c**) 0.60 s, and (**d**) 0.8 s.

The pressure oscillation and VF for FL30TB30 are shown in Figure 16. In the microtube, nucleate boiling with an evicted liquid droplet from the interface occurs at high FFR. Figure 16a shows that liquid droplets escape from the interface at an axial location of 3.4 mm and have a pressure of 104.5 kPa, which is due to the high kinetic energy of the droplets. Similar to this, the nucleate bubble that occupies two-thirds of the microtube's diameter has an extreme pressure of 101.8 kPa prior to a larger size and breaking down the liquid–vapor interface. Figure 16b shows a nucleation site with a vapor VF of 0.4 that has a pressure of 103.2 Kpa, while two small nucleating bubbles have a maximum pressure of 102.8 kPa. A single liquid droplet is seen at both 0.2 and 0.4 s, with an approximate pressure of 104.2 kPa. Figure 16c shows a very tiny micro droplets with a pressure of 105 kPa due to a small volume.

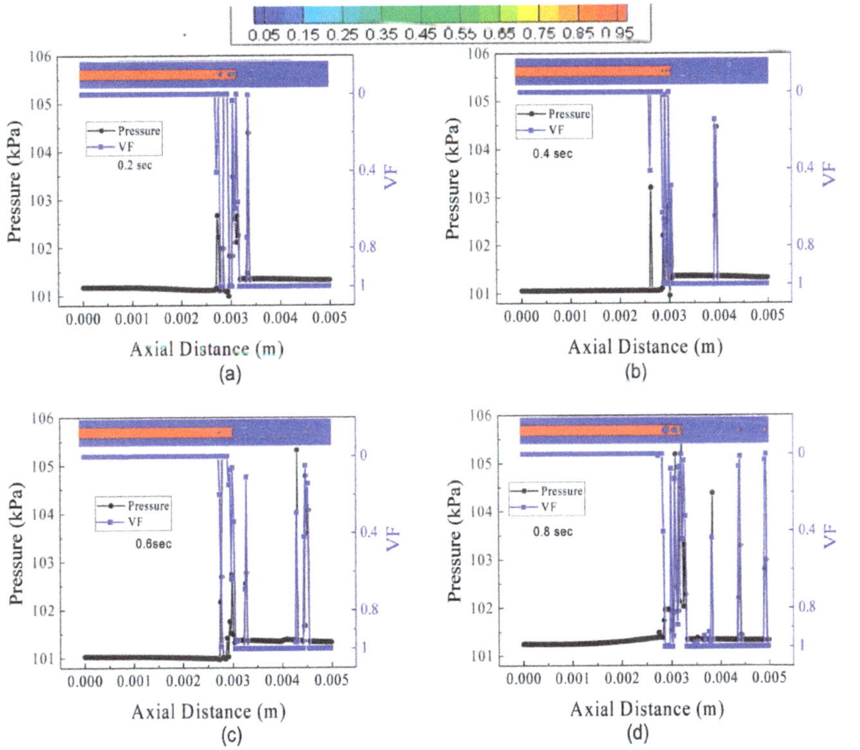

Figure 16. Pressure and volume fraction in the microtube at different intervals of time for FL30TB30: (**a**) 0.20 s, (**b**) 0.40 s, (**c**) 0.60 s, and (**d**) 0.8 s.

The pressure oscillation and VF for FL70TB70 are shown in Figure 17. The liquid droplet intensity breakout from the interface is higher than FL30TB30. The liquid slug is visible at 0.4 s and a train of the liquid droplet can be termed as Taylor bubbles. It is important to note that at 0.2 s, the nucleating bubble continues to grow and the liquid regime's axial pressure continues to decline, whereas at 0.4 s, the nucleating bubble breaks up and the liquid regime's axial pressure tends to increase. The rise in wall temperature may be the cause of the pressure drop, whilst the energy development in the nascent nucleating bubble is the cause of the increase. The increase in axial pressure can also be seen at 0.6 s, as shown in Figure 17c. A similar increase and decrease in axial pressure can also be seen in Figure 16a,d for FL30TB30; however, this increase or decrease is not as steep as in Figure 17a,c, and this may be due to an increase in the fuel flow rate for the latter case.

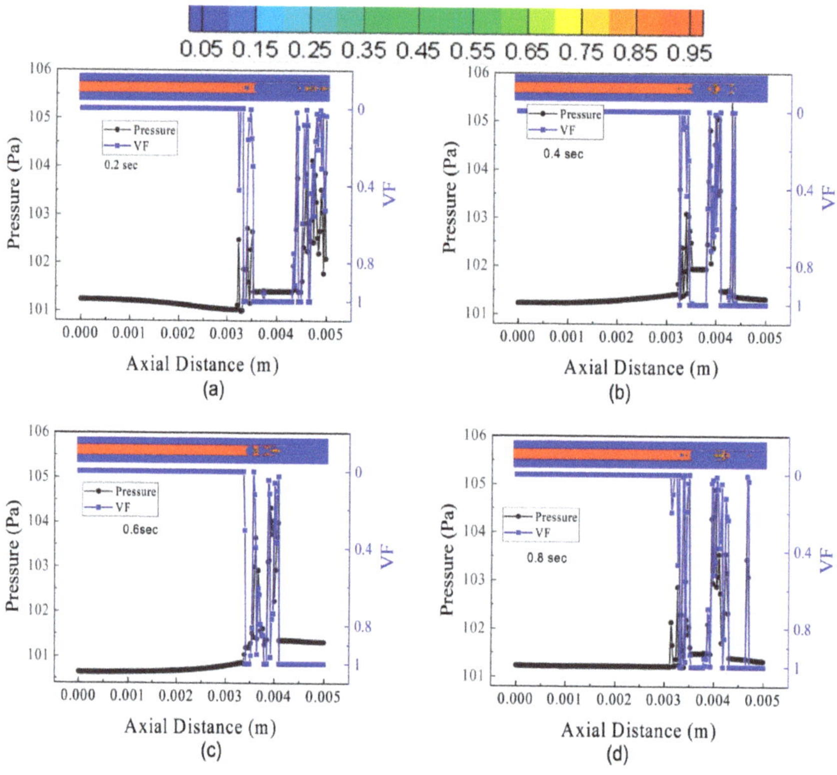

Figure 17. Pressure and volume fraction in the microtube at several intervals of time for FL70TB70: (**a**) 0.20 s, (**b**) 0.40 s, (**c**) 0.60 s, and (**d**) 0.8 s.

5.5. Fuel Flow Rate Effect on Ressure Drop

To determine the impact of fuel flow rate and wall temperature on the fluctuation frequency and amplitude, the data on pressure drop are analyzed. For each of the 13 cases listed in Table 4, the pressure drop data are presented in Figure 18a,c,e against a timescale of one second. Figure 18a demonstrates that the maximum pressure drop increases from 500 Pa to 2800 Pa when the fuel flow rate increases from 5 μL/min to 70 μL/min. This increase in extreme pressure drop is caused by greater flow resistance at an upper mass flow rate. Wang et al. [54] investigated the pressure drop fluctuation of two-phase flow in a 0.5 mm microtube and reported similar results. However, Yen et al. [52] disagreed with this assertion and came to the conclusion that the pressure drop features in microtubes remain unchanged. Figure 18c depicts that the maximum pressure drop oscillations are up to 900 Pa if the FFR is kept constant, whereas Figure 18e demonstrates that the maximum pressure drop oscillations are up to 2600 Pa if the FFR is changed and TB is kept constant. This depicts that fuel flow rate has a dominant impact on pressure decrease.

The results of the pressure drop in this part represent the entire pressure drop during the two-phase flow. As a result, the pressure drop entails three components: pressure drop owing to acceleration, which is also influenced by void fraction and density changes, pressure drop due to wall friction, and pressure drop caused by abrupt contraction and expansion in the microchannel [55]. According to Kawahara et al. [55] the frictional pressure drop accounts for the majority of the total pressure loss. The pressure reduction in the current study is mostly caused by phase shift, the presence of the nucleating bubble, and the liquid droplet.

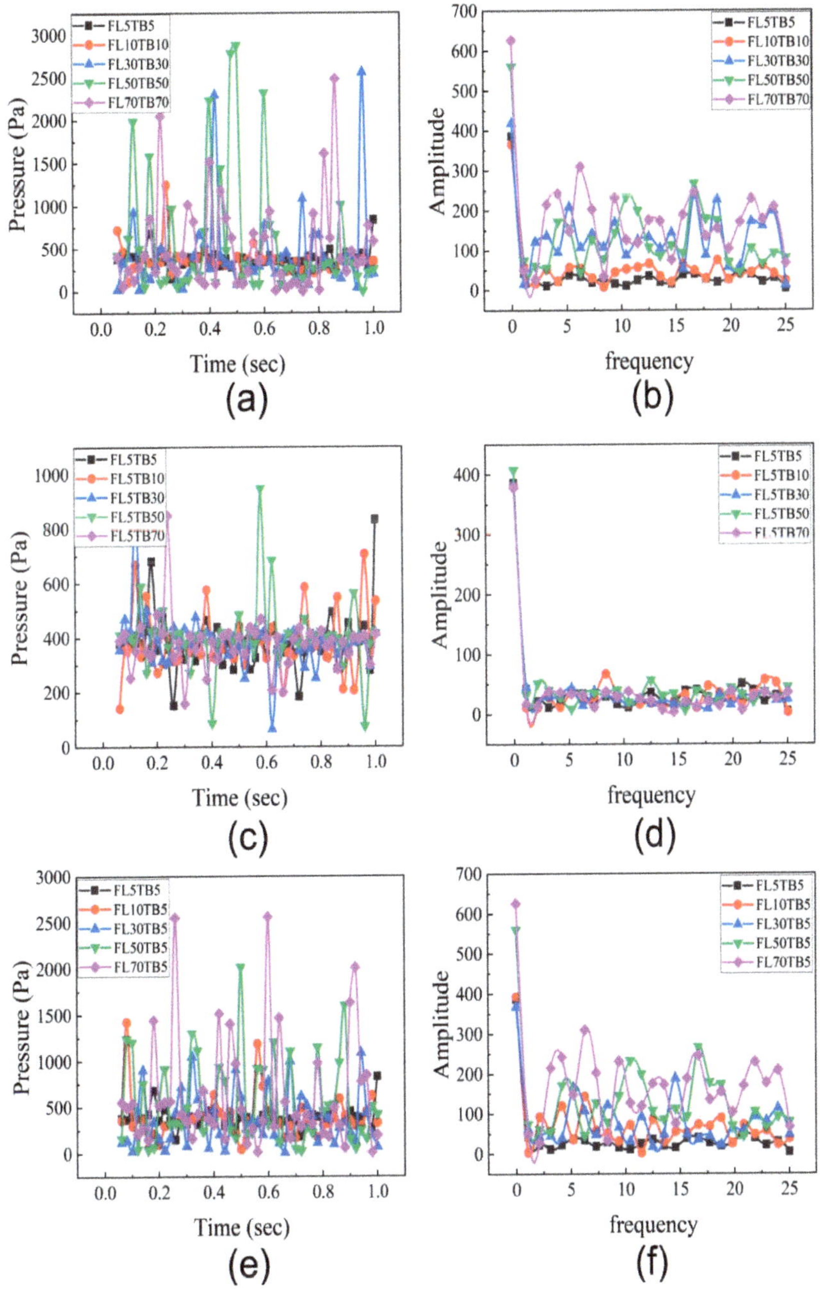

Figure 18. Pressure drop data for different flow conditions. (**a**) Pressure drop with different flow rate and different temperature boundary. (**b**) Pressure drop frequency spectrum with different flow rate and different temperature boundary. (**c**) Pressure drop with same flow rate and different temperature boundary. (**d**) Pressure drop frequency spectrum with same flow rate and different temperature boundary. (**e**) Pressure drop with different flow rate and same temperature boundary. (**f**) Pressure drop frequency spectrum with different flow rate and same temperature boundary.

The frequency spectrum of the pressure drop was obtained for all 13 cases as described in Table 4 and shown in Figure 18b,d,e. Figure 18b shows the FFT analysis for flow conditions with different flow rates and different temperature gradients. For 5 μL/min, the high amplitude of 380 is found at a low frequency of 2 Hz while at a high frequency, the amplitude is 50. For 70 μL/min, the amplitude was 620 at a low frequency, while at a higher frequency, the maximum amplitude is 300. Figure 18d shows the FFT analysis for 5 μL/min with different temperature gradients. As seen in Figure 18d, at low frequency, the amplitude is 400 for all temperature gradients, while at higher frequency, the amplitude is around 50 for all temperature gradients. This concludes that the high amplitude is due to the mass flow rate. Bogojevic et al [56] and Wang et al. [54] captured the high-frequency low-altitude fluctuation as well as the low-frequency high-amplitude fluctuation. While the high-frequency, low-amplitude fluctuations were mostly caused by bubbles and liquid droplets, the low-frequency, high-amplitude fluctuations were thought to be caused by two-phase vapor–liquid separation. This is the reason that at 5 μL/min, the amplitude fluctuation is not observed due to the absence of liquid droplets.

The pressure drop's coefficient of variation (COV) is also used to study pressure drop oscillations. This coefficient is used to determine variability and diversity of the data from the average value. It is the ratio of the mean value ΔP_{avg} to the standard deviation ΔP_{rms}. Singh et al. [57] previously employed this analysis technique. The coefficient of variation is plotted in Figure 19. The COV increases with the rise in the fuel flow rate for the first group till 50 μL/min, and then decreases, and a second-order polynomial curve is obtained. For the third group, the COV is almost similar at all flow rates and a linear fitted curve was obtained.

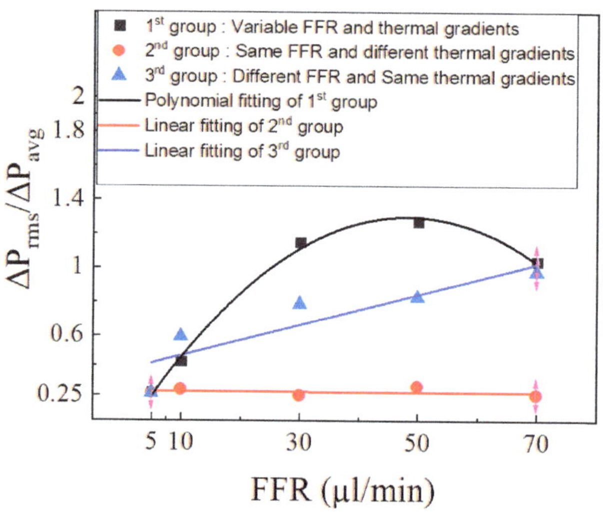

Figure 19. Coefficient of variation at different FFR.

6. Conclusions

The numerical model analyses and thoroughly discusses the flow boiling of liquid n-heptane in a microtube at various fuel flow rates. In the experimental work, nucleating bubbles, liquid bubbles, and two-phase flow patterns are all observed. To examine heat transfer, pressure oscillation, pressure drop, and evaporative flux at the meniscus, numerical simulations have been carried out at a fuel flow rate comparable to that employed in the experimental investigations. The main outcomes of this study are as follows:

1. The steady flow with a dynamic meniscus was obtained at FFR < 10 μL/min and the highest wall temperature was 1050 K. In the microtube, an unstable flow with nucleate bubbles and liquid droplets predominates at an FFR of 30 to 70 μL/min.

2. It is found that the pressure drop across the meniscus and the interface distance of the meniscus from the exit of the microtube affect the evaporative flux at the meniscus.
3. One peak of HTC was obtained when the FFR was less than 10 µL/min, but at high fuel flow rates of 30 to 70 µL/min, numerous peaks of HTC were obtained at various time steps. The location of the peak was determined based on the temperature boundary condition, whereas the peak of HTC is dependent on the fuel flow rate.
4. At a low fuel flow rate of 10 µL/min, the capillary force was found to be dynamic. The pressure oscillation was found at 30–70 µL/min. In the liquid regime, the axial pressure tends to rise during nucleating bubble breakup and decrease during bubble growth.
5. At various fuel flow rates, the oscillation of the pressure drop was also obtained in the microtube. At all flow rates, both high-frequency low-amplitude and low-frequency high-amplitude pressure fluctuations were observed. The liquid n-heptane fuel flow rate is mostly responsible for the large amplitude.
6. The fuel flow rate remained constant while the temperature boundary changed, or it remained constant while the fuel flow rate changed, although the slope of the fitted curve was larger in the latter instance.

Author Contributions: M.T.R.: conceptualization, formal analysis, software, investigation, and writing—original draft; N.A. and R.F.S.: conceptualization, formal analysis, investigation, and writing—original draft; M.B.K.: methodology, formal analysis, validation; N.A.: supervision, visualization, review and writing—original draft; M.T.R., R.F.S., and M.B.K.: data curation, writing—review and editing, proofreading. All authors have read and agreed to the published version of the manuscript.

Funding: This research receives no external fundings.

Data Availability Statement: No data available.

Conflicts of Interest: The authors declare no conflict of interest.

References

1. Ju, Y.; Maruta, K. Microscale combustion: Technology development and fundamental research. *Prog. Energy Combust. Sci.* **2011**, *37*, 669–715. [CrossRef]
2. Maruta, K. Micro and mesoscale combustion. *Proc. Combust. Inst.* **2011**, *33*, 125–150. [CrossRef]
3. Epstein, A.H. Millimeter-Scale, Micro-Electro-Mechanical Systems Gas Turbine Engines. *J. Eng. Gas Turbines Power* **2004**, *126*, 205–226. [CrossRef]
4. Gu, Y. Gasdynamic Modeling and Parametric Study of Mesoscale Internal Combustion Swing Engine/Generator Systems. Ph.D. Thesis, University of Michigan, Ann Arbor, MI, USA, 2006.
5. Dahm, W.; Mijit, J.; Mayor, R.; Qiao, G.; Benajmin, A.; Gu, Y.; Lei, Y.; Papke, M.; Wu, S. Micro internal combustion swing engine (MICSE) for portable power generation systems. In Proceedings of the 40th AIAA Aerospace Sciences Meeting & Exhibit, Reno, NV, USA, 14–17 January 2002.
6. Liu, Q.; Jiang, L.; Yang, W.; Zhao, D. Experimental study of flame propagation in single chamber of MICSE. In Proceedings of the 10th Asia-Pacific Conference on Combustion, ASPACC, Beijing, China, 19–22 July 2015.
7. Zhou, X.; Zhang, Z.; Kong, W.; Du, N. Investigations of leakage mechanisms and its influences on a micro swing engine considering rarefaction effects. *Appl. Therm. Eng.* **2016**, *106*, 674–680. [CrossRef]
8. Kim, N.I.; Kato, S.; Kataoka, T.; Yokomori, T.; Maruyama, S.; Fujimori, T.; Maruta, K. Flame stabilization and emission of small Swiss-roll combustors as heaters. *Combust. Flame* **2005**, *141*, 229–240. [CrossRef]
9. Kim, N.I.; Aizumi, S.; Yokomori, T.; Kato, S.; Fujimori, T.; Maruta, K. Development and scale effects of small Swiss-roll combustors. *Proc. Combust. Inst.* **2007**, *31*, 3243–3250. [CrossRef]
10. Chou, S.K.; Yang, W.M.; Chua, K.J.; Li, J.; Zhang, K.L. Development of micro power generators—A review. *Appl. Energy* **2011**, *88*, 1–16. [CrossRef]
11. Jiang, D.; Yang, W.; Chua, K.J.; Ouyang, J. Thermal performance of micro-combustors with baffles for thermophotovoltaic system. *Appl. Therm. Eng.* **2013**, *61*, 670–677. [CrossRef]
12. Dunn-Rankin, D.; Leal, E.M.; Walther, D.C. Personal power systems. *Prog. Energy Combust. Sci.* **2005**, *31*, 422–465. [CrossRef]
13. Hussain, T.; Markides, C.N.; Balachandran, R. Flame dynamics in a micro-channeled combustor. *AIP Conf. Proc.* **2015**, *1642*, 130–137. [CrossRef]

14. Loire, S.; Fonoberov, V.A. Combustion of methane in microchannels. In Proceedings of the ASME 2009 International Mechanical Engineering Congress and Exposition, Lake Buena Vista, FL, USA, 13–19 November 2009; American Society of Mechanical Engineers Digital Collection; American Society of Mechanical Engineers: New York, NY, USA, 2009. [CrossRef]
15. Maruta, K. Flame Chromatography: Toward Fuel Indexing Based on Multiple Weak Flames in a Meso-Scale Channel with a Prescribed Temperature Profile. In Proceedings of the ASME 2011 9th International Conference on Nanochannels, Microchannels, and Minichannels, Edmonton, AB, Canada, 19–22 June 2011; American Society of Mechanical Engineers Digital Collection; Canada. American Society of Mechanical Engineers: New York, NY, USA, 2012. [CrossRef]
16. Kinoshita, K.; Yoshimoto, T.; Kato, Y.; Kubo, N.; Takagi, T. Stability Limits, Behaviors and Characteristics of the Diffusion Flame. *Proc. Conf. Kansai Branch* **2004**, *79*, 5–15. [CrossRef]
17. Rashid, M.T.; Li, J.; Chen, X.; Wang, N. Flow Boiling and Heat Transfer of N-heptane Flow in a Microtube Heated by Concurrent Microflame. *Combust. Sci. Technol.* **2021**, *195*, 265–293. [CrossRef]
18. Leong, K.; Ho, J.; Wong, K. A critical review of pool and flow boiling heat transfer of dielectric fluids on enhanced surfaces. *Appl. Therm. Eng.* **2016**, *112*, 999–1019. [CrossRef]
19. Prajapati, Y.K.; Bhandari, P. Flow boiling instabilities in microchannels and their promising solutions—A review. *Exp. Therm. Fluid Sci.* **2017**, *88*, 576–593. [CrossRef]
20. Dalkilic, A.S. A review of flow boiling in mini and microchannel for enhanced geometries. *J. Therm. Eng.* **2018**, *4*, 2037–2074. [CrossRef]
21. Zhou, X.; Li, J.; Chen, X.; Wang, N. Effects of pulse power on evaporation characteristics of an n-heptane droplet in a microtube. *Appl. Therm. Eng.* **2023**, *231*, 120895. [CrossRef]
22. Nebuloni, S.; Thome, J.R. Numerical modeling of laminar annular film condensation for different channel shapes. *Int. J. Heat Mass Transf.* **2010**, *53*, 2615–2627. [CrossRef]
23. Rashid, M.T.; Li, J.; Chen, X.; Song, A.; Wang, N. Numerical simulation of evaporation phenomena and heat transfer of liquid Hydrocarbon in a microtube. *Int. J. Heat Mass Transf.* **2021**, *179*, 121734. [CrossRef]
24. Zhao, T.; Liao, Q. Theoretical analysis of film condensation heat transfer inside vertical mini triangular channels. *Int. J. Heat Mass Transf.* **2002**, *45*, 2829–2842. [CrossRef]
25. Wang, H.S.; Rose, J.W. A Theory of Film Condensation in Horizontal Noncircular Section Microchannels. *J. Heat Transf.* **2005**, *127*, 1096–1105. [CrossRef]
26. Wang, H.S.; Rose, J.W. Film condensation in horizontal microchannels: Effect of channel shape. *Int. J. Therm. Sci.* **2006**, *45*, 1205–1212. [CrossRef]
27. Ferrari, A.; Magnini, M.; Thome, J.R. Numerical analysis of slug flow boiling in square microchannels. *Int. J. Heat Mass Transf.* **2018**, *123*, 928–944. [CrossRef]
28. Guo, Z.; Haynes, B.; Fletcher, D. Numerical simulation of annular flow boiling in microchannels. *J. Comput. Multiph. Flows* **2016**, *8*, 61–82. [CrossRef]
29. Tiwari, N.; Moharana, M.K. Conjugate heat transfer analysis of liquid-vapor two phase flow in a microtube: A numerical investigation. *Int. J. Heat Mass Transf.* **2019**, *142*, 118427. [CrossRef]
30. Ahmed, S.A.; Zhou, S.; Tsegay, S.; Ahmad, N.; Zhu, Y. Effects of hydrogen-enriched biogas on combustion and emission of a dual-fuel diesel engine. *Energy Sources Part A Recover. Util. Environ. Eff.* **2020**, 1–16. [CrossRef]
31. Ahmed, S.A.; Zhou, S.; Tsegay, A.S.; Zhu, Y.; Malik, A.; Ahmad, N.; Legese, Z. Effects of CO_2 ratio in biogas on performance, combustion and emissions of a dual-fuel diesel engine. *Int. Energy J.* **2020**, 15–38.
32. Müller-Steinhagen, H.; Epstein, N.; Watkinson, A.P. Subcooled-boiling and convective heat transfer for heptane flowing inside an annulus and past a coiled wire: Part II—Correlation of data. *J. Heat Transfer.* **1986**, *108*, 928–933. [CrossRef]
33. Müller-Steinhagen, H.; Watkinson, A.; Epstein, N. Subcooled-boiling and convective heat transfer to heptane flowing inside an annulus and past a coiled wire: Part I—Experimental results. *J. Heat Transfer.* **1986**, *108*, 922–927. [CrossRef]
34. Hapke, I.; Boye, H.; Schmidt, J. Flow boiling of water and n-heptane in micro channels. *Microscale Thermophys. Eng.* **2002**, *6*, 99–115. [CrossRef]
35. Dhavaleswarapu, H.K.; Murthy, J.Y.; Garimella, S.V. Numerical investigation of an evaporating meniscus in a channel. *Int. J. Heat Mass Transf.* **2012**, *55*, 915–924. [CrossRef]
36. Ahmed, S.A.; Zhou, S.; Zhu, Y.; Feng, Y.; Malik, A.; Ahmad, N. Influence of Injection Timing on Performance and Exhaust Emission of CI Engine Fuelled with Butanol-Diesel Using a 1D GT-Power Model. *Processes* **2019**, *7*, 299. [CrossRef]
37. Lee, W.H. *Multiphase Trabsport Fundamentals, Reator Saftey, Applications*; Hemisphere Publishing: Washington, DC, USA, 1980; Volume 1.
38. Khan, Z.A.; Ahmad, N.; Sattar, M.; Haq, M.A.; Khan, I.; Ganie, A.H. Cell alternation algorithm for simulating bubble growth in boiling flows through volume of fluid (VOF) method in fluent. *Alex. Eng. J.* **2022**, *61*, 13051–13066. [CrossRef]
39. Fang, C.; David, M.; Rogacs, A.; Goodson, K. Volume of fluid simulation of boiling two-phase flow in a vapor-venting microchannel. *Front. Heat Mass Transf.* **2010**, *1*, 013002. [CrossRef]
40. Schrage, R.W. *A Theoretical Study of Interphase Mass Transfer*; Columbia University Press: New York, NY, USA, 1953.
41. Wu, C.; Li, J. Numerical simulation of flow patterns and the effect on heat flux during R32 condensation in microtube. *Int. J. Heat Mass Transf.* **2018**, *121*, 265–274. [CrossRef]

42. Lee, P.-S.; Garimella, S.V. Thermally developing flow and heat transfer in rectangular microchannels of different aspect ratios. *Int. J. Heat Mass Transf.* **2006**, *49*, 3060–3067. [CrossRef]
43. Bertsch, S.S.; Groll, E.A.; Garimella, S.V. A composite heat transfer correlation for saturated flow boiling in small channels. *Int. J. Heat Mass Transf.* **2009**, *52*, 2110–2118. [CrossRef]
44. Yang, Z.; Peng, X.; Ye, P. Numerical and experimental investigation of two phase flow during boiling in a coiled tube. *Int. J. Heat Mass Transf.* **2008**, *51*, 1003–1016. [CrossRef]
45. Lemmon, E.W.; Huber, M.L.; McLinden, M.O. *REFPROP: Reference Fluid Thermodynamic and Transport Properties*; NIST Standard Reference Database; Standard Reference Data Program; National Institute of Standards and Technology: Gaithersburg, MA, USA, 2007; Volume 23.
46. Wang, H.; Murthy, J.Y.; Garimella, S.V. Transport from a volatile meniscus inside an open microtube. *Int. J. Heat Mass Transf.* **2008**, *51*, 3007–3017. [CrossRef]
47. Polansky, J.; Kaya, T. Stability of an evaporating meniscus: Part I–Theoretical analysis. *Int. J. Therm. Sci.* **2016**, *107*, 209–219. [CrossRef]
48. Lie, Y.; Su, F.; Lai, R.; Lin, T. Experimental study of evaporation heat transfer characteristics of refrigerants R-134a and R-407C in horizontal small tubes. *Int. J. Heat Mass Transf.* **2006**, *49*, 207–218. [CrossRef]
49. Bertsch, S.S.; Groll, E.A.; Garimella, S.V. Effects of heat flux, mass flux, vapor quality, and saturation temperature on flow boiling heat transfer in microchannels. *Int. J. Multiph. Flow* **2008**, *35*, 142–154. [CrossRef]
50. Basu, S.; Ndao, S.; Michna, G.J.; Peles, Y.; Jensen, M.K. Flow Boiling of R134a in Circular Microtubes—Part I: Study of Heat Transfer Characteristics. *J. Heat Transf.* **2011**, *133*, 051502. [CrossRef]
51. Huang, S.-C.; Kawanami, O.; Kawakami, K.; Honda, I.; Kawashima, Y.; Ohta, H. Experimental Study of the Relation Between Heat Transfer and Flow Behavior in a Single Microtube. *Microgravity Sci. Technol.* **2008**, *20*, 193–197. [CrossRef]
52. Yen, T.-H.; Kasagi, N.; Suzuki, Y. Forced convective boiling heat transfer in microtubes at low mass and heat fluxes. *Int. J. Multiph. Flow* **2003**, *29*, 1771–1792. [CrossRef]
53. Wang, S.; Liu, P.; Zhao, H.; Zhang, Y. A novel method for calculating the dynamic capillary force and correcting the pressure error in micro-tube experiment. *Sci. Rep.* **2017**, *7*, 16590. [CrossRef] [PubMed]
54. Wang, Y.; Sefiane, K.; Wang, Z.-G.; Harmand, S. Analysis of two-phase pressure drop fluctuations during micro-channel flow boiling. *Int. J. Heat Mass Transf.* **2014**, *70*, 353–362. [CrossRef]
55. Kawahara, A.; Chung, P.-Y.; Kawaji, M. Investigation of two-phase flow pattern, void fraction and pressure drop in a microchannel. *Int. J. Multiph. Flow* **2002**, *28*, 1411–1435. [CrossRef]
56. Bogojevic, D.; Sefiane, K.; Walton, A.; Lin, H.; Cummins, G. Two-phase flow instabilities in a silicon microchannels heat sink. *Int. J. Heat Fluid Flow* **2009**, *30*, 854–867. [CrossRef]
57. Singh, S.G.; Bhide, R.R.; Duttagupta, S.P.; Puranik, B.P.; Agrawal, A. Two-Phase Flow Pressure Drop Characteristics in Trapezoidal Silicon Microchannels. *IEEE Trans. Components Packag. Technol.* **2009**, *32*, 887–900. [CrossRef]

micromachines

MDPI

Article

Experimental Investigation on the Flow Boiling of Two Microchannel Heat Sinks Connected in Parallel and Series for Cooling of Multiple Heat Sources

Zhengyong Jiang [1], Mengjie Song [1,*], Jun Shen [1], Long Zhang [1], Xuan Zhang [1] and Shenglun Lin [2]

1 Department of Energy and Power Engineering, School of Mechanical Engineering, Beijing Institute of Technology, Beijing 100081, China

2 Department of Environmental Engineering, National Cheng Kung University, Tainan 70101, Taiwan; z11208011@ncku.edu.tw

* Correspondence: mengjie.song@gmail.com; Tel.: +86-188-1172-6830

Abstract: Cooling methods for multiple heat sources with high heat flux have rarely been reported, but such situations threaten the stable operation of electronic devices. Therefore, in this paper, the use of two microchannel heat sinks is proposed, with and without grooves, labeled Type A and Type B, respectively. Experimental investigations on the flow boiling of two microchannel heat sinks connected in parallel and in series are carried out under different mass fluxes. In addition, a high-speed camera is used to observe flow patterns in the microchannels. The cold plate wall temperature (T_w), heat transfer coefficient (*HTC*), and pressure drop (*PD*) are obtained with the use of two microchannel heat sinks. The flow patterns of the bubbly flow and elongated bubbles in the microchannels are observed. The results of the analysis indicated that the T_w, *HTC*, and *PD* of the two microchannel heat sinks connected in parallel were degraded, especially when using the Type A-B parallel connection. Compared to the use of a single heat sink, the maximum decrease in *HTC* was 9.44 kW/(m^{2}K) for Type A heat sinks connected in parallel, which represents a decrease of 45.95%. The influence of the series connection on the T_w, *HTC*, and *PD* of the two heat sinks is obvious. The Type A-A series connection exerted the greatest positive effect on the performance of the two heat sinks, especially in the case of the postposition heat sink. The maximum increase in *HTC* was 12.77 kW/(m^{2}K) for the postposition Type A heat sink, representing an increase of 72.88%. These results could provide a reference for a two-phase flow-cooling complex for multiple heat sources with high heat flux.

Keywords: microchannel heat sinks; flow boiling; series connection; parallel connection; heat transfer coefficient; wall temperature; pressure drop

1. Introduction

From the chip level to the system level, there are an enormous number of high-power electronic modules that are widely used to perform multi-threaded tasks requiring high strength, such as in supercomputing [1,2], high-power lasers, automotive power battery packs [3], phased-array radar, etc. [4]. To ensure stability and reliability, the problem of cooling multiple heat sources with high heat flux needs to be addressed. As stated, eliminating high heat flux from multiple heat sources is important, but research on relevant strategies have rarely been reported [5]. At present, traditional air and liquid-flow cooling strategies are widely used to cool electronics [6]. Air cooling is typically used when the heat flux is lower than 50 W/cm^2. Liquid-flow cooling can be used to address heat flux reaching 100 W/cm^2 but with a greater degree of uniformity of temperature between the inlet and outlet [7,8]. Furthermore, the heat flux of highly integrated electronic devices can reach 1000 W/cm^2 [9]. The traditional cooling methods are not able to meet the heat dissipation of these high-heat-flux devices. Moreover, cooling problems for multiple heat sources with high heat flux are more challenging to overcome. Hence, there is a pressing need to develop and test advanced cooling techniques. One efficient method for solving high-heat-flux cooling problems is microchannel flow boiling [10,11].

Two-phase flow cooling has a higher heat transfer capacity and lower coolant flow velocity compared to single-phase flow [12,13]. This is because it mainly utilizes the latent heat of the coolant to absorb and carry away large amounts of heat. Due to the rapid evaporation of liquid, a large number of bubbles are produced, which are accompanied by variable and complex flow patterns [14,15]. In addition, the heat transfer properties of flow boiling in microchannels are strongly dependent on the flow pattern [16,17]; for example, the liquid–vapor flow distribution in the microchannels can affect the heat transfer capacity and pressure drop of flow boiling [18]. In addition, when flow boiling occurs in a long tube, there are several regions with increased superheat [10]. It has been reported that flow boiling processes are classically divided into five categories, including liquid convection, subcooled boiling, saturated boiling, transition boiling, and film boiling [19]. In addition, saturated boiling has the highest heat transfer capacity. However, as the superheating increases, there is a critical heat flux. When the heat flux exceeds the value of the critical heat flux, the heat transfer capacity of the flow boiling is suppressed. As a result, many methods have been tested with the aim of enhancing heat transfer and critical heat flux, such as modifying the geometrical structure of the microchannel.

Flow boiling in microchannel heat sinks has been widely studied, and the structure of the microchannels has been optimized to further optimize flow boiling [20,21]. Flow boiling in a copper foam fin microchannel heat sink was investigated experimentally by Fu et al. [20]. The results showed an 80% improvement in *HTC* and a 25% improvement in critical heat flux compared to the solid fin microchannel heat sink. Similarly, two microchannel heat sinks made of porous copper and solid copper were investigated by Yin et al. in flow boiling experiments [21], and the experimental results showed that the *HTC* of the porous open-microchannel was greatly improved compared with that for the solid copper open-microchannel heat sink. In addition, as reported by Li et al. [22], the critical heat flux was increased by 33.8~57.2% in a bidirectional counter-flow microchannel heat sink compared to the parallel-flow microchannel heat sink. A novel porous heat sink with reentrant microchannels was developed by Deng et al. [23], and the test results showed that the *HTC* value of the porous reentrant microchannels was 2–5 times higher than that of the solid copper microchannels. Furthermore, a bidirectional counter-flow microchannel heat sink was designed, and experimental study was performed [24,25]. The results showed that the average *HTC* of these novel microchannels was 33.5~62.0% higher than the traditional parallel-flow microchannels [24,25]. Subsequently, as reported by Jiang et al. [26], the counter-flow microchannels were modified with expanding angle, and flow boiling was performed in this heat sink at a higher heat flux of 2677 $kW \cdot m^{-2}$. Flow boiling has been extensively studied in microchannel heat sinks, and it is expected that it will be useful for cooling multiple heating sources with high heat fluxes.

Although flow boiling in microchannel heat sinks has been widely studied as a highly promising high-flux cooling method, only single-heat-source cooling has routinely been considered. Multiple-heat-source cooling problems are frequently encountered in advanced technology, and urgently need to be solved. Xu et al. [27] fabricated a cold plate with four microchannel regions made of copper to cool four heat sources. The experimental results showed that the temperature distribution of electronic devices with multiple heat sources was focused on single-phase flow. However, single-phase flow cooling was not a better method for high heat flux dissipation. Conversely, Tan et al. [28] modeled heat sources as point sources using the Dirac Delta function, and then extended this to locate the positions of the multiple heat sources in order to optimize the heat pipe performance. Similarly, an improved quasi-dynamic multiple-heat-source model was developed by Dan et al. [7] with the aim of optimizing the temperature distribution of a vapor chamber with multiple heat sources. It can be seen there are a small number of studies considering multiple heat sources with a vapor chamber in mathematical modeling. Recently, Zhang et al. [5] considered flow boiling in parallel/tandem microchannel heat sinks for cooling multiple heat sources. The experimental results showed that the phase transition heat of the upstream heat sink had a strong effect on the downstream heat sink. The connection of heat sinks is important for flow boiling with the aim of cooling multiple heat sources with high heat flux, but there is a lack of research results that would help to better understand this cooling method. As discussed, the studies on the cooling of multiple heat sources have rarely been reported, and more studies are needed to address the cooling problems related to multiple heat sources with high heat flux.

The importance of removing high heat flux from multiple heat sources has been widely reported, but research into relevant strategies is rare. Therefore, in this study, a method is proposed based on two efficient rectangular radial expanding microchannel heat sinks with and without grooves, designed previously [29], which were labeled as Type A and Type B, respectively. Figure 1a presents the rectangular radial expanding microchannel heat sink assembled with a heat source. In addition, two cold plates with and without grooves are shown in Figure 1b,c. In the present work, two microchannel heat sinks are connected in either series or parallel, and share a circulating system in order to perform experimental studies on flow boiling at a fixed heating load of 300 W. The T_w, *HTC*, and *PD* characteristics are obtained for the two heat sinks under different connection modes with different mass fluxes. The main corresponding flow patterns were observed using a high-speed camera. In addition, these characteristics are analyzed and compared in order to understand flow boiling in the radial expanding microchannels. Furthermore, the performances of the two heat sinks in different connection modes are compared in order to determine the most efficient connection mode for multiple-heat-source cooling. The results are expected to provide a reference for solving the high-heat-flux and complex multiple-heat-source cooling problems, thus improving the reliability of highly integrated electronic devices.

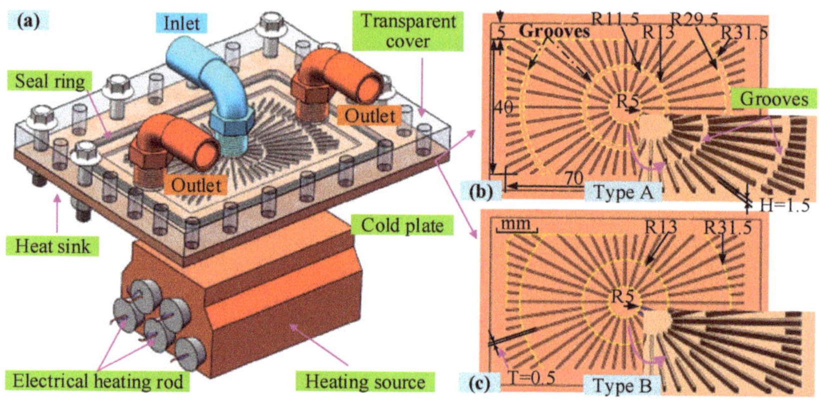

Figure 1. Diagram of the heat sink assembly with heating source and two cold plates: (**a**) the heat sink assembly with heating source; (**b**) microchannel cold plate with grooves; (**c**) microchannel cold plate without grooves.

2. Materials and Methods

2.1. Experimental System and the Connection of Two Heat Sinks

The schematic diagram of a two-phase cooling system is presented in Figure 2a. The main experimental equipment includes a peristaltic pump, a flowmeter, a test section of the heat sink with the heat source, two thermostatic water baths, a reservoir, and a filter. In the test section, two heat sinks are connected in series or parallel to experimentally investigate flow boiling and to further analyze the performance of the two heat sinks. For the two heat sinks, one cold plate with annular grooves at the downstream microchannel was labeled Type A, while the other one without grooves was labeled Type B, as shown in

Figure 1b,c, respectively. In addition, details regarding the size of the two microchannel cold plates can be found in refs [30,31]. The two heat sinks connected in parallel and series are shown in Figure 2b,c, respectively. Subsequently, the locations at which the temperature and *PD* parameters of the two heat sinks connected in parallel are measured are indicated in Figure 2b, as well as the inlet temperature of the working medium, T_{inlet}, the outlet temperature of the working medium $T_{1,\text{out}}$, $T_{2,\text{out}}$, the wall temperature of the two cold plates, T_1, T_2, and the *PD* of the two heat sinks, PD_1, PD_2. Similarly, the locations at which the parameters for two heat sinks connected in series are measured are marked in Figure 2c. Real-time temperature and *PD* data are collected at 1 s intervals using an Agilent data acquisition instrument. In addition, the location of the cold plate wall temperature measurement point is shown in Figure 2d, and there is a groove with a depth of 1 mm in the back of the cold plate for welding the thermocouple measurement points. Furthermore, a high-speed camera is positioned directly above the heat sink in order to visually record the flow states, boiling bubbles, and flow patterns in the microchannels.

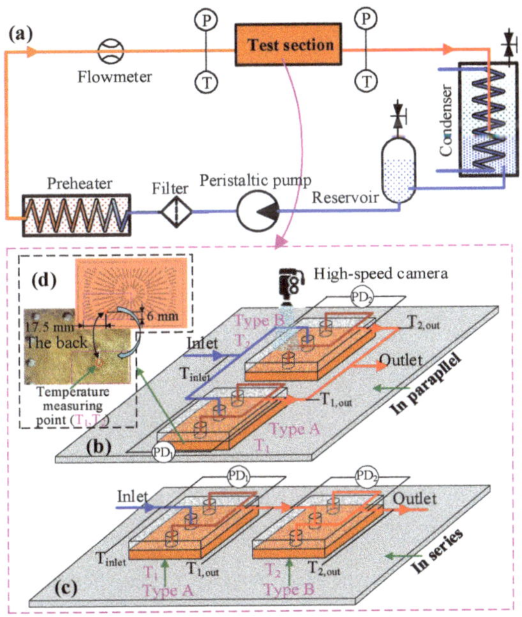

Figure 2. Diagrams of the flow boiling system and two heat sinks connected in parallel or series: (**a**) two-phase cooling system; (**b**) two cold plates connected in parallel; (**c**) two cold plates connected in series; (**d**) wall temperature measurement point.

2.2. Experimental Procedures and Conditions

The two-phase cooling system is presented in Figure 2a. The two-phase flow medium from the heat sink flows directly into an open cooling box in a short pipeline to cool the liquid and condensate steam. In order to evaluate the T_w, *HTC*, and *PD* of the two heat sinks in different connection modes at ambient atmospheric pressure, initially, the working medium consisted of deionized pure water. The heating load was set to a constant value of 300 W. The heating source was controlled separately using the voltage regulator and insulated with asbestos. The voltage ammeter was used to measure the power in real time. The water temperature at the inlet of the heat sink was kept constant at 88 °C (±0.2 °C) by a preheater, and a thermostatic water bath was used as the preheater. Subsequently, two thermocouples were used to monitor the inlet temperature at the inlet of the heat sink. In addition, 45 experimental cases were then designed under different mass fluxes with different connection modes.

As shown in Table 1, for two heat sinks connected in parallel, the main pipe volume flow rate was changed from 0.12 L/min to 0.40 L/min, with a gradient of 0.04 L/min, while the heating loads of the two heat sinks were kept constant at 300 W. In particular, the volume flow rates of two heat sinks connected in parallel were adjusted to be consistent before the phase change. Meanwhile, for the two heat sinks connected in series, the volume flow rate was changed from 0.12 L/min to 0.32 L/min and the heating loads of the two heat sinks was 300 W. In addition, the mass flux, G, was calculated using Equation (1) on the basis of the volume flow rate listed in Table 1. In addition, the T_w, HTC, and PD characteristics for the two heat sinks were then experimentally investigated under different mass fluxes. The volume flow rate was measured before liquid boiling, and was controlled by changing the rotational speed of the peristaltic pump. Therefore, the heating loads were turned off to measure the next volume flow rate. Due to the delay in the increase in the cold plate wall temperature, it generally took about 4 min for the heat sink to reach the heat transfer balance. When the trend of T_w remained unchanged, or T_w was no longer increasing, the heat sink reached the heat transfer balance. The time of the heat transfer balance must last for at least 5 min. In addition, the data of the temperature and PD were then calculated by averaging the data over 4 min.

Table 1. Forty-five experimental cases were designed at a fixed heating load of 300 W.

Types	No.	Heat Sink Type	Volume Flow Rates and Mass Fluxes
In parallel	Case 1 Case 2 Case 3	Type A–A Type A–B Type B–B	Main pipe volume flow rates (L/min): 0.16, 0.20, 0.24, 0.28, 0.32, 0.40 Corresponding mass fluxes (kg/(m^{2}s)): 46.14, 57.67, 69.20, 80.74, 92.27, 103.81, 115.34
In series	Case 4 Case 5 Case 6 Case 7	Type A–B Type A–A Type B–A Type B–B	Volume flow rates (L/min): 0.12, 0.16, 0.20, 0.24, 0.28, 0.32 Corresponding mass fluxes (kg/(m^{2}s)): 34.60, 46.14, 57.67, 69.20, 80.74, 92.27

2.3. Calculated Parameters and Uncertainties

Firstly, the averaged mass flux, G, in the expanding microchannel was calculated using Equation (1).

$$G = \frac{M}{2}\left(\frac{1}{A_{entr}} + \frac{1}{A_{\text{exit}}}\right) \tag{1}$$

where M is the mass flow rate, A_{in} is the total cross-sectional area of the channel entrance, at 3.21×10^{-5} m^2, and A_{exit} is the total cross-sectional area of the channel exit, at 2.88×10^{-4} m^2.

Then, the heat flux, q_{eff}, was calculated using Equation (2), and the HTC of the heat sink was calculated on the basis of the average cold plate wall temperature using Equation (3).

$$q_{\text{eff}} = \frac{Q}{A} \tag{2}$$

$$HTC = \frac{q_{\text{eff}}}{(T_w - T_{\text{sat}})} \tag{3}$$

where Q is the heating load, at 300 W, and A is the total efficient heat transfer area, which includes the base area and fin area. The areas of the Type A and Type B heat sinks are 4.868×10^{-3} m^2 and 5.033×10^{-3} m^2, respectively. T_w is the cold plate wall temperature measured by the thermocouple sensor, and T_{sat} is the saturated temperature in the heat sink, where the value of the T_{sat} is represented by the outlet temperature of the heat sinks.

The heat loss of the heat sink was determined, using the increase in the temperature for single-phase flow, as flow:

$$\varphi_{loss} = \left[Q - C_p M(T_{\text{out}} - T_{\text{in}})\right]/Q \tag{4}$$

where C_p is the specific heat capacity of water, at 4.2×10^3 J/(kg·k), and M is the mass flow. T_{out} is the liquid temperature at the outlet of the heat sink, and T_{in} is the inlet temperature of the liquid.

All the measuring sensors were calibrated before testing. In addition, the dimensions of the microchannel are ±0.01 mm. The uncertainty of *HTC* was obtained using Equation (5).

$$\frac{\partial HTC}{HTC} = \pm\sqrt{\left(\frac{\partial U}{U}\right)^2 + \left(\frac{\partial I}{I}\right)^2 + \left(\frac{\partial A}{A}\right)^2 + \left(\frac{\partial T_{wall}}{T_{wall}}\right)^2 + \left(\frac{\partial T_{out}}{T_{out}}\right)^2} \quad (5)$$

The maximum measured heat loss was 5.68%. The uncertainty of the calculated *HTC* was 7.68%. Similarly, the uncertainties of the other measured parameters were obtained using the same calculation method. The uncertainties for the other measurement parameters are listed in Table 2 with measuring ranges and errors.

Table 2. The uncertainties of the measured and calculated parameters.

No.	Measured Parameters	Uncertainties
1	Volume flow rate (q_v)	±0.69%
2	Heating load (Q)	±0.2%
3	Temperature (T_{in}, T_{out}, T_w)	±0.2 K
4	Pressure drop (PD)	±0.25%
5	Heat transfer coefficient (HTC)	±7.68%
6	Mass flow (M)	±0.96%
7	Dimension (A, H, D)	±0.01 mm

2.4. Comparison with Existing Correlations

The experimental *HTC*s of the two Type A and Type B heat sinks, used individually, were compared to existing correlations under different mass fluxes. As listed in Table 3, the *HTC* correlations of Kandlikar [32], Fang and Zhou [33], and Fang and Wu [34] were used. Initially, these *HTC* correlations were applied to the straight microchannels, while the flow rate and the hydraulic diameter in the expanding microchannels were calculated on the basis of the average. In addition, the *HTC* in the two expanding microchannel cold plates was calculated in three stages when the fins changed. The experimental *HTC* was compared with the predicted *HTC* calculated using the three correlations. The calculation of the mean absolute deviation (*MAD*) is shown in Table 3. It can be seen that the *MAD* of the *HTC* between the experimental and the predicted results for a single Type A heat sink is within 30%, while the smallest *MAD* was 21.67%, and was obtained using the correlation of Fang and Zhou. For a single Type B heat sink, the *MAD* is within about 20%, and the best was 6.00%, using Kandlikar's correlation. Therefore, based on existing *HTC* correlations, the experimental *HTC*s of the two heat sinks were first compared, in order to show a certain degree of accuracy under different mass fluxes at a heating load of 300 W.

Table 3. Mean absolute deviations between the experimental and predicted *HTC*.

No.	Correlations	[a] Mean Absolute Deviations	
		Type A	Type B
1	Kandlikar	24.32%	6.00%
2	Fang and Zhou	21.67%	20.05%
3	Fang and Wu	26.40%	19.28%

[a] Mean absolute deviation $= \frac{1}{N}\sum_{1}^{N}\left|\frac{HTC_{exp}-HTC_{pre}}{HTC_{exp}}\right|$, where N is the number of data.

3. Results

Based on the above experimental steps, T_w, *PD*, *HTC*, and the corresponding flow patterns were obtained at different mass fluxes. Subsequently, these characteristic parameters of T_w, *PD*, and *HTC* were analyzed. The influences of the parallel and series connections of the two heat sinks on the values of T_w, *PD*, and *HTC* were quantified. In addition, the influence of the connection modes on Type A and Type B heat sinks was compared.

3.1. Results for Two Heat Sinks Connected in Parallel

Two Type A heat sinks connected in parallel are presented in Figure 3b, with the measurement parameters marked out, where the subscript 1 and 2 differentiate the measurement data for the two heat sinks. In Figure 3a, the solid curves indicate the temperature data for two Type A heat sinks connected in parallel. In addition, the dotted line is the T_w of a single Type A heat sink. The abscissa of Figure 3a is the mass flux for a heat sink. In addition, the volume flow rate of each heat sink is adjusted to be the same before liquid boiling. Thus, the main pipe volume flow rate is evenly distributed. It can be seen that the T_w of the two heat sinks connected in parallel is similar. When the mass flux is increased from 23.08 kg/(m^{2}s) to 57.67 kg/(m^{2}s), the change in the T_w is small for two heat sinks. However, the T_w values for two heat sinks connected in parallel are higher than those for single heat sinks at the same mass flux. For example, the maximum temperature difference is 3.26 °C under a mass flux of 34.60 kg/(m^{2}s). In addition, the outlet temperature for two heat sinks connected in parallel first increases and later decreases with increasing mass flux. As analyzed, the T_w values for heat sinks utilizing the Type A-A parallel connection are degraded compared to those of single heat sinks.

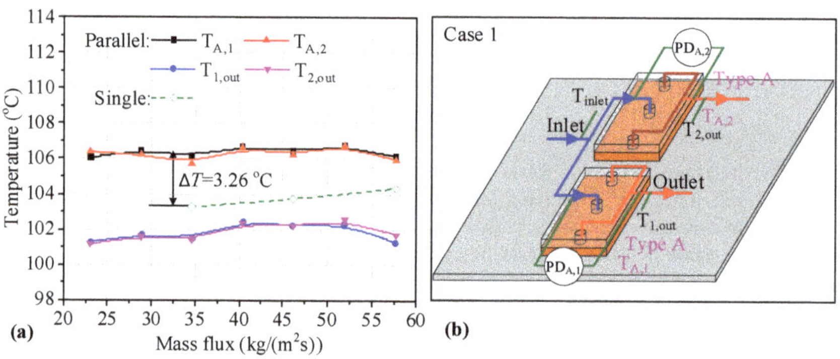

Figure 3. Temperatures of two heat sinks and measurement parameters in Case 1. (**a**) temperatures of two heat sinks; (**b**) the layout of the measurement parameters.

The *HTC* values for two Type A heat sinks connected in parallel are shown in Figure 4a. It can be seen that the *HTC* for heat sinks connected in parallel first increases and later decreases. However, the *HTC* for single heat sinks decreases with increasing mass flux. The *HTC* values for heat sinks connected in parallel are lower than those for single heat sinks under the same mass flux. As shown, the maximum difference in *HTC* values between single heat sinks and those connected in parallel is 9.17 kW/(m^{2}K). The decrease in the *HTC* is 43.32%. As shown in Figure 4b, the difference in *PD* between two Type A heat sinks connected in parallel is small, and the maximum difference occurs at a mass flux of 51.90 kg/(m^{2}s). It can be seen that the trend of *PD* values of two heat sinks connected in parallel increases slightly until the mass flux reaches 51.90 kg/(m^{2}s), and decreases sharply at the end. This may be because boiling is inhibited by higher mass flux. The *PD* values for heat sinks connected in parallel are higher than those for single heat sinks at mass flux values of 34.60 and 46.14 kg/(m^{2}s). It can be seen that the *HTC* and *PD* values for two Type A heat sinks connected in parallel are degraded.

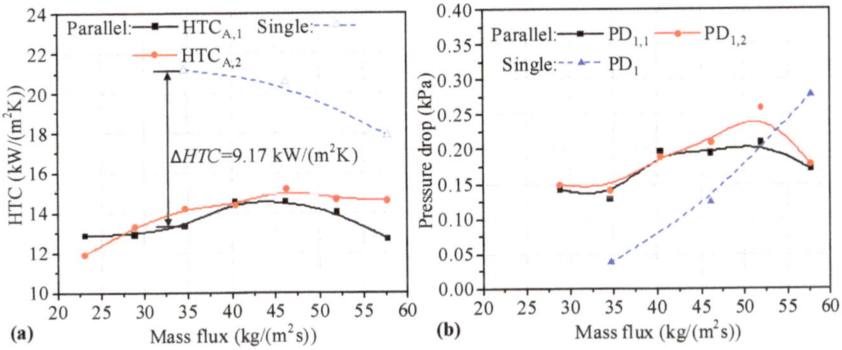

Figure 4. *HTC* and *PD* for two heat sinks utilizing the Type A-A parallel connection. (**a**) *HTC* of single and two parallel heat sinks; (**b**) *PD* of single and two parallel heat sinks.

Figure 5b depicts two heat sinks of Types A and B connected in parallel, along with the measurement parameters. The temperatures measured for two heat sinks connected in parallel are shown in Figure 5a. It can be seen that the T_w values for two heat sinks of Types A and B increase slightly with increasing mass flux. The values are higher than those of single heat sinks, with the maximum difference between single Type A heat sinks and those connected in parallel being 4.18 °C, while for the Type B heat sink the maximum difference is 3.57 °C. The T_w values for the Type A heat sink are lower than those of the Type B heat sink when connected in parallel. The outlet temperatures of the two heat sinks are similar until the mass flux reaches 51.90 kg/(m^{2}s). Meanwhile, the outlet temperature of the Type B heat sink decreases obviously. This may be because boiling is inhibited by higher mass flux, and these changes can be further explained by the flow patterns in Figure 6. It can be seen that the T_w values for two heat sinks of Types A and B connected in parallel are degraded.

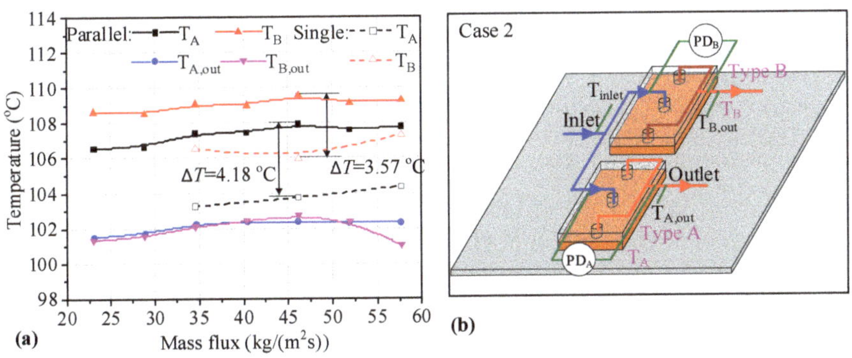

Figure 5. Temperatures of two heat sinks with the measured parameters marked in Case 2. (**a**) temperatures of two heat sinks; (**b**) the layout of the measurement parameters.

The change in the flow patterns of two heat sinks connected in parallel is shown in Figure 6. In addition, the change in the flow patterns could reflect differences in the heat transfer mechanism under different mass fluxes. It can be seen that, when the mass flux increases from 23.08 kg/(m^{2}s) to 57.67 kg/(m^{2}s), the main flow patterns of two heat sinks connected in parallel correspond to elongated bubble flow, elongated bubble with liquid flow, and bubbly flow. In addition, this change in flow pattern indicates that the heat transfer mechanism has also changed, with thin liquid film evaporation occurring at a lower mass flux and bubble nucleation at a higher mass flux. Furthermore, the change in the heat transfer mechanism leads to a change in the T_w and *HTC* in a non-monotonic relationship with mass flux. At lower mass flux, the heat transfer mechanism is the evaporation of thin liquid film, which has a higher heat transfer capacity compared to the bubble nucleation heat transfer mechanism [35]. Thus, the T_w increases with increasing mass flux when the flow rates are low. Furthermore, the evaporation decreases at high mass flux, which may lead to a decrease in the outlet temperature of the medium.

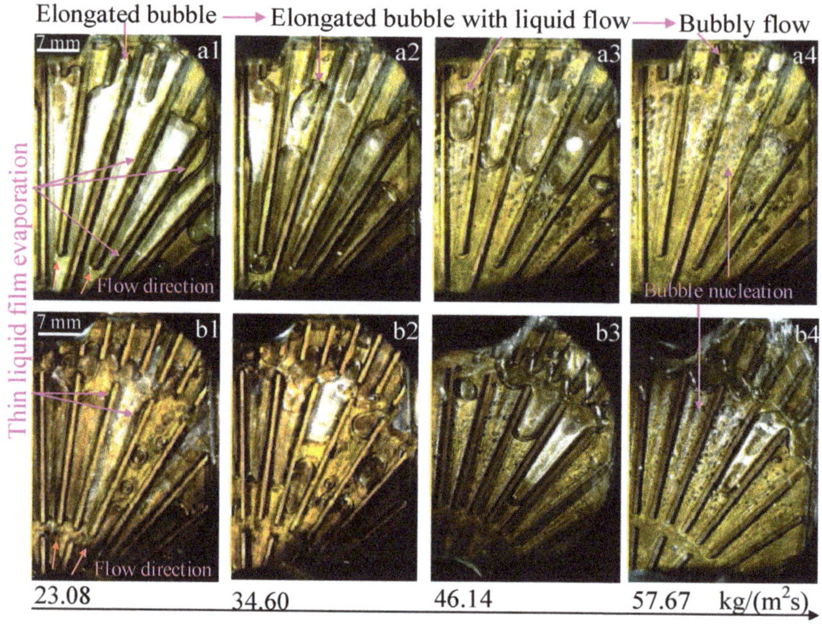

Figure 6. Flow patterns in two heat sinks under different mass fluxes in Case 2. (**a1**) flow patterns in channels without grooves at 23.08 kg/(m^{2}s); (**a2**) flow patterns in channels without grooves at 34.60 kg/(m^{2}s); (**a3**) flow patterns in channels without grooves at 46.14 kg/(m^{2}s); (**a4**) flow patterns in channels without grooves at 57.67 kg/(m^{2}s); (**b1**) flow patterns in channels with grooves at 23.08 kg/(m^{2}s); (**b2**) flow patterns in channels with grooves at 34.60 kg/(m^{2}s); (**b3**) flow patterns in channels with grooves at 46.14 kg/(m^{2}s); (**b4**) flow patterns in channels with grooves at 57.67 kg/(m^{2}s).

The *HTC* values of two heat sinks of Types A and B connected in parallel are shown in Figure 7a. It can be seen that the *HTC* values of the Type B heat sinks connected in parallel first increase and later decrease with increasing mass flux. The *HTC* values for Type A heat sinks connected in parallel decrease slightly with increasing mass flux. The *HTC* values for two heat sinks connected in parallel are lower than those of single heat sinks under the same mass flux. In addition, the difference in the *HTC* values between single Type A heat sinks and those connected in parallel is more obvious. It can be seen that the maximum difference in *HTC* values between single Type A heat sinks and those connected in parallel is 9.44 kW/(m^{2}K) at a mass flux of 46.14 kg/(m^{2}s). The decrease in *HTC* is 45.95%. For Type B heat sinks, the maximum difference in *HTC* values between single heat sinks and those connected in parallel is 2.93 kW/(m^{2}K) at a mass flux of 57.67 kg/(m^{2}s). As shown in Figure 7b, the *PD* values for two heat sinks connected in parallel are fluctuation, with one increasing and the other decreasing under the same mass flux. The *PD* change trend for two heat sinks connected in parallel shows a slight upward trend compared to single heat sinks. The *PD* for single heat sinks is lower than that of heat sinks connected in parallel at

34.60 and 51.90 kg/(m^{2}s). It can be seen that the *HTC* values for two heat sinks connected in parallel is degraded, along with *PD*, at low mass flux.

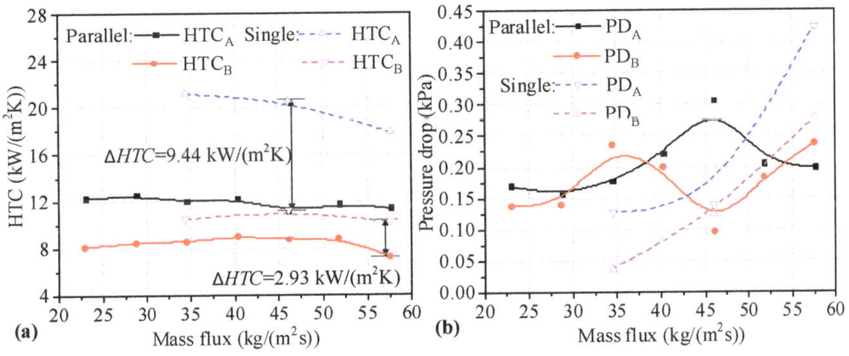

Figure 7. *HTC* and *PD* values for two heat sinks utilizing the Type A-B parallel connection. (**a**) *HTC* of single and two parallel heat sinks; (**b**) *PD* of single and two parallel heat sinks.

Boiling bubbles developed for the two heat sinks when utilizing a Type A-B parallel connection, as shown in Figure 8. The boiling bubbles grow and exit periodically in the downstream microchannels of two heat sinks. As shown in the area marked in the red box, it takes approximately 270 ms for the bubbles to nucleate and grow to completely fill in the microchannels at a heating load of 300 W and a mass flux of 46.14 kg/(m^{2}s). For the Type A heat sinks connected in parallel, the bubble first grows in the downstream annular grooves and then extends into adjacent channels. Subsequently, the bubbles coalesce in the middle channel to form a large bubble that fully fills the microchannel. For the Type B heat sink, the bubbles nucleate in downstream microchannels, and then the bubbles coalesce first in the middle microchannels. Subsequently, the coalescent bubbles first extend downstream, and upstream later. The development of the boiling bubbles in microchannels reflects the heat transfer state of two heat sinks. The main flow patterns are the bubbly flow and the elongated bubbles in the microchannel. The main heat transfer mechanism is bubble nucleation, and liquid boiling evaporation absorbs heat for bubble growth.

Figure 9b presents a schematic diagram of two Type B heat sinks connected in parallel on which the measured parameters are marked. The temperatures for two heat sinks connected in parallel are shown in Figure 9a. It can be seen that T_w values of the two Type B heat sinks are similar, and they have a slightly increasing trend with the increase in mass flux. The values are higher than those for single heat sinks, with the maximum difference between the single Type B heat sinks and those connected in parallel being 3.78 °C. The outlet temperatures of two heat sinks first increase and later decrease. This may be because boiling is inhibited by a higher mass flux. It can be seen that the T_w for two Type B heat sinks connected in parallel is degraded.

The *HTC* values of two Type B heat sinks connected in parallel are shown in Figure 10a. It can be seen that the *HTC* of heat sinks connected in parallel first increases and then decreases. The *HTC* values of the parallel heat sink are lower than those of single heat sinks under the same mass flux. It can be seen that the maximum *HTC* difference between the parallel and single heat sinks is 2.77 kW/(m^{2}K) at a mass flux of 40.37 kg/(m^{2}s). The decrease in the *HTC* is 24.72%. As shown in Figure 10b, the changing trend of $PD_{B,2}$ in the heat sinks connected in parallel shows an upward trend, reaching a peak at a mass flux of 46.14 kg/(m^{2}s), after which $PD_{B,2}$ decreases slightly. Although the change trend of $PD_{B,1}$ is fluctuant, the *PD* values in two Type B heat sinks connected in parallel are small, and the maximum value is about 0.27 kPa for $PD_{B,1}$. This may be because boiling is inhibited by the higher mass flux. The *PD* of heat sinks connected in parallel is higher than that of single heat sinks at mass flux values of 34.60 and 46.14 kg/(m^{2}s), and the difference in *PD* between single Type B heat sinks and those connected in parallel is small. However, the *PD* of single Type B heat sinks is significantly higher than that of heat sinks connected in parallel. It can be seen that the *HTC* values for two Type B heat sinks connected in parallel are degraded, but the *PD* values of two Type B heat sinks connected in parallel decrease with high mass flux.

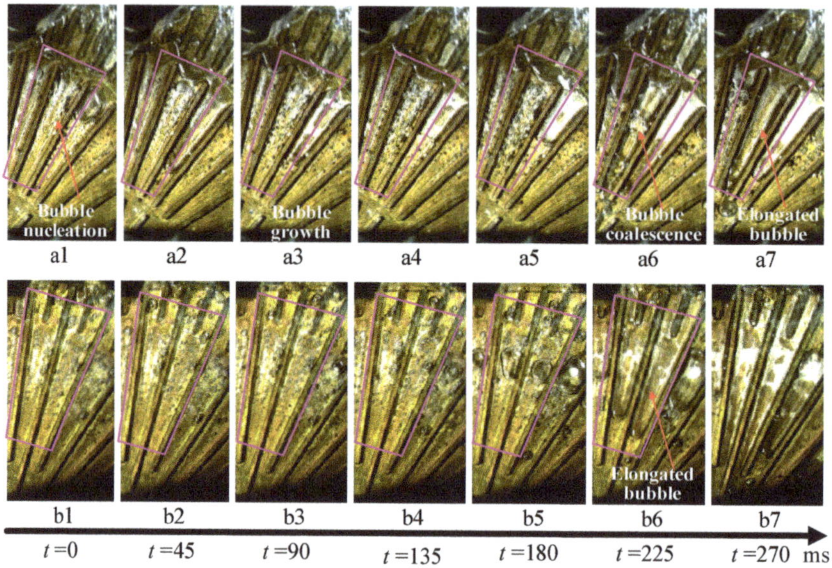

Figure 8. Main flow patterns in microchannels of the two heat sinks in Case 2. (**a1**) flow patterns in channels with grooves at t = 0 ms; (**a2**) flow patterns in channels with grooves at t = 45 ms; (**a3**) flow patterns in channels with grooves at t = 90 ms; (**a4**) flow patterns in channels with grooves at t = 135 ms; (**a5**) flow patterns in channels with grooves at t = 180 ms; (**a6**) flow patterns in channels with grooves at t = 225 ms; (**a7**) flow patterns in channels with grooves at t = 270 ms; (**b1**) flow patterns in channels without grooves at t = 0 ms; (**b2**) flow patterns in channels without grooves at t = 45 ms; (**b3**) flow patterns in channels without grooves at t = 90 ms; (**b4**) flow patterns in channels without grooves at t = 135 ms; (**b5**) flow patterns in channels without grooves at t = 180 ms; (**b6**) flow patterns in channels without grooves at t = 225 ms; (**b7**) flow patterns in channels without grooves at t = 270 ms.

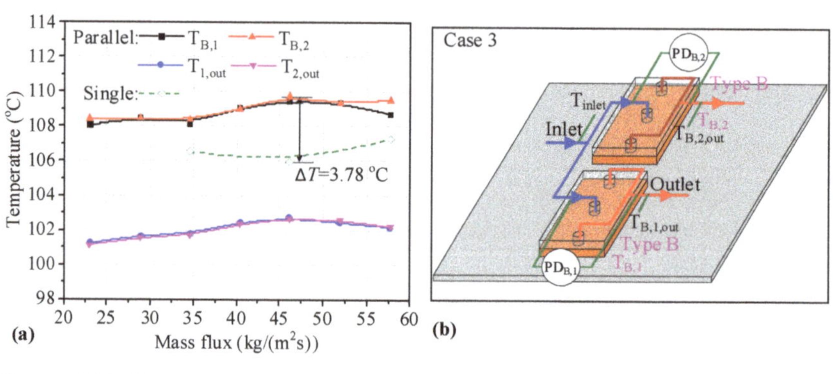

Figure 9. Main flow patterns in microchannels of two heat sinks in Case 2. (**a**) temperatures of two heat sinks; (**b**) the layout of the measurement parameters.

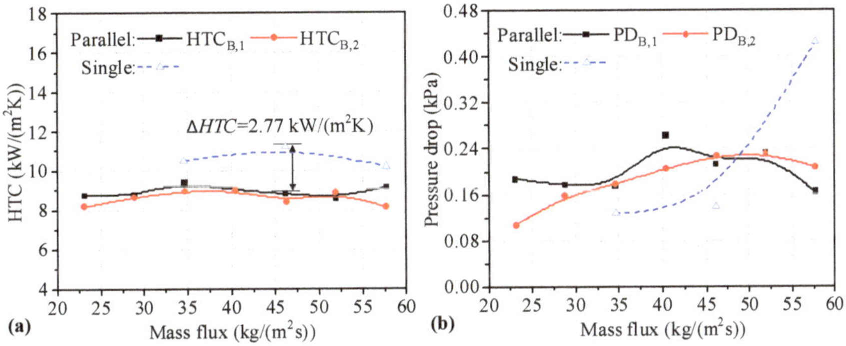

Figure 10. *HTC* and *PD* of two heat sinks in a Type B-B parallel connection. (**a**) *HTC* of single and two parallel heat sinks; (**b**) *PD* of single and two parallel heat sinks.

Table 4 lists the mean values of T_w, *HTC*, and *PD* for two heat sinks connected in parallel at different flow rates. It can be seen from the mean T_w, *HTC*, and *PD* values for the two heat sinks that the performance of heat sinks is better when utilizing the Type A-A parallel connection compared to when utilizing the Type A-B and Type B-B parallel connections. In addition, the performance of the heat sink in the Type A-B parallel connection is the worst, where the T_w of the Type B heat sink is the highest and the *HTC* is the lowest. Therefore, when two heat sinks are connected in parallel, the two heat sinks should be of the same type in order to achieve a stable performance. In addition, as described above, the T_w, *PD*, and *HTC* values of heat sinks connected in parallel are different from those of individual heat sinks, because there are mutual influences affecting the two heat sinks connected in parallel. This may be related to the fluctuation of the mass flux when phase change occurs in two heat sinks connected in parallel.

Table 4. Mean T_w, *HTC*, and *PD* of two heat sinks connected in parallels.

No.	Parallel Connections	Right Heat Sink			Left Heat Sink		
		T_w	*HTC*	*PD*	T_w	*HTC*	*PD*
Case 1	Type A–Type A	106.38	13.59	0.18	106.17	14.48	0.16
Case 2	Type A–Type B	107.33	11.86	0.20	109.02	8.45	0.18
Case 3	Type B–Type B	108.75	8.94	0.20	108.97	8.60	0.19

3.2. Results of Two Heat Sinks Connected in Series

There are four series connection modes for two Type A or B heat sinks connected in series, as listed in Table 1. The T_w values determined for the preposition and postposition heat sinks after testing and analysis are shown in Figure 11a,b, respectively. The solid curves represent the data correspond to heat sinks connected in series, and the dotted lines belong to single heat sinks. As seen in Figure 11a, for the preposition heat sinks connected in series, the T_w in Type A heat sinks connected in series always shows an upward trend with increasing mass flux, while the T_w in Type B heat sinks connected in series is fluctuant and even exhibits different change trends with different series connections. In addition, the T_w of single heat sinks first increases, until the mass flux reaches 69.20 kg/(m^{2}s), and then sharply decreases. In contrast, all the T_w values of preposition heat sinks utilizing different series connections are higher than those of single heat sinks. This may be because the flow boiling state in the two cold plates is affected by the series connection, increasing, for example, the pressure in the preposition heat sinks. Furthermore, all of the T_w values of Type A heat sinks are lower than those of the Type B heat sinks connected either in series or individually. The reason for this is that the grooves in the channels are beneficial for the outflow of large bubbles from the channel. The analysis showed that the difference in the value of T_w between the single Type A heat sinks and those connected in series is lower than that of Type B heat sinks. For preposition heat sinks connected in series, a lower T_w value was found for the Type A-A series connection compared to the other series connection modes. However, the lowest value of T_w observed for preposition heat sinks connected series is higher than that of single heat sinks. The T_w of two preposition heat sinks connected in series is degraded.

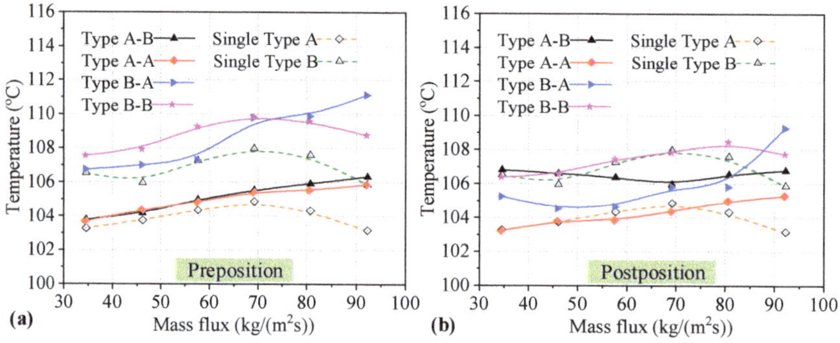

Figure 11. T_w of two heat sinks utilizing different series connections. (**a**) T_w of preposition heat sinks; (**b**) T_w of postposition heat sinks.

As shown in Figure 11b, for postposition Type A heat sinks connected in series, the T_w exhibits an upward trend with increasing mass flux in the Type A-A connection mode, while the T_w first decreases slightly, until a mass flux of 46.14 kg/(m^2s) is reached, and then increases sharply for the Type B-A connection mode. For postposition Type B heat sinks connected in series, the T_w gradually increases until the mass flux reaches 69.20 kg/(m^2s), and then it decreases slowly until the end in the Type B-B connection mode, while the T_w first decreases slightly until the mass flux reaches 69.20 kg/(m^2s), before increasing slightly until the end with the Type A-B connection mode. In contrast, the majority of the T_w values of Type A heat sinks connected in series are higher than those of single heat sinks. Only at mass flux values of 57.67 kg/(m^2s) and 69.20 kg/(m^2s) are the T_w values of heat sinks in the Type A-A connection mode lower than those of single heat sinks. The majority of T_w values of postposition Type B heat sinks are higher than those of Type A heat sinks. The majority of T_w values of postposition Type B heat sinks in series are similar to those of single heat sinks, or even lower than those of single heat sinks. For postposition heat sinks connected in series, a lower T_w value is obtained for the Type A-A series connection compared to the other series connection modes. In addition, there are some T_w values for preposition heat sinks connected in series that are lower than those of single heat sinks. It can be seen that it was possible to optimize the T_w of preposition heat sinks connected in series.

Furthermore, for the *HTC* of two heat sinks utilizing the four series connection modes, the *HTC* values of preposition and postposition heat sinks are shown in Figure 12a,b, respectively. As shown in Figure 12a, for preposition Type A heat sinks connected in series, the *HTC* of the heat sink in the Type A-B connection mode shows a downward trend with increasing mass flux, while the *HTC* of the heat sink in the Type A-A connection mode decreases slowly first, and then increases slightly towards the end. For preposition Type B heat sinks connected in series, the *HTC* values of the heat sink in Type B-A and Type B-B connection modes show a downward trend with increasing mass flux. In contrast, the majority of *HTC* values in preposition Type B heat sinks connected in series are lower than those of single heat sinks. However, the majority of *HTC* values of Type A heat sinks are higher than those of single heat sinks. In addition, all the *HTC* values of Type A heat sinks are higher than those of Type B heat sinks with different mass fluxes. The grooves in the channel cause the differences in the performance of the two heat sinks. It can be seen that the difference in HTC between the single heat sinks and those connected in series is small. For preposition heat sinks connected in series, higher HTC values were found for the Type A-B series connection compared to the other series connection modes. In addition, the highest HTC value observed for preposition heat sinks connected in series is higher than that for single heat sinks. The effect of being connected in series on the HTC of two preposition heat sinks is small, and the HTC value of the Type A heat sink is optimal in the Type A-B series connection mode.

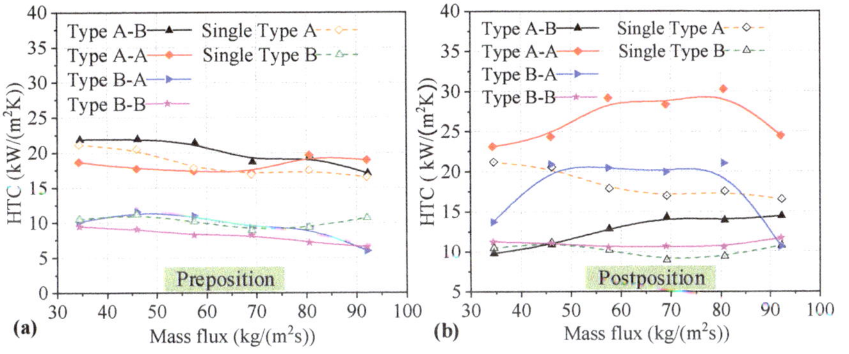

Figure 12. *HTCs* of two heat sinks utilizing different series connections. (**a**) *HTC* of preposition heat sinks; (**b**) *HTC* of postposition heat sinks.

As shown in Figure 12b, for postposition Type A heat sinks connected in series, the *HTC* of heat sinks in the Type A-A and Type B-A connection modes increases at the beginning and maintains a constant at moderate mass flux, but decreases sharply at the end. For postposition Type B heat sinks connected in series, the *HTC* in the Type A-B connection mode increases slightly until the mass flux reaches 69.20 kg/(m^2s), after which the *HTC* maintains a constant level. In addition, the *HTC* value of heat sinks in the Type B-B connection mode remains almost stable with increasing mass flux. In contrast, the majority of the *HTC* values of postposition heat sinks connected in series are higher than those of single heat sinks. In addition, the majority of *HTC* values of Type A heat sinks are higher than those of Type B heat sinks under different mass fluxes. It can be seen that a higher *HTC* value was obtained for the Type A-A series connection compared to the other series connection modes. In addition, the highest *HTC* value observed for preposition heat sinks connected in series is significantly higher than those of single heat sinks. The maximum increase in *HTC* is 12.77 kW/(m^2K), at a mass flux of 80.74 kg/(m^2s), which represents an improvement of 72.88% compared to a single Type A heat sink. The reason for the increase in *HTC* is that the heat transfer mechanism of the postposition heat sinks is mainly the thin liquid film evaporation mechanism. This mechanism is caused by the preposition heat sinks supplying a portion of the steam. The effect of being connected in series on the *HTC* of the two postposition heat sinks is obvious, and the *HTC* in two postposition heat sinks is optimal.

The change trends of *PD* in two heat sinks connected in series are shown in Figure 13. Firstly, the *PD* values for individual Type A and Type B heat sinks are similar, and have an upward trend. The *PD* of two single heat sinks increases when the mass flux changes from 34.60 kg/(m^2s) to 57.67 kg/(m^2s). When the mass flux exceeds 57.67 kg/(m^2s), the *PD* of the two single heat sinks barely changes. These trends are similar to the change trend of the *PD* for preposition Type A and B heat sinks connected in series. Conversely, for postposition heat sinks connected in series, the *PD* increased significantly compared to single heat sinks. The analysis showed that the maximum difference in *PD* between the single heat sinks and the Type B heat sinks connected in series is 0.79 kPa, which represents an increase by about 200%. This is mainly because the flow boiling state in the postposition heat sink is more dramatic and postposition heat sinks have a higher *HTC* compared to preposition heat sinks. Therefore, the series connection has a significant impact on *PD* values in postposition heat sinks. In addition, the *PD* values of two heat sinks in the Type A-A and Type B-B series connections have lower values compared to other series connection modes. The *PD* for postposition heat sinks connected in series is increased significantly compared to single heat sinks, but the *HTC* for postposition heat sinks connected in series is higher.

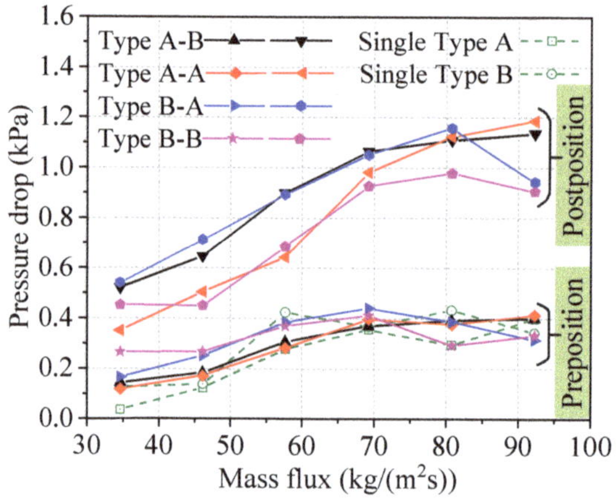

Figure 13. *PD*s of two heat sinks utilizing different series connections.

3.3. Comparison of Two Heat Sinks Connected in Series and Parallel

The mean differences in the T_w, *HTC*, and *PD* between single heat sinks and two heat sinks connected in series or parallel at mass fluxes of 34.60, 46.14, and 57.67 kg/(m^{2}s) are presented in Table 5. Firstly, the mean difference in the T_w for two heat sinks connected in parallel is about +2.18~+3.91 °C. This shows that the T_w of two heat sinks when using different parallel connections is obviously degraded. In addition, the T_w in the Type A-B parallel connection is the worst. The mean difference in the *HTC* for two heat sinks connected in parallel is about -1.50~-8.40 kW/(m^{2}K). This shows that the *HTC* of two heat sinks utilizing different parallel connections is degraded; in particular, this is most obvious for the Type A heat sink. Meanwhile, the mean difference in the *PD* for two heat sinks connected in parallel is small. This shows that the effect of parallel connection on the *PD* for heat sinks is small.

Subsequently, the mean differences in T_w between preposition heat sinks in different series connections increase, which shows that the series connection degrades the T_w of preposition heat sinks connected in series. However, the effect on the T_w is small compared to when connected in parallel. For the postposition heat sinks connected in series, the T_w is optimal in the Type A-A series connection. In addition, the mean *HTC* differences between the preposition and postposition heat sink exhibit a small increase in the Type A-B series connection. The *HTC* in the Type A-A series connection shows the most obvious improvement. This shows that the *HTC* of the postposition heat sink can be optimized when connected in series, and that the effect of the series connection on the *HTC* of the preposition heat sink is small. The mean difference in the *PD* of the preposition heat sinks increases slightly when connected in series, but the increase is obvious for the postposition heat sinks. In addition, the difference in *PD* between the preposition and postposition heat sinks is up to 5~6 times larger. The series connection degrades the *PD* of two heat sinks utilizing different connection modes, especially in the case of postposition heat sinks. On the basis of an analysis of the mean difference, the Type A-A series connection has the greatest positive effect on the performance of two heat sinks connected in series, especially in the case of postposition heat sinks at moderate mass flux. However, the T_w, *HTC*, and *PD* characteristics of the two heat sinks are been degraded when connected in parallel. Thus, the Type A-A series connection of two heat sinks is recommended for cooling multiple heat sources. In addition, when two heat sinks are connected in parallel, the two heat sinks should be of the same type in order to achieve stable performance.

Table 5. Mean differences in T_w, *HTC*, and *PD* under mass flux of 34.60, 46.14, and 57.67 kg/(m^{2}s).

No.	Parallel Connections							
	Left Heat Sink			Heat Sink Types		Right Heat Sink		
	T_w	*HTC*	*PD*	Left	Right	T_w	*HTC*	*PD*
Case 1	+2.43	−5.81	+0.02	Type A–Type A		+2.46	−6.32	+0.03
Case 2	+2.69	−2.43	−0.04	Type A–Type B		+3.91	−8.40	+0.08
Case 3	+2.61	−2.12	−0.03	Type B–Type B		+2.18	−1.50	−0.05
	Series Connections							
No.	Preposition Heat Sink			Heat Sink Types		Postposition Heat Sink		
	T_w	*HTC*	*PD*	Pre.	Post.	T_w	*HTC*	*PD*
Case 4	+0.52	+1.88	+0.06	Type A–Type B		+0.02	+0.56	+0.46
Case 5	+0.51	−1.90	+0.04	Type A–Type A		−0.15	+5.68	+0.35
Case 6	+0.43	+0.29	+0.04	Type B–Type A		+1.03	−1.50	+0.57
Case 7	+1.66	−1.67	+0.07	Type B–Type B		+0.15	+0.36	+0.30

The mean difference in $Tw = \frac{1}{N}\sum_{1}^{N}(T_{series} - T_{single})$°C, and the mean differences in *HTC* kW/(m^{2}K) and *PD* kPa are calculated using the same method.

4. Conclusions

Experimental investigation of flow boiling with two microchannel heat sinks connected in series and parallel for the cooling of multiple heat sources was performed. The following four conclusions were drawn in this study:

(1) An experimental investigation on flow boiling of two microchannel heat sinks connected in series and parallel was conducted, and the T_w, *HTC*, and *PD* characteristics of two microchannel heat sinks utilizing different connection modes were obtained. The bubbly flow and elongated bubble flow patterns were observed. In addition, the effect on these characteristics of being connected in series versus in parallel was then analyzed.

(2) The parallel connection significantly affected the T_w, *HTC*, and *PD* of the two heat sinks. The majority of these characteristics were degraded when connected in parallel. For example, compared to single heat sinks, the maximum increase in T_w was 4.18 °C for the Type A heat sinks connected in parallel, and the maximum decrease in *HTC* was 45.95%. The *PD* of two heat sinks connected in parallel was higher than that of a single heat sink.

(3) The influence of being connected in series on the T_w for two Type A and Type B heat sinks is obvious. The T_w in preposition heat sinks connected in series was degraded, but some of the T_w in postposition heat sinks connected in series could be optimized. The *PD* of the postposition heat sinks connected in series was significantly increased, but the *HTC* was optimized. For example, the maximum increase in *HTC* was 72.88% for the postposition heat sinks connected in series.

(4) The performance of two heat sinks in different parallel connections was degraded compared to that of a single heat sink. Meanwhile, the Type A-A series connection had the most positive effect in terms of the performance of the two heat sinks, especially in the case of the postposition heat sink with moderate mass flux. Thus, the Type A-A series connection is recommended for cooling multiple heat sources.

Author Contributions: Methodology, Z.J., M.S., J.S., L.Z., X.Zm and S.L.; investigation, Z.J., M.S., J.S., L.Z., X.Z. and S.L.; writing—original draft preparation, Z.J. and M.S. All authors have read and agreed to the published version of the manuscript.

Funding: This research was funded by the Chinese Association of Refrigeration (No. KT202004), the National Natural Science Foundation of China (No. 52076013), Beijing Municipal Science & Technology Commission (No. 3212024).

Data Availability Statement: The research data for this article can be obtained from the corresponding author "Mengjie Song" on reasonable request.

Conflicts of Interest: The authors declare no conflict of interest.

References

1. Cai, Y.H.; Li, Z.T.; Zhai, J.M.; Tang, Y.; Yu, B.H. Experimental investigation on a novel multi-branch heat pipe for multi-heat source electronics. *Int. J. Heat Mass Transf.* **2017**, *104*, 467–477. [CrossRef]
2. Elliott, J.W.; Lebon, M.T.; Robinson, A.J. Optimising integrated heat spreaders with distributed heat transfer coefficients: A case study for CPU cooling. *Case Stud. Therm. Eng.* **2022**, *38*, 102354. [CrossRef]
3. Li, W.; Klemeš, J.J.; Wang, Q.W.; Zeng, M. Efficient thermal management strategy of Li-ion battery pack based on sorption heat storage. *Energy Convers. Manag.* **2022**, *256*, 115383. [CrossRef]
4. Moon, S.; Choi, K.; Lee, J.; Kim, H. Application of aluminum flat heat pipe for dry cooling near the hot spot of a radar array with a multiscale structure. *Appl. Therm. Eng.* **2020**, *169*, 114894. [CrossRef]
5. Zhang, Y.X.; Wang, J.T.; Liu, W.; Liu, Z.C. Heat transfer and pressure drop characteristics of R134a flow boiling in the parallel/tandem microchannel heat sinks. *Energy Convers. Manag.* **2017**, *148*, 1082–1095. [CrossRef]
6. Wiriyasart, S.K.; Naphon, P. Heat spreading of liquid jet impingement cooling of cold plate heat sink with different fin shapes. *Case Stud. Therm. Eng.* **2020**, *20*, 100638. [CrossRef]
7. Siva, V.M.; Pattamatta, A.; Das, S.K. Effect of flow maldistribution on the thermal performance of parallel microchannel cooling systems. *Int. J. Heat Mass Transf.* **2014**, *73*, 424–428. [CrossRef]
8. Dan, D.; Li, W.F.; Zhang, Y.J.; Xie, Y. A quasi-dynamic model and thermal analysis for vapor chambers with multiple heat sources based on thermal resistance network model. *Case Stud. Therm. Eng.* **2022**, *35*, 102110. [CrossRef]
9. Krishnan, S.; Garimella, S.V.; Chrysler, G.M.; Mahajan, R.V. Towards a thermal Moore's law. *IEEE Trans. Adv. Packag.* **2007**, *30*, 462–474. [CrossRef]
10. Song, M.J.; Mao, N.; Xu, Y.J. Chapter 18—Instability control of two-phase flow in microchannel heat exchangers. In *Advanced Analytic and Control Techniques for Thermal Systems with Heat Exchangers*; Pekař, L., Ed.; Academic Press: Cambridge, MA, USA, 2020; pp. 387–410.
11. Cheng, X.; Wu, J.J. Unravelling the role of inlet valve in flow boiling characteristics in microchannel: Flow reversal suppression. *Int. Commun. Heat Mass Transf.* **2022**, *139*, 106438. [CrossRef]
12. Radwan, A.; Ahmed, M. The influence of microchannel heat sink configurations on the performance of low concentrator photovoltaic systems. *Appl. Energy* **2017**, *206*, 594–611. [CrossRef]
13. Zhang, X.; Wu, X.M. Time and Frequency Characteristics of Pressure Fluctuations during Subcooled Nucleate Flow Boiling. *Heat Transf. Eng.* **2018**, *39*, 642–653. [CrossRef]
14. Triplett, K.A.; Ghiaasiaan, S.M.; Abdel-Khalik, S.I. Gas-liquid two-phase flow in microchannels Part I: Two-phase flow patterns. *Int. J. Multiph. Flow* **1999**, *25*, 377–394. [CrossRef]
15. Serizawa, A.; Feng, Z.; Kawara, Z. Two-phase flow in microchannels. *Exp. Therm. Fluid Sci.* **2002**, *26*, 703–714. [CrossRef]
16. Cheng, L.; Ribatski, G.; Thome, J.R. Gas-liquid two-phase flow patterns and flow pattern maps: Fundamentals and applications. *ASME Appl. Mech. Rev.* **2008**, *61*, 050802. [CrossRef]
17. Cheng, L. Chapter 3—Microscale flow patterns and bubble growth in microchannels. In *Microchannel Phase Change Heat Transfer*, 1st ed.; Saha, S.K., Ed.; Elsevier: Amsterdam, The Netherlands, 2015; pp. 91–140.
18. Cheng, L. Chapter 4—Flow boiling heat transfer with models in microchannels. In *Microchannel Phase Change Heat Transfer*, 1st ed.; Saha, S.K., Ed.; Elsevier: Amsterdam, The Netherlands, 2015; pp. 141–191.
19. Cheng, L.X.; Xia, G.D. Fundamental issues, mechanisms and models of flow boiling heat transfer in microscale channels. *Int. J. Heat Mass Transf.* **2017**, *108*, 97–127. [CrossRef]
20. Fu, K.; Gao, W.H.; Xu, X.H.; Liang, X.G. Flow boiling heat transfer and pressure drop characteristics of water in a copper foam fin microchannel heat sink. *Appl. Therm. Eng.* **2023**, *218*, 119295. [CrossRef]
21. Yin, L.F.; Sun, M.M.; Jiang, P.X.; Dang, C.; Jia, L. Heat transfer coefficient and pressure drop of water flow boiling in porous open microchannels heat sink. *Appl. Therm. Eng.* **2023**, *218*, 119361. [CrossRef]
22. Li, Y.; Wu, H.Y. Experiment investigation on flow boiling heat transfer in a bidirectional counter-flow microchannel heat sink. *Int. J. Heat Mass Transf.* **2022**, *187*, 122500. [CrossRef]
23. Deng, D.X.; Tang, Y.; Liang, D.J.; He, H.; Yang, S. Flow boiling characteristics in porous heat sink with reentrant microchannels. *Int. J. Heat Mass Transf.* **2014**, *70*, 463–477. [CrossRef]

24. Zong, L.X.; Xia, G.D.; Jia, Y.T.; Liu, L.; Ma, D.D.; Wang, J. Flow boiling instability characteristics in microchannels with porous-wall. *Int. J. Heat Mass Transf.* **2020**, *146*, 118863. [CrossRef]
25. Yang, F.; Li, W.; Dai, X.; Li, C. Flow boiling heat transfer of HFE-7000 in nanowire-coated microchannels. *Appl. Therm. Eng.* **2016**, *93*, 260–268. [CrossRef]
26. Jiang, X.C.; Zhang, S.W.; Li, Y.J.; Pan, C. High performance heat sink with counter flow diverging microchannels. *Int. J. Heat Mass Transf.* **2020**, *162*, 120344. [CrossRef]
27. Xu, S.; Wang, W.J.; Guo, Z.K.; Hu, X.L.; Guo, W. A multi-channel cooling system for multiple heat source. *Therm. Sci.* **2016**, *20*, 1991–2000. [CrossRef]
28. Tan, B.K.; Huang, X.Y.; Wong, T.N.; Ooi, K.T. A study of multiple heat sources on a flat plate heat pipe using a point source approach. *Int. J. Heat Mass Transf.* **2000**, *43*, 3755–3764. [CrossRef]
29. Song, M.J.; Dang, C.B.; Hihara, E. Experimental investigation on the heat transfer characteristics of novel rectangle radial microchannel heat exchangers in two-phase flow cooling system for data centers. *J. Therm. Anal. Calorim.* **2020**, *141*, 199–211.
30. Jiang, Z.J.; Song, M.J.; Dang, C.B.; Jiang, Y.Y.; Wan, M.P.; Chao, C.Y.H. Two mathematical models of flow boiling and flow instability in rectangular expanding microchannel heat exchangers and structure optimization. *Appl. Therm. Eng.* **2023**, *228*, 120483. [CrossRef]
31. Song, M.J.; Jiang, Z.Y.; Dang, C.B.; Jiang, Y.Y.; Shen, J.; Luo, X.Y. Mathematical modeling investigation on flow boiling and high efficiency heat dissipation of two rectangular radial microchannel heat exchangers. *Int. J. Heat Mass Transf.* **2022**, *190*, 122736. [CrossRef]
32. Kandlikar, S.G.; Balasubramanian, P. An extension of the flow boiling correlation to transition, laminar, and deep laminar flows in minichannels and microchannels. *Heat Transf. Eng.* **2004**, *3*, 86–93. [CrossRef]
33. Fang, X.D.; Zhou, Z.R.; Wang, H. Heat transfer correlation for saturated flow boiling of water. *Appl. Therm. Eng.* **2015**, *76*, 147–156. [CrossRef]
34. Fang, X.D.; Wu, Q.; Yuan, Y.L. A general correlation for saturated flow boiling heat transfer in channels of various sizes and flow directions. *Int. J. Heat Mass Transfer* **2017**, *107*, 972–981. [CrossRef]
35. Zhang, P.; Wang, T.; Jiang, Y.Y.; Guo, C.H. Measurement of transient liquid film and its effect on flow boiling heat transfer in non-circular microchannels. *Int. J. Therm. Sci.* **2023**, *184*, 108004. [CrossRef]

micromachines

Article

Experimental Investigations upon Ultrasound Influence on Calefaction of AdBlue in Selective Catalytic Reduction Systems (SCR)

Claudiu Marian Picus, Ioan Mihai * and Cornel Suciu *

Faculty of Mechanical Engineering, Automotive and Robotics, Stefan cel Mare University, 720229 Suceava, Romania; claudiu.picus@usm.ro
* Correspondence: mihai.i@usm.ro (I.M.); suciu@usm.ro (C.S.); Tel.: +40-735173288 (I.M.); +40-742007820 (C.S.)

Abstract: The present paper intends to provide an analysis of how the process of calefaction occurs in a selective catalytic reduction (SCR) system and the mechanisms by which the deposition of AdBlue crystals on a hot surface evolve. Experimentally, two aluminium samples heated to 200 °C were used, over which AdBlue droplets with different atomisation rates were dropped, maintaining the same dynamic flow parameters, in order to observe the influence of temperature effects on the degree of deposition of crystallised sediment on the surface. The authors proposed the use of calefaction in an ultrasonic environment to prevent deposition and to increase droplet fragmentation by a break-up process. To prove the performance of this method one sample was subjected to a normal flow regime while a second sample was exposed to ultrasound. Both samples were assembled on a magnetostrictive concentrator operating at a frequency of 20 kHz. The obtained results indicated that the sample exposed to ultrasound demonstrated lower urea crystallisation compared to the sample that was not exposed to this treatment. Thus, it can be seen that the proposed method of injecting AdBlue into an ultrasonic zone gives the desired results.

Keywords: AdBlue; heat transfer; crystallization; SCR; droplets; microchannels

1. Introduction

In the current situation of global challenges associated with air pollution and exhaust emissions from trucks and marine shipping with very heavy-duty diesel engines, SCR technology and the use of the additive AdBlue have become increasingly important for reducing nitrogen oxide (NO_x) emissions. AdBlue is now commonly used as an emission reducing agent in the SCR process, where it reacts with nitrogen oxides to convert NO_x to nitrogen and water. However, the improvement of the efficiency and performance of SCR systems for cars, trucks, and marine ships remains a priority. One of the most important issues in this area is the efficient heating and vaporisation of AdBlue before entering the catalyst, especially during the time between engine warm-up and operating temperature. In modern configurations, AdBlue is injected into the exhaust manifolds of diesel engines above SCR systems to remove pollutants and neutralise NO_x. We proposed to study how the calefaction process takes place. It consists of an intense vaporisation of AdBlue droplets on the surface of a liquid located near a heated solid body, in this case the intake manifold of a combustion engine. Vaporisation takes place throughout the volume and on surfaces, the influence of vaporisation in the ultrasonic field is studied in this article. AdBlue is composed of 32.5% urea and 67.5% distilled water, so when the water evaporates, the urea tends to crystallise and causes deposits on the exhaust system walls and at the inlet of the SCR microchannels, hindering their correct operation and performance. Many authors [1–6] have studied the crystallization of urea from a droplet of evaporating aqueous solution placed on a heated flat surface, observing macroscopically the morphology of crystal growth under different conditions. It was seen that, after a certain period of time, increasing concentrations of urea in the droplet caused crystal growth. Moreover, thermal imaging indicated an exothermic character of the process through the significant release of heat on the surface where solidification occurred. Other researchers [7–11], have studied methods to prevent crystallization with constructive techniques such as gas mixed with AdBlue solution in order to improve gas atomized liquid droplet homogenization and to increase the performance of the selective catalytic reduction (SCR) system. Several studies can be found in the literature that investigate the use of ultrasonic vibrations in catalytic reactions, with numerous applications in green energy production [12–16]. In this context, the present work intended to find solutions to reduce urea crystallization especially in the microchannels of the SCR system using an ultrasonic process. The effect of vibrations on crystal deposition during evaporation of atomized AdBlue droplets in high temperature gaseous media and in direct contact with a heated disc was also considered during the presented investigations. The mechanism of the crystallisation process and the microscopic deposition of the formed crystal, in the presence and absence of vibrations, were examined in particular. The characteristics of the crystals formed, and their rate of formation depend on the working temperature of the gaseous media, which can be controlled by efficient preheating methods in the immediate period after the engine is started up to operating temperature. The use of this method helps to avoid residue deposition on SCR system components, which has a direct impact on SCR catalyst performance and durability.

2. Determination of the Temperature Distribution of the Ultrasonic Generator Disc and the Droplets on Its Surface

However, as discussed, vaporisation of AdBlue droplets at a high temperature surface can lead to the appearance and formation of crystal deposits. These deposits on the inlet face of the SCR catalyst can block or even obstruct the inlets of the SCR microchannels; in this case it will lose its exhaust gas cleaning capacity. This paper presents a method to control and reduce the deposits using the effect of an ultrasonic generator which produces vibrations through a magneto-strictive concentrator to a disc. AdBlue droplets are injected on the disc. The vibrations of the disc provide a much more accentuated dispersion of the droplets already atomised by injection, with a direct effect on the deposits. The method ensures a significant quantitative reduction of solid deposits when combined with a precise temperature control of the gaseous medium. The vibrations produced have a fundamental role in improving heat transfer within the droplets, but also on mass transfer.

2.1. A Description of the Processes That Take Place When Injecting Adblue onto a Disc with and without Vibration

A two-phase solution is considered in this case, where the turbulence created by vibrations in the boundary layer and in the bulk result in a decrease in thermal resistance, amplifying the rate of heat transfer. The equation used in the case of solid-fluid particles for a vibrating disc is given by Strouhal's homo-iconicity criterion, denoted Sh:

$$Sh = 0.746 Re^{0.5} Sc^{0.33} \left(\frac{A}{r}\right)^{1/6}, \tag{1}$$

In Equation (1) the following notations are made: Reynolds criterion ($Re = 2\omega Ar/\nu_f$), ω—the pulsation, A—the amplitude, r—radius, ν_f—the vibration frequency, Sc—the Schmidt criterion ($Sc = \eta/\rho D$). Equation (1) was used to calculate the temperature in the droplet as a function of the distance from the centre of the droplet. The temperature is assumed to change according to an exponential function and an attenuation factor given by $\exp(-rs^2)$ where, s—distance from the centre of the droplet. The larger the s, the smaller is the attenuation factor, while the temperature at the initial distance (denoted T_p) decreases progressively. The temperature inside the droplet can be calculated with the following relationship:

$$T_d = T_i + (T_p - T_i) exp(-rs^2), \tag{2}$$

where: T_d—represents the temperature at a certain distance s inside the droplet, T_i—represents the initial temperature in the droplet, T_p—represents the temperature at the initial distance $s = 0$ (which can be considered as the temperature of the disc), r—is the radius of the droplet, s—is the distance from the centre of the droplet where the temperature T_d is measured. In order to have a more accurate view of the vaporization phenomenon, the finite conductivity model was considered, which does not consider the study of phenomena occurring inside the droplet, but rather the thermal diffusion in the gas [17]. Convection in the droplet is neglected and concentration and temperature gradients are taken into account [18,19]. The temperature evolution is represented using the energy conservation equation below:

$$\frac{\partial T}{\partial t} = \frac{\lambda_i}{\rho C_p r^2} \frac{\partial}{\partial r}\left(r^2 \frac{\partial T}{\partial r}\right), \tag{3}$$

where: T—temperature, t—time, λ_i—heat conduction coefficient, ρ—density, C_p—specific heat at constant pressure, r—radius. In order to analyse the evolution of the exothermic effect of the droplet, the parameters in Table 1 were used by considering the temperature variation under conditions of whether the disc is vibrated or not.

Table 1. Parameters used in the mathematical model.

Parameters	Values	Units
AdBlue droplet temperature (T_i)	25	[°C]
Disc temperature (T_{disc})	200	[°C]
Frequency (f)	20,000	[Hz]

2.2. A Description of the Processes That Take Place When Injecting AdBlue onto a Disc with and without Vibration

The evolution of disc temperature T_{disc} at time t on the surface of the disc was determined for two cases, that is without and in the presence of ultrasound [17]. The equation used for the first case is as follows:

$$T_{disc} = T_i + (T_p - T_i)(1 - exp(-t)). \tag{4}$$

The mathematical calculation of the temperature variation over time when using the magneto-strictive concentrator for generating vibration is T_{dv} below:

$$T_{dv} = T_i + (T_p - T_i)(1 - exp(-t) + Asin(\omega t)), \tag{5}$$

where: ω—is the pulsation expressed by the relation $\omega = 2\pi f$, f—frequency. The calculation results obtained in MATLAB R11b are those in Figure 1a,b, considering a transient regime where at $t = 0$ the disc temperature is that of the ambient.

Figure 1a,b shows that there is a temperature difference for the two cases. In contrast to the first case (without vibrations Figure 1a), we remark that in Figure 1b, the field marked with yellow indicates an increase in the temperature ΔT, resulting from vibrations of frequency 20 kHz. In the analysis of the 3D temperature distribution of the heated disc, the temperature of the disc was calculated for each point on the radius, knowing the values of the heat flux Φ, with the following relationship:

$$T_{disc} = T_{di} + \frac{\Phi}{4k_d}\left(\frac{1 - r^2}{2}\right)^2, \tag{6}$$

where: T_{di}—initial disc temperature, k_d—disc conductivity, r—disc radius.

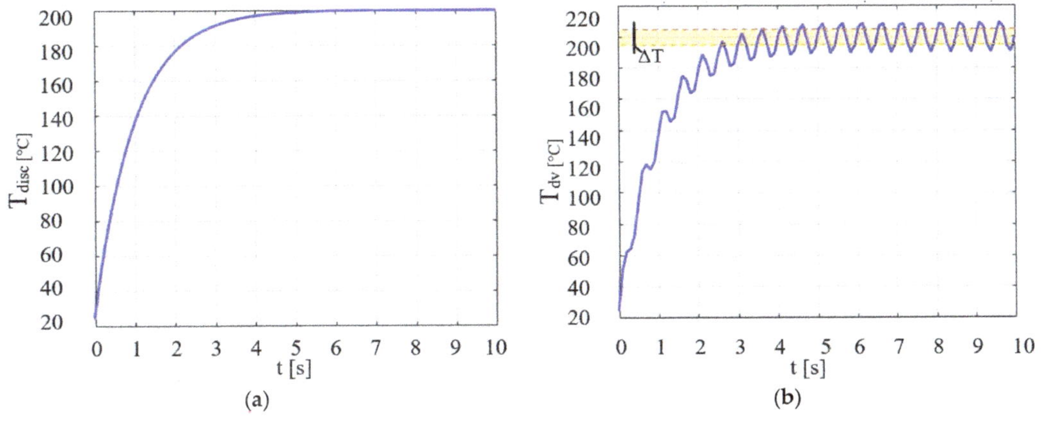

Figure 1. Temperature evolution of the evaporation process: (**a**) without the influence of vibration and (**b**) case with the effect of vibration.

The calculation results obtained in a MATLAB code are shown in Figure 2.

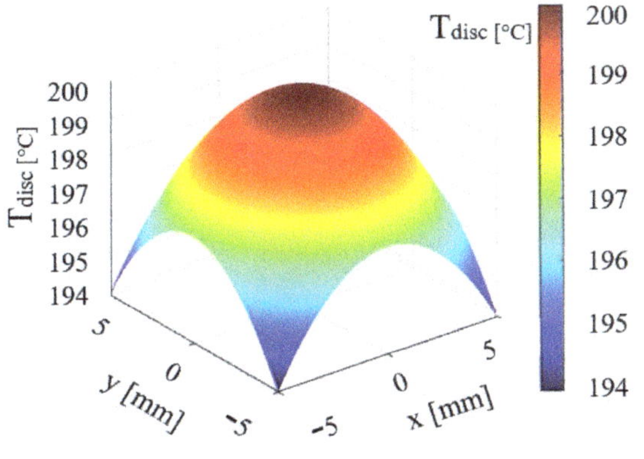

Figure 2. Temperature distribution on the disc surface.

When the disc reaches the maximum temperature and the AdBlue drops are injected and impinge on the disc surface, a heat transfer takes place. Thus, in the case of a sudden surface energy pulse, E = lim(qst) [kJ/m^2], and as time t approaches zero, at the surface the instantaneous temperature response is given by Equation (7):

$$T_m - T_0 = \frac{E}{\rho C_p \sqrt{\pi a t}} exp\left(\frac{x^2}{4at}\right), \quad (7)$$

where: T_m—temperature at the calculated moment, T_0—initial temperature, E—surface energy, ρ—density, C_p—specific heat at constant pressure, a—coefficient of thermal diffusivity, t—time, x—heat propagation distance. Considering the initial temperature of the disc as 200 °C and considering the thermal inertia when the droplet contacts the hot surface of the disc, a cooling of the disc of about 34 °C occurs. Figure 3 represents the temperature change on the radius of the disc if the liquid droplet remains on the surface for 1 to 20 ms if no vibrations are applied.

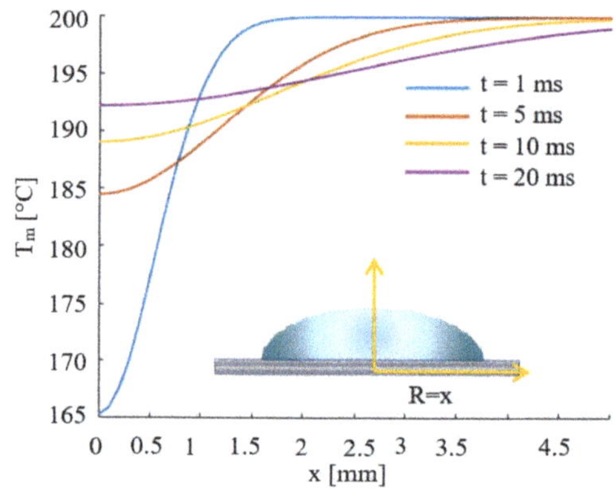

Figure 3. Temperature change of the disc surface at the moment of impact with an AdBlue droplet in a non-vibrating media.

The calculations show that after 1 ms the disc temperature in the centre of the droplet has the lowest value 165 °C which can be explained by the large amount of liquid in the centre of the droplet. With increasing distance x from the centre of the droplet to the extremity of the droplet, the disc temperature increases to the maximum value of 200 °C, as the amount of heated liquid decreases progressively. At the centre of the droplet when $x = 0$, after 5 ms the temperature in the centre of the droplet reaches 184 °C, after 10 ms it reaches 188 °C, and at 20 ms it reaches 192 °C. On the other hand, at the droplet tip ($x = 3.7$ mm) temperatures at the calculated moment are close to the initial disc temperature, their values being between 198 °C and 200 °C. For the case where vibrations are applied to the disc, the following equation was used:

$$T_m - T_0 = \frac{E}{\rho C_p \sqrt{\pi a t}} exp\left(\frac{2x(j)^2}{4atsin(2\pi f t)}\right), \quad (8)$$

The results obtained in a Matlab code are shown in Figure 4.

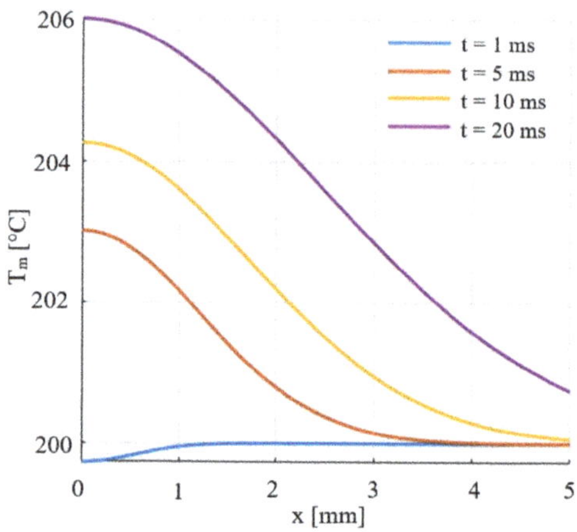

Figure 4. Variation of disc surface temperature at the time of impact with an AdBlue droplet in a vibrating media.

In Figure 4 we see that the blue line corresponding to the first millisecond of contact between the AdBlue droplet and the vibrating disc represents the case of a pulsating process. Compared to the previous variant, where the droplet–disc contact was all over the surface, in this case the contact times are much shorter. In Figure 4 it can be seen, due to the vibrations, how a much more pronounced heat transfer takes place. In contrast to the previous case, even if we move away from the centre of the droplet, due to the phenomenon of enhanced diffusive heat transfer, the cooling of the disc is more pronounced in the centre of the droplet, which is explained by the mass transfer. On the other hand, the heat transfer is more intense than in the previous case at the periphery of the droplet, which can be explained by the pronounced increase in molecular agitation.

2.3. Determination of the Evaporation Rate of AdBlue Droplets and Modelling This Phenomenon in ANSYS

The evaporation rate of atomized droplets on a high temperature quasi-isothermal cylindrical surface is calculated using the equations below. Analysis of the results obtained, allows a better understanding of the behaviour of AdBlue droplets during vaporization under the specific conditions of the experiment.

Equation (9) represents the variation of diameter of the droplet as a function of media temperature and time:

$$D_i = Dexp\left(\frac{2m_d i T_a t(j)^2}{D^2}\right), \tag{9}$$

Equation (10) represents the temperature variation on the disc surface:

$$T_{disc} = T_{di} + \alpha(T_a - T_p)exp(-T(j)), \tag{10}$$

Considering Equations (8) and (9) it can be written that the evaporation rate E_r is according to Equation (11):

$$E_r(j) = \left(\frac{m_d}{m_d + \alpha}\right)\pi {D_i}^2 C(T_a - T_p)\left(\frac{1 + A}{1000}\right), \tag{11}$$

where: E_r—evaporation rate, m_d—initial mass of the droplet, α—mass transfer coefficient, D_i—initial diameter of the AdBlue droplet, C—AdBlue concentration, T_a—ambient temperature, T_{disc}—disc temperature, A—oscillation amplitude.

Figure 5a,b presents the droplet evaporation rate as a function of time at different ambient temperatures, without the use of vibrations.

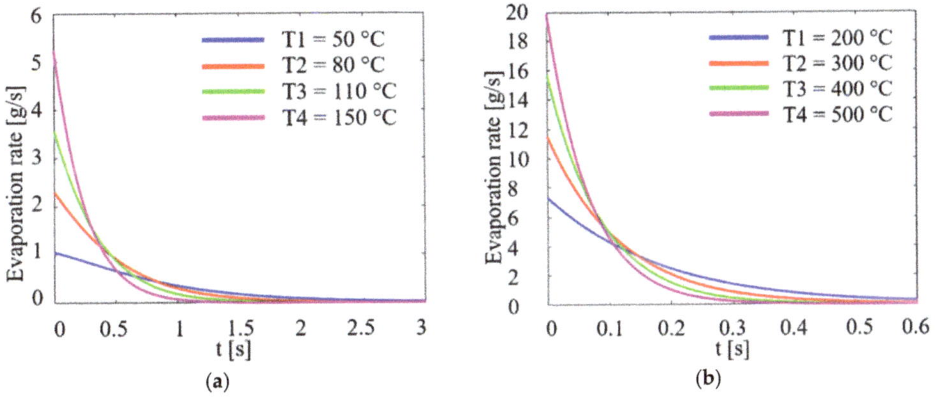

Figure 5. Evaporation rate of a liquid droplet as a function of time and temperature, without vibration, in a transient regime: (**a**) from temperature values from 50 °C to 150 °C, (**b**) from 200 °C to 500 °C.

Analysing Figure 5a,b it can be seen that the evaporation rate increases with the ambient temperature, Ta, for both variants. If the disc is not subjected to vibrations, a very low evaporation rate is observed at 50 °C, with a value of about 1 g/s, and a maximum value at 500 °C not exceeding 20 g/s. In the case of the theoretical presence of vibrations on the test disc obtained by means of a magnetostrictive concentrator, calculations using a Matlab code obtained the following in Figure 6a,b.

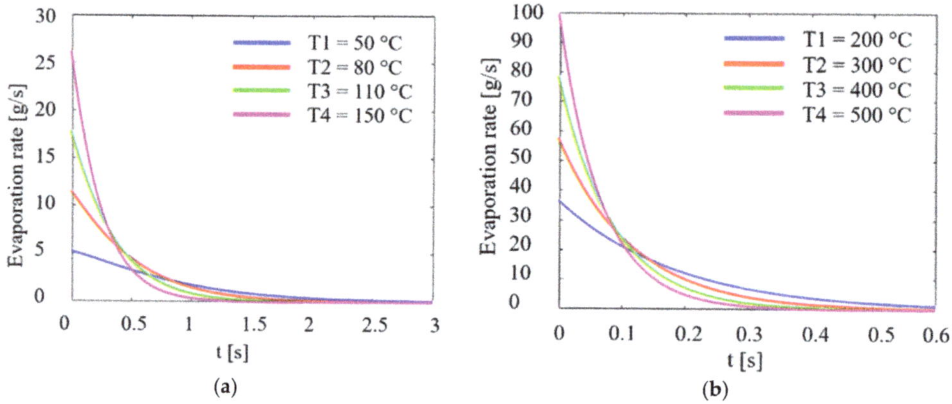

Figure 6. Evaporation rate of a liquid droplet as a function of time and temperature, in transient regime. (**a**) from temperature values from 50 °C to 150 °C, (**b**) from 200 °C to 500 °C, with vibration.

In the situation where vibrations are present, as shown in Figure 6a,b, the evaporation rate is 5 g/s at a temperature of 50 °C, indicating slow evaporation (blue line, Figure 6a, and at a maximum temperature of 500 °C, the evaporation rate changes rapidly to 100 g/s.

Analysis of the figures shows that the evaporation rate increases with the ambient temperature in both cases and that vibrations can significantly influence this parameter. It was found that the highest evaporation rate is obtained at the beginning of atomization, this decreasing with time.

In to establish how droplet atomization proceeds on a high temperature surface and to visualize the temperature evolution of AdBlue micro-droplets, a simulation was performed in the ANSYS programming environment. This simulation considered droplets in contact with a surface subjected to vibrations generated by the magneto-strictive concentrator. The result of the simulation in Figure 7 provided a prediction of the temperature evolution and dynamics of the AdBlue micro-droplets during interaction with the magneto-strictive concentrator.

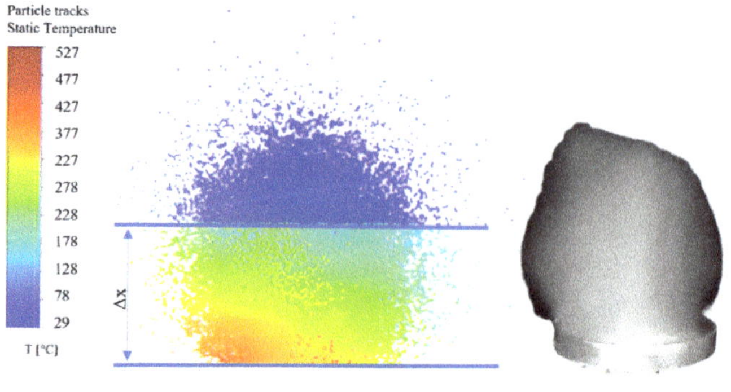

Figure 7. Ansys simulation of droplet evolution on impact with a vibrating disc at a frequency of 20 kHz.

During the vaporization phenomenon induced by the vibrations generated by the magnetostrictive concentrator in contact with the AdBlue droplets, a deep crumbling of the already atomized droplets is observed due to the intensification of thermal and mass diffusion in the system. In Figure 7 the prediction of the dispersion and temperature evolution of AdBlue droplets in a controlled environment was performed for a distance Δx of 10 mm from the vibrating disc. A pronounced heat transfer is observed in the 10 mm area, and subsequently the particles reach the ambient temperature relatively quickly. On the right side of the figure an experiment performed by the authors is shown for AdBlue droplets diffused into the air using an ultrasound generator, illustrating the closeness of the results with those from ANSYS.

3. Experimental Results and Interpretation

3.1. Description of the Experimental Laboratory Test Bench

In order to investigate the evaporation process and the evolution of deposits resulting from the evaporation of water from the AdBlue solution, laboratory tests were carried out on droplets in contact with a hot surface with or without using ultrasound. The experiments focused on the impact of ultrasonic vibrations on the microscopic deposition of AdBlue crystals on two disc-shaped samples. Figure 8 shows the equipment used for the experiments.

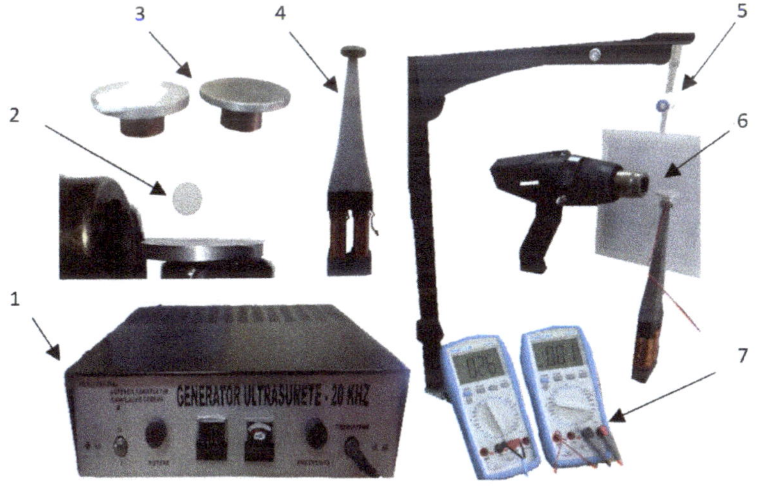

Figure 8. Experimental equipment for the study of AdBlue substance calefaction process.

The significance of the notations in Figure 8 is as follows: 1—frequency generator, 2—AdBlue drop (illustrative picture), 3—disc-shaped samples, 4—magnetostrictive concentrator, 5—stalagmometer, 6—hot air blower, 7—thermometers. The AdBlue droplet 2 flow is controlled by the stalagmometer 5 over the surface of the disc-shaped samples 3. The disc-shaped samples 3 are rigidly attached to the tip of the magnetostrictive concentrator 4 which generates vibrations due to the frequency induced by the frequency generator 1 which can generate 20 kHz. The AdBlue droplet 2 flows into the air perpendicular to the disc 4 samples followed by diffusion with the hot air flow at 200 °C generated by the blower 6. The heating to any temperature of the samples 3 and hence the ambient environment in the area is controlled by temperature sensors via thermometers 7. The results of these tests can be used to better understand the calefaction phenomenon between a heated flat surface and AdBlue droplets in order to optimise the efficiency of the catalytic emission reduction process.

The process that takes place during the conducted experiments is illustrated graphically in the schematic diagram shown in Figure 9.

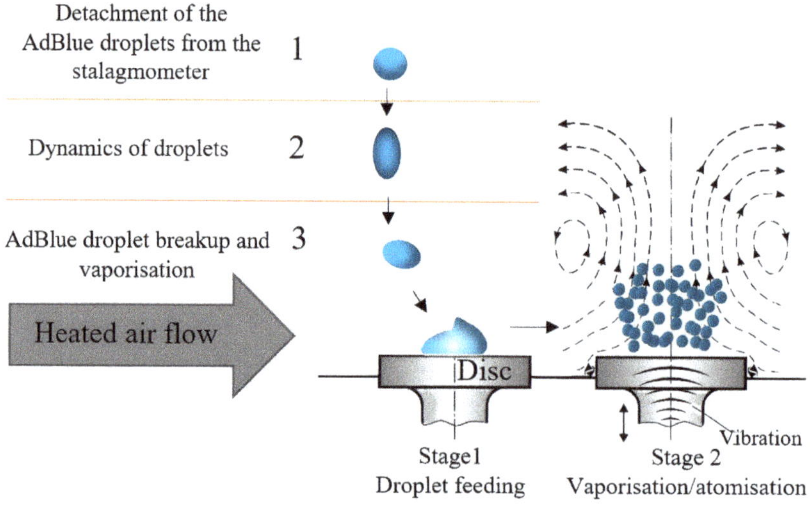

Figure 9. Schematic diagram of the experimental process.

Stage 1 of the process starts with the detachment of an AdBlue droplet from the stalgmometer. First, the droplet falls perpendicularly towards the disc surface, as its dynamics is only influenced by gravity. In the second portion of the trajectory, the droplet follows a slightly curved path because its trajectory is influenced by both gravity and the heated air flow. Once the droplet reaches the heated surface, it suffers a "break-up" phenomenon and fragments into multiple smaller droplets.

Stage 2 of the process consists of the intensification of the fragmentation phenomenon under the influence of ultrasonic vibrations. This leads to obtaining atomized spherical AdBlue nanodroplets that are spread over the whole surface of the disc, thus facilitating the vaporisation process and the transfer of mass and heat.

3.2. *Use of High-Speed Recording to Highlight the Evolution of AdBlue Droplets in the Calefaction Process*

For detailed analysis of the results, fast 240 fps shooting was performed using trouble shooter equipment produced by Fastec Imaging Corporation Model TSHRCS. The disc of element 4 in the figure is heated to a temperature of 200 °C by means of a stream of hot air. Two samples are used, one of which is subjected to vibrations generated by a magnetostrictive concentrator. After the experiments, the physical properties and structure of the AdBlue crystals are evaluated with and without vibrations. This research contributes to the understanding of the crystallisation process of AdBlue under controlled vibration conditions and provides new insight into improving the efficiency of SCR systems for cars, trucks, industries, and marine ships. For this study, aspects such as air flow velocity, pressure, ambient temperature, and geometrical characteristics of the disc on the magnetostrictive concentrator were considered. Using computational analysis methods and appropriate mathematical models, results were obtained that provided information on the temperature distribution and heat transfer in the system under analysis. Regarding the thermal aspect, an aerotherm was used to heat the vibrating disc and the environment, which provided heat to both the air and the vibrating disc at 200 °C. Then, a droplet of AdBlue having a temperature of 25 °C and a mass of 100 mg was allowed to flow through a capillary tube until the droplet had direct contact with the surface of the hot disc. By heating the disc and the adjacent space, a favourable environment was created for rapid vaporisation of the water in the solution, causing residual deposits to be formed. During the experiments, important dynamic parameters such as temperature, mass and evaporation rate were monitored and recorded. This information was collected in order to obtain accurate and relevant data, allowing a detailed analysis of the behaviour of the AdBlue droplet and its interaction with the hot disc. Figure 10 shows the results of a high-speed recording to analyse the heating process of a droplet dropped in the direction of the red arrow on the high temperature disc. The vaporization process takes place instantaneously, with the droplet jumping up and down from the surface due to the breaking of contact.

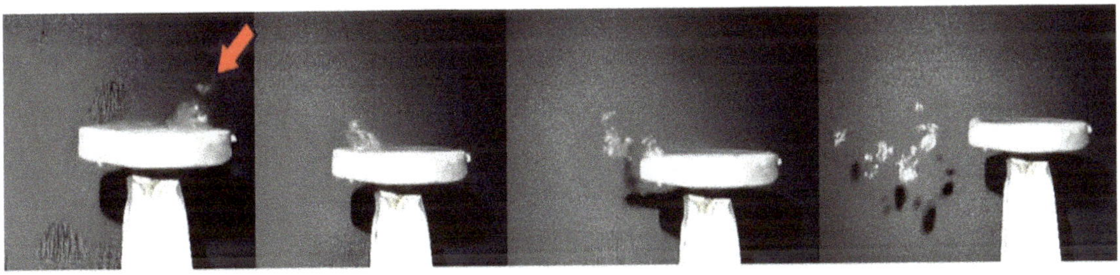

Figure 10. Dynamic evolution of an AdBlue drop without vibration.

In this stage of the experiment (Figure 9), the disc is not subjected to vibrations, in which case the "break-up" phenomenon of the droplet can be observed when it comes into contact with the heated surface. This "break-up" phenomenon refers to the dispersion of the droplet into smaller fragments due to impact with the hot surface. In the second phase of the experiment, the disc is subjected to vibrations, and thus a much stronger impact on the vaporisation and break-up phenomenon occurs. The calefactory mechanism is totally different in this case, the vibrations induced by the magnetorestrictive generator intensifying the break-up process. Figure 11 shows that the mass and temperature transfer in the case of vaporisation is significantly higher compared to the previous situation where the droplet is not subjected to vibrations.

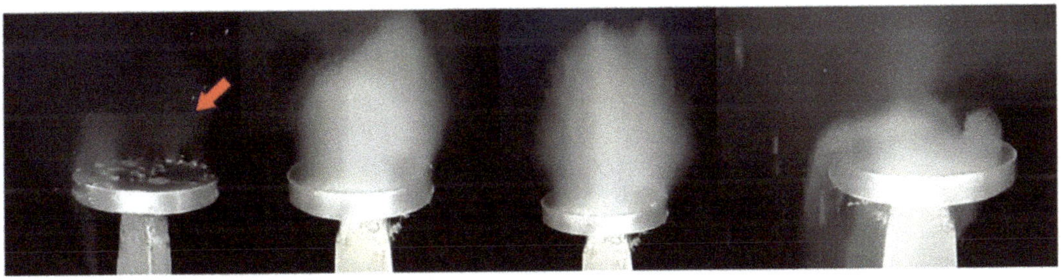

Figure 11. The process of calefaction of an AdBlue droplet in the presence of vibrations with a frequency of 20 kHz.

On contact with the surface of the strongly heated disc, a certain part of the droplet does not vaporise instantly but breaks up into relatively large droplets. At this stage the vibrations help to disperse the droplet into nanodroplets and subsequently spread them over the surface of the disc, thus facilitating the vaporisation process and the transfer of mass and heat.

3.3. Application of the IR Technique to the Calefaction of AdBlue Droplets

In addition, an infrared (IR) camera was used to better highlight the differences mentioned between the use and non-use of ultrasound, and the results obtained are shown in Figures 12 and 13. This technique allows thermal visualization of the process, which helps to distinguish the degree of evaporation and vaporization for the same quantity of AdBlue. With the IR camera, it is possible to observe how the heat is dispersed in the droplets as well as how the vaporisation process develops.

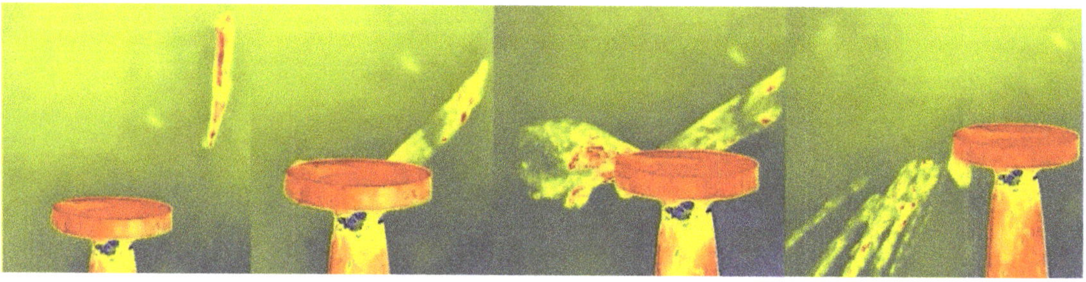

Figure 12. Vapour evolution for a droplet of AdBlue without vibration.

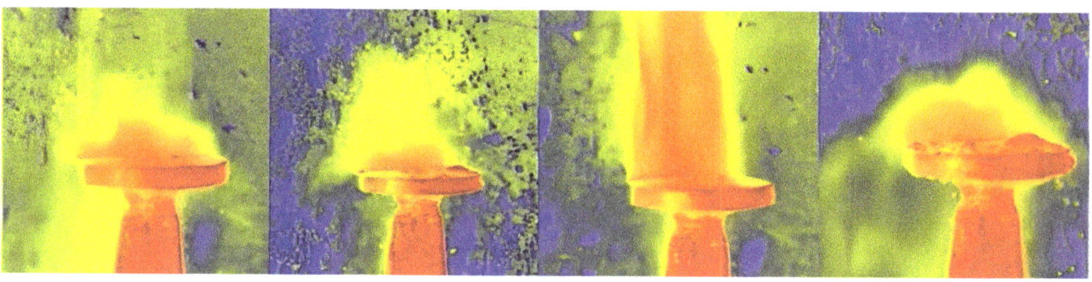

Figure 13. Vapour evolution of an AdBlue droplet in the condition of 20 kHz vibrations.

The figures show the totally different evolution of the AdBlue droplet break-up phenomenon, showing that the use of a magneto-restrictive concentrator produces fragmentation thousands of times amplified compared to the normal injection case.

3.4. Evidence of Crystal Deposition in the Case of Calefaction Phenomenon

The crystallised deposits which result from thermal decomposition of AdBlue aqueous solution, can have a negative impact on the efficient operation of selective catalytic reduction (SCR) systems. The appearance of salts through the calefaction process can lead to clogging of part of the SCR microchannels while also partially obstructing the flow of flue gas through the system. Deposition can lead to increased emission of NO_X. In this context, the study of the evaporation and vaporization of the AdBlue droplet revealed the formation of deposits in different amounts for the two samples, as illustrated in Figure 14 where the colour images were obtained with an IR camera code FLIR TG165. These deposits can interfere with the gas and chemical flow of SCR installations and affect the overall performance of the system.

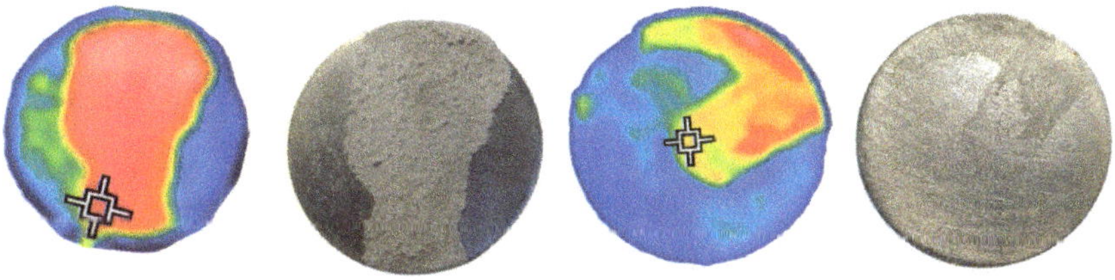

(**a**) (**b**)

Figure 14. Experimental AdBlue deposition: (**a**) without vibration (**b**) with vibration.

Investigations revealed an intensification of mass transfer processes in the presence of vibrations in AdBlue droplets, leading to a reduction of deposits on the sample surface. In the case of the evaporation phenomenon, the deposition of solid urea creates a rough region, marked in red in Figure 14. The liquid film tends to move towards the yellow marked area during evaporation. Furthermore, experiments showed that the deposition rate in the areas subjected to vibration was much lower than in samples that were not influenced by vibration, as shown in Figure 14b. To evaluate the deposition of AdBlue salts, measurements of the resulting surface microtopography were performed using a MarSurf CWM 100 confocal microscope and interferometer produced by Mahr GmbH, Gottingen, Germany (Figure 15).

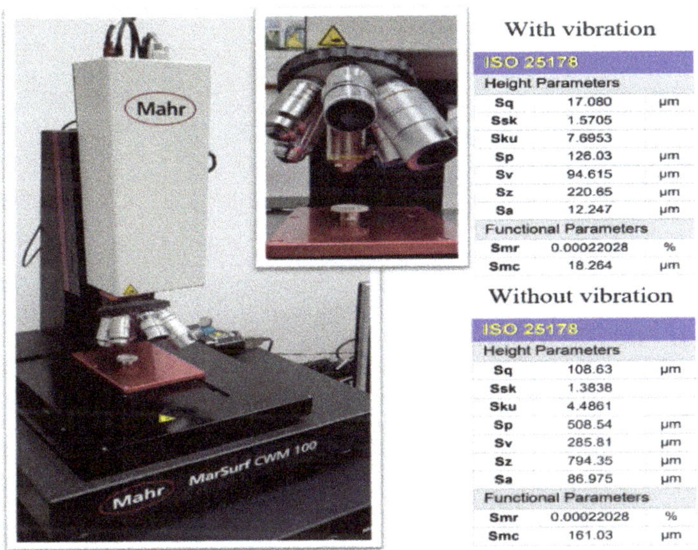

Figure 15. MarSurf CWM 100 confocal microscope.

Measurements of the surface microtopography were made in the central region of the samples, in 2 × 1.5 mm^2 areas, considering both longitudinal and transverse directions. In order to quantitatively determine crystal deposition on the surfaces of the two samples (with or without vibration), quantitative analyses were performed at selected points of interest on each piece. These points were placed from right to left on the part surfaces using a random sampling method. The areas of interest, illustrated in Figure 16, were identified based on specific criteria, such as roughness level and deposit distribution, in order to obtain a more precise assessment of the amount and distribution of deposits on the surfaces.

Figure 16. AdBlue residue deposits in absence (**left**) and presence (**right**) of vibrations, with indication of investigated areas.

The analysis of these areas of interest involved the use of high-resolution microscopy with a Mahr CWM 100 confocal microscope, using specific analytical methods to process the results. The use of the confocal microscope, allowed a detailed analysis of the height of the deposits for the two samples investigated, as shown in Figure 17a,b.

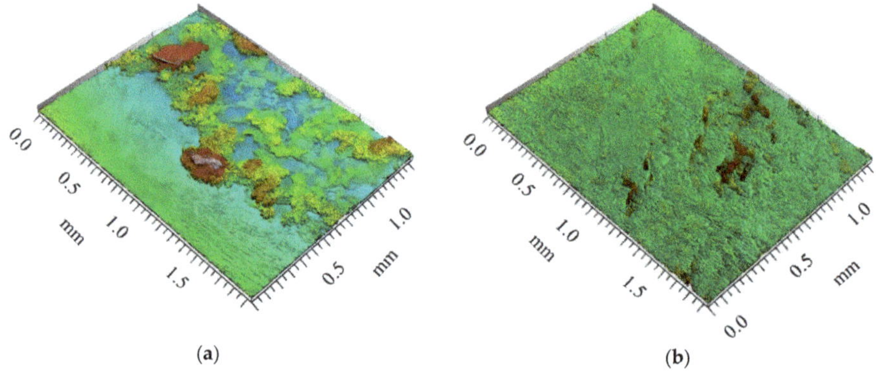

Figure 17. 3D profiles of urea deposition in AdBlue calefaction obtained with Mahr CWM 100 confocal microscope: (**a**) without vibration, (**b**) with vibration.

From the microscope measurement reports, it can be seen that there are significant differences between the two cases in terms of the height of the deposition profile. In case (a), where no ultrasound was used, the height of the deposition profile is much higher compared to case (b), where ultrasound was used. This suggests that the use of ultrasound has a positive effect in reducing the height of the deposits, thus contributing to the partial or total removal of the crystals formed by calefaction. Where no vibration was used, it can be seen from Figure 18 that a significant accumulation of solid deposits occurs, with heights of the order of 370 μm.

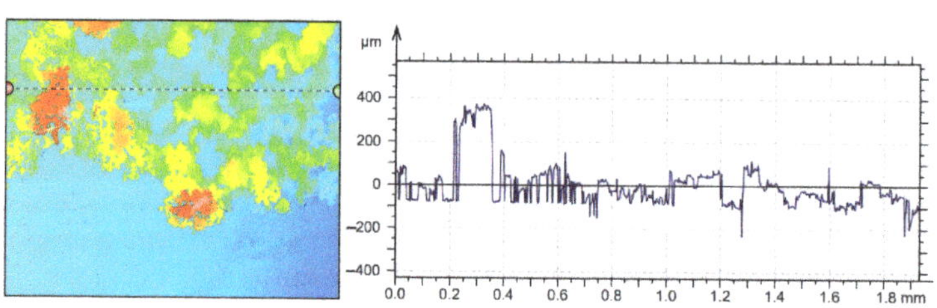

Figure 18. Deposition profile of AdBlue urea on calefaction in non-vibration sample.

These deposits are more concentrated in certain areas of the analysed surface compared to the case where vibrations were used. The distribution of residual solid particles does not cover the entire studied surface to the same extent as when using vibration. These crystals accumulate in larger concentrated areas, which can create significant variations in deposition between different regions of the surface. Analysing the profile of the samples in the case of vibration shown in Figure 19, a constant distribution of asperity peaks can be observed over approximately 80% of the surface.

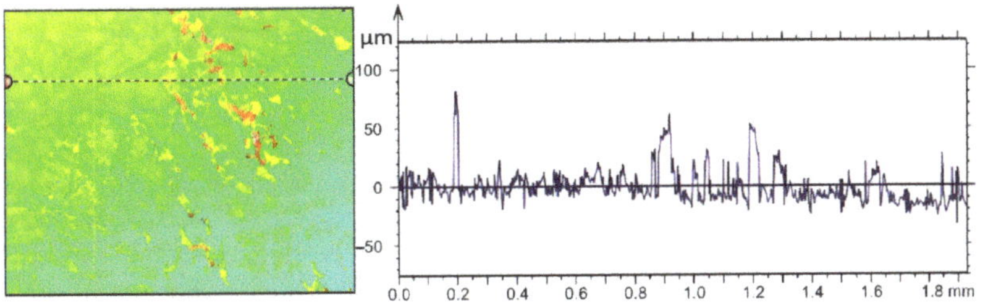

Figure 19. Deposition profile of AdBlue urea on calefaction in vibration test.

Small clusters of crystals are also found concentrated in certain areas of the surface, where differences of up to 50 µm can be observed. These differences represent a significant variation in heights in those areas and can have a significant impact on the remaining deposition. Using confocal microscopy, it was possible to determine the roughness parameters of the profiles, Ra and Rz, Figure 20.

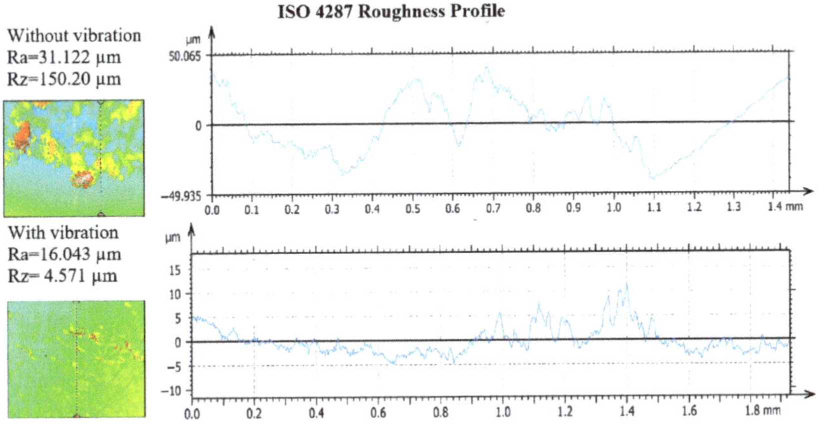

Figure 20. Roughness profiles of AdBlue, urea deposits by calefaction Ra and Rz.

These parameters are essential in evaluating surface roughness properties. For the analysed profile, an average roughness value Ra of 16.043 µm was obtained for the case where there were vibrations, respectively, 31.122 µm for the sample that was not subjected to vibrations. Ra roughness values are important in surface characterization because they reflect the absolute heights of the elevations and depressions on the surface. Thus, lower values of Ra indicate better surface quality with reduced roughness and more uniform texture.

4. Conclusions

The present article discusses the importance of eliminating all or part of the urea formations and deposits on the microchannels of SCR systems. The practice of injecting AdBlue into SCR systems is commonly used and is still being improved upon. One method is to heat the exhaust gases in the engine warm-up mode until the operating temperature is reached. During this period the emissions are much higher. The present paper showed that injecting AdBlue into the gas stream at high temperatures can cause deposits to form through calefaction under specific conditions. These deposits are undesirable in the operation of SCR systems, so the authors propose the following two routes:

- injecting the AdBlue solution into an ultrasonic field
- gas temperature control with an external induction heating system (not part of this study).

The experiments carried out and the results obtained have undoubtedly shown that major differences in the heating phenomenon occur in the presence or absence of a vibration-generating ultrasonic field. The images obtained in Figures 10 and 12 without vibrations and Figures 11 and 13 with vibrations show clearly how the AdBlue droplets go through an advanced process of fragmentation through a break-up effect; in such a case the vaporization process will also be much more efficient. This increase in vaporization intensity is attributed to the increased convective surface area of the atomized spherical AdBlue droplets and the increased velocity difference of the ultrasound relative to the gaseous medium in which it travels. In our opinion, with minimal investment, the atomization and reduction of NO_x when injecting AdBlue into a gaseous medium gives satisfactory results. In addition, even in the case of a calefaction process, AdBlue droplets in quasi-periodic contact with an extremely hot surface can lead to deposits in the form of crystals. This phenomenon is undesirable. Controlling the temperature at which AdBlue is injected is essential as low and very high temperatures will help avoid this phenomenon. The authors therefore propose controlling the AdBlue injection parameters according to the temperature in the SCR system. Microscopy measurements showed that deposits can approach 400 μm in just a few cycles of droplet injection, so that over time millimetre-sized plates can form at the SCR catalyst inlet surface with fundamental consequences. In conclusion the method of AdBlue injection in a controlled ultrasonic environment is effective and justified by increasing droplet break-up and avoiding crystal deposition on adjacent surfaces.

Author Contributions: Conceptualization, I.M.; methodology, I.M.; software, I.M.; validation, I.M., C.S. and C.M.P.; formal analysis, I.M.; investigation, I.M., C.S. and C.M.P.; resources, I.M. and C.S; data curation, I.M.; writing—original draft preparation, I.M.; writing—review and editing, C.S.; visualization, I.M., C.S. and C.M.P.; supervision, I.M.; project administration, I.M.; funding acquisition, I.M. All authors have read and agreed to the published version of the manuscript.

Funding: This work was funded by Ministry of Public Works, Development and Administration within REGIONAL OPERATIONAL PROGRAM 2014–2020, PRIORITY AXIS 1: PROMOTING TECHNOLOGICAL TRANSFER, PRIORITY OF INVESTMENTS 1.1, OPERATION B—Supporting science and technology parks, under contract no. 6592/24.12.2020 (SMIS: 128730), project title: ESTABLISHMENT OF THE EAST EUROPEAN BORDER SCIENTIFIC AND TECHNOLOGICAL PARK IN SIRET.

Data Availability Statement: Some or all data, models, or code generated or used during the study are available from the corresponding author by request.

Conflicts of Interest: The authors declare no conflict of interest.

References

1. Schmid, J.; Zarikos, I.; Terzis, A.; Roth, N.; Weigand, B. Crystallization of urea from an evaporative aqueous solution sessile droplet at sub-boiling temperatures and surfaces with different wettability. *Exp. Therm. Fluid Sci.* **2018**, *91*, 80–88. [CrossRef]
2. Eakle, S.; Kroll, S.; Yau, A.; Gomez, J.; Henry, C. *Investigation of Urea Derived Deposits Composition in SCR Systems and Their Potential Effect on Overall PM Emissions (No. 2016-01-0989)*; SAE Technical Paper; SAE International: Warrendale, PA, USA, 2016. [CrossRef]

3. Weeks, C.L.; Ibeling, D.R.; Han, S.; Ludwig, L.; Ayyappan, P. Analytical Investigation of Urea Deposits in SCR System. *SAE Int. J. Engines* **2015**, *8*, 1219–1239. [CrossRef]
4. Smith, H.; Lauer, T.; Mayer, M.; Pierson, S. Optical and numerical investigations on the mechanisms of deposit formation in SCR systems. *SAE Int. J. Fuels Lubr.* **2014**, *7*, 525–542. [CrossRef]
5. Munnannur, A.; Chiruta, M.; Liu, Z.G. *Thermal and Fluid Dynamic Considerations in Aftertreatment System Design for SCR Solid Deposit Mitigation (No. 2012-01-1287)*; SAE Technical Paper; SAE International: Warrendale, PA, USA, 2012. [CrossRef]
6. Zheng, G.; Fila, A.; Kotrba, A.; Floyd, R. Investigation of urea deposits in urea SCR systems for medium and heavy-duty trucks (No. 2010-01-1941). In Proceedings of the SAE 2010 Commercial Vehicle Engineering Congress, Chicago, IL, USA, 5 October 2010; Volume 1.
7. Park, K.; Hong, C.H.; Oh, S.; Moon, S. Numerical prediction on the influence of mixer on the performance of urea-SCR system. *Int. J. Mech. Aerosp. Ind. Mechatron. Manuf. Eng.* **2014**, *8*, 998–1004.
8. Habchi, C.; Nicolle, A.; Gillet, N. Numerical study of deposits formation in SCR systems using urea-water solution injection. *J. Mater. Sci. Nanotechnol.* **2018**, *6*, 201.
9. Jeong, S.; Kim, H.; Kim, H.; Kwon, O.; Park, E.; Kang, J. Optimization of the urea injection angle and direction: Maximizing the uniformity index of a selective catalytic reduction system. *Energies* **2020**, *14*, 157. [CrossRef]
10. Mehdi, G.; Zhou, S.; Zhu, Y.; Shah, A.H.; Chand, K. Numerical investigation of SCR mixer design optimization for improved performance. *Processes* **2019**, *7*, 168. [CrossRef]
11. Bouaziz, M.; Daïf, A. Etude numérique et expérimentale de l'évaporation simultanée de plusieurs gouttes de carburants. Influence de la pression, des convections et de interaction mécanique. *Perpignan Via Domitia Perpignan* **1998**. Available online: https://pascal-francis.inist.fr/vibad/index.php?action=getRecordDetail&idt=192736 (accessed on 15 May 2023).
12. Li, S.; Zhao, Z.; Li, J.; Liu, H.; Liu, M.; Zhang, Y.; Su, L.; Pérez-Jiménez, A.I.; Guo, Y.; Yang, F.; et al. Mechanically induced highly efficient hydrogen evolution from water over piezoelectric SnSe nanosheets. *Small* **2022**, *18*, 2202507. [CrossRef] [PubMed]
13. Li, S.; Zhao, Z.; Liu, M.; Liu, X.; Huang, W.; Sun, S.; Jiang, Y.; Liu, Y.; Zhang, J.; Zhang, Z. Remarkably enhanced photocatalytic performance of Au/$AgNbO_3$ heterostructures by coupling piezotronic with plasmonic effects. *Nano Energy* **2022**, *95*, 107031. [CrossRef]
14. Kuna, E.; Behling, R.; Valange, S.; Chatel, G.; Colmenares, J.C. Sonocatalysis: A Potential Sustainable Pathway for the Valorization of Lignocellulosic Biomass and Derivatives. *Top. Curr. Chem. (Z)* **2017**, *375*, 41. [CrossRef] [PubMed]
15. Wang, M.; Wang, B.; Huang, F.; Lin, Z. Enabling PIEZOpotential in PIEZOelectric Semiconductors for Enhanced Catalytic Activities. *Angew. Chem. Int. Ed. Engl.* **2019**, *58*, 7526–7536. [CrossRef] [PubMed]
16. Machado, I.V.; Santos, J.R.N.D.; Januario, M.A.P.; Corrêa, A.G. Greener organic synthetic methods: Sonochemistry and heterogeneous catalysis promoted multicomponent reactions. *Ultrason. Sonochem.* **2021**, *78*, 105704. [CrossRef] [PubMed]
17. Incropera, F.P.; DeWitt, D.P.; Bergman, T.L.; Lavine, A.S. *Fundamentals of Heat and Mass Transfer*; Wiley: New York, NY, USA, 1996; Volume 6, pp. 408–409.
18. Raghavan, V. Numerical modelling of evaporation and combustion of isolated liquid fuel droplets: A review. *J. Indian Inst. Sci.* **2019**, *99*, 5–23. [CrossRef]
19. Godsave, G.A.E. Studies of the combustion of drops in a fuel spray—The burning of single drops of fuel. *Symp. (Int.) Combust.* **1953**, *4*, 818–830. [CrossRef]

micromachines

MDPI

Article

Assessment of Vapor Formation Rate and Phase Shift between Pressure Gradient and Liquid Velocity in Flat Mini Heat Pipes as a Function of Internal Structure

Ioan Mihai *, Cornel Suciu * and Claudiu Marian Picus

Faculty of Mechanical Engineering, Automotive, and Robotics, Stefan cel Mare University, 720229 Suceava, Romania; claudiu.picus@usm.ro

* Correspondence: mihai.i@usm.ro (I.M.); suciu@usm.ro (C.S.); Tel.: +40-735-173-288 (I.M.); +40-742-007-820 (C.S.)

Abstract: Flat mini heat pipes (FMHPs) are often used in cooling systems for various power electronic components, as they rapidly dissipate high heat flux densities. The main objective of the present work is to experimentally investigate whether differences in the rate of vapor formation occur on an internal structure containing trapezoidal microchannels and porous sintered copper powder material. Several parameters, such as hydraulic diameter and fluid velocity through the material, as a function of the internal structure porosity, were determined by calculation for a steady state regime. Reynolds number was determined as a function of porosity, according to Darcy's law, and the Nusselt number was calculated. Since the flow is Darcy-type through the porous medium inside the FMHP, the Darcy friction factor was calculated using five methods: Colebrook, Darcy–Weisbach, Swamee–Jain, Blasius, and Haaland. After experimental tests, it was found that when the porous and trapezoidal microchannel layers are wetted at the same time, the vaporization progresses at a faster rate in the porous material, and the duration of the process is shorter. This recommends the use of such an internal structure in FMHPs since the manufacturing technology is simpler, the materials are cheaper, and the heat flux transport capacity is higher.

Keywords: microchannels; mini flat MHP's; pressure gradient; porous medium

1. Introduction

To date, a multitude of micro flat heat pipes has been developed with different internal structures for cooling various systems, including power electronics. So far, mainly three types of heat pipes have been made: grooved, sintered, or composited heat pipes (including FMHPs). A classification of heat pipes [1], first by construction principle and by the structure of the internal capillary layer, is shown in Figure 1.

In addition to the relatively low cost of FMHPs, the low dimensions, and high heat flux dissipation power, the literature [2] shows that FMHPs have various typical applications of microelectronic devices. According to [3], electronic devices work at temperatures usually between $-5 \div 65$ °C, but recently, the maximum heat flux in these components is increasing more and more, reaching maximum temperatures of $70 \div 80$ °C. Increases above a certain threshold can cause functional instability or even destruction of electronic devices. To avoid this, one of the cooling methods uses thermal tubes. A great diversity of thermal tubes has appeared, as shown by [2,4–6], who mentions the use of FMHPs in electronics cooling, space thermal control, aircraft devices, traction drives, audio amplifiers, in cooling of closed cabinets in harsh environmental conditions, space applications under vacuum conditions [4,7] nanofluidic technologies in medicine, process engineering, avionics applications, solar photovoltaic/loop-heat-pipe, (PV/LHP) heat pump system [8] electronic applications, solar units, and aircraft [9] self-driving car, smart TV, smart grid, aerospace, radar, camera, computer, cellphone [10], solar energy systems, heat recovery systems, air conditioning systems, cooling of energy storage and electronic equipment, industrial applications and space apparatus.

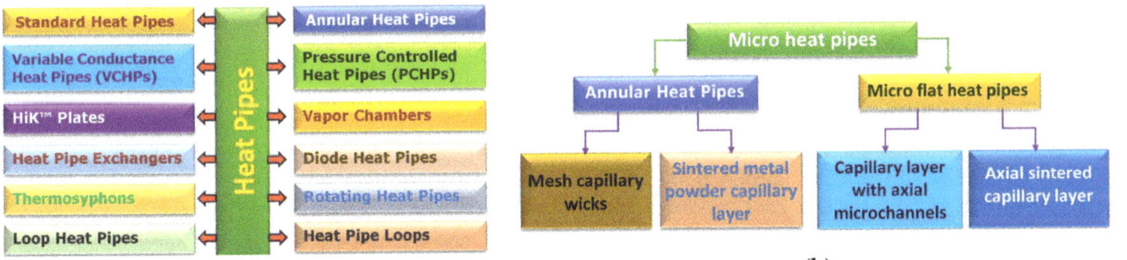

Figure 1. Heat pipes classification: (**a**) by construction principle and (**b**) by the structure of the internal capillary layer.

Similarly, Ref. [11] states that MHPs are of interest for use in implanted neural pacemakers, sensors and pumps, electronic wristwatches, active transponders, self-powered temperature displays, and temperature warning systems. Recently, thermal tubes have also found applications in the fields of Energy, Global Warming & Environment, and Healthcare [12].

According to M V Pukhovoy et al. [13], with increasing heat dissipation power of power electronic components, which often reach heat flux densities of the order of 1000 W/cm^2, the standard dimensions of microelectronic elements are reduced to 7 nm. Devices made of silicon carbide or gallium nitride have been developed, with chip element sizes reaching 6 ÷ 9 nm. In these specific cases, heat flux densities in a range of 100 ÷ 300 W/cm^2 are achieved for surface areas as small as 4 mm^2 and 10 W/cm^2 for 1 cm^2 [2,12–15]. In the case of FMHP [2], for copper-water coupling, heat fluxes of the order of 40 W/cm^2 applied to the evaporator wall in a vertical position can be achieved. This paper combines analytical analysis for a number of functional parameters of FMHP with experimental studies of their behavior when using a porous wick structure with trapezoidal microchannels. The porosity of the material denoted by Φ, which in this case is part of a packed bed, according to [16], represents the ratio of the volume occupied by the pores to the total volume of the assembly. The choice of the inner capillary layer for the FMHP depends on several factors, but the most important is the compatibility of the material with the properties of the working fluid. It is well known that the role of the capillary layer is to generate an internal capillary pressure which, together with the adhesion forces, leads to the movement of the working fluid through the FMHP. The choice of the type of capillary layer is secondly aimed at its ability to evenly distribute the working fluid in the vaporization zone. Providing both functions requires the use of capillary layers with different shapes, especially for FMHPs where operation requires horizontal placement. In this case, the return of the condensed liquid to the evaporator is no longer produced gravitationally, the major role of the capillary layer being to distribute the working liquid circumferentially. Another characteristic of the capillary layer that needs to be optimized is its thickness. The heat transport capacity of an FMHP increases as the thickness of the capillary layer increases up to an optimum maximum value. However, the increase in the radial thermal resistance of the capillary layer due to the increase in thickness is contrary to the increase in transport capacity and leads to a reduction in the maximum heat flux from the evaporator. The total thermal resistance of the vaporizer also depends on the conductivity of the working fluid in the capillary layer. Another necessary property to be fulfilled by the capillary layer is6 its compatibility with the wettability of the working fluid, which must allow constructive adaptability to the shape of the inner wall of the FMHP. In the case of FMHP, there are several materials that can be used for inner capillary layers. In order to achieve the most optimal heat transport characteristics, a compromise between the thickness of the capillary layer size and the structure of the inner pores (sintered capillary layers) is necessary. In this way, the inhomogeneity of the porous capillary layer assimilated to the pore size will result in a total thermal conductivity for the whole assembly, which depends equally on the working fluid, the material from which it is made, and the number of pores per unit volume. Experiments carried out by Peter, H.J. de Bock et al. [17] have shown that the use of sintered capillary layers, in combination with micro-channels distributed longitudinally along the direction of vapor travel in certain operating situations, can reduce the drying of the vaporization zone. However, even in this case, situations may arise where the variation of the heat flux may be large enough to cause a blockage of the flow of accumulated condensate from the condenser to the vaporizer. In two-phase heat transfer, the heat transfer performance of an FMHP is strongly influenced by the working fluid fill ratio and the space left free for vapor circulation. If the free space through which the vapor flows during heat transfer is greatly reduced, the un-evaporated liquid remaining in the capillary layer tends to accumulate in the corners and at the edges of the FMHP, thus reducing the overall thermal resistance of the system. Under the same conditions, there is a reduction in the radius of curvature of the liquid meniscus in the condenser. This has been experimentally found by F. Lefèvre et al. [18], who showed that if the space through which the vapor flow decreases by 1 mm, the heat transfer capacity of the FMHP decreases by up to 35%. An important condition in the design of an FMHP is the capillary boundary. The threshold at the capillary limit is reached when vapor and liquid exceed the thickness of the capillary layer, the internal pressure prevents the movement of vapor and liquid relative to each other, and if the absorption capacity of a given porous material can no longer provide for fluid circulation within its pores. The maximum thickness of the sintered capillary layer is strictly related to the size of the voids in the layer or, more precisely, to the size of the microspheres making up the layer. In the case of FMHP, this capillary boundary phenomenon prevents the movement of freshly condensed liquid from the condenser to the vaporization zone. As in the other cases presented above, this also leads to the "drying out" of the vaporization zone during the heat transfer that takes place through the micro-tube.

2. Computational Details

2.1. *Influence of Polysynthetic Medium Porosity on FMHP's Surface Layer Heat Transfer*

The porous material obtained and used in our own laboratory for FMHP is sintered copper powder. The simplest method to study heat transfer through porous media is that of heat transfer under the assumption that the fluid diffuses into the porous medium and remains at rest, being traversed by a heat flux ϕ [19,20]. If, by initial boundary conditions, the temperature of the fluid (index "l") and that of the solid (index "s") in the transient regime is considered, the heat transfer equation for the two media will be:

$$(\rho c)_{l\_s}\frac{\partial T}{\partial t} = \nabla(h_{l\_s}\nabla T). \tag{1}$$

In Equation (1), the following notations are made, ρ—density of the working fluid, c—specific heat, T—temperature, t—time, ∇—Nabla differential operator, h—heat conduction coefficient. At the fluid (index "fl")—porous material (index "ps") interface, the $(\rho c)_{l\_s}$ term can be defined as follows:

$$(\rho c)_{l\_s} = \phi\rho_{fl}c_{fl} + (1-\phi)\rho_{ps}c_{ps}, \tag{2}$$

and represents an average value of the specific heat for the two environments.

This approach can only be applied for the condition when the transient thermal regime is short and when the ratio between the thermal conductivity of the liquid and the solid is much different from unity. If the solid and the liquid are not in thermal equilibrium, a system consisting of two equations is considered. For ease of calculation, a rigid medium with a fixed structure is considered. The liquid in the porous medium is assumed to have constant viscosity, density, and thermal conductivity. Since the liquid is considered incompressible, by applying an external temperature, the force it exerts on the porous medium can be characterized by the energy equation [21]:

$$\rho_l c_l \frac{\partial T}{\partial t} + \rho_l c_l \nabla\left(\vec{u}\,T\right) = h_l \nabla^2 T, \tag{3}$$

where $\vec{u}$ is the velocity vector in the direction of fluid flow. Of course, in this case, heat transfer cannot be minimized by the viscosity of the interacting fluid media. This can only be considered when the flow is predominantly in one direction (unidirectional flow) or for boundary layer flow.

In the analysis of convective heat transfer through porous media, it should be taken into account that the most important transformations occur at the interface of these media with the wall of the FMHP. The general equation of convective heat transfer [21] from the solid medium (FMHP wall) to the working fluid diffusing into the porous medium can be written as:

$$q_{s\_l} = h_{s\_l} a_{s\_l} \left(\overline{T_s} - \overline{T_l}\right). \tag{4}$$

In Equation (4), the area of the solid-liquid interface has been denoted by a_{s_l} and was determined to be $a_{s\_l} = 0.015\ \text{m}^2$, while h_{s_l} represents the convective solid-liquid heat transfer coefficient. An analysis of heat transfer in porous media shows that the two forms of heat transfer, convective and conductive, cannot be separated as the interface between the two media determines this. These two effects, convective and conductive, cause a volumetric force to arise, determined by the porosity coefficient of the medium, the hydraulic diameter of the FMHP, and the macroscopic parameters of the media.

The hydraulic diameter D_h, can be determined by the relation [22]:

$$D_h = \frac{4\,D_m\,\Phi_p}{C_s\left(1 - \Phi_p\right)}. \tag{5}$$

In Equation (5), the hydraulic diameter of the microfluid flow channels was noted as $D_h = 4R_h$, where R_h is the hydraulic radius, $D_m = 0.456 \times 10^{-3}$m is the grain diameter, porosity was considered $\Phi_p = 0.25 \div 0.75$, and C_s is the shape constant of the cross-section, which is determined as [22]:

$$C_s = \frac{m_{fl}\mu_{fl}}{\gamma_W {R_h}^{2\,i\,a_{po}}}. \tag{6}$$

In Equation (6), the fluid flow rate is denoted by m_{fl} and the fluid viscosity by μ_{fl}.

The calculations were performed using a code developed under Mathcad14.

In order to determine the flow regime estimated with the Reynolds number, it is necessary to determine the flow velocity of the working fluid through the porous material [23].

$$w_l = \frac{\Phi_p}{\mu_{fl}}\left(\frac{D_h^2}{16C_k}\right)\left(\frac{\Delta P}{L_{hp}}\right)\left(\frac{L_{hp}}{L_e}\right)^2. \tag{7}$$

The notations in the fluid velocity relation [23] have the following meanings: μ_{fl}—fluid viscosity, D_h—hydraulic diameter, C_k—Kozeny constant, ΔP—pressure difference, L_{hp}—medium length, L_e—actual flow length. In microthermal tubes, it is also possible to consider the phenomenon of convective diffusion, which occurs due to free convection and occurs as a result of density differences induced by changes in pressure and inner temperature. Diffusion will be analyzed by determining the velocity of fluid movement through porous media. The evolution of the hydraulic diameter and the fluid flow velocity through the porous medium in dependence on the porosity of the polysynthetic medium is shown in Figure 2.

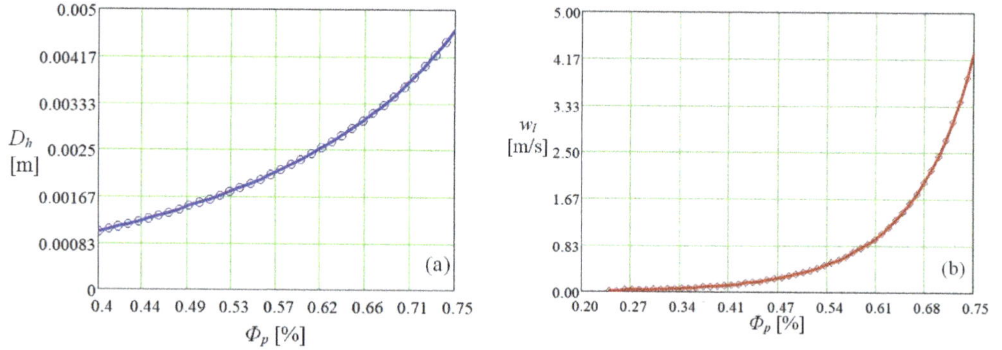

Figure 2. Variation of hydraulic diameter (**a**) and liquid velocity through a porous material (**b**) as a function of porosity.

The increase in porosity obviously leads to an increase in the hydraulic diameter (case a) and the flow velocity of the working fluid (b), which can be seen in Figure 2.

Knowing the average value for grain diameter $D_g = 1.78 \times 10^{-3}$m, obtained by measurements, one can calculate the dynamic specific surface area [22] with the relation:

$$a_{vd} = \frac{C_s}{D_g} \tag{8}$$

The Reynolds number will be determined using Equation (9) [22].

$$Re_h = \frac{\rho_{fl}\, w_l \frac{4\Phi_p}{a_{vd}(1-\Phi_p)}}{\mu_{fl}}. \tag{9}$$

Using the developed code, the Reynolds number variation as a function of the porosity of the polysynthetic material was determined, and the values obtained by calculation are shown in Figure 3.

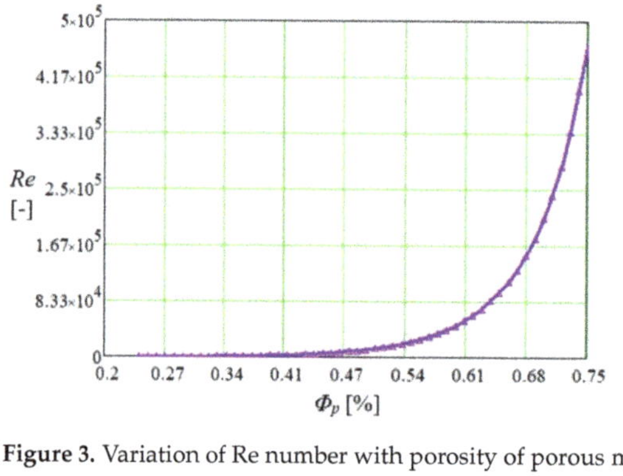

Figure 3. Variation of Re number with porosity of porous material in an FMHP.

The heat exchange process takes place through two different mechanisms. The Nusselt number is determined for the case of fluid diffused in the porous copper medium and for fluid flowing through polysynthetic material. The total heat transfer coefficient of the fluid flowing through the porous medium formed by copper microspheres expressed by interfacial Nusselt number Nu_{s_l} [24], can be given as a function of Prandtl and Reynolds numbers, as:

$$Nu_{s\_l} = \begin{vmatrix} Nu^*_{s\_l}\left(1 + \frac{aPrRe_h}{b+Nu^*_{s\_l}Pr^{1-m}Re_h^{1-n}}\right) \; if \; m = 0.5 \; and \; n > 0.5 \\ \frac{h_{s\_l}d_p}{k_m} \; otherwise, \end{vmatrix} \tag{10}$$

where Pr is the Prandtl number and Re_h is the Reynolds number based on Darcy velocity and hydraulic diameter. Stagnant Nusselt number, $Nu^*_{s\_l}$, can be written as a function of Darcy's laws as: $Nu^*_{sol,lic} = \frac{\sigma_{s\_l}}{\alpha_{hA}\sigma_{s|\_l}+\alpha_{hB}}$, where α_{hA} and α_{hB} represent the dimensionless conduction layer thickness in fluid and solid phases as normalized by the particle diameter, respectively σ, the solid to fluid thermal conductivity ratio. It was adopted $\sigma_{s\_l} = 394$ W/mK, this value falling between the recommended values of $1 < \sigma < 10^3$. In Equation (10), a and b represent coefficients whose ratio is a constant, $m = 0.5$, if $Pr \ll 1$ or $m = 0.33$ if $Pr \gg 1$ and $n > 0.5$ for large Reynolds numbers.

In the graph shown in Figure 4, the correlation between the Nusselt number and porosity was represented when the Prandtl number changes for convective transfer between the two capillary media. It should be noted that Petukhov [25] indicates values of the Prandtl number for a fully developed turbulent pipe flow ranging from 0.5 to 2000 while the Nusselt equation deduced by Gnielinski [22,26–28] for turbulent as well as transition regions for a smooth tube is valid only if $0.6 < Pr < 10$.

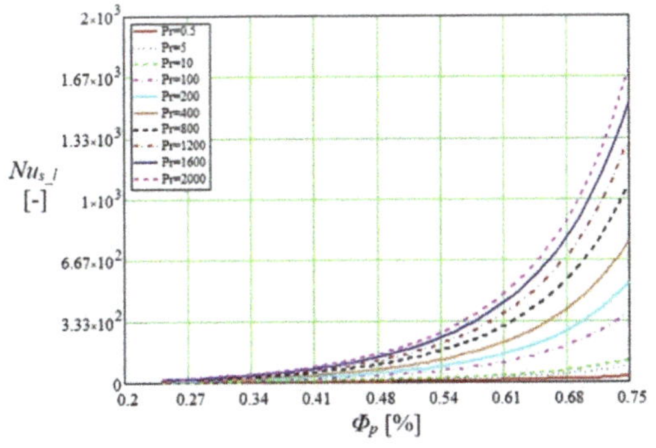

Figure 4. Nusselt representation as a function of porosity and Prandtl number values.

The amount of liquid carried through the two media (copper microsphere surface and trapezoidal microchannels) is not the same. After the entire amount of liquid has evaporated from the microchannels, the liquid in the polysynthetic capillary layer continues to evaporate. The evaporation of the fluid is caused by conductive heating of the polysynthetic porous layer by the heat from the walls of the flat thermal micro-tube. The fluid moves through the capillary layer and starts to wet the inner porous cavities of the porous medium, creating a convective-type phenomenon. In this case, for porous media, the Nusselt number can be determined as a function of Reynolds number and/or Prandtl number by empirical relations proposed by various authors in different forms [13,26–36], at which the flow regime is turbulent.

$$Nu_{Gn} = \frac{\frac{(0.79ln(Re_h)-1.64)^{-2}}{8}(Re_h - 1000)Pr}{1 + 12.7\left(\frac{(0.79ln(Re_h)-1.64)^{-2}}{8}\right)^{\frac{1}{2}}\left(Pr^{\frac{2}{3}} - 1\right)} \tag{11}$$

The Reynolds number determined by the mathematical model described above was used for the calculations. Gnielinski [37] notes the Darcy friction [36] $f = (0.79ln(Re_h) - 1.64)^{-2}$.

The variation of the Nusselt number is important for the study of heat transfer by the mechanisms described above. Starting from this consideration, a fluid Prandtl number of 0.5 to 2000 was used in calculations in the relation proposed by Gnielinski Nu_{Gn_A} [25], which provides a very accurate correlation (index A = accurate) for turbulent flow in porous media.

$$Nu_{Gn\_A} = \frac{\frac{(0.79ln(Re_h)-1.64)^{-2}}{8}(Re_h - 1000)Pr}{1 + 12.7\left(\frac{(0.79ln(Re_h)-1.64)^{-2}}{8}\right)^{\frac{1}{2}}\left(Pr^{\frac{2}{3}} - 1\right)}\left[1 + \left(\frac{D_h}{L_{HP}}\right)^{\frac{2}{3}}\right]\left(\frac{Pr}{Pr_w}\right)^{0.11} \quad (12)$$

The Prandtl number at the wall was denoted by Pr_w, knowing that $0.6 < Pr < 10^5$ for Reynolds values $2300 < Re < 10^6$. The results of the calculations performed for the Nusselt number with the Mathcad code using the two criterial relations are shown in Figure 5.

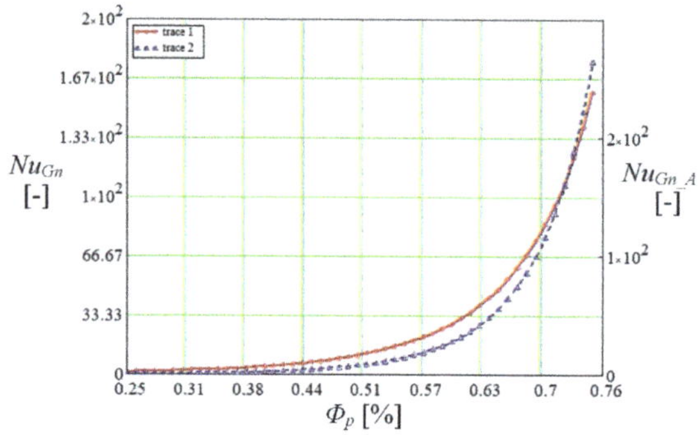

Figure 5. Variation of Nusselt number by normal (using Equation (11)) and accurate (using Equation (12)) method with porosity of porous layer.

Nusselt calculated with the normal method as specified in the article is obtained with a calculation code that uses Equation (11), and the improved one uses Equation (12). The data obtained by calculation show different values of the Nusselt number in the area of the average porosity, but the curves are the same, and the heat exchange is favored if the porosity increases. It can be appreciated that in a micro heat pipe with a composite structure, in the capillary layer made of copper microspheres, the liquid vaporizes much faster than the liquid embedded in the porous material. In composite structures, however, there is an advantage (which will also be seen in the experimental determinations) in that in the porous capillary layer made of sintered copper powder, the vaporization process starts faster and continues after the completion of this process in the trapezoidal microchannels.

Thanks to the composite structure of FMHP, an additional amount of vapor is produced by the described mechanism. Even though the structure of the FMHP is more complex, the described phenomenon prevents drying of the vaporization zone in case of high heat fluxes and, at the same time, enhances heat transfer.

2.2. Determination of the Phase Difference between the Pressure Gradient and Liquid Velocity in the Boundary Layer of an FMHP

The analyzed thermal micro tube has in its structure two porous media, one made of sintered copper powder (obtained in the laboratory) and the other of trapezoidal microchannels. By displacing the liquid in the two porous media, transient inertial forces arise between the liquid and the porous structure. The transient inertia force will be all the more important as there will be a phase difference between the oscillation velocity and the pressure gradient. If the velocity amplitude for the fluid displacement through the boundary layer is taken into account [24], the maximum pressure gradient coefficient $\hat{f}$ can be written as:

$$\hat{f} = \left| \frac{\varphi_{St}}{Re_h} + \frac{\varphi_B}{Re_h^{\frac{1}{2}}} \sqrt{\frac{2.64}{\pi}} + \frac{2.67\varphi_I}{\pi} + \frac{\varphi_M}{Re_h^{\frac{1}{2}}} \sqrt{\frac{id_{hid}}{A_{fld}}} + (1+\varphi_V)\frac{id_{hid}}{A_{fld}} \right|. \quad (13)$$

In Equation (13), φ_{St} represents the Stokes frictional force, φ_B is the frictional force due to the advection boundary layer, φ_I is the inviscid form drag, φ_M represents the Basset memory viscous force due to the transient boundary layer, φ_V is the inviscid virtual mass force, and A_{fld} denotes the amplitude of the fluid displacement of the superficial flow which has the expression $A_{fld} = \Phi\overline{A}$ where $\overline{A}$ represents the intrinsic average of the fluid displacement in the pores. The offset between pressure and velocity is determined as:

$$\theta_{p\_w} = \frac{1}{tan\left(\frac{(1+\varphi_V)\frac{d_h}{A_{fl}} + \frac{\varphi_M}{\sqrt{2Re_h}}\sqrt{\frac{d_h}{A_{fl}}}}{\frac{\varphi_S}{Re_h} + \frac{\varphi_B}{\sqrt{Re_h}}\sqrt{\frac{2.64}{\pi}} + 2.67\frac{\varphi_I}{\pi} + \frac{\varphi_M}{\sqrt{2Re_h}}\sqrt{\frac{d_h}{A_{fl}}}} \right)} \quad (14)$$

The relation given in Equation (14) shows that the pressure gradient of an oscillating flow in a porous medium depends on the Reynolds number as well as on the ratio of the hydraulic diameter d_h A_{fld}. In order to represent the phase shift between the pressure gradient and the fluid velocity on the boundary layer, the values $\varphi_{St} = 109$ N, $\varphi_B = 5$ N, $\varphi_V = 1.7$ N, $\varphi_I = 1$ N and $\varphi_M = 25$ N, were considered. The variation of the pressure gradient and the phase shift between the vapor pressure and the vapor displacement velocity depending on the porosity of the polysynthetic material obtained in Mathcad can be seen in Figure 6a,b.

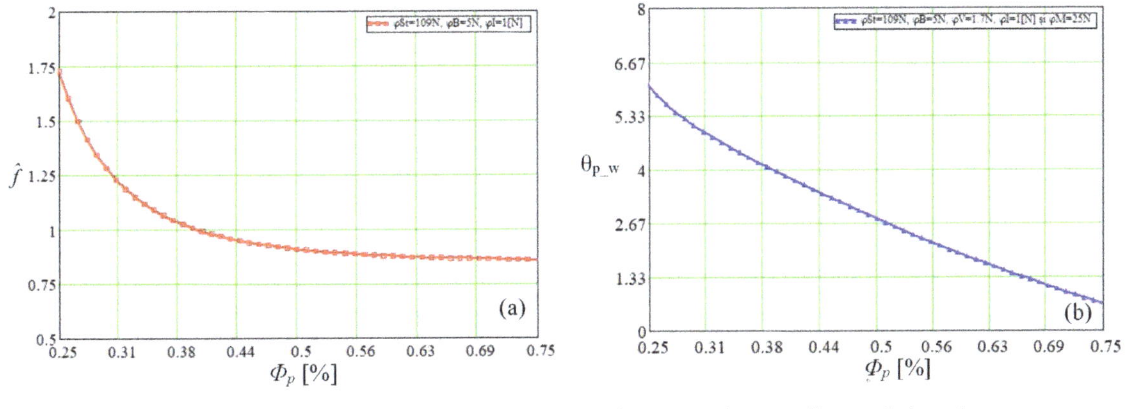

Figure 6. Maximum pressure gradient coefficient (**a**) and pressure-velocity gradient on the boundary layer (**b**) as a function of porosity.

As the evaporation of the liquid develops more and more and, the Reynolds number increases, and the pressure-velocity phase shift decreases, the amplitude of the boundary layer oscillations is reduced. In fact, the evaporation process in porous media is mainly due to convective heat transfer. Oscillations occurring in the boundary layer at the separation between two media made of different materials are caused by the nucleation phenomenon occurring in the liquid during the heat transfer process. The nucleation development will be visualized in the experimental part of this paper obtained by filming the vaporization zone for both the porous medium and the trapezoidal microchannels. One of the important factors in explaining the phenomenon is determined by the memory of the Basset viscous forces due to boundary layer transit. Vapor bubbles, which are produced inside the heated liquid, start to rise toward the boundary layer. The variable speed of the vapor bubbles, as they move through the liquid inside the FMHP, causes pressure on the boundary layer explained by the Basset viscous forces, which leads to the oscillation phenomenon, a process that diminishes as the liquid evaporates completely. At the moment of phase change, nucleation disappears, and thermodynamic equilibrium is reached. The study of oscillations due to the pressure gradient for the liquid boundary layer between the two porous media is necessary for the study of the vaporization occurring in the two layers. The different structure of the two layers leads to different behavior during heat transfer. By vaporizing the liquid in the first capillary layer, vapor bubbles of a certain temperature pass through the boundary layer, causing the second capillary layer to heat up. This results in a movement of the vapor bubbles from and towards the two layers, causing oscillations in the boundary layer, which are attributed to the existence of a pressure gradient.

2.3. Darcy Flow through a Porous Medium Inside a Horizontal Micro Heat Pipe

For the study of flow and heat transfer phenomena, it is of interest to determine the friction factor of the liquid flowing through the two porous media. This can be investigated using Colebrook's equation. Solving this equation is quite difficult [38–41] as it requires the use of iterations. Colebrook's equation allows the determination of the Darcy friction factor (also known as Darcy–Weisbach), which is necessary to calculate the pressure drop or pressure loss due to wall roughness. Over time several equations and methods of solving them have been developed. The Colebrook–White equation has the form:

$$\frac{1}{\sqrt{f}} = -2log\left(\frac{\phi}{3.7D_h} + \frac{2.51}{Re_h\sqrt{f}}\right) \tag{15}$$

Different methods are used to solve this equation [38–41], among which we mention the use of a dedicated calculator. The method solves the Colebrook Equation by Hand in Excel, using a root-finding VBA subroutine, using the Moody diagram, by approximation, or by the direct method. For relation (15), the programming code proposed by [39] was adapted to the values of the thermal microtube with a porous structure. For a comparative study of the Darcy friction factor, a code built in Mathcad14 for Equation (16), presented by Jukka Kiijärvi [41], was used.

$$\begin{aligned} f_{DW} &= \frac{2\Delta p\, D}{L_{hp}\rho_{fl}V_{fl}^2};\; f_{SJ} = 0.25\left[log\left(\frac{e/D}{3.7} + \frac{5.74}{{Re_h}^{0.9}}\right)\right]^{-2}; \\ f_{Bl} &= \frac{0.316}{Re_h^{0.25}};\; f_{Ha} = \frac{1}{-1.8\,log\left[\left(\frac{e/D}{3.7}\right)^{1.11} + \frac{6.9}{Re_h}\right]}. \end{aligned} \tag{16}$$

The meanings of the notations in Equation (16) are: f—Darcy friction factor, DW—Darcy–Weisbach, Δp—pressure loss, D—inner diameter of the heat pipe, L_{hp}—length of the heat pipe or part of the pipe, ρ_{fl}—density of the fluid, V_{fl}—flow velocity of the fluid, SJ—Swamee–Jain, e—the roughness of pipe, Bl—Blasius, Ha—Haaland.

Figure 7 plots the variation of friction coefficients for porous media made of copper and polysynthetic material as a function of material porosity (Figure 7a) and Reynolds number (Figure 7b).

The results shown in Figure 7 were obtained using two computational codes, one made by the authors in Mathcad and one proposed by Praks, P.; Brkić, D in Matlab [38]. As it can be easily seen from Figure 7, the curves have the same trend for the case of Haaland, Swamee Jain, Blasius, and Colebrook's equations, but in terms of value, there are differences of a few percent. On the other hand, the model proposed by Darcy–Weisbach, in our opinion, gives values outside the tolerance field since the shape deviations are high and significant differences in value appear. In the paper, we have predicted the relationship between heat transfer and the friction factor of fluid flow through a porous medium. The results show that the rate of heat transfer through a porous structure in FMHPs increases with increasing Reynolds number and that there is a direct link between the Nusselt number and friction factor. The Darcy number decreased with increasing Reynolds number. This paper shows how the rate of vapor generation changes with changes in the internal structure of FMHPs.

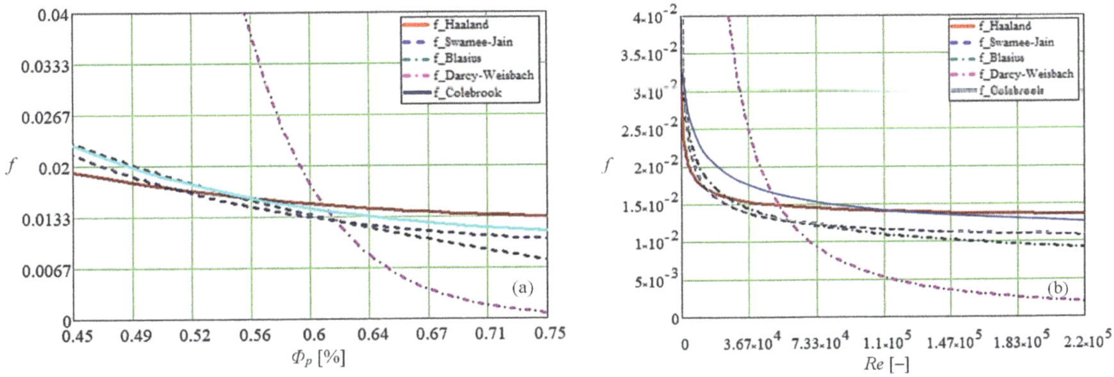

Figure 7. Friction coefficient through porous media as a function of porosity (**a**) and Reynolds number (**b**).

3. Experimental Set-Up and Methodology

For experimental determinations, a set-up shown in Figure 8 was used, containing the following elements: 1—electronic module for controlling and monitoring FMHP parameters, 2—keypad for prescribing the voltage and current consumed by the heating resistors, 3—general on/off button, 4—vaporization zone heater voltage and current display, 5—cooler on/off button, 6—condensing zone cooling control system, 7—display for displaying the temperature read by thermocouples, 8—on/off switch thermocouple module, 9—data logger and connection with PC, 10—PC, 11—thermocouples, 12—heating block, 13—micro-thermal tube, 14—cooler, 15—electronic heater control module.

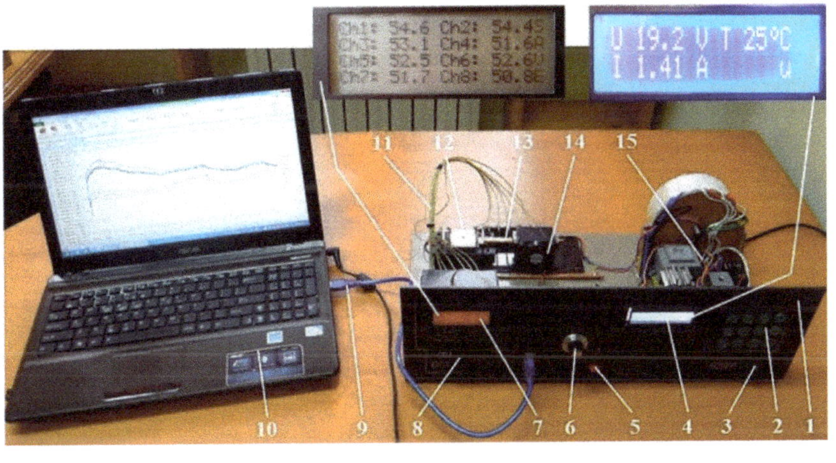

Figure 8. Experimental set-up for the study of FMHPs.

To study and visualize the behavior of the working fluid during the supply of liquid from the condensate, the initiation of vaporization, the development of the vaporization itself, and the displacement of the vapor, microscopic filming was carried out. The FMHP structure in the vaporization zone is visible in Figure 9 (made in the laboratory) and is composed of a porous medium made of sintered copper metal powder and trapezoidal copper microchannels. It is important to know that the filling of the inner structure with distilled water was performed at the same time. Over the porous layer and over the trapezoidal microchannels of the FMHP, a quartz glass was tightly applied to the top, then the tube was partially evacuated. A heating source consisting of electrical resistors fed from an adjustable power supply was placed under the micro heat pipe. Knowing the power parameters of the electrical resistors, the heat flux density can be determined by calculation.

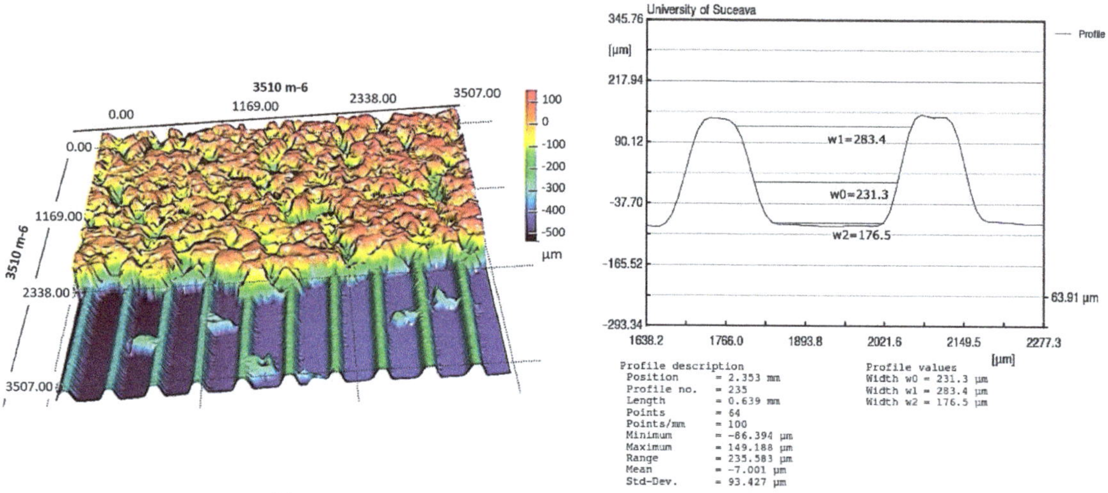

Figure 9. Sample measurement of trapezoidal microchannels: (**a**) 3D surface microtopography; (**b**) dimensional details for the microchannel transverse cross-section.

A series of laboratory tests were carried out to verify the behavior of a micro heat pipe with a porous layer of sintered copper powder and trapezoidal microchannels. From the studies and observations, we found that the permeability of the porous capillary layer increases with increasing pore size. However, for homogeneous capillary layers, there is an optimal pore size that is relatively simple to determine. A trade-off is created between permeability and pore size. Porous capillary layers used in FMHP can have low performance if the pore sizes are between 80 ÷ 150 µm for horizontal operation where the working fluid displacement should not overcome the gravitational force and high-performance capillary layers where the pore sizes are between 30 ÷ 80 µm for vertical or inclined operation where pumping of the working fluid against gravity must be ensured. To transport the condensed working fluid from the condenser to the vaporizer, a capillary layer of a porous material made in the laboratory by sintering copper powder has been incorporated into the internal structure of the FMHP. As it is well known, if the liquid has to rise against gravity, it is recommended to use a capillary layer with low porosity. The maximum capillary pressure inside an FMHP (which is initially partially evacuated) is determined by the average pore size of the sintered capillary layer, decreasing as it increases. Sintered structures used in capillary layers provide higher capillary pressure than grooved or trapezoidal microchannel structures. However, the permeability of sintered capillary layers is lower. Therefore, there will be pressure losses due to friction. Both micro-channels and sintered layers must be made of the same material as the FMHP wall and be fixed to the inner walls of the FMHP in order to achieve the best thermal contact. In the case of the porous material, we have concluded from experimental determinations that it should be chosen with a relatively high degree of porosity in order to retain as much liquid as possible inside the pores. The absorption of liquid into the larger pores of the material ensures an increase in the capillary radius. The choice of material should be made in relation to the temperature at which the FMHP is to be used so that once the vaporization temperature of the working fluids is reached, the material does not change its physical properties. Also, the change in physical properties must not lead to the interaction of the material with the working fluids used.

4. Experimental Results and Discussion

The dimensions of the trapezoidal microchannels were obtained by laser profilometry using a NanoFocus µScan laser profilometer, a sample measurement of a microchannel is given in Figure 9a, and the details of the internal structure of the FMHP used in the experiments was obtained by the same method and can be seen in Figure 9b.

In the following paragraphs, reference is made to the polysynthetic material used in practice in the FMHP structure since its absorption capacity, internal structure, and compression deformation have superior quality parameters to other materials. Preliminary tests proved that the porous material retained its absorbent qualities when using distilled water, acetone as well as methanol. Figure 10a,b shows a microscopic image of two cases, namely the structure of the sintered material made of dry sintered copper powder (case a) and wetted with distilled water (case b).

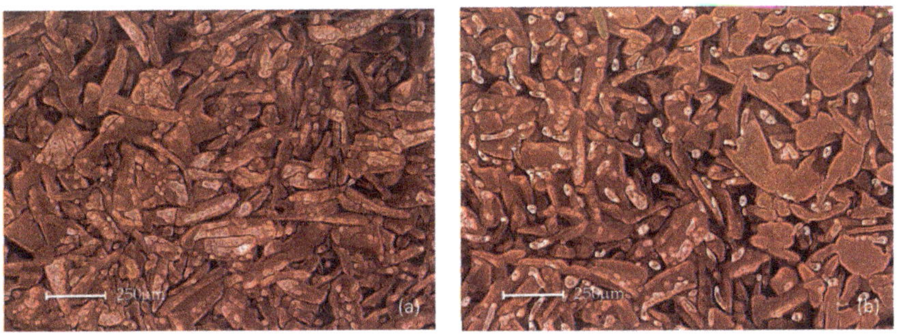

Figure 10. Porous sintered copper powder material: dry (**a**) and wetted with condensation liquid (**b**).

In order to know the quantities used in calculations, laser profilometry was used to measure the surface of the porous material, both in a dry and fully wetted state. Figure 11 shows the results first scanned for the dry porous material.

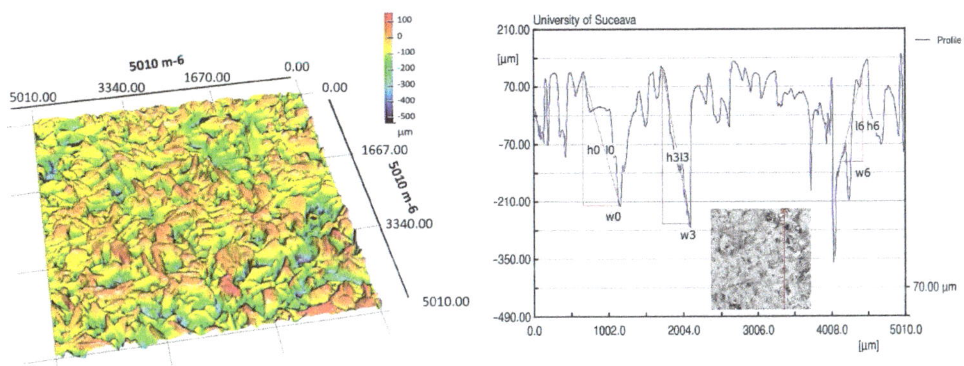

(**a**) (**b**)

Figure 11. 3D dry porous layer images (**a**) and sample with dimensions (**b**) obtained by laser profilometry.

In order to observe how the pore size changes when the material is wetted with the working fluid, it was necessary to scan it by laser profilometry, results illustrated in Figure 12 providing a series of physical parameters specific to this case.

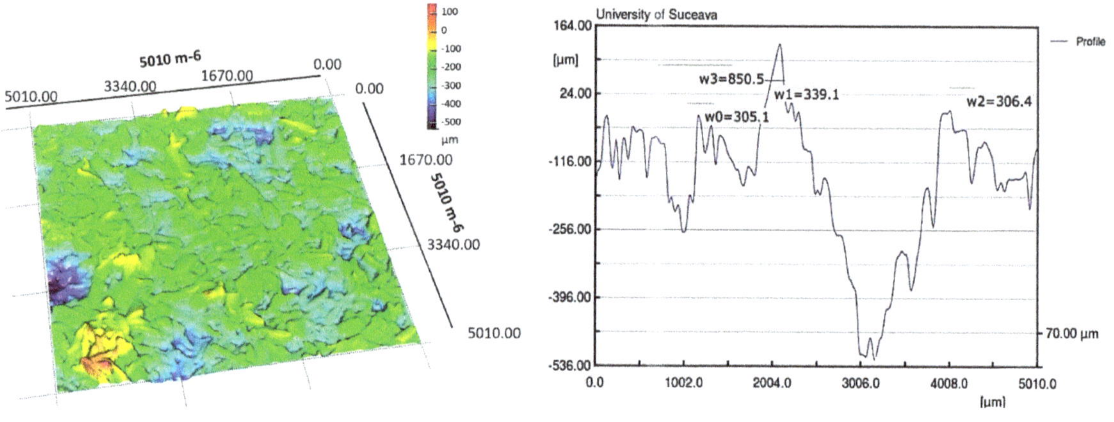

(**a**) (**b**)

Figure 12. Images of the wetted porous layer in 3D (**a**) and the dimensions of the formed microchannels (**b**), obtained by laser profilometry.

For each case, the equipment creates a report from which an extract of the most important parameters for the two cases is shown in Table 1.

Table 1. Polysynthetic layer microchannel profile values measured by laser profilometry.

Case \ Measured Parameter	Length [mm]	Points –	Minimum [µm]	Maximum [µm]	Mean [µm]	Std-Dev. [µm]
Dry porous material	5.010	501	−401.945	148.977	13.208	88.085
Wetted porous material	5.010	501	−522.357	126.796	156.629	135.783

The experimental data obtained clearly show a change in the mean values of the microchannel parameters when using a porous material wetted with condensate due to the influence of the liquid weight on the pores. The values for pore area, diameter or size of microspheres or copper microparticles permeability, respective pore thread diameter of the porous material, and porosity were assumed, as shown in Table 2.

Table 2. Considered values for capillary layers.

Material	Pore Area [m^2]	Microsphere Diameter/Average Values of Copper Powder Particles [m]	Porosity [%]
Capillary layer of sintered copper microspheres (average values)	$(5 \div 15) \times 10^{-8}$	$(52 \div 78) \times 10^{-6}$	$35 \div 85$
Porous sintered copper powder material (obtained by the authors)	$(98 \div 195) \times 10^{-9}$	$(10 \div 32) \times 10^{-6}$	$25 \div 75$

The experiments allowed to film at the same time two categories of internal structures (wick structure) of the FMHP. The results obtained can be seen in Figure 13, which includes a sample of the filmed frames.

In all the selected frames, it can be seen how in the first heating step and reaching the vaporization temperature, the vapor formed in the porous zone migrates to the trapezoidal microchannel zone. The velocity of fluid or vapor movement can be determined by calculations using the distance of fluid or vapor movement or development obtained on the basis of the micrometric scale of the images and the time on the basis of a specialized image processing editor (determined for the beginning and end of the processes).

Figure 14 shows extracts of detail frames for the porous zone obtained from the microscopy images.

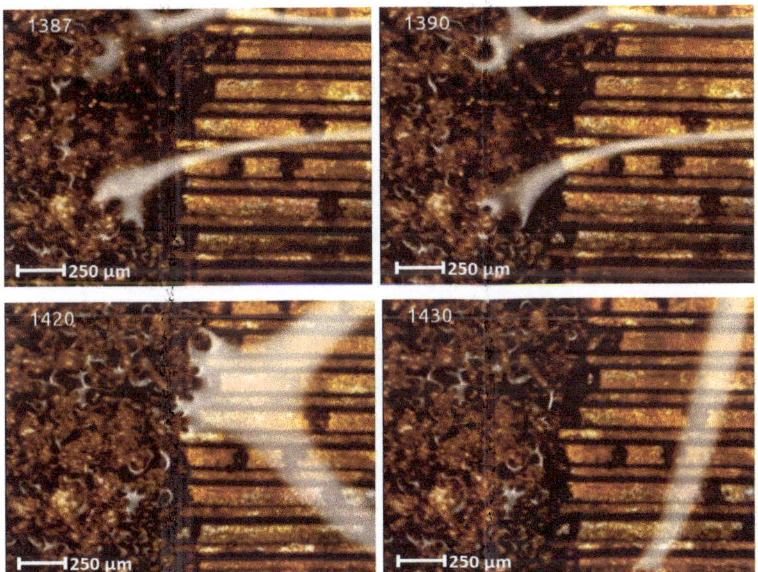

Figure 13. Vaporization zone with porous material and trapezoidal microchannels.

The porous layer in the evaporation zone of frame 1350 in Figure 12 shows is initially dry, in which case, after liquid supply, the liquid pressure inside will be equal to the vapor pressure. As liquid displacement through the capillary layer (visible in frame 1360 right) takes place, the pressure inside the FMHP decreases, a phenomenon mainly due to friction of the working fluid with the internal structure of the capillary, gravity, and depending on the mass of liquid displaced. This reduction in capillary pressure is sufficient to cause the fluid to start boiling (nucleation) once condensate starts to reach the zone (frames 1370 to 1390). The pressure in the capillary layer is important because it causes nucleation to begin in the pore layer and at the wall of the FMHP. As this pressure decreases, the risk of boiling the liquid increases. There are several factors that can substantially influence the pressure in the capillary layer. Temperature is the determining factor as it affects the physical properties of the working fluid in the FMHP and is, in turn, controlled by the applied heat flux. In turn, the degree of permeability of the capillary layer influences the total pressure drop. Since the pressure loss in the capillary layer decreases with the length of the FMHP, the most likely place where fluid boiling can occur is at the top of the vaporization zone, which can be visualized in frames 1380 and 1390. If liquid boiling is avoided in this section, then this phenomenon will not occur along the entire length of the capillary layer. The occurrence of a pressure loss gradient is mainly attributed to the porosity of the capillary layer. For lower porosities, it has been found that the pressure increases faster than when the porosity tends towards maximum values. At porosities tending towards maximum values, more liquid accumulates in the porous material, which leads to a certain incompressibility of the porous material, in which case the capillary pressure loss is lower.

Figure 14. Frames with vapor formation and displacement in a porous medium in an FMHP.

If the liquid vaporizes, the pressure in the vaporizer drops very much, but after leaving the vaporizer, the pressure in the adiabatic zone does not drop very much, and then the pressure in the condensation zone increases slightly due to the temperature reduction. An important influence on nucleation and boiling is the Reynolds number, which, when increasing, the capillarity of the polysynthetic starter decreases due to the increase in pressure by changing the vapor velocity, which causes the axial hydrostatic pressure to have an increasing tendency. An important condition for high capillarity is that the capillary layer has a high degree of permeability. This can be determined by its internal structure through the size of the pore size. A structure with the smallest pore size has a high degree of permeability, and as the porosity decreases, the capillary pressure decreases. The vaporization process continues for a long period (1400 to 1435 frames) with the development of new nucleation centers. After the nucleation phenomenon has fully developed, a flow of vapor formed through microchannels begins (frames 1440 to 1465) and then continues until a new wave of cold liquid condensate appears.

In order to study the vaporization phenomenon in a porous environment in comparison with that in a trapezoidal microchannel environment, movies were taken (an extract of frames is the one in Figure 15) on the same FMHP. The study allows a comparative analysis of vaporization and differentiation of the nucleation and vapor development mode for the two cases.

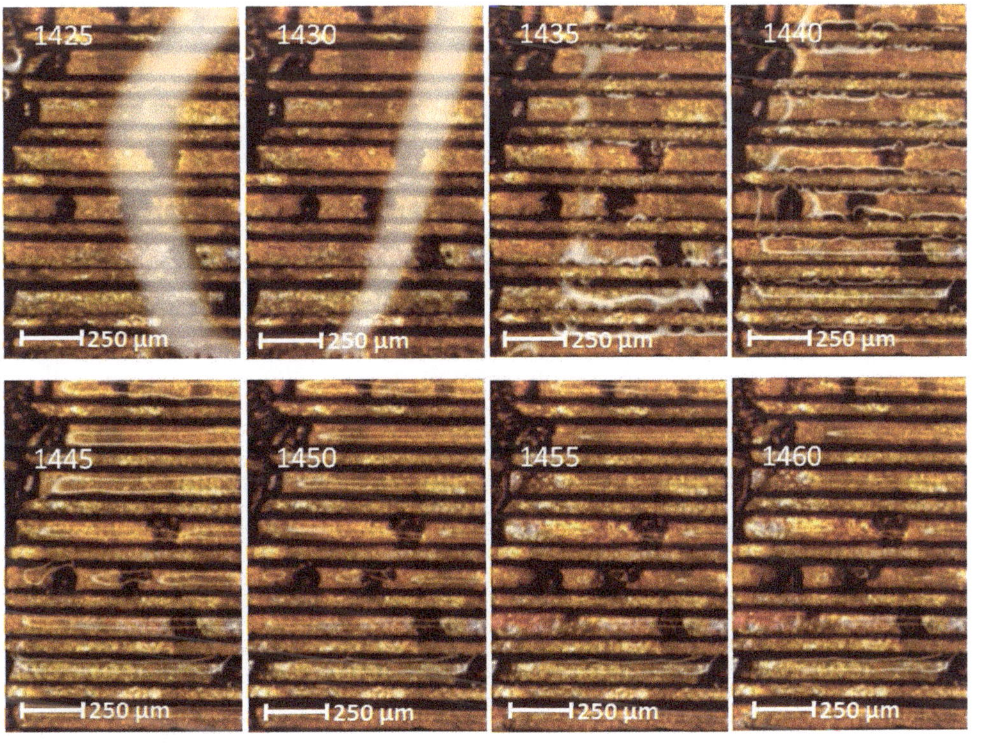

Figure 15. Frames with vapor formation and displacement in a porous medium in a micro heat pipe.

If, in the first case, corresponding to the porous layer, the vaporization started at frame 1360 and ended at frame 1465, in the case of trapezoidal channels, there is a phase shift compared to the case of micro trapezoidal channels. Although the liquid in the condensate covered the trapezoidal microchannels and the porous material at the same time, there is a delay from frame 1360 (porous layer) to 1435 (trapezoidal microchannel layer) where it is observed for the second case how nucleation starts and how the vaporization process begins to propagate.

5. Conclusions

In the present paper, various parameters, such as the hydraulic diameter and fluid velocity through the material as a function of the porosity of the internal structure, were determined by calculation for a steady state regime. Reynolds number was determined as a function of porosity, according to Darcy's law, and the Nusselt number was calculated. Both are of interest in the study of heat exchange in an FMHP. The calculation of the Nusselt number was performed using the normal (using Equation (11)) and improved method (using Equation (12)) as a function of porosity and Prandtl number values. The determination of the phase difference between the pressure gradient and the liquid velocity in the boundary layer of a FMHP allows the analysis of how flow through the porous layer takes place. Since the flow is Darcy-type through the sintered copper powder porous medium inside the FMHP, the Darcy friction factor was calculated using five methods: Colebrook, Darcy–Weisbach, Swamee–Jain, Blasius, and Haaland. The results obtained were analyzed as a function of the change in porosity and Reynolds number.

The experiments were carried out on a surface structure composed of trapezoidal microchannels with an average base of 176.5 μm and a peak of 283.4 μm (see Figures 9, 13 and 15) and porous sintered copper powder material (see Figures 10–12) with an average depth of 210 μm which is relevant for a comparative study as they have close average values. The internal structure of the porous material has a porosity ranging from 55% to 80% at a thickness that, in turn, ranges from 0.8 mm to 0.38 mm.

The heating system used provides a heat flux on the evaporator side of up to 140 W, which, used at maximum values, indicated that the vaporization process is in strict dependence on the internal structure of the FMHP. Theoretical calculations showed that when the maximum value of the porosity of the FMHP material is reached, the hydraulic diameter increases to 4.63×10^{-3} m, the Reynolds number to 4.585×10^{5}, and the Nusselt number to 1.67×10^{3}, which favors the vaporization process by intensifying the convective phenomenon (if the Nusselt number increases, then the convective heat coefficient has the same tendency). The calculations showed that there is a phase difference between the pressure gradient and liquid velocity in the boundary layer of an FMHP.

The dimensional data used as input in the conducted calculations was obtained by measuring the surface microtopography by laser profilometry.

Several experimental tests were conducted by filming under the microscope the fluid feeding process of the internal structure of the FMHP. The condensed liquid was introduced into the FMHP, which was heated in the vaporization zone, followed by a nucleation process followed by vapor development. Knowing the time between frames and the space traveled by the vapor, it is possible to determine the vapor formation rate for both the porous medium and the trapezoidal microchannels. Using the same method, it was possible to estimate the velocity phase shift for the two interior structures, the porous one and the one with trapezoidal microchannels.

It was experimentally shown that for the porous medium structure versus trapezoidal microchannels, the vaporization proceeds faster from 5.44 ms to 5.86 ms versus a delayed onset at 5.74 ms to the end of vaporization at 5.84 ms. This shows that the vaporization process proceeds more efficiently in the porous sintered copper powder material layers subjected to this study. The explanation lies in the fact that the porous material performs better at high temperatures as it has a lower thermal resistance compared to the case of fluid flow through trapezoidal microchannels. This is due to the smaller thickness of the fluid film in relation to the contact surface, which in this case is larger in the porous material than in the case of trapezoidal channels. The study carried out experimentally proved that under certain operating conditions, it is possible that a porous material in an FMHP can transport heat more efficiently than a trapezoidal microchannel system.

In the present paper, an analysis was made on the heat transfer phenomena inside the FMHP, their limits, as well as the phase shifts that occur in the vaporization process depending on their inner structure.

Author Contributions: Conceptualization, I.M.; methodology, I.M.; software, I.M.; validation, I.M., C.S. and C.M.P.; formal analysis, I.M.; investigation, I.M., C.S. and C.M.P.; resources, I.M. and C.S; data curation, I.M.; writing—original draft preparation, I.M.; writing—review and editing, C.S.; visualization, I.M., C.S. and C.M.P.; supervision, I.M.; project administration, I.M.; funding acquisition, I.M. All authors have read and agreed to the published version of the manuscript.

Funding: This work was funded by Ministry of Public Works, Development and Administration within REGIONAL OPERATIONAL PROGRAM 2014–2020, PRIORITY AXIS 1: PROMOTING TECHNOLOGICAL TRANSFER, PRIORITY OF INVESTMENTS 1.1, OPERATION B—Supporting science and technology parks, under contract no. 6592/24.12.2020 (SMIS: 128730), project title: ESTABLISHMENT OF THE EAST EUROPEAN BORDER SCIENTIFIC AND TECHNOLOGICAL PARK IN SIRET.

Data Availability Statement: Some or all data, models, or codes generated or used during the study are available from the corresponding author by request.

Conflicts of Interest: The authors declare no conflict of interest.

References

1. Available online: https://www.1-act.com/resources/heat-pipe-fundamentals/different-types-of-heat-pipes/ (accessed on 12 May 2023).
2. Zaghdoudi, M.C.; Maalej, S.; Mansouri, J.; Sassi, M.B.H. Flat Miniature Heat Pipes for Electronics Cooling: State of the Art, Experimental and Theoretical Analysis. *Int. J. Mech. Mechatron. Eng.* **2011**, *5*, 714–737. [CrossRef]
3. Xin, F.; Lyu, Q.; Tian, W.; Xin, F.; Lyu, Q.; Tian, W. Visualization and Heat Transfer Performance of Mini-Grooved Flat Heat Pipe Filled with Different Working Fluids. *Micromachines* **2022**, *13*, 1341. [CrossRef]
4. Nesterov, D.A.; Derevyanko, V.A.; Suntsov, S.B. Development of T-shaped flat heat pipes for cooling electronic equipment. In Proceedings of the 20th IHPC and 14th IHPS, Gelendzhik, Russia, 7–10 September 2021; pp. 1–10.
5. Yang, K.S.; Cheng, Y.C.; Jeng, M.S.; Chien, K.H.; Shyu, J.C. An Experimental Investigation of Micro Pulsating Heat Pipes. *Micromachines* **2014**, *5*, 385–395. [CrossRef]
6. Song, W.; Xu, Y.; Xue, L.; Li, H.; Guo, C. Visualization Experimental Study on Silicon-Based Ultra-Thin Loop Heat Pipe Using Deionized Water as Working Fluid. *Micromachines* **2021**, *12*, 1080. [CrossRef]
7. Chen, X.; Ye, H.; Fan, X.; Ren, T.; Zhang, G.Q. A review of small heat pipes for electronics. *Appl. Therm. Eng.* **2016**, *96*, 1–17. [CrossRef]
8. Rangasamy, S.; Raghavan, R.R.V.; Elavarasan, R.M.; Kasinathan, P. Energy Analysis of Flattened Heat Pipe with Nanofluids for Sustainable Electronic Cooling Applications. *Sustainability* **2023**, *15*, 4716. [CrossRef]
9. Zhang, Z.; Wang, X.; Yan, Y. A review of the state-of-the-art in electronic cooling, e-Prime-Advances in Electrical Engineering. *Electron. Energy* **2021**, *1*, 100009.
10. Riffat, S.; Ma, X. Recent developments in heat pipe technology and applications: A review. *Int. J. Low Carbon Technol.* **2007**, *2*, 162–177. [CrossRef]
11. Vasiliev, L.L. Micro and miniature heat pipes-electronic components coolers. In Proceedings of the VI Minsk International Seminar "Heat Pipes, Heat Pumps, Refrigerators", Minsk, Belarus, 12–15 September 2005; pp. 74–86.
12. Mochizuki, M.; Nguyen, T.; Mashiko, K.; Saito, Y.; Nguyen, T.; Wuttijumnong, V. A review of heat pipe application including new opportunities. *Front. Heat Pipes FHP* **2011**, *2*, 013001. [CrossRef]
13. Pukhovoy, M.V.; Kunts, K.A.; Spesivtsev, S.E.; Kabov, O.A. Maximum heat fluxes and features of heat transfer mechanisms with boiling during jet impingement cooling of electronics. In *Journal of Physics: Conference Series*; IOP Publishing: Bristol, UK, 2021; Volume 1867, p. 012036.
14. Zeghari, K.; Louahlia, H. Flat miniature heat pipe with sintered porous wick structure: Experimental and mathematical studies. *Int. J. Heat Mass Transf.* **2020**, *158*, 120021–120058. [CrossRef]
15. Lv, L.C.; Li, J. Micro Flat Heat Pipes for Microelectronics Cooling: Review. *Recent Pat. Mech. Eng.* **2013**, *6*, 169–184. [CrossRef]
16. Ahmed, A.A.; Gennish, R.; Bilrrazig, A.; Jolgam, S.A. Darcy Flow and Heat Transfer Through Porous Media Inside a Horizontal Fixed Packed Bed. *Univ. Bull.* **2021**, *4*, 109–122.
17. de Bock, H.P.; Varanasi, K.; Chamarthy, P.; Deng, T.; Kulkarni, A.; Rush, B.M.; Russ, B.A.; Weaver, S.E.; Gerner, F.M. Experimental investigation of micro/nano heat pipe wick structures. In Proceedings of the ASME International Mechanica Engineering Congress and Exposition, Boston, MA, USA, 31 October–6 November 2008; pp. 1–6.
18. Lefèvre, F.; Lips, S.; Rullière, R.; Conrardy, J.B.; Raynaud, M.; Bonjour, J. Flat plate heat pipes: From obsevation to the modeling of the capillary structure. *Front. Heat Pipes* **2012**, 1–9. [CrossRef]
19. Lienhard, J.H., IV; Lienhard, J.H. *A Heat Transfer Textbook*, 5th ed.; Phlogiston Press: Cambridge, MA, USA, 2019; 771p.
20. Heat Conduction–Chapther, P373588.tex 1/2/2007, 1–42. Available online: https://booksite.elsevier.com/samplechapters/9780123735881/9780123735881.pdf (accessed on 10 March 2023).

21. Lu, X. Fluid Flow and Heat Transfer in Porous Media Manufactured by a Space Holder Method. Ph.D. Thesis, University of Liverpool, Liverpool, UK, 2020; 284p.
22. Chang-Hoon, S. Application of the effective diameters of porous media to the non-Darcy low analyses. *Sci. Rep.* **2022**, *12*, 5321.
23. Chang-Hoon, S. Permeability variation analysis using the superficial diameter correlation with porosity change. *Phys. Fluids* **2021**, *33*, 053108.
24. Hsu, C.T. Heat conduction in porous media. In *Handbook of Porous Media*, 1st ed.; Vafai, K., Ed.; Marcel Dekker: New York, NY, USA, 2000; pp. 171–199.
25. Petukhov, B.S. Heat Transfer and Friction in Turbulent Pipe Flow with Variable Physical Properties. In *Advances in Heat Transfer*; High Temperature Institute, Academy of Science of the USSR: Moscow, Russia, 1970; pp. 503–564.
26. Ammar, S.M.; Park, C.W. Validation of the Gnielinski correlation for evaluation of heat transfer coefficient of enhanced tubes by non-linear regression model: An experimental study of absorption refrigeration system. *Int. Commun. Heat Mass Transf.* **2020**, *118*, 104819. [CrossRef]
27. Nouri-Borujerdi, A. A new approach to thermo-fluid behavior through porous layer of heat pipes. *Sci. Iran. B* **2018**, *25*, 1236–1242. [CrossRef]
28. Nojoomizadeh, M.; Karimipour, A.; Firouzi, M.; Afrand, M. Investigation of permeability and porosity effects on the slip velocity and convection heat transfer rate of Fe_3O_4/water nanofluid flow in a microchannel while its lower half filled by a porous medium. *Int. J. Heat Mass Transf.* **2018**, *119*, 891–906. [CrossRef]
29. Rashidian, S.; Tavakoli, M.R. Using Porous Media to Enhancement of Heat Transfer in Heat Exchangers. *Int. J. Adv. Eng. Manag. Sci. IJAEMS* **2017**, *3*, 1051–1064. [CrossRef]
30. Harmen; Adriansyah, W.; Abdurrachim; Pasek, A.D. Theoretical investigation of heat transfer correlations for supercritical organic fluids. In *AIP Conference Proceedings, International Conference on Thermal Science and Technology (ICTST), Bali, Indonesia, 17–19 November 2017*; AIP Publishing: Melville, NY, USA, 2017; pp. 020011-1–020011-13.
31. Kemerli, U.; Özgür, D.; Öztürk, A.; Kahveci, K. Effect of High-Prandtl Number on Microscale Flow and Heat Transfer. In Proceedings of the World Congress on Mechanical, Chemical, and Material Engineering (MCM 2015), Barcelona, Spain, 20–21 July 2015; pp. 319-1–319-9.
32. Zulfattah, Z.M.; Rafeq, S.A.; Shukri, Z.M.; Ridhwan, J.; Hanafi, M.H.M.; Norain, I.; Nazri, Z.H.; Shakinah, M.I.N.; Najib, A.M. Gnielinski Method in Calculating the Heat Transfer Coefficient for Metallic Solar Tower Absorber. *Procedia Eng.* **2013**, *68*, 293–298. [CrossRef]
33. Ji, W.T.; Zhang, D.C.; He, Y.L.; Tao, W.Q. Prediction of fully developed turbulent heat transfer of internal helically ribbed tubes–An extension of Gnielinski equation. *Int. J. Heat Mass Transf.* **2012**, *55*, 1375–1384. [CrossRef]
34. Delgado, J.M.P.Q. (Ed.) *Heat and Mass Transfer in Porous Media, Bock*; Springer: Berlin/Heidelberg, Germany; Dordrecht, The Netherlands; London, UK; New York, NY, USA, 2012; 263p.
35. Hsu, C.T. Thermal dispersion in a porous medium. *Int. J. Heat Mass Transf.* **1990**, *33*, 1587–1597. [CrossRef]
36. Vafai, K.; Tien, C.L. Boundary and inertia effects on flow and heat transfer in porous media. *Int. J. Heat Mass Transf.* **1981**, *24*, 195–203. [CrossRef]
37. Cengel, A.Y. *Heat Transfer–A practical Approach, Boock*, 2nd ed.; Prentice-Hall: Hoboken, NJ, USA, 2003; 853p.
38. Praks, P.; Brkić, D. Advanced Iterative Procedures for Solving the Implicit Colebrook Equation for Fluid Flow Friction. *Adv. Civ. Eng.* **2018**, *2018*, 5451034. [CrossRef]
39. Clamond, D. Accurate and Robust Computation of the Darcy-Weisbach Friction Factor F according to the Colebrook Equation File colebrook.m, MATLAB Central File Exchange. Retrieved, 2023. Available online: https://www.mathworks.com/matlabcentral/fileexchange/21990-colebrook-m (accessed on 20 March 2023).
40. ** Colebrook-White Equation. Available online: https://engineerexcel.com/colebrook-white-equation/ (accessed on 4 June 2023).
41. Lunowa, J.K. Darcy Friction Factor Formulae in Turbulent Pipe Flow. *Fluid Mech. Pap.* **2011**, *110727*, 1–11.

micromachines

MDPI

Article

Numerical Investigation of Fluid Flow and Heat Transfer in High-Temperature Wavy Microchannels with Different Shaped Fins Cooled by Liquid Metal

Tingfang Yu [1], Xing Guo [1], Yicun Tang [1,*], Xuan Zhang [2,*], Lizhi Wang [3] and Tao Wu [3]

[1] Department of Energy and Power Engineering, School of Advanced Manufacturing, Nanchang University, Nanchang 330031, China; yutingfang@ncu.edu.cn (T.Y.); xingg@email.ncu.edu.cn (X.G.)
[2] Department of Energy and Power Engineering, School of Mechanical Engineering, Beijing Institute of Technology, Beijing 100081, China
[3] Yangzhou Collaborative Innovation Research Institute Co., Ltd., Yangzhou 225006, China; lychee@mail.ustc.edu.cn (L.W.); wutao_921006@163.com (T.W.)
* Correspondence: tyc@ncu.edu.cn (Y.T.); xuan.zhang@bit.edu.cn (X.Z.)

Abstract: The microchannel heat sink has been recognized as an excellent solution in high-density heat flux devices for its high efficiency in heat removal with limited spaces; however, the most effective structure of microchannels for heat dissipation is still unknown. In this study, the fluid flow and heat transfer in high-temperature wavy microchannels with various shaped fins, including the bare wavy channel, and the wavy channel with circular, square, and diamond-shaped fins, are numerically investigated. The liquid metal-cooled characteristics of the proposed microchannels are compared with that of the smooth straight channel, with respect to the pressure drop, average Nusselt number, and overall performance factor. The results indicate that the wavy structure and fin shape have a significant effect on the heat sink performance. Heat transfer augmentation is observed in the wavy channels, especially coupled with different shaped fins; however, a large penalty of pressure drops is also found in these channels. The diamond-shaped fins yield the best heat transfer augmentation but the worst pumping performance, followed by the square-, and circular-shaped fins. When the *Re* number increases from 117 to 410, the *Nu* number increases by 61.7% for the diamond fins, while the Δp increases as much as 7.5 times.

Keywords: microchannel; heat transfer augmentation; wavy channel; fin shape; liquid metal

1. Introduction

The microchannel heat sink is encountered in many industrial applications such as chip cooling, aerospace engineering, micro-electromechanical systems, and microfluidic control devices [1–4]. Due to the advantage of a high heat transfer coefficient and an enlarged heat transfer area, the microchannel heat sink is recognized as an excellent heat removal device with limited spaces. Tuckerman and Peace [5] introduced the first microchannel heat sink, which was fabricated on a silicon wafer and used water as the liquid coolant, where the heat flux that was dissipated was 790 W/cm^2. Since then, much attention has been focused on the improvement of the cooling performance for the microchannel heat sink [6–11]. Steinke and Kandlikar [12] made a systematic review of the single-phase flow friction factors in microchannels, and they compared the experimental data to discover the deviations in the literature. Smakulski and Pietrowicz [13] compared the cooling techniques in microchannel heat sinks, by considering the coolant type, pressure drop, maximal heat flux, and heat transfer coefficient. They found that spray cooling could enhance the heat transfer significantly more than normal, single-phase flow cooling.

The microchannel configurations have a direct effect on the performance of the heat sink, and many novel designed microchannels have been proposed in the literature, which includes secondary branches, different shaped cross-sectional shapes, various inlets and outlets, different flow passage types, and porous fillings. Xia et al. [14] investigated the effects of different inlet and outlet positions, and channel shapes on the performance of microchannel heat sinks, and they found that the I-type flow passage with a rectangular-shaped channel achieved the optimal performance. Gong et al. [15] numerically investigated the effects of metallic, porous fins on the microchannel heat sink performance, and they concluded that the porous fins could deteriorate the heat transfer because the bulk flow velocity decreased, but that the pressure drop was lowered because shear stress at the solid–liquid interface was reduced.

The coolant type also influences the microchannel heat sink performance greatly due to excellent thermal properties. Some commonly used coolants, such as water, ethyl alcohol, and ethylene glycol, cannot keep up with the growing demand of the cooling capacity for high-density heat flux devices; therefore, many researchers have paid attention to new kinds of coolants, such as nanofluids and liquid metals [16–21]. Nanofluids can achieve a better heat performance because of their enlarged heat transfer area brought about by the numerous nanoparticles. Salman et al. [17] investigated the influences of nanoparticle and base fluid types on the heat dissipation capacity of microchannel heat sinks, and they found that SiO_2 nanoparticles suspended in ethylene glycol achieved the largest Nusselt number, followed by ZnO, CuO, and Al_2O_3. Ho et al. [18] investigated the heat transfer of a microchannel heat sink with nanofluids experimentally, and concluded that the thermal resistance was cut down by 12.61%, and the convective heat transfer coefficient increased by 14.43%.

Liquid metal is considered the most appropriate coolant for high-temperature cooling situations for its high thermal conductivity, electrical conductivity, and mobility [22]. Wu et al. [23] numerically investigated a silicon carbide-fabricated heat sink with different shaped cross-sectional microchannels, and cooled by different liquid metal fluids, including sodium, potassium, a sodium–potassium alloy, and lithium. They found that the liquid lithium with circular-shaped microchannels had the best heat dissipation performance among all the microchannels studied. Ghoshal et al. [24] investigated a GaIn alloy loop-cooled heat sink, and found that the mean heat transfer coefficient could achieve 20 $W/cm^2 \cdot K$, with the thermal resistance lowered to 0.22 K/W. Tawk et al. [25] studied a rectangular-shaped channel cooler with a gallium alloy as the coolant, and concluded that the liquid metal could efficiently enhance the heat removal for high-density heat flux electronic devices.

From the literature review, it can be concluded that novel configuration designs of microchannel heat sinks can achieve better heat dissipation performances. In addition, the proper coolant type has a significant influence on the heat transfer and fluid flow characteristics in the microchannels; however, the most efficient structure of microchannels for heat dissipation is still unknown. In this paper, the laminar fluid flow and heat transfer in high-temperature wavy microchannels with various shaped fins are numerically investigated, including the bare wavy channel, and the wavy channel with circular-, square-, and diamond-shaped fins, while liquid lithium is chosen as the coolant. Special attention is paid to the comparison between the proposed microchannel heat sinks with the smooth straight channel, according to the pressure drop, average friction coefficient, average Nusselt number, and the overall performance factor. The effects of the fin relative size and fin relative height on the fluid flow and heat transfer characteristics of the microchannel heat sink are also investigated.

2. Theoretical Model and Numerical Procedure

2.1. Physical Model

The 3D-microchannel heat sink studied in this work is illustrated schematically in Figure 1a, whose length is L, width is W, and height is H. The smooth, straight channel heat sink has six parallel microchannels with rectangular cross-sectional shapes. The microchannel has a length of L, width of W_c, and height of H_c, while the thickness between the adjacent channels is W_w. The microchannels are fabricated on a silicon carbide material, and the base thickness is H_w. The coordinate system is also shown in Figure 1a. The coolant fluid flows toward the positive x-direction with a uniform inflow velocity. At the bottom surface, the microchannel heat sink is heated by a uniform and constant heat flux. Figure 1b shows the four new microchannels proposed in this study, including the wavy channel without fins, and the wavy channel with circular-, square-, and diamond-shaped fins. The wavy channel is sinusoidal shaped with a wavelength of L_t and amplitude of W_t. The different shaped fins have the same hydraulic diameter D_h and fin height H_b. The geometric parameters are listed in Table 1. The computational domain of the physical model and the relevant boundary conditions are depicted in Figure 1c, which only includes one microchannel due to the symmetric feature and the consideration of saving computation resources.

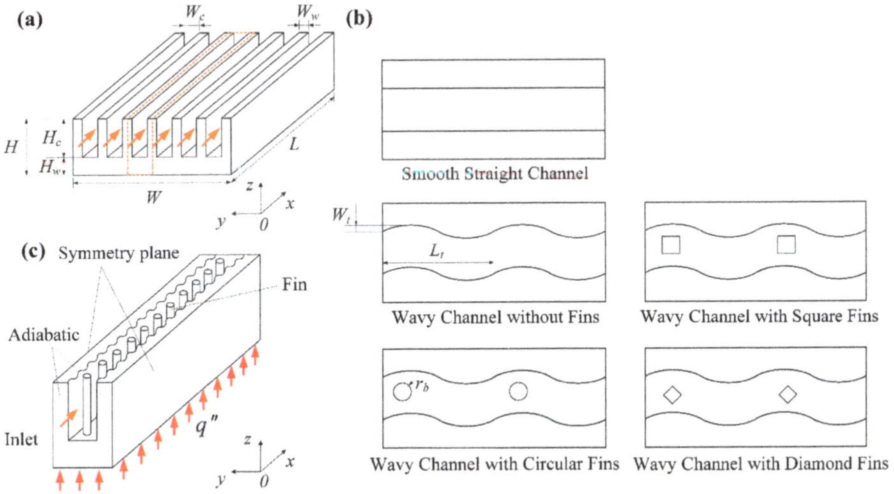

Figure 1. (**a**) Physical model; (**b**) microchannels with different geometric configurations; (**c**) computational domain and boundary conditions.

Table 1. Geometric sizes used in the present study.

Geometric Parameters	L	W	H_c	W_c	W_w	H_w	L_t	W_t	r_b	H_b
Value (mm)	20	20	3	0.6	0.4	2	0.3	0.1	0.1	3

2.2. Mathematical Model

The following assumptions are made in order to develop the mathematical model:

(1) The flow is laminar and steady state, and viscous dissipation is neglected;
(2) The radiative heat transfer, surface tension, and gravitational force are not considered;
(3) The density, dynamic viscosity, specific heat capacity, and thermal conductivity of the fluid are functions of the temperature, and the thermal properties of the solid are supposed to be constant.

The laminar assumptions for the fluid flow in wavy channels are indicated by Sui et al. [26], in which the wavy microchannel with a rectangular-shaped cross-section was investigated, and they found that the flow state was laminar in wavy channels when Re < 800. In the present study, the Re number ranges from 117 to 410; therefore, the laminar assumption can be accepted in this regard.

Based on the assumptions above, the mass conservation equation can be written as:

$$\frac{\partial}{\partial x_i}\left(\rho_f u_i\right) = 0,\ i,j = 1,2,3 \tag{1}$$

The momentum equation can be expressed by:

$$\frac{\partial}{\partial x_i}\left(\rho_f u_i u_j\right) = -\frac{\partial p}{\partial x_j} + \frac{\partial}{\partial x_i}\left[\mu_f\left(\frac{\partial u_j}{\partial x_i} + \frac{\partial u_i}{\partial x_j}\right)\right],\ i,j = 1,2,3 \tag{2}$$

The energy equation for the fluid can be given by:

$$\frac{\partial}{\partial x_i}\left(\rho_f u_i c_{p,f} T_f\right) = \frac{\partial}{\partial x_i}\left(\lambda_f \frac{\partial T_f}{\partial x_i}\right),\ i,j = 1,2,3 \tag{3}$$

The energy equation for the solid is as follows:

$$\frac{\partial}{\partial x_i}\left(\lambda_s \frac{\partial T_s}{\partial x_i}\right) = 0,\ i,j = 1,2,3 \tag{4}$$

where i, j = 1, 2, 3 represent the x, y, and z directions, respectively. u_i (i = 1, 2, 3) stands for the velocity component in the x, y, and z direction, respectively. T depicts the temperature, and p is the static pressure. ρ stands for the fluid density, μ is the dynamic viscosity, and c_p and λ are the specific heat capacity and thermal conductivity, respectively. The subscripts f and s denote the fluid and solid, respectively. The thermal properties of the silicon carbide is specified as follows: ρ_s = 3220 kg/m^3, $c_{p,s}$ = 800 J/(kg·K), and λ_s = 80 W/(m·K). The thermal properties of the liquid lithium are defined as follows [23]:

$$\rho_f = 278.5 - 0.04657T + 274.6(1 - T/3500)^{0.467} \tag{5}$$

$$c_{p,f} = 4754 - 0.925T + 2.91 \times 10^{-4}T^2 \tag{6}$$

$$\lambda_f = 22.28 + 0.05T - 1.243 \times 10^{-5}T^2 \tag{7}$$

$$\mu_f = \exp[-4.16435 - 0.6374\ln(T) + 292.1/T] \tag{8}$$

Boundary conditions:

(1) At the channel inlet (x = 0), the inlet velocity and fluid temperature are specified as:

$$u = u_{\text{in}},\ v = 0,\ w = 0;\ T_f = T_{\text{in}} \tag{9}$$

where u_{in} is set to be 0.1, 0.15, 0.2, 0.25, 0.3, and 0.35 m/s, and the Reynolds number is 117.1, 175.7, 234.3, 292.8, 351.4, and 410.0, respectively. The inlet temperature is T_{in} = 600 K.

(2) At the channel outlet (x = L), the pressure is defined as the atmospheric pressure:

$$p_{\text{out}} = p_{\text{atm}} \tag{10}$$

where p_{atm} uses the gauge pressure at the channel outlet, and p_{atm} = 0 Pa.

(3) At the solid–fluid interface, non-slip and coupled boundaries are applied:

$$u = 0, v = 0, w = 0;\ T_f = T_s;\ -\lambda_s\frac{\partial T_s}{\partial n} = -\lambda_f\frac{\partial T_f}{\partial n} \tag{11}$$

where n denotes the normal direction.

(4) At the bottom surface (z = 0), a uniform and constant heat flux q'' is applied:

$$-\lambda_s \frac{\partial T_s}{\partial z} = q'' \tag{12}$$

where q'' is set to be 200 W/cm^2 in this study.

(5) At the symmetric planes ($y = 0$, and $y = W_c + 2W_w$), symmetric boundary conditions are used:

$$-\lambda_s \frac{\partial T_s}{\partial y} = 0 \tag{13}$$

(6) For other surfaces, the adiabatic boundaries are used:

$$-\lambda_s \frac{\partial T_s}{\partial n} = 0;\ -\lambda_f \frac{\partial T_f}{\partial n} = 0 \tag{14}$$

2.3. Numerical Work

The meshing procedure for the computational domain was performed in the commercial software ICEM CFD 19.0, which utilizes unstructured hexahedral cells to discretize the whole model. Then, the mesh was imported to the software ANSYS Fluent 19.0 for further setting. The coupled pressure-velocity field was decoupled using the SIMPLEC algorithm, which is more stable and can obtain a faster convergence. The convection and diffusion terms in the PDEs were discretized with the second order upwind and central difference scheme, respectively. The convergence criterions for the related parameters were specified as 10^{-5}, except for the energy equation, which was specified as 10^{-7}.

Grid independence testing was performed to obtain mesh-free results. The pressure drop and average bottom temperature were used as the parameters in this study, with the inlet velocity set as 0.25 m/s. Four grids were generated for the smooth straight channel, 0.54, 0.87, and 1.22 million grids were compared with 2.58 million grids, and the pressure drop deviated by 6.99%, 6.15%, and 0.35%, respectively. In addition, the average bottom temperature deviated by 0.019%, 0.018%, and 0.001%, respectively; therefore, 1.22 million grids were chosen for calculation in the study. Furthermore, 0.75, 1.03, and 1.67 million grids were compared with 2.57 million grids for the wavy channel with diamond fins, and the pressure drop deviated by 8.36%, 4.41%, and 2.07%, while the average bottom temperature deviated by 0.034%, 0.022%, and 0.007%, respectively; thus, 1.67 million grids were used for further study. Using the same method, the final grids for the bare wavy channel, and the wavy channel with circular-, and square-shaped fins were 1.37, 1.63, and 1.64 million, respectively.

2.4. Data Reduction

The Reynolds number is a basic parameter for evaluating the flow regime, which is calculated from:

$$Re = \frac{\rho_f u_{\text{in}} D_h}{\mu_f} \tag{15}$$

where u_{in} is the inlet flow velocity. ρ_f, and μ_f stand for the density and dynamic viscosity of the fluid, respectively. D_h depicts the hydraulic diameter, which is defined as:

$$D_h = \frac{2H_c W_c}{H_c + W_c} \tag{16}$$

The mean convective heat transfer coefficient is given by:

$$h = \frac{q'' A_b}{A_{\text{con}}\left(T_w - T_f\right)} \tag{17}$$

where T_w is the average temperature of the heating surface, and T_f stands for the average temperature of the fluid. A_b and A_{con} are the areas of the heating surface, and the contacting surface between the solid and fluid, respectively.

The average Nusselt number and average friction coefficient are calculated from:

$$Nu = \frac{hD_h}{\lambda_f} \tag{18}$$

$$f = \frac{D_h \Delta p}{2L\rho_f u_{\text{in}}^2} \tag{19}$$

where λ_f represents the fluid thermal conductivity, and Δp is the static pressure difference between the microchannel inlet and outlet.

The overall performance factor is calculated from [27]:

$$\eta = \frac{Nu/Nu_0}{(f/f_0)^{1/3}} \tag{20}$$

where Nu_0 and f_0 stand for the heat transfer and fluid flow parameters of the smooth straight channel, respectively.

The field synergy number is an effective parameter to distinguish the synergistic effect between the velocity and temperature field. In this paper, the field synergy number is adopted to evaluate the heat transfer enhancement in the newly-proposed microchannels, which is given by [28]:

$$F_c = \iiint_{V_c} \left(U_{\text{ave}} \cdot \nabla T_{f,\text{ave}}\right) dV_{\text{ave}} = \frac{Nu}{RePr} \tag{21}$$

where U_{ave} stands for the average velocity vector, $T_{f,\text{ave}}$ is the average fluid temperature, and V_{ave} depicts the fluid volume. V_c is the control volume. Pr stands for the Prandtl number, which is defined as $Pr = \mu_f \cdot c_{p,f} / \lambda_f$.

Bejan [29] used the concept of the volume entropy generation rate to evaluate the irreversible heat loss for the energy quality occurring in the microchannel heat sinks, which consists of two kinds of entropy generation in the fluid flow and heat transfer process, i.e., the friction entropy generation rate and the heat transfer entropy generation rate. The definitions of these two parameters are calculated from:

$$S_{\text{gen},\Delta p} = \iiint S'''_{\text{gen},\Delta p} dV = \frac{m}{\rho_f T_f} \Delta p \tag{22}$$

$$S_{\text{gen},\Delta T} = \iiint S'''_{\text{gen},\Delta T} dV = \frac{q'' A_{\text{con}}}{T_{\text{con}} T_f} \left(T_{\text{con}} - T_f\right) \tag{23}$$

2.5. Model Validation

In order to validate the reliability of the numerical model, the present numerical results were compared with the experimental data in the literature. Figure 2 illustrates the comparisons between the present numerical results and the experimental data from Chai et al. [30], in terms of the average Nusselt number and the product of average friction coefficient and Reynolds number. It was found that the present numerical results were in reasonable agreement with the experimental data, where the maximum deviation was 12.77% for the average Nu number, and 3.86% for the product of the average friction coefficient and *Re* number. The relative discrepancy for the average *Nu* number might have resulted from the radiative and convective heat transfer between the heat sink and the surroundings. The comparison implies that the numerical procedure can further be performed to investigate the fluid flow and heat transfer in the newly-proposed microchannel heat sinks.

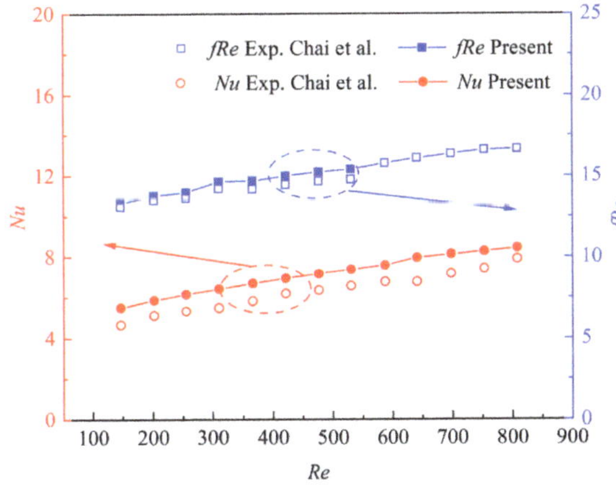

Figure 2. Comparisons of the present numerical results with experimental data from Ref. [30].

3. Results and Discussion

3.1. Fluid Flow Characteristics

To observe the velocity, streamline, pressure, and temperature distribution in the microchannels, a horizontal plane parallel to the bottom surface with a distance $z = 2.5$ mm was extracted. Figure 3 shows the velocity and streamline distributions at $z = 2.5$ mm in the microchannels for $Re = 292.8$. The plots are zoomed at the x locations in the range of $7.5 < x < 9.5$ mm. Figure 3 indicates that the velocity distribution presented a parabolic type flow in the smooth straight channel, with the streamlines in parallel with each other; however, the velocity distributions in the four new microchannels showed obvious differences with that of the smooth straight channel. For the wavy microchannel without fins, the fluid in the middle of the microchannel had a higher velocity, while the fluid near the sidewalls had a lower velocity. The velocity contour presented a periodic and wavy distribution, and the streamlines were parallel with each other in the middle of the channel, but periodic vortexes were found in the cavities on the sidewalls. The vortex sizes in the cavities were almost the same because of the equal-sized cross-sections in the wavy microchannel.

For the wavy microchannel with different shaped fins, the velocity and streamline distributions were more complex than the bare wavy channel. The fins in the middle of the channel brought about a strong blocking effect to the fluid flow, which separated the fluid into two parts, where one part flowed into the left side of the fins and the other flowed into the right side. Due to the asymmetry of the cross-sectional areas at the two sides of the fins, the separated fluids flowed around the fin with different velocities; thus, additional vortexes were formed behind the fins. The size of the additional vortexes depended on the fin shapes, where the diamond-shaped fin brought about the largest additional vortexes, followed by the square-shaped fin. The circular-shaped fin brought almost no additional vortexes because of its streamlined structures. The vortexes were also found in the cavities on the sidewalls for the wavy channel with different shaped fins. The vortex size was larger at the location where the fluid velocity was smaller. The different shaped fins could bring about a strong flow disturbance and fluid mixing; thus, the flow boundary layers in the channels were interrupted and redeveloped periodically along the flow direction.

Figure 4 illustrates the pressure distribution at $z = 2.5$ mm in the microchannels for $Re = 292.8$. It was found that the pressure in the smooth straight channel was obviously smaller than those in the wavy channels. The wavy channel with fins showed a slightly higher pressure drop than the smooth straight channel due to the periodic cavities on the sidewalls. For the wavy channel with different shaped fins, the pressure drop increased rapidly because of the blocking effect induced by the fins. The diamond-shaped fin brought about the highest pressure drop, followed by the square- and circular-shaped fins. The circular-shaped fins had a minimal blocking effect on the fluid because of their streamlined structures; however, the diamond-shaped fins greatly changed the direction of the main stream fluid, which had the maximum blocking effect on the fluid.

To have a quantitative comprehension of the blocking effects caused by the different fins, the changes in the Δp with the Re number is depicted in Figure 5a. The results show that the pressure drops increased with the Re number for all the microchannels in this study. In the range of $117.1 < Re < 410.0$, the pressure drops increased from 35.3 Pa to 157.3 Pa for the smooth straight channel, from 71.8 Pa to 409.2 Pa for the wavy channel without fins, from 318.7 Pa to 2089.9 Pa for the wavy channel with circular-shaped fins, from 545.2 Pa to 3909.1 Pa for the square-shaped fins, and from 592.6 Pa to 5040.9 Pa for the diamond-shaped fins. The pressure drop for the diamond-shaped fin increased drastically with the Re number, which was far more than the other microchannels.

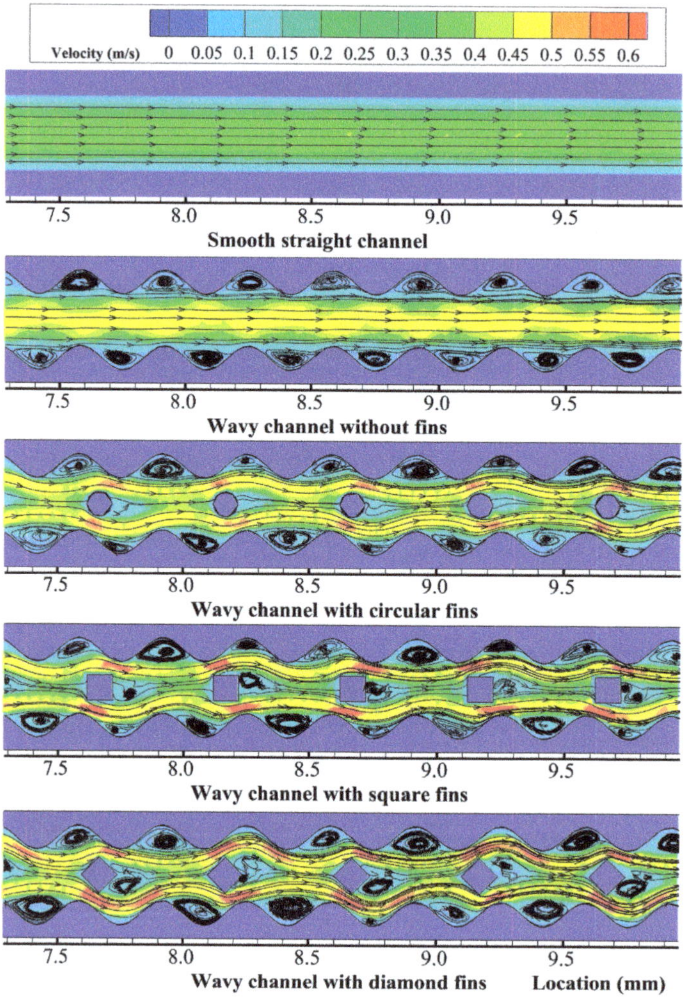

Figure 3. Velocity and streamline distributions at $z = 2.5$ mm in the microchannels for $Re = 292.8$.

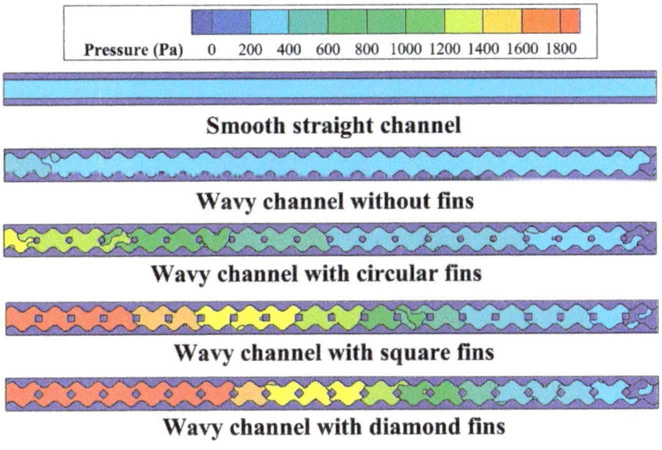

Figure 4. Pressure distribution at $z = 2.5$ mm in the microchannels for $Re = 292.8$.

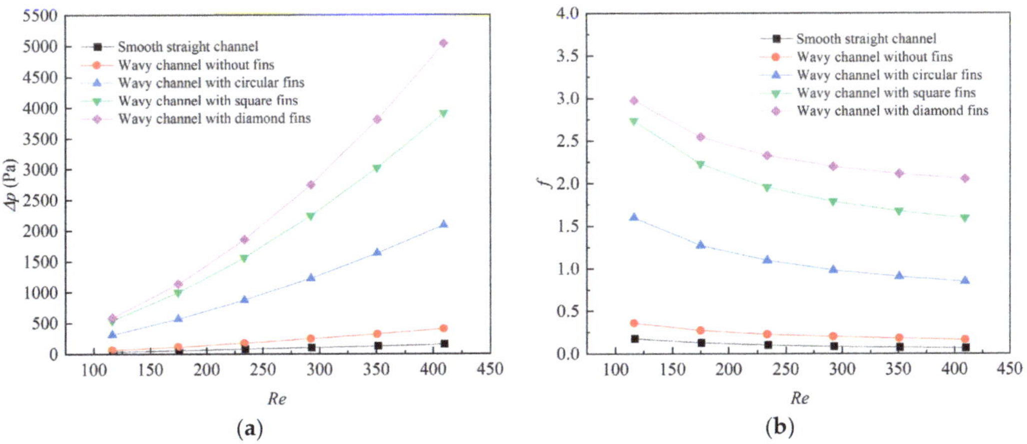

Figure 5. Changes of (**a**) pressure drop (Δp); (**b**) average friction coefficient (f) with Re.

The changes in the average friction coefficient with the Reynolds number are demonstrated in Figure 5b. The average friction coefficient was calculated from Equation (19). It can be found that the average friction coefficient declined with the Re number for all the cases. This can be explained by Equation (19), in which the average friction coefficient is strongly affected by the inlet velocity with a quadratic relation; thus, the average friction coefficient is smaller at high Re numbers. In the range of $117.1 < Re < 410.0$, the average friction coefficient declined from 0.178 to 0.064 for the smooth straight channel, from 0.361 to 0.167 for the wavy channel without fins, from 1.601 to 0.852 for the wavy channel with circular-shaped fins, from 2.739 to 1.593 for the square-shaped fins, and from 2.978 to 2.054 for the diamond-shaped fins.

3.2. Heat Transfer Characteristics

The temperature distribution at $z = 2.5$ mm in the microchannels for $Re = 292.8$ is shown in Figure 6. It can be found that the fluid temperature increased along the flow direction for all the microchannels. The figure clearly depicts the temperature difference at the solid–fluid interface, which reveals the temperature gradient and the thermal conductivity difference between the solid and the fluid phase. The temperature for the wavy channel with different shaped fins showed as slightly lower than the smooth straight channel, and the reason for this can be explained by the enlarged heat transfer area for the wavy microchannels and different fins. Moreover, the fins could have brought about periodic interruptions for the thermal boundary layers in the microchannels, and the local heat transfer was enhanced.

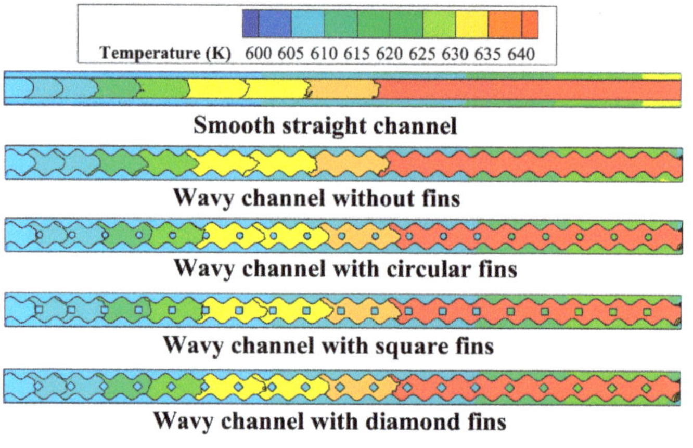

Figure 6. Temperature distributions at $z = 2.5$ mm in the microchannels for $Re = 292.8$.

The variations of the average Nusselt number is illustrated in Figure 7a. It can be found that the average *Nu* number increased with the *Re* number for all the cases. When the *Re* number increased from 117.1 to 410.0, the average *Nu* number increased from 20.09 to 27.30 for the smooth straight channel, from 22.79 to 28.29 for the wavy channel without fins, from 23.41 to 35.76 for the wavy channel with circular-shaped fins, from 23.51 to 38.29 for the square-shaped fins, and from 23.97 to 38.76 for the diamond-shaped fins. The results indicate that the wavy channel with different fins had a significant impact on the heat transfer. The diamond-shaped fin achieved the highest heat transfer augmentation among all the microchannels studied in this work.

Analyses were further carried out to investigate the overall performance factor for the different newly-proposed microchannels, which considered not only the heat transfer augmentation but also the pressure drop penalty. The overall performance factors under different *Re* numbers are given in Figure 7b, in which the overall performance factors were compared with the smooth straight channel. It can be found that the wavy channel with different fins achieved a more deteriorative overall performance factor than that of the wavy channel without fins, and the reason for this is that the blocking effects brought about by the different shaped fins were more severe than the heat transfer augmentation. Interestingly, the overall performance factor for the wavy channel with different shaped fins reached a peak at *Re* 240, which was induced by the relative dominance between the friction drag and the heat transfer augmentation. The heat transfer augmentation dominated at lower *Re* numbers but the friction drag greatly dominated at higher *Re* numbers. The overall performance factor ranged from 44.7% to 47.7% for the wavy channel with diamond-shaped fins, which was the lowest among all the microchannels.

The field synergy number *Fc* under different *Re* numbers is demonstrated in Figure 7c. It is plain to see that the field synergy number declined with the *Re* numbers for all the microchannels. The field synergy number declined from 5.0 to 1.8 for the smooth straight channel, from 5.7 to 1.8 for the wavy channel without fins, from 5.8 to 2.3 for the wavy channel with circular-shaped fins, from 5.8 to 2.5 for the square-shaped fins, and from 5.9 to 2.5 for the diamond-shaped fins. Among all the microchannels studied in this work, the diamond-shaped fin brought about the highest *Fc*. The reason for this can be explained by the synergy angle between the velocity vector and the temperature gradient. The diamond-shaped fin possessed the smallest angle due to the shape structure among all the microchannels; thus it achieved the highest field synergy number.

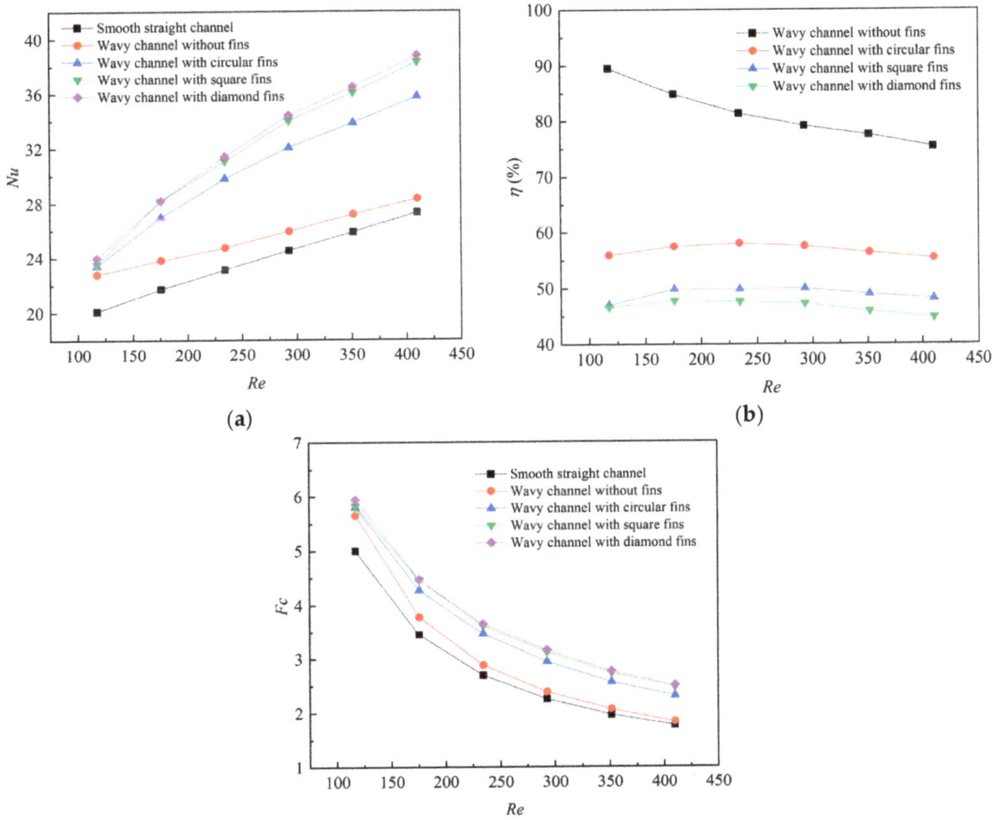

Figure 7. Variations of (**a**) average Nusselt number (*Nu*); (**b**) overall performance factor (η); (**c**) field synergy number (*Fc*) with Reynolds numbers.

The changes in the friction and heat transfer entropy generation rate with the *Re* number are shown in Figure 8. In Figure 8a, the friction entropy generation rate increases with the *Re* number for all the microchannels. The wavy channel with diamond-shaped fins achieved the largest friction entropy generation rate, then followed by the square- and circular-shaped fins. This can be explained by the strong blocking effects brought about by the fins. In Figure 8b, the heat transfer entropy generation rate for the wavy channel with different shaped fins decreases with the *Re* number, and the reason for this can be explained by the reduced temperature difference between the fluid and solid when the *Re* number was larger. The heat transfer entropy generation rate, however, for the smooth straight channel and wavy channel without fins increased with the *Re* number at first, then decreased at higher *Re* numbers, and the reason for this can also be attributed to the temperature difference variation between the solid and fluid, which was increased at first due to the shortened retention time for the heat transfer, but was decreased for the strong fluid mixing at higher *Re* numbers in the microchannels.

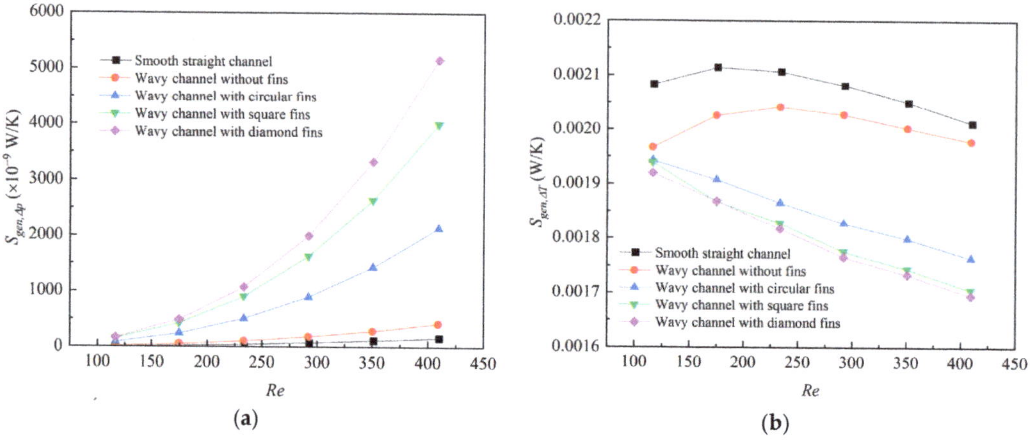

Figure 8. Changes of (**a**) friction entropy generation rate ($S_{\text{gen},\Delta p}$); (**b**) heat transfer entropy generation rate ($S_{\text{gen},\Delta T}$) with Reynolds number.

3.3. *Effect of the Fin Relative Size*

The fin relative size plays an important role in the heat transfer and fluid flow characteristics due to the constricted cross-sectional areas brought about by the fins. It is interesting to investigate the influence of the fin relative size on the comprehensive performance of the microchannel heat sinks. In this work, the fin height was maintained at 3 mm, while the fin relative size ratio was set to be 0.2, 0.4, 0.6, and 0.8, respectively. The wavy channel with square-shaped fins was chosen as the calculated microchannel, and the fin sizes were 0.04, 0.08, 0.12, and 0.16 mm, respectively. They were compared with the original microchannel, with the fin relative size set as 1.0, and the fin size as 0.2 mm.

Figure 9 demonstrates the changes in the average friction coefficient, average *Nu* number, and the overall performance factor with the fin relative size. Figure 9a shows that the average friction coefficient increased with the fin relative size, which can be explained by the constricted flow area and the increasing flow velocity and blocking effect brought about by the fins. It can also be noted that the average friction coefficient decreased with the *Re* number for a certain fin size, which can be explained by Equation (19), where the friction coefficient declined with the fluid velocity.

In Figure 9b, the average *Nu* number tends to increase with the fin relative size, and the average *Nu* number is larger under a larger *Re* number. This is reasonable because the average heat transfer coefficient was higher when the fluid flowed at a higher speed, and the heat dissipation was more reinforced. In Figure 9c, the overall performance factor decreased with fin relative size, because the average friction coefficient increased more rapidly than the heat transfer enhancement, leading to a declined overall performance factor.

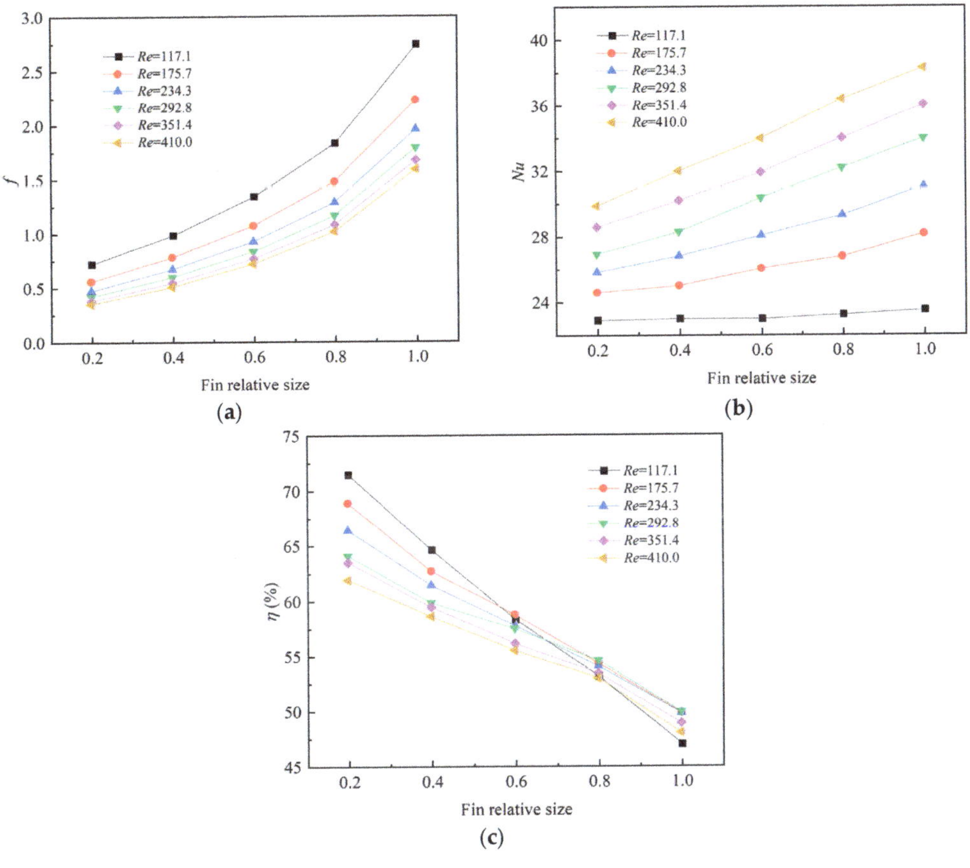

Figure 9. Changes of (**a**) average friction coefficient (f); (**b**) average Nusselt number (Nu); (**c**) overall performance factor (η) with fin relative size.

3.4. Effect of the Fin Relative Height

The fin relative height also has a significant impact on the heat transfer and fluid flow of the microchannel heat sinks due to an enlarged heat transfer area, along with the accompanying blocking effects by the fins. In this work, the fin size was kept at 0.2 mm, and the fin relative height ratio was set to be 0.2, 0.4, 0.6, and 0.8, respectively. The wavy channel with square-shaped fins was chosen as the calculated microchannel, and the fin heights were 0.6, 1.2, 1.8, and 2.4 mm, respectively. They were compared with the original microchannel, with the fin relative size set as 1.0, and the fin size as 3.0 mm.

Figure 10 depicts the variations in the average friction coefficient, average Nu number, and the overall performance factor with the fin relative height. In Figure 10a, it obviously shows that the average friction coefficient increased with the fin relative height, and the reason for is that the blocking effect was more intensive with a higher fin height. In Figure 10b, it is interesting to note that the average Nu number decreases with the fin relative height at first, then increases rapidly with the fin height. This can be explained by the relative dominance between the friction drag and the enlarged heat transfer area. When the fin relative height ratio was lower than 0.4, the friction drag dominated while the heat transfer was deteriorated with a larger fin height; however, the heat transfer augmentation dominated when the fin relative height ratio exceeded 0.4, and the average Nu number increased with the fin relative height ratio due to the enlarged heat transfer area. In Figure 10c, the overall performance factor decreased with the fin relative height at first, then rose quickly with the fin relative height, which was inconsistent with the average *Nu* number. This implies that the fin height should be paid special attention to when designing new microchannel heat sinks.

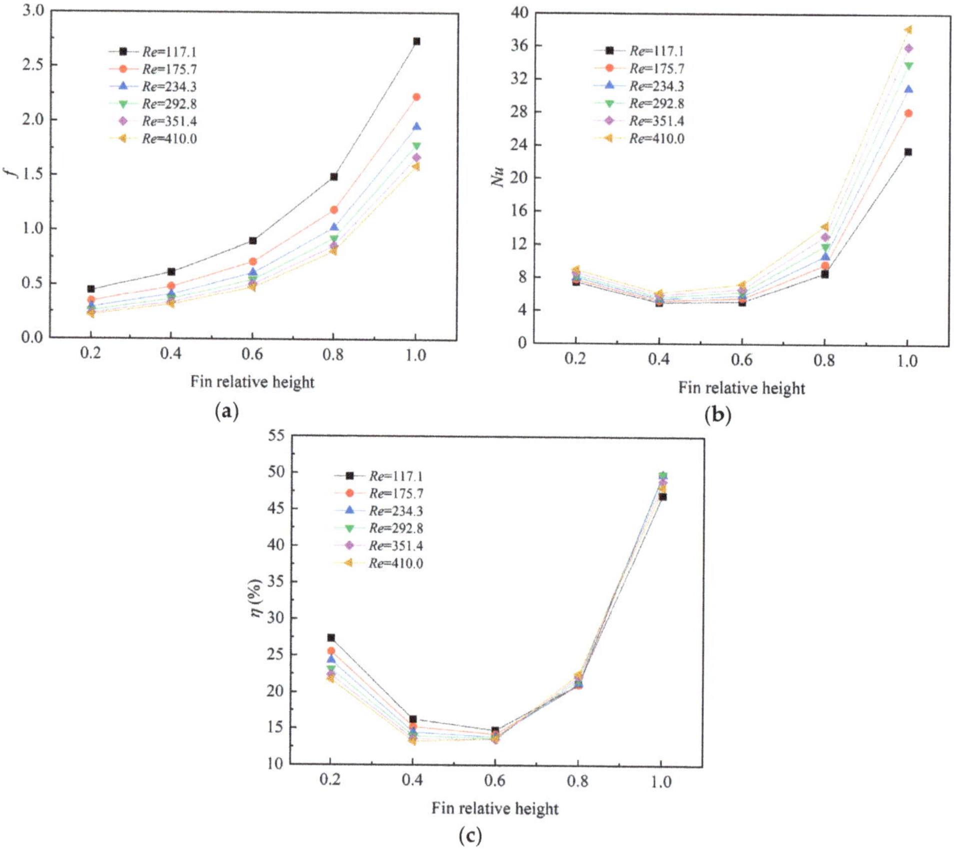

Figure 10. Variations of (**a**) average friction factor (f); (**b**) average Nusselt number (Nu); (**c**) overall performance factor (η) with fin relative height.

4. Concluding Remarks

In this paper, numerical studies were conducted on the heat transfer and laminar flow in high-temperature wavy microchannels with different shaped fins, including the bare wavy channel, and the wavy channel with circular-, square-, and diamond-shaped fins, while liquid lithium was chosen as the coolant. The pressure drop, average friction coefficient, average Nusselt number, and the overall performance factor of the proposed microchannels for different Reynolds numbers were compared with that of the smooth, straight channel. The main conclusions are summarized as follows:

(1) The presence of wavy channels and different shaped fins has an important impact on the heat transfer and fluid flow characteristics. As the *Re* number increases from 117.1 to 410.0, the average Nusselt number increases by 61.7% for the wavy channel with different shaped fins, while the pressure drop increases by 7.5 times.

(2) Among the four newly-proposed microchannels, the wavy channel with diamond-shaped fins brings about the best heat transfer enhancement, followed by the square- and circular-shaped fins. The pressure drop brought about by the diamond-shaped fins is far more than the other microchannels in this study.

(3) The overall performance factor decreases with the fin relative size for the four investigated microchannels, and the reason for this is that the average friction coefficient increases more rapidly than the heat transfer enhancement. The constricted flow cross-sectional area leads to stronger friction drag effects with larger fin relative sizes.
(4) For the four investigated microchannels, the overall performance factor decreases with the fin relative height at first, then increases drastically. The friction drag effect dominates when the fin relative height ratio is less than 0.4, but a heat transfer enhancement prevails when the fin relative height ratio exceeds 0.4, which is induced by the enlarged heat transfer area.
(5) The wavy channel with different shaped fins can lead to a strong heat transfer enhancement, but also induces a higher pumping power penalty. These configurations should be paid special attention to where the pumping power should be strictly restricted.

Author Contributions: Conceptualization, Y.T. and X.Z.; methodology, T.Y. and Y.T.; software, X.G.; validation, X.G. and T.W.; formal analysis, L.W.; investigation, X.Z.; resources, Y.T. and X.Z.; data curation, L.W.; writing—original draft preparation, T.Y. and X.G.; writing—review and editing, Y.T. and X.Z.; visualization, T.W.; supervision, Y.T. and X.Z.; project administration, Y.T.; funding acquisition, Y.T. All authors have read and agreed to the published version of the manuscript.

Funding: This research was funded by the Scientific Research Foundation of Nanchang University, grant number No. 28770602.

Data Availability Statement: The data presented in this study are available on request from the corresponding author. The data are not publicly available due to privacy.

Conflicts of Interest: The authors declare no conflict of interest.

Nomenclature

A_b	Bottom heating surface area, m^2
A_{con}	Convective heat transfer area, m^2
c_p	Specific heat capacity, J/(kg·K)
D_h	Hydraulic diameter, m
e	Relative error, %
f	Average friction factor
H	Height of the microchannel heat sink, m
h	Convective heat transfer coefficient, W/(m^2·K)
H_b	Fin height, m
H_c	Height of microchannel, m
H_w	Base thickness, m
L	Length of the microchannel heat sink, m
L_t	Wavelength of wavy channel, m
Nu	Nusselt number
p	Static pressure, Pa
q	Heat flux, W/cm^2
r_b	Radius of circular fin, m
Re	Reynolds number
T	Temperature, K
u	Velocity, m/s
W	Width of the microchannel heat sink, m
W_c	Width of microchannel, m
W_t	Amplitude of wavy channel, m
W_w	Wall thickness, m
Δp	Pressure drop, Pa

Greek symbols

η	Overall performance factor
λ	Thermal conductivity, W/(m·K)
μ	Dynamic viscosity, kg/(m·s)
ρ	Density, kg/m^3

Subscripts

b	Bottom
c	Channel
con	Contact
f	Fluid
in	Inlet
max	Maximum
out	Outlet
s	Solid
w	Wall

References

1. Miner, A.; Ghoshal, U. Cooling of high-power-density microdevices using liquid metal coolants. *Appl. Phys. Lett.* **2004**, *85*, 506–508. [CrossRef]
2. Gu, L.D.; Min, J.C.; Wu, X.M.; Yang, L.J. Airside heat transfer and pressure loss characteristics of bare and finned tube heat exchangers used for aero engine cooling considering variable air properties. *Int. J. Heat Mass Transf.* **2017**, *108*, 1839–1849. [CrossRef]
3. Yang, Y.; Du, J.; Li, M.; Li, W.; Wang, Q.; Wen, B.; Zhang, C.; Jin, Y.; Wang, W. Embedded microfluidic cooling with compact double H type manifold microchannels for large-area high-power chips. *Int. J. Heat Mass Transf.* **2022**, *197*, 123340. [CrossRef]
4. Fathi, M.; Heyhat, M.M.; Targhi, M.Z.; Bigham, S. Porous-fin microchannel heat sinks for future micro-electronics cooling. *Int. J. Heat Mass Transf.* **2023**, *202*, 123662. [CrossRef]
5. Tuckerman, D.B.; Pease, R.F.W. High-performance heat sinking for VLSI. *IEEE Electron. Device Lett.* **1981**, *2*, 126–129. [CrossRef]
6. Jiang, P.X.; Wang, Z.; Ren, Z.P.; Wang, B.X. Experimental research of fluid flow and convection heat transfer in plate channels filled with glass or metallic particles. *Exp. Therm. Fluid Sci.* **1999**, *20*, 45–54. [CrossRef]
7. Wang, X.Q.; Mujumdar, A.S.; Christopher, Y. Thermal characteristics of tree-shaped microchannel nets for cooling of a rectangular heat sink. *Int. J. Therm. Sci.* **2006**, *45*, 1103–1112. [CrossRef]
8. Hung, T.C.; Yan, W.M.; Li, W.P. Analysis of heat transfer characteristics of double-layered microchannel heat sink. *Int. J. Heat Mass Transf.* **2012**, *55*, 3090–3099. [CrossRef]
9. Alihosseini, Y.; Targhi, M.Z.; Heyhat, M.M.; Ghorbani, N. Effect of a micro heat sink geometric design on thermo-hydraulic performance: A review. *Appl. Therm. Eng.* **2020**, *170*, 114974. [CrossRef]
10. Kong, D.; Kim, Y.; Kang, M.; Song, E.; Hong, Y.; Kim, H.S.; Rah, K.J.; Choi, H.G.; Agonafer, D.; Lee, H. A holistic approach to thermal-hydraulic design of 3D manifold microchannel heat sinks for energy-efficient cooling. *Case Stud. Therm. Eng.* **2021**, *28*, 101583. [CrossRef]
11. Ji, X.; Yang, X.; Zhang, Y.; Zhang, Y.; Wei, J. Experimental study of ultralow flow resistance fractal microchannel heat sinks for electronics cooling. *Int. J. Therm. Sci.* **2022**, *179*, 107723. [CrossRef]
12. Steinke, M.E.; Kandlikar, S.G. Single-phase liquid friction factors in microchannels. *Int. J. Therm. Sci.* **2006**, *45*, 1073–1083. [CrossRef]
13. Smakulski, P.; Pietrowicz, S. A review of the capabilities of high heat flux removal by porous materials, microchannels and spray cooling techniques. *Appl. Therm. Eng.* **2016**, *104*, 636–646. [CrossRef]
14. Xia, G.D.; Jiang, J.; Wang, J.; Zhai, Y.L.; Ma, D.D. Effects of different geometric structures on fluid flow and heat transfer performance in microchannel heat sinks. *Int. J. Heat Mass Transf.* **2015**, *80*, 439–447. [CrossRef]
15. Gong, L.; Li, Y.; Bai, Z.; Xu, M. Thermal performance of micro-channel heat sink with metallic porous/solid compound fin design. *Appl. Therm. Eng.* **2018**, *137*, 288–295. [CrossRef]
16. Lee, J.; Mudawar, I. Assessment of the effectiveness of nanofluids for single-phase and two-phase heat transfer in micro-channels. *Int. J. Heat Mass Transf.* **2007**, *50*, 452–463. [CrossRef]
17. Salman, B.H.; Mohammed, H.A.; Munisamy, K.M.; Kherbeet, A.S. Three-dimensional numerical investigation of nanofluids flow in microtube with different values of heat flux. *Heat Transf.—Asian Res.* **2015**, *44*, 599–619. [CrossRef]
18. Ho, C.J.; Peng, J.K.; Yang, T.F.; Rashidi, S.; Yan, W.M. On the assessment of the thermal performance of microchannel heat sink with nanofluid. *Int. J. Heat Mass Transf.* **2023**, *201*, 123572. [CrossRef]
19. Luo, M.; Liu, J. Experimental investigation of liquid metal alloy based mini-channel heat exchanger for high power electronic devices. *Front. Energy* **2013**, *7*, 479–486. [CrossRef]
20. Bo, G.; Ren, L.; Xu, X.; Du, Y.; Dou, S. Recent progress on liquid metals and their applications. *Adv. Phys.—X* **2018**, *3*, 1446359. [CrossRef]

21. Kalkan, O. Multi-objective optimization of a liquid metal cooled heat sink for electronic cooling applications. *Int. J. Therm. Sci.* **2023**, *190*, 108325. [CrossRef]
22. Yang, X.H.; Liu, J. Advances in liquid metal science and technology in chip cooling and thermal management. *Adv. Heat Transf.* **2018**, *50*, 187–300. [CrossRef]
23. Wu, T.; Wang, L.Z.; Tang, Y.C.; Yin, C.; Li, X.K. Flow and heat transfer performances of liquid metal based microchannel heat sinks under high temperature conditions. *Micromachines* **2022**, *13*, 95. [CrossRef] [PubMed]
24. Ghoshal, U.; Grimm, D.; Ibrani, S.; Johnston, C.; Miner, A. High-performance liquid metal cooling loops. In Proceedings of the IEEE 21st Annual Semiconductor Thermal Measurement and Management Symposium, San Jose, CA, USA, 15–17 March 2005; pp. 16–19. [CrossRef]
25. Tawk, M.; Avenas, Y.; Lebouc, A.K.; Petit, M. Numerical and experimental investigations of the thermal management of power electronics with liquid metal mini-channel coolers. *IEEE Trans. Ind. Appl.* **2013**, *49*, 1421–1429. [CrossRef]
26. Sui, Y.; Lee, P.; Teo, C. An experimental study of flow friction and heat transfer in wavy microchannels with rectangular cross-section. *Int. J. Therm. Sci.* **2011**, *50*, 2473–2482. [CrossRef]
27. Webb, R.L. Performance evaluation criteria for use of enhanced heat transfer surfaces in heat exchanger design. *Int. J. Heat Mass Transf.* **1981**, *24*, 715–726. [CrossRef]
28. Guo, Z.Y.; Li, D.Y.; Wang, B.X. A novel concept for convective heat transfer enhancement. *Int. J. Heat Mass Transf.* **1998**, *41*, 2221–2225. [CrossRef]
29. Bejan, A. *Entropy Generation Minimization: The Method of Thermodynamic Optimization of Finite-Size Systems and Finite-Time Processes*; CRC Press: Boca Raton, MA, USA, 1996. [CrossRef]
30. Chai, L.; Xia, G.D.; Wang, L.; Zhou, M.Z.; Cui, Z.Z. Heat transfer enhancement in microchannel heat sinks with periodic expansion-constriction cross-sections. *Int. J. Heat Mass Transf.* **2013**, *62*, 741–751. [CrossRef]

micromachines

MDPI

Article

Heat Transfer Enhancement Using Al_2O_3-MWCNT Hybrid-Nanofluid inside a Tube/Shell Heat Exchanger with Different Tube Shapes

Maissa Bouselsal [1], Fateh Mebarek-Oudina [1,2,*], Nirmalendu Biswas [3] and Abdel Aziz I. Ismail [4]

1 Department of Physics, Faculty of Sciences, University of 20 Août 1955-Skikda, Skikda 21000, Algeria
2 Laboratoire des Matériaux et Génie Energétique (LMGE), University of 20 Août 1955-Skikda, Skikda 21000, Algeria
3 Department of Power Engineering, Jadavpur University, Salt Lake, Kolkata 700106, India
4 Mechanical Engineering Department, College of Engineering and Islamic Architecture, Umm Al-Qura University, Makkah P.O. Box 5555, Saudi Arabia
* Correspondence: f.mebarek_oudina@univ-skikda.dz; Tel.: +213-668305488

Abstract: The high demand for compact heat exchangers has led researchers to develop high-quality and energy-efficient heat exchangers at a lower cost than conventional ones. To address this requirement, the present study focuses on improvements to the tube/shell heat exchanger to maximize the efficiency either by altering the tube's geometrical shape and/or by adding nanoparticles in its heat transfer fluid. Water-based Al_2O_3-MWCNT hybrid nanofluid is utilized here as a heat transfer fluid. The fluid flows at a high temperature and constant velocity, and the tubes are maintained at a low temperature with various shapes of the tube. The involved transport equations are solved numerically by the finite-element-based computing tool. The results are presented using the streamlines, isotherms, entropy generation contours, and Nusselt number profiles for various nanoparticles volume fraction $0.01 \leq \varphi \leq 0.04$ and Reynolds numbers $2400 \leq \mathrm{Re} \leq 2700$ for the different shaped tubes of the heat exchanger. The results indicate that the heat exchange rate is a growing function of the increasing nanoparticle concentration and velocity of the heat transfer fluid. The diamond-shaped tubes show a better geometric shape for obtaining the superior heat transfer of the heat exchanger. Heat transfer is further enhanced by using the hybrid nanofluid, and the enhancement goes up to 103.07% with a particle concentration of 2%. The corresponding entropy generation is also minimal with the diamond-shaped tubes. The outcome of the study is very significant in the industrial field and can solve many heat transfer problems.

Keywords: heat exchanger; hybrid nanofluid; tube shapes; entropy generation; heat transfer

1. Introduction

Solar energy has become the energy of the times and the only solution for the poor to replace petroleum products. Heat exchangers are used in diverse fields of application such as in water heating systems in solar collectors, solar thermal energy storage systems, and other compact heat exchanger applications. The efficiency of heat exchangers is related to metallic and non-metallic materials such as aluminum, iron, copper, steel, bronze, ceramic and glass PTFE, and graphite. The heat transfer coefficient varies from one material to another. Therefore, by increasing the heat transfer coefficient, the heat exchange capacity increases. This is accomplished by modifying the geometric shape of the heat exchangers. The importance of heat exchangers differs from field to field with increasing demand; the purpose of using them is to save energy and reduce the cost. Thus, the heat exchanger is an essential element of any energy management policy and therefore of environmental protection. Indeed, a large part of the energy used in the various sectors passes at least once through a heat exchanger, which makes it a key device for thermal engineers and essential in energy management [1,2]. In general, heat exchange between the two fluids is accomplished by the heat exchangers [3] and the same versatile application in various industries for heating or cooling [4].

With the development of science and technology, energy guidance has become a major goal for scientists due to their urgent need to provide natural energy sources. Where many companies and factories compete to rationalize energy consumption through regenerative heat exchangers, it is a weapon of the times that saves large amounts of wasted energy. Thanks to it, it can capture as much waste heat as possible and use it to heat the cold gases that are usually heated in thermal furnaces and engines that use fossil fuels to warm up. Thus, these motors use less energy and therefore consume valuable fossil fuel. Underground heat exchangers have also been used underground to take advantage of geothermal heat from the earth in cooling and heating fluids for industrial, residential, and agricultural uses.

These devices are specially designed to activate the vital act of various thermal and chemical processes through the transfer of heat from one liquid to another at different temperatures. The design of these heat exchangers with reliable operation of the systems is mainly adopted for the transfer of heat in spite of the problems of this reactor such as cooling of the components of the system to reuse and avoid burning it and to intensify these exchanges to improve the production rate. To improve the heat transfer using convection as a cooling method, the researchers conducted numerical and experimental tests about the nature of the systems in which they occur and the properties of the fluids involved (physico-chemical properties).

There are different types of heat exchangers, and thus, the selection of the right type for a particular process is very important because incorrect selection leads to the malfunctioning of the system operation or sometimes failure of the equipment. Amongst the different types of heat exchangers [3,5], tube and shell types are most used in several process industries [6]. This is due to the ease of changing and cleaning these parts; it also enables high-pressure applications. It is used in many fields such as oil refineries, chemical processing, the electrical industry, hydrocarbon processing units, feed water heaters, condensing units, and others. Therefore, there is a continual demand for the development of compact heat exchangers depending on the application. To meet these demands, researchers are continually paying attention to further development in heat exchange efficacy. Apart from the geometric modification of the heat transfer surfaces/unit, there is a promising technique for further improvement in the efficacy of heat transfer by adding nano-sized particles (single type or multi types) in the heat transfer fluids. The mixture of such nanoparticles and heat transfer fluid is termed nanofluid/hybrid nanofluids. Studies have shown that adding nanoparticles to a carrier liquid matrix can increase the heat transfer compared to a pure assembled fluid, or that hybrid nanofluids have particular physical and chemical properties such as high thermal conductivity [7–11].

Actual experiments and their analysis are usually expensive as well as time-consuming. Therefore, wide variations for the parametric analysis of the prototypes are also not suitable [12,13]. Therefore, computational analysis is preferable to give a clear study of these thermal devices. There is a vast pool of literature on heat exchangers and heat transport improvement analysis [14]. Zaversky et al. [15] analyzed accurate modeling of a heat exchanger. Shahdad and Fazel [16] presented the role of perforated fins instead of simple fins for improving the convective coefficient of heat transfer and the Nusselt number. Xie et al. [17] examined the enhanced heat transfer using honeycomb tubes over smooth tubes. Matos et al. [18,19] compared the heat transfer of elliptical and circular tubes (12 numbers) under various Re values. Nouri-Borujerdi and Lavasani [20,21] examined thermo-fluid flow characteristics of simple cam-shaped conduits. Mohanty et al. [22] performed a 2D analysis in an elliptical circular tube-based heat exchanger. Dharmaiah et al. [23] numerically studied a magnetic dipole effect on a radiative ferromagnetic liquid. Li et al. [24] investigated the heat transfer as well as frictional losses in an elliptical tube-based heat exchanger.

Bouris et al. [25] studied the frictional losses in an elliptical tube-based heat exchanger and found that an increase in the heat transfer area causes a higher pressure drop. Moawed [26] conducted an experimental investigation on the forced convective heat transfer over the surface of a spiral tube-based heat exchanger. Rosen and Dincer [27] found that lower temperature differences enhance the heat exchanger performance. Khan et al. [28] analyzed the forced convective phenomena in elliptical-shaped tubes and found that the transfer of heat is more with a growing air-water flow. Harris and Goldschmidt [29] investigated the consequence of the angle of attack on the overall thermal performance in a heat exchanger with an elliptical-shaped tube. Teo et al. [30] numerically analyzed a finned heat exchanger with elliptical-shaped tubes. They observed 30% enhanced heat transfer in comparison to circular tubes. Li et al. [31] carried out calculations on thermal behavior with elliptical-shaped tubes with an axial ratio of 0.3, 0.5, and 0.8.

Kumar and Jhinge [32] conducted experimental work on tube and shell heat exchangers containing split baffles at different angular orientations of the baffles: 0°, 15°, 30° and 45° from the horizontal. Raj and Ganne [33] investigated the impact of various baffle tilt angles on the heat transfer characteristics of a tube/shell heat exchanger for three different tilt angles of the baffle: 0°, 10° and 20°, with a deflector cut of 36%. Zhang et al. [34] performed a shell/tube heat exchanger simulation with overlapping non-continuous intermediate helical baffles (30°, 40°, and 50° helix angles). Sivarajan et al. [35] performed a numerical simulation of a continuous spiral baffle tube/shell heat exchanger; their results indicated that the spiral baffle tube/shell heat exchanger has better flow performance and better heat transfer than the conventional shell and tube baffle heat exchanger. Kwon et al. [36] showed that with the use of ZnO nanoparticles and Al_2O_3, the end results are that the coefficient of heat transfer increases to 30% at a concentration of 6% Al_2O_3. Albadr et al. [37] studied heat transfer through a heat exchanger using the nanofluid Al_2O_3 at different concentrations. The forced convection heat transfer coefficient is slightly higher than that of the base liquid at the same inlet temperatures and mass flow rate. Very recently, Asadi et al. [38] carried out an experimental study of the heat transfer performance of Al_2O_3-MWCNT hybrid nanofluid (oil based) and found an enhanced heat transfer. Ghazanfari et al. [39] numerically analyzed the overall performance of a twisted-tube-based shell-type heat exchanger utilizing Al_2O_3 nanoparticles and found a 20% enhanced heat transfer and an increase in the pressure drop up to 40%.

A detailed review of the performance improvement in heat exchangers could be found in refs [40–42].

The vast pool of literature clearly indicates the usefulness of heat exchanger analysis due to the diverse fields of application in critical thermal systems. Off-course design of a compact heat exchanger (shell and tube type) is a critical task for the designer for maximizing the heat transfer and lowering the pressure drop. Both of these criteria significantly affect the equipment sizing as well as its cost. Therefore, designing high-quality and energy-efficient heat exchangers at a lower cost than the conventional (circular tube bundle) ones is the key target for the manufacturer. To address this requirement, the present study focuses on the improvements of the tube/shell heat exchanger to maximize the efficiency either by altering the geometrical shape of the tubes and/or by adding nanoparticles in its heat transfer fluid. According to the literature, this work has never been the subject of a previous study. The aim of this work is to find the ideal shape of the tubes for better heat transfer using a hybrid nanofluid and to compare it with a conventional fluid, H_2O. In this study, four different shapes (circular, square, rectangular, and diamond) of the tube bundles are examined under a wide parametric range of nanoparticles with a volume fraction $0.01 \leq \varphi \leq 0.04$ and Reynolds numbers of $2400 \leq Re \leq 2700$. For maximizing the heat transfer, the entropy generation rate is also analyzed systematically. Water-based Al_2O_3-MWCNT hybrid nanofluid is utilized here as a heat transfer fluid.

2. Geometry Description and Mathematical Model

A schematic representation of a conventional tube/shell heat exchanger (in two-dimensional forms) is shown in Figure 1. Figure 2 shows the actual view of a tube/shell heat exchanger section from RA_2K SONATRACH Skikda, Algeria. An enlarged view of a tube and its modified shape (circular, square, rectangular, and diamond) are also shown in Figure 2. The circular shape is the basic shape and is taken as a reference for the comparison of betterment in heat transfer 9, if any). The computational domain consists of half of a tube bundle and a quarter portion of the two tubes. The material of the tube bundles is considered to be aluminum. The water-based Al_2O_3-MWCNT hybrid-nanofluid flows through the gaps in between the tube bundles and carries the heat from the heated fluid flowing through the tube of different shapes. The purpose of this study is to model the shell and tube heat exchanger to improve the heat transfer of the cooling fluid using the Al_2O_3-MWCNT hybrid nanofluid.

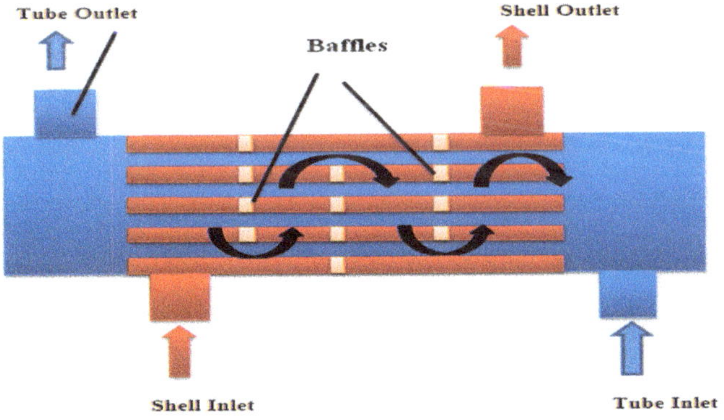

Figure 1. Schematic representation of a conventional tube/shell heat exchanger.

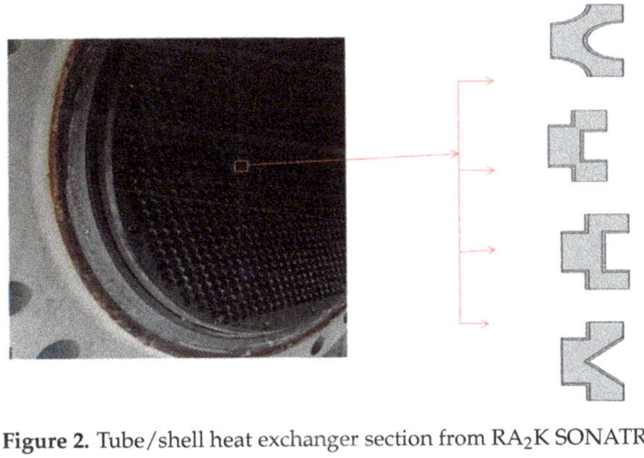

Figure 2. Tube/shell heat exchanger section from RA_2K SONATRACH Skikda, Algeria.

The leading equations for the continuity, momentum, and energy are solved numerically. The k-ε turbulent flow model is used to solve this mathematical model [43–45], and the expression is given as:

$$\frac{\partial u}{\partial t} + \rho(u \cdot \nabla)u = \nabla \cdot [-pI + K] + F \tag{1}$$

$$\frac{\partial \rho}{\partial t} + \nabla \cdot (\rho u) = 0 \tag{2}$$

where

$$K = (\mu + \mu_\tau)(\nabla u + (\nabla u)^\tau) - \frac{2}{3}(\mu + \mu_\tau)(\nabla \cdot u)I - \frac{2}{3}\rho kI \quad (3)$$

$$\rho c_p\left(\frac{\partial T}{\partial t} + u \cdot \nabla T\right) + \nabla \cdot (q + q_r) = \beta T\left(\frac{\partial p}{\partial t} + u \cdot \nabla p\right) + \tau : \nabla u + Q \quad (4)$$

$$\beta = -\frac{1}{\rho}\frac{\partial \rho}{\partial T}$$

Q contains heat sources other than viscous dissipation ($\mathrm{W/m^3}$)

$$\mathrm{Q_p} = \beta \mathrm{T}\left(\frac{\partial p}{\partial \mathrm{t}} + u \cdot \nabla p\right)$$

The second term represents viscous dissipation in the fluid:

$$Q_{vd} = \tau : \nabla u$$

For the turbulent kinetic energy, we can write the following:

$$\rho\frac{\partial k}{\partial t} + \rho(u \cdot \nabla)k = \nabla \cdot \left[\left(\mu + \frac{\mu_\tau}{\sigma_k}\right)\nabla k\right] + p_k - \rho \mathcal{E} \quad (5)$$

For dissipation:

$$\rho\frac{\partial \mathcal{E}}{\partial t} + \rho(u \cdot \nabla)\mathcal{E} = \nabla \cdot \left[\left(\mu + \frac{\mu_\tau}{\sigma_\mathcal{E}}\right)\nabla \mathcal{E}\right] + C_{\mathcal{E}1}\frac{\mathcal{E}}{k}p_K - C_{\mathcal{E}2}\rho\frac{\mathcal{E}^2}{k} \quad (6)$$

where $\mathcal{E} = ep$

For turbulent viscosity:

$$\mu_\tau = \rho C_\mu \frac{k^2}{\mathcal{E}} \quad (7)$$

The coefficients for the turbulent viscosity, turbulent kinetic energy, and dissipation are fixed: $C_{\mathcal{E}1} = 1.44, C_{\mathcal{E}2} = 1.92, \delta_k = 1, C_\mu = 0.09, \delta_\mathcal{E} = 1.3$

The turbulent kinetic energy production can be expressed as:

$$P_K = \mu_\tau\left[\nabla u : (\nabla u + (\nabla u)^\tau) - \frac{2}{3}(\nabla \cdot u)^2\right] - \frac{2}{3}\rho k \nabla \cdot u \quad (8)$$

where ρ: fluid density, p: pressure, u: fluid flow velocity, I: identity matrix, μ: dynamic viscosity of the fluid, τ: stress tensor, ∇: del operator, k: turbulent kinetic energy, ep: turbulent dissipation rate.

The velocity of the flowing fluid is nondimensionalized using the Reynolds number:

$$\mathrm{Re} = \frac{\rho u D}{\mu} \quad (9)$$

The heat transfer characteristics are evaluated using the Nusselt number (Nu) through the local as well as global heat transfer:

$$\mathrm{Nu_{log}} = \frac{k_{hnf}}{k_{bf}}\frac{\partial \theta}{\partial y} \quad (10)$$

$$\mathrm{Nu_{avg}} = \frac{1}{L}\int_0^L Nu_{log} dL \quad (11)$$

Modeling of hybrid nanofluid:

In this study, the working fluid is taken as Al_2O_3-MWCNT hybrid-nanofluid, which is basically a suspension of Al_2O_3 and MWCNT nanoparticles in the base fluid (water) following the mixture rules [45]. The mixture of Al_2O_3 and MWCNT nanoparticles (known as hybrid nanoparticles) is chosen for the present thermal–hydrodynamic study. In general, the thermal conductivity as well as the electrical conductivity of Al_2O_3 are less compared to MWCNT. Al_2O_3, however, suffers from inferior conductivities compared to that of MWCNT, and due to the combined (hybrid) preparation of MWCNT and Al_2O_3 nanoparticles, superior conductivities (much higher than the base liquid) are obtained. In this connection, it is worth mentioning that the flow of hybrid nanofluids in the heat exchanger devices is a relatively newer area; therefore, more research works can give a greater understanding of it, which is required for real-life applications [45]. On the other hand, many recent works have considered MWCNT-Al_2O_3-water hybrid nanofluid, indicating a wide field of its applications and reliability. It motivates us to choose MWCNT-Al_2O_3-water hybrid nanofluid as a working fluid for this study. The hybrid nanoparticles' volume fraction is expressed using the symbol 'φ'. The properties of the hybrid nanofluid are estimated using the standard correlations, which are expressed as

$$
\begin{aligned}
\varphi &= \varphi_{Al_2O_3} + \varphi_{MWCNT} \\
\rho_{np} &= \frac{\varphi_{Al_2O_3}\rho_{Al_2O_3} + \varphi_{MWCNT}\rho_{MWCNT}}{\varphi} \\
(C_p)_{np} &= \frac{\varphi_{Al_2O_3}(C_p)_{Al_2O_3} + \varphi_{MWCNT}(C_p)_{MWCNT}}{\varphi} \\
\beta_{np} &= \frac{\varphi_{Al_2O_3}\beta_{Al_2O_3} + \varphi_{MWCNT}\beta_{MWCNT}}{\varphi} \\
k_{np} &= \frac{\varphi_{Al_2O_3}k_{Al_2O_3} + \varphi_{MWCNT}k_{MWCNT}}{\varphi} \\
\sigma_{np} &= \frac{\varphi_{Al_2O_3}\sigma_{Al_2O_3} + \varphi_{MWCNT}\sigma_{MWCNT}}{\varphi}
\end{aligned}
\tag{12}
$$

Hybrid Nanofluid:

$$
\begin{aligned}
\sigma_{hnf} &= (1-\varphi)\sigma_{bf} + \varphi\sigma_{np} \\
\rho_{hnf} &= (1-\varphi)\rho_{bf} + \varphi\rho_{np} \\
(\rho\beta)_{hnf} &= (1-\varphi)(\rho\beta)_{bf} + \varphi(\rho\beta)_{np} \\
(\rho C_p)_{hnf} &= (1-\varphi)(\rho C_p)_{bf} + \varphi(\rho C_p)_{np} \\
\alpha_{hnf} &= \frac{k_{hnf}}{(\rho C_p)_{hnf}} \\
\frac{k_{hnf}}{k_{bf}} &= \frac{k_{np} + (n-1)k_{bf} - (n-1)\left(k_{bf} - k_{np}\right)\varphi}{k_{np} + (n-1)k_{bf} + \left(k_{bf} - k_{np}\right)\varphi} \\
\mu_{hnf} &= \frac{\mu_{bf}}{(1-\varphi)^{2.5}} \\
\frac{\sigma_{hnf}}{\sigma_{bf}} &= 1 + \frac{3\left(\sigma_{np} - \sigma_{bf}\right)\varphi}{\left(\sigma_{np} + 2\sigma_{bf}\right) - \left(\sigma_{np} - \sigma_{bf}\right)\varphi}
\end{aligned}
\tag{13}
$$

where the total entropy and efficiency are defined as:

$$S_{tot} = \frac{k}{(T_{avg})^2}\left[\left(\frac{\partial T}{\partial x}\right)^2 + \left(\frac{\partial T}{\partial y}\right)^2\right] + \frac{\mu_{nf}}{T_{avg}}\left[\frac{\mathcal{E}_p}{k}(u^2+v^2) + 2\left(\frac{\partial u}{\partial x}\right)^2 + 2\left(\frac{\partial v}{\partial y}\right)^2 + \left(\frac{\partial u}{\partial y} + \frac{\partial v}{\partial x}\right)^2\right] + \frac{\sigma_{nf}}{T_{avg}}\beta^2 v \tag{14}$$

With

$$T_{avg} = \frac{T_{inlet} + T_{tube}}{2}$$

$$\mathcal{E}p = \frac{q}{q_{max}} = \frac{q}{C_{min}(\theta_{inlet} - \theta_{tube})}$$

$$q = UA\Delta T_{ml}$$

$$\Delta T_{ml} = \frac{\Delta T_1 - \Delta T_2}{ln(\Delta T_1/\Delta T_2)}$$

C_{min} = minimum heat capacity rate
q = heat transfer rate, kW
A = total heat transfer surface area of the heat exchanger and total heat transfer surface area of all matrices of a regenerator, m^2
U = overall heat transfer coefficient, W/m^2K
Boundary conditions
The boundary conditions adopted for this study are:
1—symmetry (thermal insulation) at the region borders.
2—fixed inlet temperature T = 285 K with speed $v = -0.01$ m/s.
3—fixed temperature at the inner surfaces of the tube T = 278 K.
4—transport dominated by convection at the exit.
5—tube material: aluminum
The thermo-physical properties of the used hybrid nanofluids are presented in Table 1.

Table 1. Thermo-physical proprieties of the Al_2O_3-MWCNT/water [44,46].

	ρ (kg/m³)	k (W/m k)	C_p (J/kg k)	σ (s/m)
Water	997.1	0.613	4179	5.5×10^{-6}
MWCNT	2100	3000	711	10^{-7}
Al_2O_3	3950	36.96	785.02	

3. Numerical Methodology

The evolved Equations (1)–(7) of the chosen geometry are numerically solved utilizing the boundary conditions. The conservation equations are discretized over the computing domain following the Galerkin finite element method (FEM). The computing domain is split into an appropriate smaller mesh structure. The discretization of the computational domain is achieved through the function space adopting the piecewise cubic polynomials. The appropriate mesh elements are distributed in a manner such that all of the related hydro-thermal boundary layers nearer to the walls are captured correctly. The element size is selected through the mesh independence study. The elemental equations are computed through an iterative process (with convergence criteria of 10^{-8} for obtaining the final solution) and solved data are stored for post-processing, which are utilized for generating the local and global data.

3.1. Grid Test

Before conducting the extensive analysis of the present problem, an extensive grid independence study is carried out considering the circular tube bundle. Here, six different mesh sizes, extra coarse, very coarse, coarse, normal, fine, and very fine, are utilized with number elements of 1398, 3501, 5487, 13,108, 32,128 and 42,948. The comparison is made through the average Nusselt number, which is shown in Figure 3 and Table 2. From Figure 3 and Table 2, it is observed that the last three values of the average Nusselt are close to each other, so the middle value (with a fine mesh structure) is chosen to save the computational time to generate the correct results.

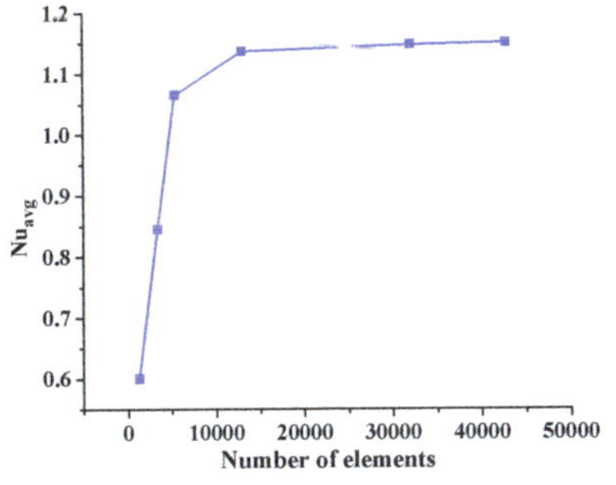

Figure 3. Grid test, variation of the average via element number.

Table 2. Mesh comparison.

Mesh	Element Number	Average Nusselt Number
Extra coarse	1398	0.6011
Very coarse	3501	0.84613
Coarse	5487	1.0666
normal	13,108	1.13758
Fine	32,128	1.14863
Very fine	42,948	1.15084

3.2. Validation

After finalizing the mesh structure, a validation study is conducted considering the published work of Roy and Mondal [47]. The comparison of the present findings and the published work of Roy and Mondal [47] is shown in Figure 4 using the isotherms. The comparison shows a close match between the two results [47].

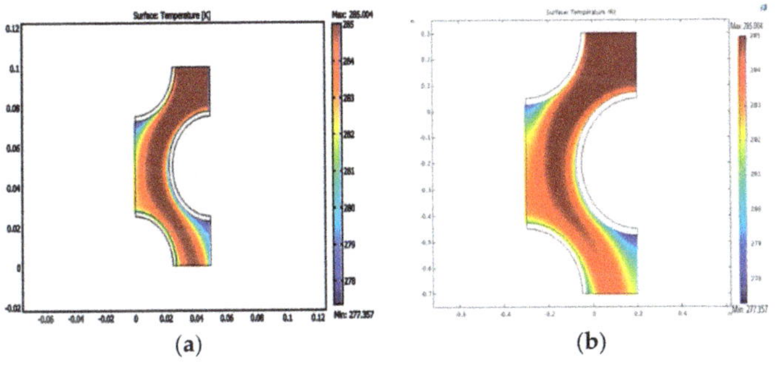

Figure 4. Validation of our results with ref. [47]. (**a**) Results of ref. [47]; (**b**) our results.

4. Results and Discussion

The present study focuses on the improvements of the tube/shell heat exchanger to maximize the efficiency either by altering the geometrical shape of the tubes and/or by adding nanoparticles in its heat transfer fluid. Water-based Al_2O_3-MWCNT hybrid nanofluid is utilized here as a heat transfer fluid. The results are presented using the streamline, isotherms, and entropy contours, and the average Nusselt number for two main parameters: the Reynolds number ($2400 \leq Re \leq 2700$) and the volume fraction

($0.01 \leq \varphi \leq 0.04$), to study the heat transfer performance of the heat exchanger. The shape (circular, square, rectangular, and diamond) of the heat exchanger tube is modified from a conventional circular for obtaining the best shape.

4.1. Effect of Nanofluid Volume Fraction

The effect of the increasing hybrid nanofluid volume fraction ($0.01 \leq \varphi \leq 0.04$) on the heat transfer is shown in Figure 5, at Re = 2600 for the four different shapes of the tube bundle. From the figure, it is observed that the Nu_{avg} rises monotonically with the increasing φ for the circular, rectangular, and diamond-shaped tubes, whereas there is no change in the Nu_{avg} for the square-shaped tubes. With the growth in the φ, the effective thermal conductivity of the heat transfer fluid (compared to the base fluid) increases, which leads to the improvement in the heat transfer rate. It should be mentioned that these properties cause an enhancement in the thermal conducting properties of the working fluid [45–54]. Therefore, the average Nu is proportional to the nanoparticle concentrations, and such a correlation contributes to the enhanced convective transfer. Further improvement in the Nu_{avg} is accomplished due to the modification in the tube shape, which allows more removal of the heat from the tube bundles. However, a diamond-shaped tube only offers superior performance in heat transfer compared to all other shaped tubes.

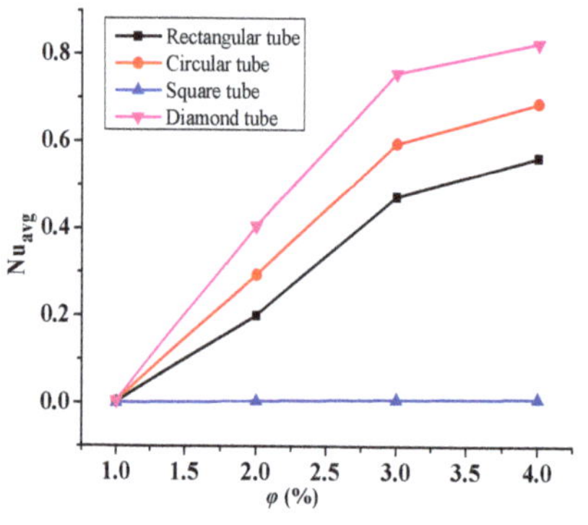

Figure 5. Variation of the Nu_{avg} with the increasing φ(%) for Re = 2600.

4.2. Reynolds Number Effect

The reason for the improvement in the Nu_{avg} could be realized from the isotherm and streamline contours, as shown in Figures 6 and 7, respectively, for the different time steps (t = 0 to 100 s) with Al_2O_3-MWCNT/water hybrid nanofluid at Re = 2600 and φ = 0.02. Here, the effects of the Reynolds number ($2400 \leq Re \leq 2700$) on the streamlines, isotherms, and entropy production are also analyzed. As the fluid flows through the complex path, the fluid flows through the narrow gaps due to the thermal buoyancy caused by the temperature gradient between the inlet wall and the outlet walls of the shell/tube heat exchanger. It is pertinent to mention that, at the initial time, the flows begin to move from the inlet (of the flow passage) and then move to the next complex and narrow passage in between the tube bundle gaps. The isotherms replicate the thermal conditions (temperature) across the chamber. As the fluid is heated from below, the concentration of the isotherm lines is clustered at the bottom and fades as it spreads vertically toward the outlet of the flow passage. Heat is absorbed by the working fluid from the bottom and dissipates as it rises to an outlet as it begins to spread upwards with the passage of time. At the lower flow regime (Re = 2600), the convection fluxes are very weak, and uniformly distributed temperature lines occupy the entire domain. Therefore, the thermal conduction modes dictate the thermal behavior. The values of the isothermal lines gradually increase from the hot wall to the cold wall. This topology is the thermal stratification, where thermal diffusion causes heat transfer. With the progress in time, the isotherm contours distribute uniformly over the flow passage, and such uniformity is more with the diamond-shaped tube bundle compared to all other shapes of the tubes. From the streamline contours (Figure 7), it is observed that there are multi-cellular vortices within the flow path. The position, size, and shape of these vortices change with progress in time. In fact, the results reveal that over time there is a linear growth in the flow functions, which is due to the flow through the complex narrow gaps. Figure 7 shows that over time, the eddies in the vicinity of the entrance wall become larger and stronger, and it soon heads upwards, allowing strong free convection, which strengthens the flow velocity and accelerates the fluid flow. With the increasing Re, as the flow passes through the inlet for increased Re values, the isotherms were blown to the edge. Furthermore, in the case of rectangular and square tubes, the fluid faces more resistance in its flow path due to the presence of the shape corners at the inlet and outlet of the flow domain. However, the circular and diamond fluid faces lower resistance, which leads to a higher rate of heat removal.

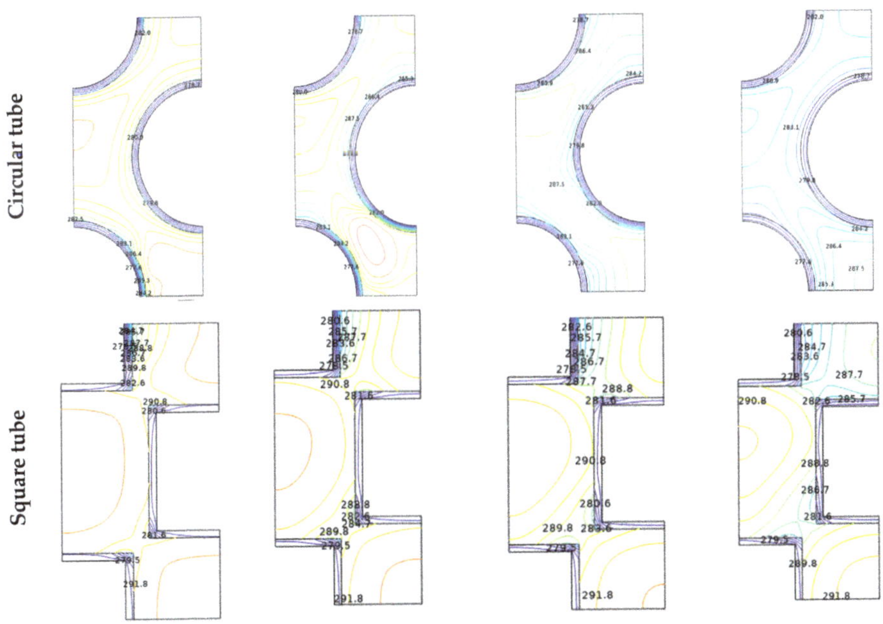

Figure 6. *Cont.*

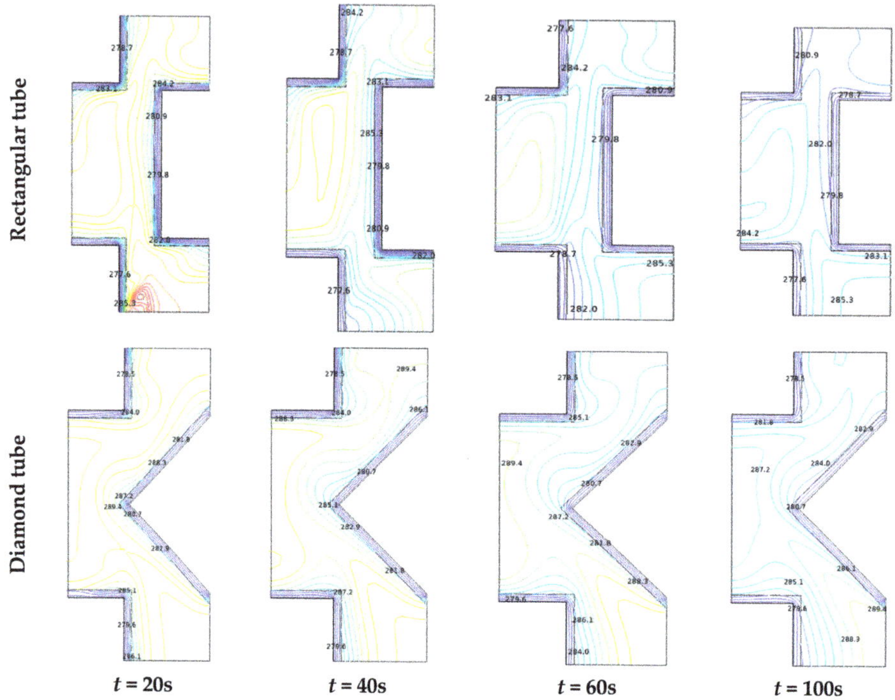

Figure 6. Isotherms in a tube/shell heat exchanger with different shape of manufacturing tubes for Re = 2600, $\varphi = 0.02$.

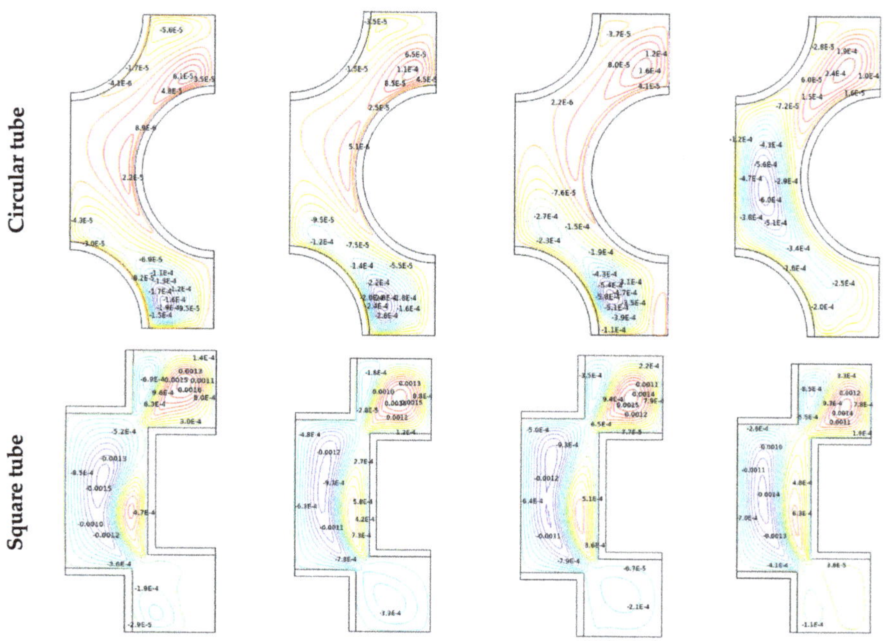

Figure 7. *Cont.*

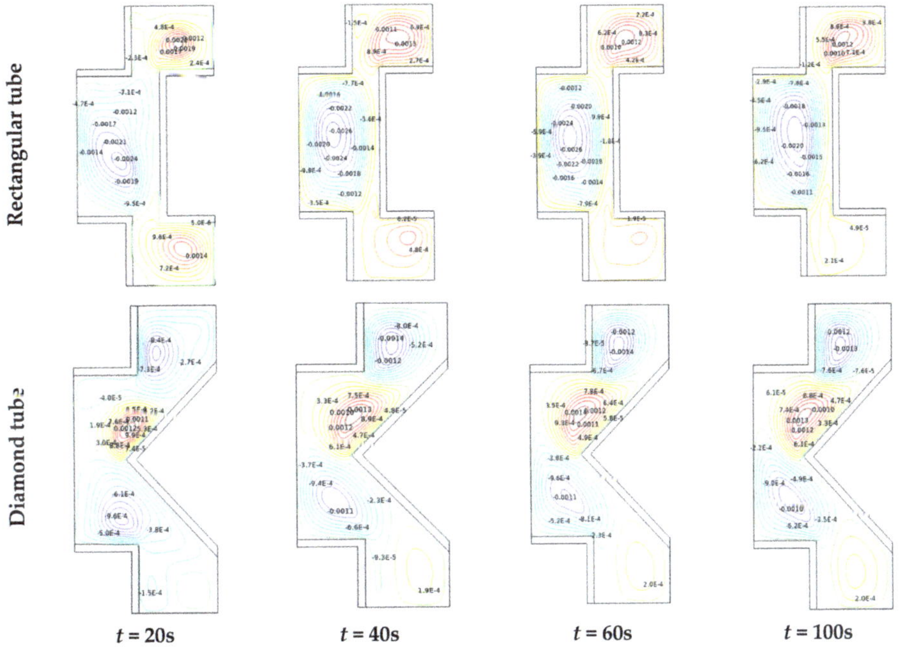

Figure 7. Streamlines in a tube/shell heat exchanger with different shape of manufacturing tubes for Re = 2600, $\varphi = 0.02$.

For understanding the overall impact of the different shaped tube bundles on the overall heat transfer, the Nu_{avg} is plotted for the varying Re, which is shown in Figure 8. The heat transmission efficiency of Al_2O_3-MWCNT/water hybrid nanofluid improves as the flow velocity increases, which enhances the thermal convection effect. The higher fluid Re corresponds to the higher heat transfer. Furthermore, the diamond-shaped tube bundles correspond to the superior heat transfer, and this is true for any Re values.

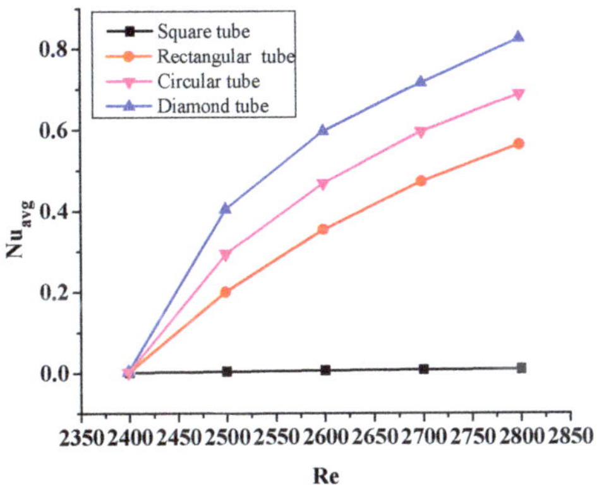

Figure 8. Variation of the Nu_{avg} as a function of change in the Re at $\varphi = 0.02$.

4.3. The Effect of Tube Shape

By examining the results of streamlines and isotherms as shown in Figures 6 and 7, the flow velocity is lessened with the square-, circular- and rectangular-shaped tubes due to the flow separation compared to the diamond-shaped tubes, which increases the transfer of heat since it provides the largest flux function values. In fact, the vortex is distributed between the inlet wall and the outlet wall, providing a higher area for the flowing fluid, thus enhancing the rate of heat transfer. The variation in the temperature also goes a long way in improving buoyancy forces, which stimulate natural convection.

Additionally, Figures 5 and 8 show that the pipe's diamond shape improves the heat transfer by exhibiting the maximum peak of average Nu values in comparison to the other shapes. Furthermore, the geometric characteristics of the diamond shape and the uniform area available around it, which allow hybrid nanofluid dispersion. This helps to improve the Nu_{avg} number and thus thermal convection.

4.4. Entropy Generation

In this section, entropy generation is calculated (Figure 9), and low entropy is observed at the hot wall cones, as a result of weaker isotherms close to the region. Therefore, more circulation of fluid is essential. The distribution is also bi-cellular, except the shape of the circulating cells, and it follows an oval shape. This implies an enhanced induced flow. Therefore, thermal convection dominates, which is indicated by the isotherms. The latter indicates enhanced convection, which favors the mechanism of convection over that of conduction. Moreover, Figure 10 shows that the diamond tube shape improves the heat exchanger's effectiveness by exhibiting the highest decreases in entropy compared to the other shapes.

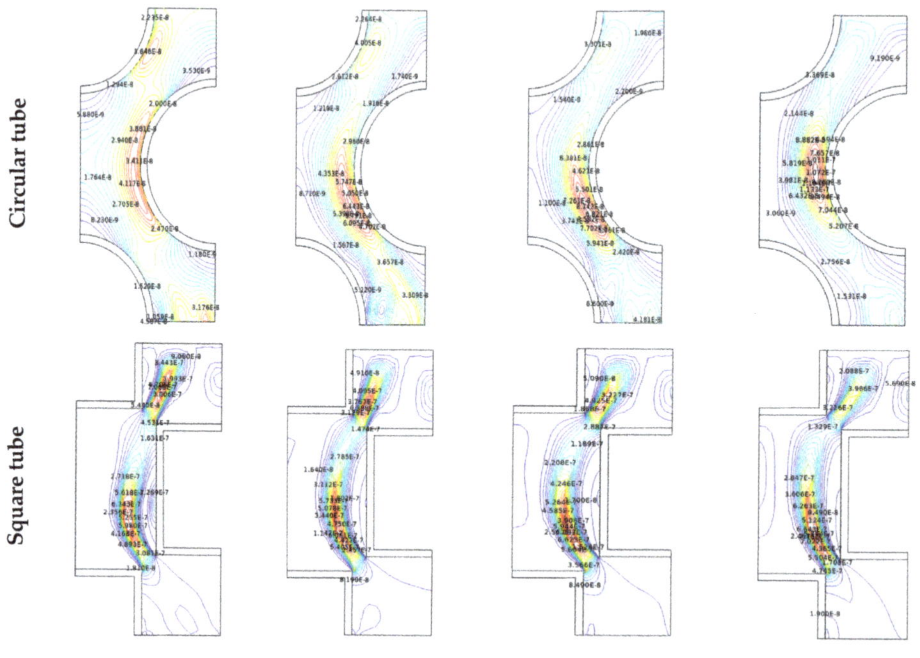

Figure 9. *Cont.*

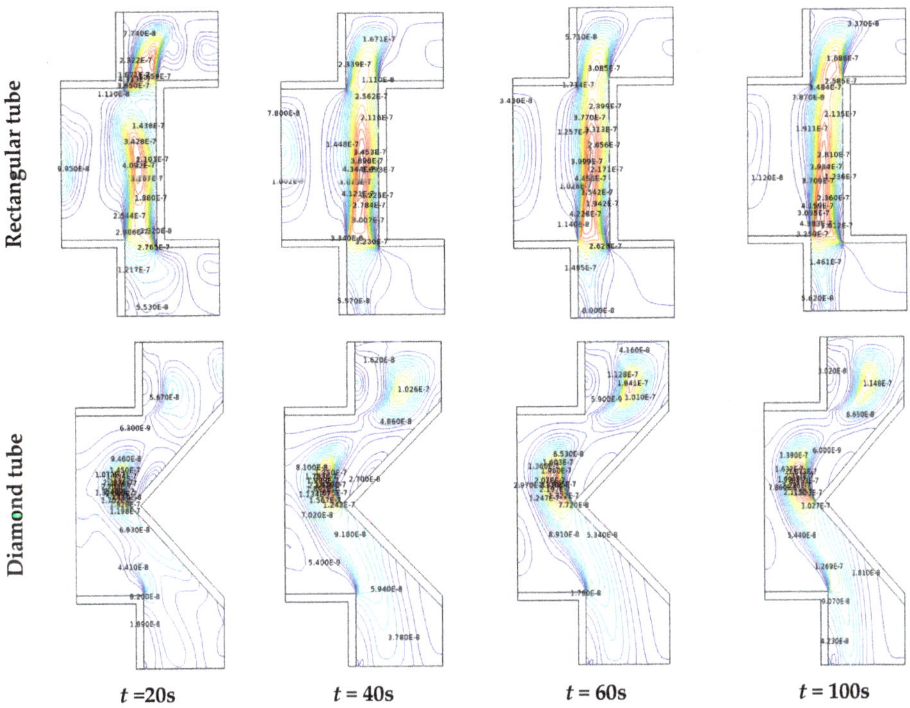

Figure 9. Entropy in a tube/Shell heat exchanger with different shape of manufacturing tubes for Re = 2600 and $\varphi = 0.02$.

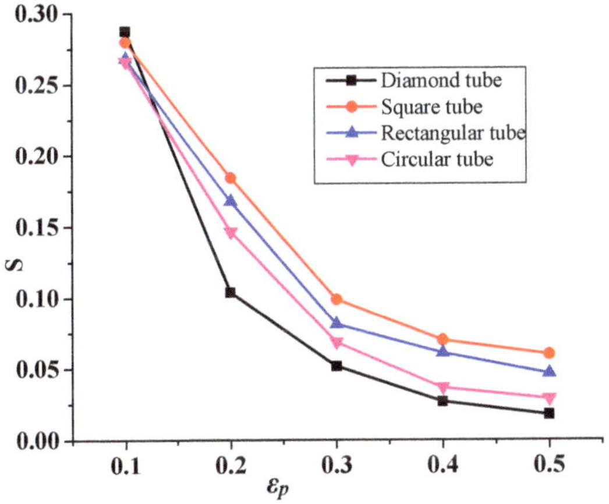

Figure 10. Entropy generation (S) variation as a function of effectiveness of the heat exchanger (ε_p) for Re = 2600 and $\varphi = 0.02$.

5. Conclusions

The present study focuses on the improvements of the tube/shell heat exchanger to maximize the efficiency either by altering the geometrical shape of the tubes and/or by adding nanoparticles in its heat transfer fluid. Water-based Al_2O_3-MWCNT hybrid nanofluid is utilized here as a heat transfer fluid. The analysis is carried out for two main parameters: the Reynolds number and the volume fraction, to study the heat transfer performance of the heat exchanger. The shape (circular, square, rectangular, and diamond) of the heat exchanger tube is modified from a conventional circular for obtaining the best shape. The following conclusions were revealed:

- The heat transfer rises with the growing fluid velocity (function of Re).
- The heat transfer rises with growing nanoparticle concentration. Compared to the circular tubes, the heat transfer enhancement goes up to 103.07% with a particle concentration of 2%.
- The diamond-shaped tube only offers superior performance in heat transfer compared to all other shaped tubes.
- The diamond tube shape improves the heat exchanger performance by exhibiting the most decreases in entropy compared to the other shapes.

In general, the outcome of the study is very significant in the industrial field and can solve many heat transfer problems. However, the present study could be extended for further research to incorporate three-dimensional models, other geometries, and flow-controlling parametric ranges and boundary conditions for unsteady or turbulent fluid flow, which are limited in the present study.

Author Contributions: Conceptualization, F.M.-O.; Methodology, F.M.-O. and A.A.I.I.; Software, F.M.-O.; Validation, F.M.-O.; Formal analysis, F.M.-O.; Investigation, F.M.-O.; Resources, F.M.-O. and A.A.I.I.; Data curation, F.M.-O.; Writing—original draft, M.B. and F.M.-O.; Writing—review & editing, F.M.-O., N.B. and A.A.I.I.; Visualization, F.M.-O.; Supervision, F.M.-O.; Project administration, F.M.-O.; Funding acquisition, F.M.-O. All authors have read and agreed to the published version of the manuscript.

Funding: This research received no external funding.

Institutional Review Board Statement: Not applicable.

Informed Consent Statement: Not applicable.

Data Availability Statement: Data are available on request.

Acknowledgments: The authors would like to thank the Deanship of Scientific Research at Umm Al-Qura University for supporting this work by Grant Code: (22UQU4240002DSR15).

Conflicts of Interest: The authors declare no conflict of interest.

Abbreviations

Nomenclature	
g	Acceleration due to gravitational ($m \cdot s^{-2}$)
H	Heat exchanger height (m)
c_p	Specific heat capacity at constant pressure ($J/(kg \cdot K)$)
Nu	Nusselt number (average)
p	pressure (Pa)
Pr	Prandtl number
Re	Reynolds number
t	time (s)
T	temperature (K)
q	heat flux by conduction (W/m^2)
$\mathbf{q_r}$	heat flux by radiation (W/m^2)
u, v	components of velocity ($m\ s^{-1}$)
x, y	Cartesian coordinates (m)
Greek symbols	
α	thermal diffusivity ($m^2 \cdot s^{-1}$)
β	coefficient of thermal expansion (K^{-1})
μ	dynamic viscosity ($N\ m^{-2} \cdot s^{-1}$)
ν	kinematic viscosity ($m^2 \cdot s^{-1}$)
ρ	fluid density ($kg\ m^{-3}$)
φ	hybrid nanoparticles volume fraction (%)

ψ	Dimensionless stream function
θ	Dimensionless temperature
τ	viscous stress tensor (Pa)
∇	del operator

References

1. Bontemps, A.; Garrigue, A.; Goubier, C.; Huetz, J.; Marvillet, C.; Mercier, P.; Vidil, R. Echangeurs de Chaleur. In *Définitions et Architecture Générale. France: Techniques de l'Ingénieur, Traité Génie Energétique.* 1994; p. B2340. Available online: https://www.techniques-ingenieur.fr/base-documentaire/archives-th12/archives-thermique-industrielle-tiabec/archive-1/echangeurs-de-chaleur-b2340/architecture-generale-de-l-echangeur-b2340niv10002.html (accessed on 15 May 2023).
2. Finkbeiner, F.; Gonard, T.; Filiol, B. *Echangeurs Thermiques: Enjeux, Marchés, Technologie et Politique d'Innovation*; Editions Européennes Thermique et Industrie (EETI): Paris, France, 1993.
3. Labbadlia, O.; Laribi, B.; Chetti, B.; Hendrick, P. Numerical study of the influence of tube arrangement on the flow distribution in the header of shell and tube heat exchangers. *Appl. Therm. Eng.* **2017**, *126*, 315–321. [CrossRef]
4. He, Z.; Fang, X.; Zhang, Z.; Gao, X. Numerical investigation on performance comparison of non-Newtonian fluid flow in vertical heat exchangers combined helical baffle with elliptic and circular tubes. *Appl. Therm. Eng.* **2016**, *100*, 84–97. [CrossRef]
5. Qiu, Y.; Li, M.-J.; Wang, W.-Q.; Du, B.-C.; Wang, K. An experimental study on the heat transfer performance of a prototype molten-salt rod baffle heat exchanger for concentrated solar power. *Energy* **2018**, *156*, 63–72. [CrossRef]
6. Yang, J.-F.; Zeng, M.; Wang, Q.-W. Numerical investigation on shell-side performances of combined parallel and serial two shell pass shell-and-tube heat exchangers with continuous helical baffles. *Appl. Energy* **2015**, *139*, 163–174. [CrossRef]
7. Ayub, Z.H.; Yang, D.; Khan, T.S.; Al-Hajri, E.; Ayub, A.H. Performance characteristics of a novel shell and tube heat exchanger with shell side interstitial twisted tapes for viscous fluids application. *Appl. Therm. Eng.* **2018**, *134*, 248–255. [CrossRef]
8. Biswas, N.; Manna, N.K.; Chamkha, A.J.; Mandal, D.K. Effect of surface waviness on MHD thermo-gravitational convection of $Cu-Al_2O_3-$water hybrid nanofluid in a porous oblique enclosure. *Phys. Scr.* **2021**, *96*, 105002. [CrossRef]
9. Mandal, D.K.; Biswas, N.; Manna, N.K.; Gorla, R.S.R.; Chamkha, A.J. Hybrid nanofluid magnetohydrodynamic mixed convection in a novel W-shaped porous system. *Int. J. Numer. Meth. Heat Fluid Flow* **2023**, *33*, 510–544. [CrossRef]
10. Horvat, A.; Leskovar, M.; Mavko, B. Comparison of heat transfer conditions in tube bundle cross-flow for different tube shapes. *Int. J. Heat Mass Transf.* **2006**, *49*, 1027–1038. [CrossRef]
11. Mebarek-Oudina, F.; Preeti; Sabu, A.S.; Vaidya, H.; Lewis, R.W.; Areekara, S.; Mathew, A.; Ismail, A.I. Hydromagnetic flow of magnetite-water nano-fluid utilizing adapted Buongiorno model. *Int. J. Mod. Phys. B* **2023**, *2450003*. [CrossRef]
12. Fazelpour, F.; Vafaeipour, M.; Rahbari, O. CFD simulation of hydrodynamics of gas-solid two-phase flow for different geometries of solid particles. In Proceedings of the 5th International Congress on Energy and Environment Engineering and Management, Lisbon, Portugal, 17–19 July 2013.
13. Skoglund, T.; Årzén, K.-E.; Dejmek, P. Dynamic object-oriented heat exchanger models for simulation of fluid property transitions. *Int. J. Heat Mass Transf.* **2006**, *49*, 2291–2303. [CrossRef]
14. Raza, J.; Mebarek-Oudina, F.; Lund, L.A. The flow of magnetised convective Casson liquid via a porous channel with shrinking and stationary walls. *Pramana—J. Phys.* **2022**, *96*, 229. [CrossRef]
15. Zaversky, F.; Sánchez, M.; Astrain, D. Object-oriented modeling for the transient response simulation of multi-pass shell-and-tube heat exchangers as applied in active indirect thermal energy storage systems for concentrated solar power. *Energy* **2014**, *65*, 647–664. [CrossRef]
16. Shahdad, I.; Fazelpour, F. Numerical analysis of the surface and geometry of plate fin heat exchangers for increasing heat transfer rate. *Int. J. Energy Environ. Eng.* **2018**, *9*, 155–167. [CrossRef]
17. Xie, S.; Liang, Z.; Zhang, L.; Wang, Y. A numerical study on heat transfer enhancement and flow structure in enhanced tube with cross ellipsoidal dimples. *Int. J. Heat Mass Transf.* **2018**, *125*, 434–444. [CrossRef]
18. Matos, R.; Laursen, T.; Vargas, J.; Bejan, A. Three-dimensional optimization of staggered finned circular and elliptic tubes in forced convection. *Int. J. Therm. Sci.* **2004**, *43*, 477–487. [CrossRef]
19. Matos, R.; Vargas, J.; Laursen, T.; Saboya, F. Optimization study and heat transfer comparison of staggered circular and elliptic tubes in forced convection. *Int. J. Heat Mass Transf.* **2001**, *44*, 3953–3961. [CrossRef]
20. Nouri-Borujerdi, A.; Lavasani, A. Experimental study of forced convection heat transfer from a cam shaped tube in cross flows. *Int. J. Heat Mass Transf.* **2007**, *50*, 2605–2611. [CrossRef]
21. Nouri-Borujerdi, A.; Lavasani, A.M. Pressure loss and heat transfer characterization of a cam-shaped cylinder at different orientations. *J. Heat Transf.* **2008**, *130*, 124503. [CrossRef]
22. Mohanty, R.L.; Swain, A.; Das, M.K. Thermal performance of mixed tube bundle composed of circular and elliptical tubes. *Therm. Sci. Eng. Prog.* **2018**, *5*, 492–505. [CrossRef]
23. Dharmaiah, G.; Mebarek-Oudina, F.; Balamurugan, K.S.; Vedavathi, N. Numerical Analysis of the Magnetic Dipole Effect on a Radiative Ferromagnetic Liquid Flowing over a Porous Stretched Sheet. *Fluid Dyn. Mater. Process.* **2023**. [CrossRef]
24. Li, B.; Feng, B.; He, Y.-L.; Tao, W.-Q. Experimental study on friction factor and numerical simulation on flow and heat transfer in an alternating elliptical axis tube. *Appl. Therm. Eng.* **2006**, *26*, 2336–2344. [CrossRef]

25. Bouris, D.; Papadakis, G.; Bergeles, G. Numerical evaluation of alternate tube configurations for particle deposition rate reduction in heat exchanger tube bundles. *Int. J. Heat Fluid Flow* **2001**, *22*, 525–536. [CrossRef]
26. Moawed, M. Experimental study of forced convection from helical coiled tubes with different parameters. *Energy Conv. Manag.* **2011**, *52*, 1150–1156. [CrossRef]
27. Rosen, M.; Dincer, I. Exergy methods for assessing and comparing thermal storage systems. *Int. J. Energy Res.* **2003**, *27*, 415–430. [CrossRef]
28. Khan, M.G.; Fartaj, A.; Ting, D.S.-K. An experimental characterization of cross-flow cooling of air via an in-line elliptical tube array. *Int. J. Heat Fluid Flow* **2004**, *25*, 636–648. [CrossRef]
29. Harris, D.K.; Goldschmidt, V.W. Measurements of the overall heat transfer from combustion gases confined within elliptical tube heat exchangers. *Exp. Therm. Fluid Sci.* **2002**, *26*, 33–37. [CrossRef]
30. Tao, Y.; He, Y.; Wu, Z.; Tao, W. Three-dimensional numerical study and field synergy principle analysis of wavy fin heat exchangers with elliptic tubes. *Int. J. Heat Fluid Flow* **2007**, *28*, 1531–1544. [CrossRef]
31. Li, Z.; Davidson, J.; Mantell, S. Numerical simulation of flow field and heat transfer of streamlined cylinders in cross flow. In *ASME 2005 Summer Heat Transfer Conference Collocated with the ASME 2005 Pacific Rim Technical Conference and Exhibition on Integration and Packaging of MEMS, NEMS, and Electronic Systems 2005*; American Society of Mechanical Engineers: New York, NY, USA, 2005; pp. 531–541.
32. Kumar, N.; Jhinge, P.K. Effect of segmental baffles at different orientation on the performances of single pass shell and tube heat exchanger. *Int. J. Ing. Trends Technol.* **2014**, *15*, 423–428.
33. Raj, T.K.; Ganne, S. Shell side numerical analysis of a shell and tube heat exchanger considering the effects of baffle inclination angle on fluid flow. *Therm. Sci.* **2012**, *16*, 1165–1174. [CrossRef]
34. Zhang, J.; He, Y.; Tao, W. 3D numerical simulation on shell-and-tube heat exchangers with middle-overlapped helical baffles and continuous baffles—Part II: Simulation results of periodic model and comparison between continuous and noncontinuous helical baffles. *Int. J. Heat Mass Transf.* **2009**, *52*, 5381–5389. [CrossRef]
35. Sivarajan, C.; Rajasekaran, B.; Krishnamohan, N. Comparision of numerical heat transfer in conventional and helically baffled heat exchanger. *IJERA* **2012**, *2*, 1278–1282.
36. Kwon, Y.H.; Kim, D.; Li, C.G.; Lee, J.K. Heat transfer and pressure drop characteristics of Nano fluids in a plate heat exchanger. *J. Nano Sci. Nanotechnol.* **2011**, *11*, 5769–5774. [CrossRef] [PubMed]
37. Albadr, J.; Tayal, S.; Alasadi, M. Heat transfer through heat exchanger using Al_2O_3 nanofluid at different concentration. *Case Stud. Therm. Eng.* **2013**, *1*, 38–44. [CrossRef]
38. Asadi, A.; Asadi, M.; Rezaniakolaei, A.; Rosendahl, L.A.; Afrand, M.; Wongwises, S. Heat transfer efficiency of Al_2O_3-MWCNT/thermal oil hybrid nanofluid as a cooling fluid in thermal and energy management applications: An experimental and theoretical investigation. *Int. J. Heat Mass Transf.* **2018**, *117*, 474–486. [CrossRef]
39. Ghazanfari, V.; Imani, M.; Shadman, M.M.; Amini, Y.; Zahakifar, F. Numerical study on the thermal performance of the shell and tube heat exchanger using twisted tubes and Al_2O_3 nanoparticles. *Prog. Nucl. Energy* **2023**, *155*, 104526. [CrossRef]
40. Multiphysics, C. *Comsol Multiphysics User Guide, Version 4.3 a*; COMSOL, AB: Stockholm, Sweden, 2012; pp. 39–40.
41. Gourari, S.; Mebarek-Oudina, F.; Makinde, O.D.; Rabhi, M. Numerical Investigation of Gas-Liquid Two-Phase Flows in a Cylindrical Channel. *Defect Diffus. Forum* **2021**, *409*, 39–48. [CrossRef]
42. Salari, M.; Malekshah, E.H.; Malekshah, M.H. Natural convection in a rectangular enclosure filled by two immiscible fluids of air and Al_2O_3-water nanofluid heated partially from side walls. *Alex. Eng. J.* **2018**, *57*, 1401–1412. [CrossRef]
43. Waini, I.; Ishak, A.; Pop, I. Hybrid nanofluid flow and heat transfer over a nonlinear permeable stretching/shrinking surface. *Int. J. Numer. Methods Heat Fluid Flow* **2019**, *29*, 3110–3127. [CrossRef]
44. Chabani, I.; Mebarek-Oudina, F.; Vaidya, H.; Ismail, A.I. Numerical analysis of magnetic hybrid Nano-fluid natural convective flow in an adjusted porous trapezoidal enclosure. *J. Magn. Magn. Mater.* **2022**, *564*, 170142. [CrossRef]
45. Mebarek-Oudina, F.; Chabani, I. Review on Nano-fluids applications and heat transfer enhancement techniques in different enclosures. *J. Nanofluids* **2022**, *11*, 155–168. [CrossRef]
46. Khan, U.; Mebarek-Oudina, F.; Zaib, A.; Ishak, A.; Abu Bakar, S.; Sherif, E.M.; Baleanu, D. An exact solution of a Casson fluid flow induced by dust particles with hybrid nanofluid over a stretching sheet subject to Lorentz forces. *Waves Random Complex Media* **2022**. [CrossRef]
47. Roy, P.; Mondal, A. Analysis of Shell and Tube heat exchanger design using Comsol Multiphysics. *Int. J. Sci. Appl. Res.* **2016**, *3*, 70–82.
48. Mebarek-Oudina, F.; Fares, R.; Aissa, A.; Lewis, R.W.; Abu-Hamdeh, N. Entropy and convection effect on magnetized hybrid nano-liquid flow inside a trapezoidal cavity with zigzagged wall. *Int. Comm. Heat Mass Trans.* **2021**, *125*, 105279. [CrossRef]
49. Mebarek-Oudina, F.; Fares, R.; Choudhari, R. Convection Heat Transfer of MgO-Ag/Water Magneto-Hybrid Nanoliquid Flow into a Special Porous Enclosure. *Alg. J. Renew. Energ. Sust. Develop.* **2020**, *2*, 84–95.
50. Shafiq, A.; Mebarek-Oudina, F.; Sindhu, T.N.; Rassoul, G. Sensitivity analysis for Walters' B nanoliquid flow over a radiative Riga surface by RSM. *Sci. Iran.* **2022**, *29*, 1236–1249. [CrossRef]
51. Mebarek-Oudina, F. Convective Heat Transfer of Titania Nanofluids of different base fluids in Cylindrical Annulus with discrete Heat Source. *Heat Transf.-Asian Res.* **2019**, *48*, 135–147. [CrossRef]

52. Ramesh, K.; Mebarek-Oudina, F.; Ismail, A.I.; Jaiswal, B.R.; Warke, A.S.; Lodhi, R.K.; Sharma, T. Computational analysis on radiative non-Newtonian Carreau nanofluid flow in a microchannel under the magnetic properties. *Sci. Iran.* **2023**, *30*, 376–390.
53. Ali, A.; Mebarek-Oudina, F.; Barman, A.; Das, S.; Ismail, A.I. Peristaltic transportation of hybrid nano-blood through a ciliated micro-vessel subject to heat source and Lorentz force. *J. Therm. Anal. Calorim.* **2023**, *36*, 101510. [CrossRef]
54. Mebarek-Oudina, F.; Chabani, I. Review on Nano Enhanced PCMs: Insight on nePCM Application in Thermal Management/Storage Systems. *Energies* **2023**, *16*, 1066. [CrossRef]

micromachines

MDPI

Article

Mixed Convection Hybrid Nanofluid Flow Induced by an Inclined Cylinder with Lorentz Forces

Farizza Haniem Sohut [1], Umair Khan [1,2], Anuar Ishak [1,*], Siti Khuzaimah Soid [3] and Iskandar Waini [4]

[1] Department of Mathematical Sciences, Faculty of Science and Technology, Universiti Kebangsaan Malaysia, Bangi 43600, Malaysia; p106827@siswa.ukm.edu.my (F.H.S.); umairkhan@iba-suk.edu.pk (U.K.)
[2] Department of Mathematics and Social Sciences, Sukkur IBA University, Sukkur 65200, Pakistan
[3] School of Mathematical Sciences, College of Computing, Informatics and Media, Universiti Teknologi MARA, Shah Alam 40450, Malaysia; khuzaimah@tmsk.uitm.edu.my
[4] Faculty of Mechanical and Manufacturing Engineering Technology, Universiti Teknikal Malaysia Melaka, Hang Tuah Jaya, Durian Tunggal, Melaka 76100, Malaysia; iskandarwaini@utem.edu.my
* Correspondence: anuar_mi@ukm.edu.my

Abstract: Hybrid nanofluids may exhibit higher thermal conductivity, chemical stability, mechanical resistance and physical strength compared to regular nanofluids. Our aim in this study is to investigate the flow of a water-based alumina-copper hybrid nanofluid in an inclined cylinder with the impact of buoyancy force and a magnetic field. The governing partial differential equations (PDEs) are transformed into a set of similarity ordinary differential equations (ODEs) using a dimensionless set of variables, and then solved numerically using the bvp4c package from MATLAB software. Two solutions exist for both buoyancy opposing ($\lambda < 0$) and assisting ($\lambda > 0$) flows, whereas a unique solution is found when the buoyancy force is absent ($\lambda = 0$). In addition, the impacts of the dimensionless parameters, such as curvature parameter, volume fraction of nanoparticles, inclination angle, mixed convention parameter, and magnetic parameter are analyzed. The results of this study compare well with previously published results. Compared to pure base fluid and regular nanofluid, hybrid nanofluid reduces drag and transfers heat more efficiently.

Keywords: magnetohydrodynamic (MHD); mixed convection; inclined cylinder; stagnation point; hybrid nanofluid; dual solutions; stability analysis

1. Introduction

A nanofluid is a solid-liquid mixture containing nanoparticles and a base liquid. According to Daungthongsuk and Wongwises [1], the thermal conductivity of a fluid is boosted by mixing nanoparticles with a regular fluid. Because thermal conductivity is the most important issue in heat transfer problems, many mathematical models for nanofluids have been developed by researchers such as Buongiorno [2] and Tiwari and Das [3]. In addition, issues surrounding the improvement of heat transfer are also important for other cases such as employing microchannel heat sink (MCHS) with V-ribs and electrohydrodynamic (EHD) using water-based nanofluid. The heat transfer performance of a MCHS is enhanced due to the periodically arranged V-ribs that interrupt the thermal boundary, which produce chaotic convention and augment the heat transfer area [4]. Wang et al. [5] mentioned that the major processes involved in increasing heat transfer in microchannels are chaotic convection caused by the electrophoretic effect and thermophoretic effect. Although nanofluids have extremely efficient qualities and properties for industrial and engineering operations, researchers are still searching for fluid variants by studying several of their physical aspects such as the type, shape, and volume fraction of nanoparticles. Hybrid nanofluid exhibits significant developments in its thermal and rheological properties when compared to regular nanofluid, particularly in improving the heat conductivity of base fluids. The upgraded features of nanofluids result from the addition of two or more types of nanoparticles to a common fluid that is called a hybrid nanofluid and serves the purpose of increasing the heat transfer rate. Suresh et al. [6] mentioned that researchers have used a variety of nanoparticles including carbon nanotubes, metallic particles (Cu, Al, Fe, Au, and Ag), and non-metallic particles (Al_2O_3, CuO, Fe_3O_4, TiO_2, and SiC). Furthermore, if the nanoparticle materials are properly chosen, they can positively enhance each other and significantly improve the thermal conductivity of the nanofluid. Based on the characteristics of metallic and non-metallic nanoparticles, it can be expected that the addition of metallic nanoparticles such as copper (Cu) into a nanofluid composed on a basis of Al_2O_3 nanoparticles can enhance the thermophysical properties of this mixture. It has been determined that most hybrid nanofluids studied have a higher thermal conductivity than nanofluid, but Guan et al. [7] further explored the reason why hybrid nanofluids have high thermal conductivity. The nanolayer densities and diffusion coefficients were calculated for various hybrid nanofluids to explain the underlying mechanism of the enhancement in the thermal properties. The results showed that a hybrid nanofluid with Cu-Ag 50%/Ar can obtain the greatest improvement in thermal conductivity compared to liquid Ar, Cu/Ar and Ag/Ar nanofluids.

Waini et al. [8] reported that a hybrid nanofluid, also known as a base fluid with multiple types of nanoparticles, had improved thermal properties. Highly effective thermal conductivity, enhanced heat transfer, the synergistic effect of nanoparticles, stability, and good aspect ratio are the advantages of a hybrid nanofluid. Lower operating costs and higher energy efficiency are the best definitions of enhanced thermal conductivity of a hybrid nanofluid [9]. A hybrid nanofluid mixed convection flow has applications in various modern manufacturing processes, such as metal extrusion, wire drawing, welding, and heat pipe, etc. [10]. Suresh et al. [11] investigated how an Al_2O_3-Cu/H_2O hybrid nanofluid affected heat transfer. They identified that the hybrid nanofluid had a greater friction coefficient than that of an Al_2O_3/H_2O nanofluid. Alshuhail et al. [12] reviewed the thermal efficiency of hybrid nanofluids in solar energy applications. Ranga et al. [13] reported that for Cu-CuO/H_2O hybrid nanofluid at 5% volume fraction, the collector efficiency increased by 2.175%. An experimental analysis conducted by Farajzadeh et al. [14] found that the collector efficiency increased by 26% on using Al_2O_3-TiO_2/H_2O hybrid nanofluids.

The mathematical analysis of the dual solutions of the boundary layer flow along moving surfaces has a practical application in the engineering scenario. It enables the determination of the most practical, stable, and physically acceptable solutions. Early research on Blasius flow for non-unique solutions by considering various flow situations and boundary conditions was conducted by Steinheuer [15]. He performed an investigation on boundary layer flow over a semi-infinite long flat plate with moving but impermeable soil to give a non-unique solution. Later, Klemp and Acrivos [16] reported that non-unique solutions exist in the case where the plate and the free stream move with the reverse flow. The flow field differs from the ordinary boundary layer in that an inviscid collision area exists around the site of detachment, where the reverse boundary-layer flow is turned in the direction of the mainstream. In addition, other scholars such as Bognár [17], Hussaini and Lakin [18], and Ishak and Bachok [19] reported similar problems with the moving flat plate for Newtonian fluid flow. It was discovered that the existence of solutions for a semi-infinite plate is dependent on the ratio of the plate surface velocity to the free stream velocity. It has been demonstrated that a solution exists only if this parameter does not exceed a specific critical value, and numerical computations have been performed to demonstrate that this solution is non-unique. However, the existence of non-unique solutions also occurs for the Blasius boundary conditions problem when the surface is at rest, as reported by Khashi'ie et al. [20], Waini et al. [21], Sohut et al. [22], etc.

A stagnation point (SP) is a point on a solid body which directly faces the stream and where the streamlines are separated. On the other hand, the flow of a fluid in the neighborhood of an SP is called stagnation point flow (SPF). The exploration of SPF is a hot topic in fluid mechanics because of its functions and applications in industries. The stagnation point flow, also commonly known as the Hiemenz flow, was established by

Hiemenz [23] and can be accurately analyzed using the Navier–Stokes equation. He was the first researcher to find an outcome for the two-dimensional SPF model over a horizontal plate. Hiemenz [23] solved the Navier–Stokes equation numerically by transforming it into a regular differential equation using a suitable similarity transformation. Many cases that involved SPF have also been discovered by many scholars. For instance, Chiam [24] explored the SPF past a stretching sheet. Wang [25] examined both axisymmetric and two-dimensional stagnation flows toward a shrinking sheet. He found that the flow structure is complicated due to the non-alignment of the stagnation flow on the shrinking sheet. A few years later, Lok et al. [26] continued the investigation by including the effect of MHD. However, due to the shrinkage effect, the streamlines are not always parallel, but a reverse flow is designed or originated near the surface. Furthermore, Awaludin et al. [27] investigated the SPF over a shrinking as well as a stretching sheet. They found that the solution for the stretching case is unique, but two solutions were found for the shrinking case.

Many industrial applications and natural phenomena involve mixed convection flows (MCFs), which involve a combination of free and forced convection flows, such as electronic equipment cooled by fans, flows in the ocean, underground cable systems, etc. In MCF, the existence of dual solutions has recently become a subject of discussion among scholars. For example, Ishak et al. [28] explored the magnetohydrodynamic MCF on a vertical continuous porous wall using a finite difference method. They reported that both opposing and assisting flows produce dual solutions. The existence of the solutions is dependent on the buoyancy parameter, where the range increases with suction effect at the boundary. Later, Aman et al. [29] continued the study by adding the slip effect at the boundary. Abbas et al. [30] studied the MHD mixed convection flow between concentric cylinders and reported that the mixed convection increases the velocity on the heated cylinder, but it decreases on the cold cylinder. The combined effects of MHD and slip velocity on the mixed convection flow around a circular cylinder was investigated by Ullah et al. [31]. They reported that an increase in slip factor enhances velocity as well as temperature. The MHD flow of a hybrid nanofluid past a vertical plate was considered by Zainal et al. [32], who found two branches of solutions in a certain range of the mixed convection strength. The temporal stability study verified that only one of the solutions was stable. Wang et al. [33] explored the influence of magnetic and electric fields on heat transfer to analyze the thermal conductivity and heat transfer enhancement of nanofluids, respectively, and the chaotic convention. They found that the applied magnetic and electric fields significantly affect the heat transfer performance of the nanofluid.

The study of fluid flow over an inclined cylindrical surface has significance for applications in industrial and engineering processes, e.g., magnetohydrodynamic (MHD) power generators, the polymer industry, and gas turbines [34]. Studies on the flow over a horizontal or a vertical cylinder have been explored widely; however, studies involving an inclined cylinder remain uncommon. An analysis of stratified stagnation flow induced by an inclined cylinder with mixed convection and a magnetic field was presented by Hayat et al. [35]. Bilal et al. [36] studied the Darcy Forchheimer flow of a hybrid nanofluid (consisting of carbon nanotubes) through the impermeable inclined cylinder using a homotopy analysis method. They found that the higher the inclination angle, the slower the fluid flow. Rehman et al. [37] extended the study to include logarithmic and parabolic curve fitting analysis and stated that analysis is the first step and can be of great help to previous studies. In addition, the study of MHD mixed convection on an inclined cylinder has also attracted the attention of researchers for the case of nanofluid flow, such as Dhanai et al. [38], who examined the effects of slip boundary conditions, Brownian motion, thermophoresis, and viscous dissipation over an inclined cylinder. They discovered that heat transfer increases with the augmentation of the mass transfer parameter but the opposite occurs for the thermal slip parameter. Moreover, Gupta and Sharma [39] studied a similar problem in the presence of thermal radiation. They used the differential transform method (DTM) with Padé approximation in their study and found that the DTM can overcome traditional perturbation limits, assumptions and restrictions.

Based on the previous investigations, hybrid nanofluid flow over an inclined cylinder has not been extensively studied. Binary hybrid nanofluid flow is a popular topic in fluid dynamics inspired by the above literature. The current investigation intends to discover and analyze the impact of magnetohydrodynamic on hybrid nanofluid flow over an inclined cylinder. This study is the extension of what has been investigated by researchers, such as Rehman et al. [37] and Devi and Devi [9], which considers binary hybrid Al_2O_3-Cu/H_2O nanofluid. The stimulus of the governing parameters is generally represented via several graphs and tables, and the quantitative outcomes obtained are validated for the limited cases using the results of previous studies. Furthermore, the current study reveals that the stability investigation proves the physical outcome. These duality and stability results were also the main objectives of the current work.

2. Description of the Mathematical Model

Consider the buoyancy effects on a magnetohydrodynamic SPF of a steady incompressible hybrid nanofluid past an inclined cylindrical surface, as shown in Figure 1. The coordinates (x, r) are measured in the corresponding axial and radial directions of the cylinder. It is assumed that the temperature of the cylindrical surface, T_w, is higher than the surrounding temperature, T_∞. In the current study, the thermophysical properties of the hybrid nanofluid are taken to be uniform. Meanwhile, copper nanoparticles are combined with the base fluid to produce Cu/H_2O nanofluid, which is then combined with the alumina nanoparticles, Al_2O_3, to create the required hybrid nanofluid, Al_2O_3-Cu/H_2O.

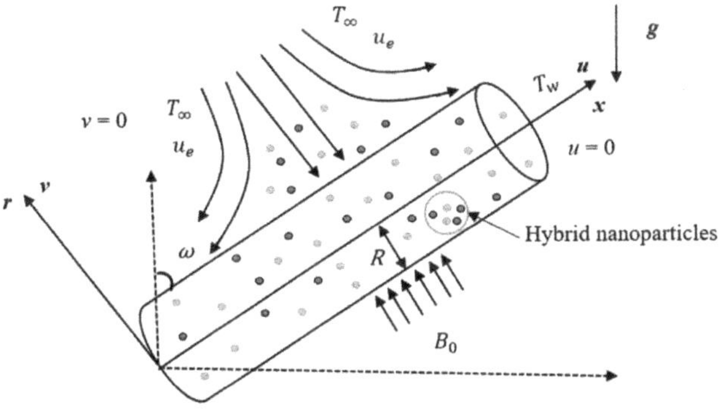

Figure 1. Physical configuration.

The boundary layer equations comprising hybrid nanofluid can take place as follows [9,37]:

$$\frac{\partial(ru)}{\partial x} + \frac{\partial(rv)}{\partial r} = 0 \tag{1}$$

$$\begin{aligned}&\rho_{hnf}\left(u\frac{\partial u}{\partial x} + v\frac{\partial u}{\partial r} - u_e\frac{\partial u_e}{\partial x}\right) = \mu_{hnf}\frac{1}{r}\frac{\partial}{\partial r}\left(r\frac{\partial u}{\partial r}\right) - \sigma_{hnf}{B_0}^2(u - u_e)\\ &+(\rho\beta)_{hnf}(T - T_\infty)g\cos\omega\end{aligned} \tag{2}$$

$$(\rho C_p)_{hnf}\left(u\frac{\partial T}{\partial x} + v\frac{\partial T}{\partial r}\right) = k_{hnf}\frac{1}{r}\frac{\partial}{\partial r}\left(r\frac{\partial T}{\partial r}\right) \tag{3}$$

subject to the boundary conditions

$$\begin{aligned} r = R: &\quad u = 0, \quad v = 0, \quad T = T_w,\\ r \to \infty: &\quad u \to u_e = ax, \quad T \to T_\infty. \end{aligned} \tag{4}$$

Here, v and u are the elements of velocity along the r and x axes, respectively, B_0 is the uniform magnetic field, g is the gravity acceleration, ω is the inclination angle, T is the temperature of the hybrid nanofluid, and a is a positive constant.

Where

$$\mu_H = \frac{\mu_{hnf}}{\mu_f},\ \rho_H = \frac{\rho_{hnf}}{\rho_f},\ \sigma_H = \frac{\sigma_{hnf}}{\sigma_f},\ k_H = \frac{k_{hnf}}{k_f},\ (\rho C_p)_H = \frac{(\rho C_p)_{hnf}}{(\rho C_p)_f},\ (\rho\beta)_H = \frac{(\rho\beta)_{hnf}}{(\rho\beta)_f} \tag{5}$$

In Table 1, $\phi_{hnf} = \phi_1 + \phi_2$ is the hybrid nanoparticle volume fraction, where ϕ_f corresponds to the regular liquid. In addition, ϕ_1 and ϕ_2 indicate aluminum oxide (Al_2O_3) and copper (Cu) nanoparticle volume fractions, and their solid components are symbolized by the subscripts Al and C, respectively. In addition, μ, ρ, k, C_p, σ, and β refer to the absolute viscosity, density, TCN, specific heat capacity, electrical conductivity and thermal expansion coefficient, respectively. The subscripts *hnf* and *f*, respectively, stand for the requisite posited hybrid nanofluid and the regular fluid. Table 2 provides the values of the thermophysical characteristics of the nanoparticles and water (regular fluid).

Table 1. Thermophysical correlations of Al_2O_3–Cu/H_2O (see Takabi and Salehi [40], Ganguly et al. [41] and Waini et al. [42]).

Properties	Hybrid Nanofluid
Density	$\rho_{hnf} = \phi_1\rho_{Al} + \phi_2\rho_C + \left(1 - \phi_{hnf}\right)\rho_f$
Dynamic viscosity	$\mu_{hnf} = \mu_f(1 - \phi_1 - \phi_2)^{-2.5}$
Thermal conductivity	$k_H = \frac{\frac{\phi_1 k_{Al}+\phi_2 k_C}{\phi_1+\phi_2}+2k_f+2(\phi_1 k_{Al}+\phi_2 k_C)-2(\phi_1+\phi_2)k_f}{\frac{\phi_1 k_{Al}+\phi_2 k_C}{\phi_1+\phi_2}+2k_f-(\phi_1 k_{Al}+\phi_2 k_C)+(\phi_1+\phi_2)k_f}$
Electrical conductivity	$\sigma_H = \frac{\frac{\phi_1\sigma_{Al}+\phi_2\sigma_C}{\phi_1+\phi_2}+2\sigma_f+2(\phi_2\sigma_{Al}+\phi_1\sigma_C)-2(\phi_1+\phi_2)\sigma_f}{\frac{\phi_1\sigma_{Al}+\phi_2\sigma_C}{\phi_1+\phi_2}+2\sigma_f-(\phi_1\sigma_{Al}+\phi_2\sigma_C)+(\phi_1+\phi_2)\sigma_f}$
Heat capacity	$(\rho C_p)_{hnf} = \left(1 - \phi_{hnf}\right)(\rho C_p)_f + \phi_1(\rho C_p)_{Al} + \phi_2(\rho C_p)_C$
Thermal expansion	$(\rho\beta)_{hnf} = \left(1 - \phi_{hnf}\right)(\rho\beta)_f + \phi_1(\rho\beta)_{Al} + \phi_2(\rho\beta)_C$

Table 2. Thermophysical characteristics of the regular fluid and hybrid nanoparticles (see Waini et al. [42] and Oztop and Abu-Nada [43]).

Physical Properties	Water	Cu	Al_2O_3
$C_p(J/KgK)$	4179	385	765
$\beta \times 10^{-5}(K^{-1})$	21	1.67	0.85
$k(W/mK)$	0.613	400	40
$\sigma(S/m)$	0.05	5.96×10^7	3.69×10^7
$\rho(kg/m^3)$	997.1	8933	3970
Pr	6.2		

In order to get similarity solutions, the dimensionless variables (6) are introduced [37,44]:

$$\psi = \left(u_e v_f x\right)^{1/2} R f(\eta), \quad \eta = \frac{r^2 - R^2}{2R}\sqrt{\frac{u_e}{v_f x}}, \quad \theta(\eta) = \frac{T - T_\infty}{T_w - T_\infty} \tag{6}$$

where prime denotes differentiation with respect to η, and ψ is the stream function defined as $u = r^{-1}\partial\psi/\partial r$ and $v = -r^{-1}\partial\psi/\partial x$, which identically satisfies the continuity Equation (1). The temperature at the surface of the cylinder is assumed to be $T_w = T_\infty + bx$, where b is a constant.

By utilizing the similarity variables (6), Equation (1) is identically satisfied, whereas Equations (2)–(4) are transformed into the following ODEs:

$$\frac{\mu_H}{\rho_H}[(1 + 2K\eta)f''' + 2Kf''] + ff'' + 1 - f'^2 - \frac{\sigma_H}{\rho_H}M^2(f' - 1) + \frac{(\rho\beta)_H}{\rho_H}\lambda\theta\cos\omega = 0 \tag{7}$$

$$\frac{1}{\Pr}\frac{k_H}{(\rho C_p)_H}[(1+2\eta K)\theta''+2K\theta']+\theta' f-\theta f'=0 \tag{8}$$

$$\begin{array}{c} f(0)=0, \quad f'(0)=0, \quad \theta(0)=1 \\ f'(\eta)\to 1, \quad \theta(\eta)\to 0 \quad \text{as} \quad \eta\to\infty \end{array} \tag{9}$$

where M stands for magnetic parameter, Pr is Prandtl number, K is the curvature parameter, and λ is the mixed convection parameter (constant). Further, the buoyancy term is equal to the ratio of the local Grashof number, Gr_x, and the square of the local Reynolds number, ${\mathrm{Re}_x}^2$. The dimensionless quantities are defined as:

$$M=\sqrt{\frac{\sigma_f {\beta_0}^2}{\rho_f a}}, \quad \Pr=\frac{v_f}{\alpha_f}, \quad K=\frac{1}{R}\sqrt{\frac{v_f}{a}}, \quad \lambda=\frac{Gr_x}{{\mathrm{Re}_x}^2}(=\text{const.}), \quad Gr_x=\frac{g\beta_f(T_w-T_\infty)x^3}{{v_f}^2} \tag{10}$$

The skin friction coefficient and local Nusselt number (heat transfer rate) are given by:

$$C_f=\frac{\mu_{hnf}}{\rho_f {u_e}^2(x)}\left(\frac{\partial u}{\partial r}\right)_{r=R}, \quad Nu_x=-\frac{x k_{hnf}}{k_f\left(T_f-T_\infty\right)}\left(\frac{\partial T}{\partial r}\right)_{r=R} \tag{11}$$

which, after applying (5) and (11), then become the following:

$${\mathrm{Re}_x}^{1/2}C_f=\mu_H f''(0), \quad {\mathrm{Re}_x}^{-1/2}Nu_x=-k_H\theta'(0) \tag{12}$$

where $\mathrm{Re}_x=u_e x/v_f$ is the local Reynolds number.

3. Stability Analysis

Several researchers [45–47] have demonstrated the temporal stability of the numerical solutions. The purpose being to verify the stability of the solutions as time evolves. In the present study, two solutions for Equations (7)–(9) are found within a certain range of the physical parameters. It is worth investigating the time stability of the solutions and determining which one of the two solutions is reliable, and which one is not reliable, over time. For the working procedure, the following dimensionless time variable $\tau=at$ is introduced, as proposed by Merkin [48] and Weidman et al. [49]. The unsteady or time-dependent form of Equations (2) and (3) are considered, and then they are reduced to:

$$\begin{array}{l} \rho_{hnf}\left(\frac{\partial u}{\partial t}+u\frac{\partial u}{\partial x}+v\frac{\partial u}{\partial r}-u_e\frac{\partial u_e}{\partial x}\right)=\mu_{hnf}\frac{1}{r}\frac{\partial}{\partial r}\left(r\frac{\partial u}{\partial r}\right)-\sigma_{hnf}{B_0}^2(u-u_e) \\ +(\rho\beta)_{hnf}(T-T_\infty)g\cos\omega \end{array} \tag{13}$$

$$(\rho C_p)_{hnf}\left(\frac{\partial T}{\partial t}+u\frac{\partial T}{\partial x}+v\frac{\partial T}{\partial r}\right)=k_{hnf}\frac{1}{r}\frac{\partial}{\partial r}\left(r\frac{\partial T}{\partial r}\right) \tag{14}$$

where t stands for time. Based on Equation (6), the dimensionless transformations are written as follows:

$$\begin{array}{l} \psi=\left(u_e v_f x\right)^{1/2}Rf(\eta,\tau), \quad \eta=\frac{r^2-R^2}{2R}\sqrt{\frac{u_e}{v_f x}}, \quad u=ax\,f'(\eta,\tau), \\ v=-\frac{R}{r}\sqrt{av_f}f(\eta,\tau), \quad \theta(\eta,\tau)=\frac{T-T_\infty}{T_w-T_\infty}. \end{array} \tag{15}$$

Substituting Equation (15) into Equations (13) and (14) results in

$$\begin{array}{l} \frac{\mu_H}{\rho_H}\left[(1+2K\eta)\frac{\partial^3 f}{\partial\eta^3}+2K\frac{\partial^2 f}{\partial\eta^2}\right]-\frac{\partial^2 f}{\partial\eta\partial\tau}+1+f\frac{\partial^2 f}{\partial\eta^2}-\frac{\sigma_H}{\rho_H}M^2\left(\frac{\partial f}{\partial\eta}-1\right)-\left(\frac{\partial f}{\partial\eta}\right)^2 \\ +\frac{(\rho\beta)_H}{\rho_H}\lambda\theta\cos\omega=0, \end{array} \tag{16}$$

$$\frac{1}{\Pr}\frac{k_H}{(\rho C_p)_H}\left[(1+2K\eta)\frac{\partial^2\theta}{\partial\eta^2}+2K\frac{\partial\theta}{\partial\eta}\right]-\frac{\partial\theta}{\partial\tau}+\frac{\partial\theta}{\partial\eta}f-\frac{\partial f}{\partial\eta}\theta=0, \tag{17}$$

subjected to:

$$\begin{array}{llll} f(0,\tau)=0, & \frac{\partial f}{\partial \eta}(0,\tau)=0, & \theta(0,\tau)=1 & \\ \frac{\partial f}{\partial \eta}(\eta,\tau)\to 1, & \theta(\eta,\tau)\to 0 & \text{as} & \eta\to\infty. \end{array} \tag{18}$$

To verify the temporal stability of the dual solutions, we follow Weidman et al. [49] by introducing the following perturbations:

$$f(\eta,\tau)=f_0(\eta)+e^{-\gamma\tau}F(\eta,\tau), \quad \theta(\eta,\tau)=\theta_0(\eta)+e^{-\gamma\tau}G(\eta,\tau). \tag{19}$$

Here, the functions F and G are considered small compared to the steady solutions f_0 and θ_0, and γ stands for an unknown eigenvalue. Substituting (19) into Equations (16)–(18) results in

$$\begin{array}{l} \frac{\mu_H}{\rho_H}\left[(1+2K\eta)\frac{\partial^3 F}{\partial \eta^3}+2K\frac{\partial^2 F}{\partial \eta^2}\right]+f_0\frac{\partial^2 F}{\partial \eta^2}+F\frac{\partial^2 f_0}{\partial \eta^2}+\left(\gamma-2\frac{\partial f_0}{\partial \eta}\right)\frac{\partial F}{\partial \eta}-\frac{\sigma_H}{\rho_H}M^2\frac{\partial F}{\partial \eta} \\ +\frac{(\rho\beta)_H}{\rho_H}\lambda G\cos\omega=0, \end{array} \tag{20}$$

$$\frac{1}{\Pr}\frac{k_H}{(\rho C_p)_H}\left[(1+2K\eta)\frac{\partial^2 G}{\partial \eta^2}+2K\frac{\partial G}{\partial \eta}\right]-\theta_0\frac{\partial F}{\partial \eta}+f_0\frac{\partial G}{\partial \eta}+F\frac{\partial \theta_0}{\partial \eta}+G\left(\gamma-\frac{\partial f_0}{\partial \eta}\right)=0, \tag{21}$$

$$\begin{array}{llll} F(0,\tau)=0, & \frac{\partial F}{\partial \eta}(0,\tau)=0, & G(0,\tau)=0, & \\ \frac{\partial F}{\partial \eta}(\eta,\tau)\to 0 & G(\eta,\tau)\to 0 & \text{as} & \eta\to\infty. \end{array} \tag{22}$$

By rendering $\tau=0$, the stability of the steady flow solutions f_0 and θ_0 can be investigated, and hence the initial decay or growth of the solutions can be identified by $F=F_0(\eta)$ and $G=G_0(\eta)$, respectively. The linearized eigenvalue equations are:

$$\begin{array}{l} \frac{\mu_H}{\rho_H}[(1+2K\eta)F_0'''+2KF_0'']+f_0F_0''+F_0f_0''+(\gamma-2f_0')F_0'-\frac{\sigma_H}{\rho_H}M^2F_0' \\ +\frac{(\rho\beta)_H}{\rho_H}\lambda G_0\cos\omega=0, \end{array} \tag{23}$$

$$\frac{1}{\Pr}\frac{k_H}{(\rho C_p)_H}[(1+2K\eta)G_0''+2KG_0']-\theta_0F_0'+f_0G_0'+F_0\theta_0'+G_0(\gamma-f_0')=0, \tag{24}$$

subject to the boundary conditions

$$\begin{array}{llll} F_0(0)=0, & F_0'(0)=0, & G_0(0)=0, & \\ F_0'(\eta)\to 0 & G_0(\eta)\to 0 & \text{as} & \eta\to\infty. \end{array} \tag{25}$$

Following Harris et al. [50], the condition $F'_0(\eta)\to 0$ as $\eta\to\infty$ in Equation (25) is relaxed, and is substituted by the new boundary condition $F''_0(0)=1$, without loss of generality. This replacement is performed in order to obtain the smallest eigenvalue γ_1 from the infinite set of eigenvalues $\gamma_1<\gamma_2<\gamma_3\ldots$. The flow solution is considered stable only when the generated smallest eigenvalue is positive ($\gamma_1>0$), and this positive smallest eigenvalue eventually approaches zero as $\lambda\to\lambda_c$, indicating that the solution is stable as the applied perturbation decreases with time.

4. Results and Discussion

The system of Equations (7)–(9) was numerically solved using the bvp4c package available from MATLAB R2021a software. It is necessary to set an initial guess at the initial mesh point, and to set the step size, $\Delta\eta$, in order to obtain the specified accuracy. In the present study, we set the boundary layer thickness as $\eta_\infty=10$ and the step size as $\Delta\eta=0.01$ with the relative error tolerance set to 10^{-10}. The initial guesses are dependent on the parameter values applied for computation. The validity of the numerical results was checked by examining the related profiles. Moreover, the numerical solutions were compared with the available results for the particular case of this study.

Several representing values of parameters were selected in such a way as to examine the behaviors of the heat transport and flow dynamics characteristics. For instance, the present problem comprised several influential parameters such as curvature parameter K, mixed convection parameter λ, magnetic parameter M, inclination angle ω, and the solid volume fraction of nanoparticles ϕ_1, ϕ_2. For clarity, subscripts 1 and 2, which represent Al_2O_3 and Cu, respectively, were introduced. In addition, the effects of these parameters on fluid motion, skin friction coefficient, temperature and heat transfer profiles are shown in several graphs and tables. For the purpose of the computations, we used the following default values: $M = 0.1$, $\omega = 30°$, $\Pr = 6.2$, $K = 0.2$ and $\phi_1 = \phi_2 = 0.01$. Furthermore, the values of $f''(0)$ and $-\theta'(0)$, which signify the skin friction coefficient and the heat transfer rate, were compared with those obtained previously by Lok et al. [51] and Ishak et al. [28], which show a good agreement. This validation is highlighted for the case of assisting flow, $\lambda = 1$ in the absence of hybrid nanoparticles $\phi_1 = \phi_2 = 0$, curvature parameter $K = 0$, inclination angle $\omega = 0$, and magnetic parameter $M = 0$, for different values of Prandtl number Pr, as presented in Tables 3 and 4.

Table 3. The output values of the skin friction coefficient for several values of Pr.

Pr	Lok et al. [51] (Keller-Box)	Ishak et al. [28] (Keller-Box)		Current Study (bvp4c)	
	First Solution	First Solution	Second Solution	First Solution	Second Solution
0.7	1.706376	1.7063	1.2387	1.706323	1.238728
1.0	-	1.6755	1.1332	1.675437	1.133192
7.0	1.517952	1.5179	0.5824	1.517913	0.582401
10.0	-	1.4928	0.4958	1.492839	0.495779
20.0	1.448520	1.4485	0.3436	1.448483	0.343640
40.0	1.410094	1.4101	0.2111	1.410058	0.211101
50.0	-	1.3989	0.1720	1.398930	0.172048
60.0	1.390311	1.3903	0.1413	1.390274	0.141292

Table 4. The output values of the local Nusselt number for several values of Pr.

Pr	Lok et al. [51] (Keller-Box)	Ishak et al. [28] (Keller-Box)		Current Study (bvp4c)	
	First Solution	First Solution	Second Solution	First Solution	Second Solution
0.7	0.764087	0.7641	1.0226	0.764063	1.022631
1.0	-	0.8708	1.1691	0.870779	1.169126
7.0	1.722775	1.7224	2.2191	1.722382	2.219194
10.0	-	1.9446	2.4940	1.944617	2.494029
20.0	2.458836	2.4576	3.1646	2.457590	3.164608
40.0	3.103703	3.1011	4.1080	3.101093	4.108024
50.0	-	3.3415	4.4976	3.341458	4.497588
60.0	3.555404	3.5514	4.8572	3.551406	4.857187

Figures 2–9 represent the deviations of the drag forces or the skin friction coefficient and the heat transfer rate for $\lambda < 0$ (opposing flow) and $\lambda > 0$ (assisting flows) with mixed convection parameter λ for several values of the non-dimensional parameters. Dual solutions are possible for both assisting and opposing flows, as shown in the figures, but the solution is unique for the case of forced convection flow ($\lambda = 0$). Two solutions exist in the range $\lambda > \lambda_c$, and a unique solution is seen at $\lambda = \lambda_c$, but there is no solution when $\lambda < \lambda_c$. Here, λ_c denotes the bifurcation point or the critical value of λ for which the solutions are in existence. This critical value λ_c is a unique value where the first solution meets the second solution, after which the boundary layer separates from the surface, and no solution is found using the boundary layer approximations.

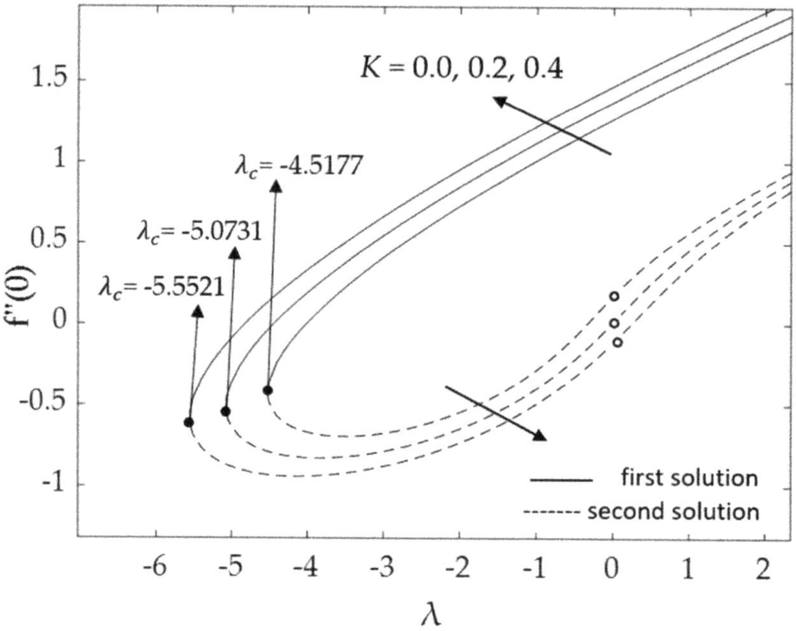

Figure 2. Deviations of $f''(0)$ for distinct values of K.

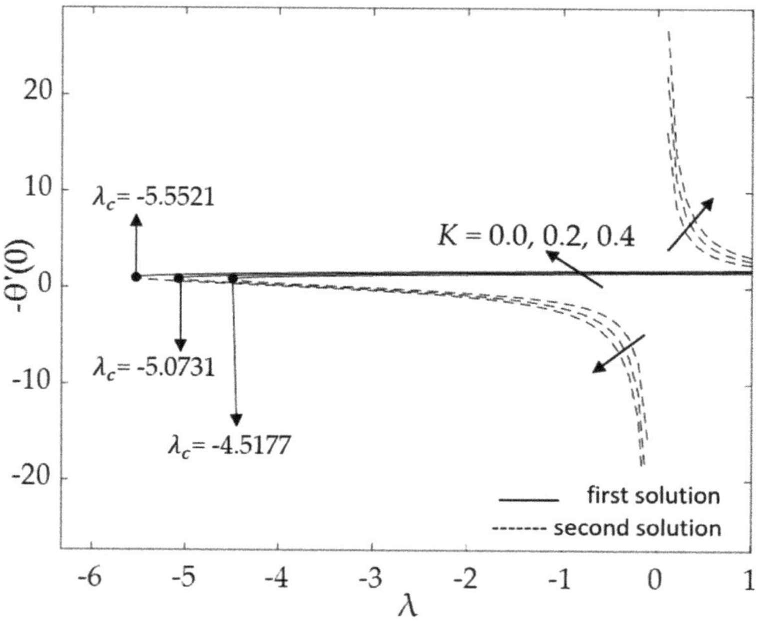

Figure 3. Deviations of $-\theta'(0)$ for distinct values of K.

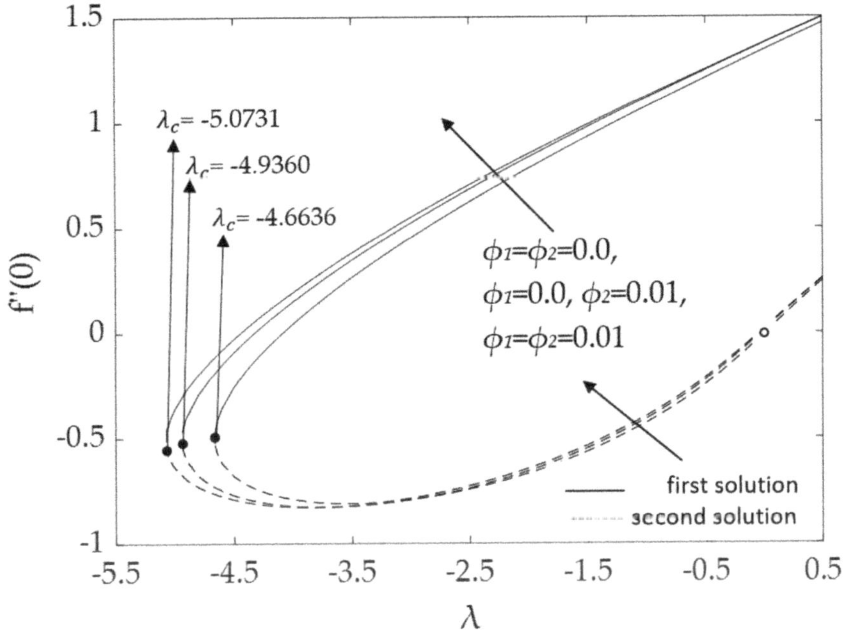

Figure 4. Deviations of $f''(0)$ for distinct values of ϕ_1, ϕ_2.

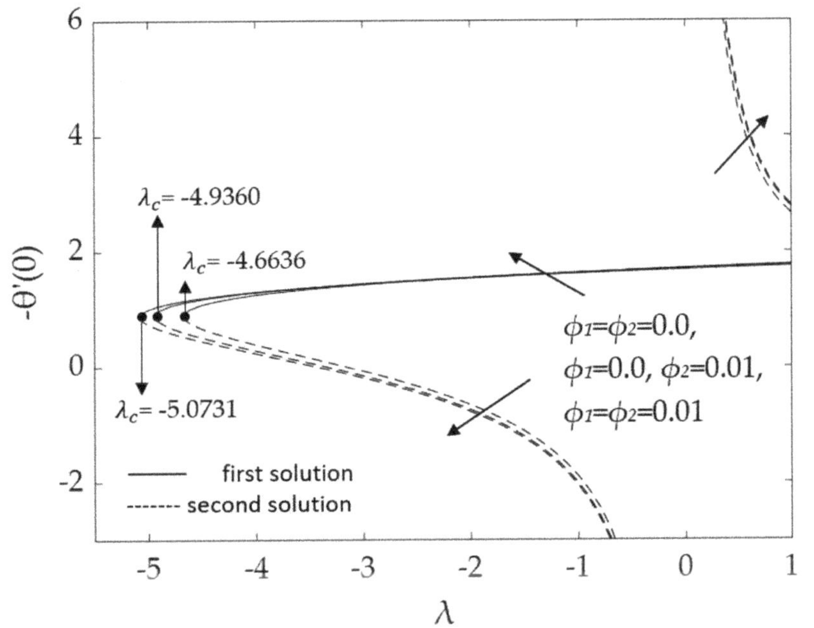

Figure 5. Deviations of $-\theta'(0)$ for distinct values of ϕ_1, ϕ_2.

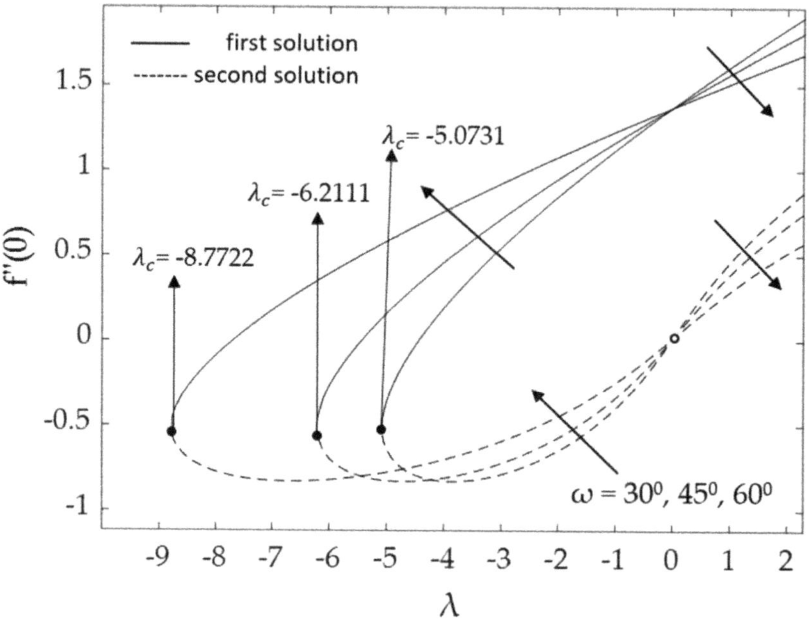

Figure 6. Deviations of $f''(0)$ for distinct values of ω.

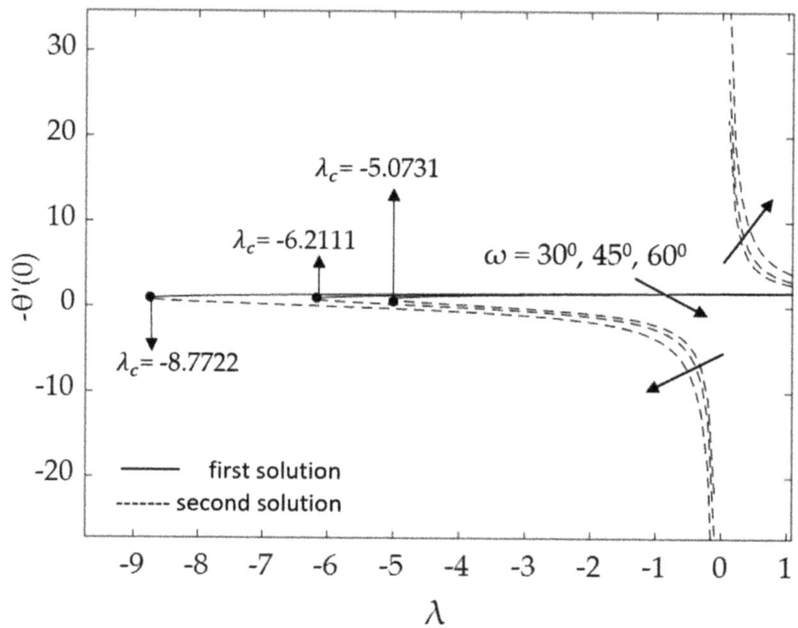

Figure 7. Deviations of $-\theta'(0)$ for some values of ω.

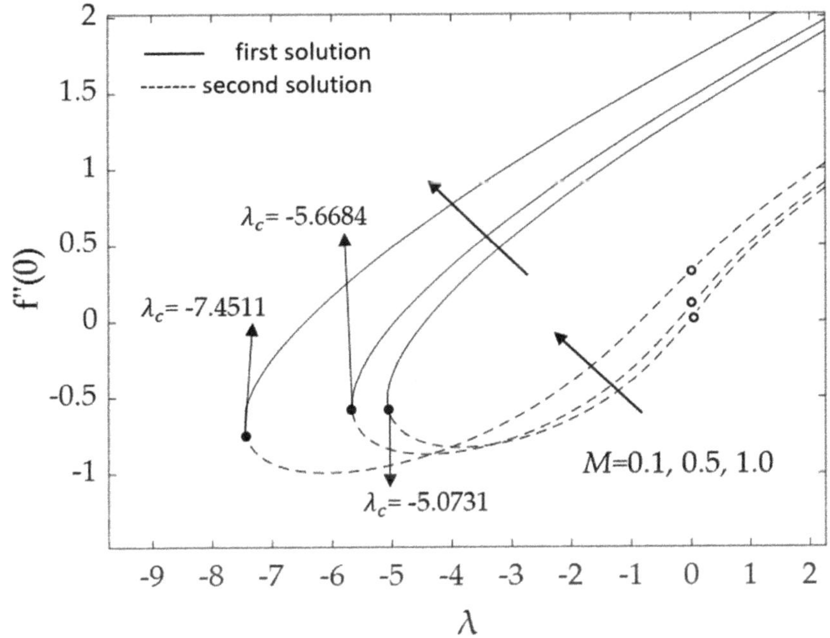

Figure 8. Deviations of $f''(0)$ for some values of M.

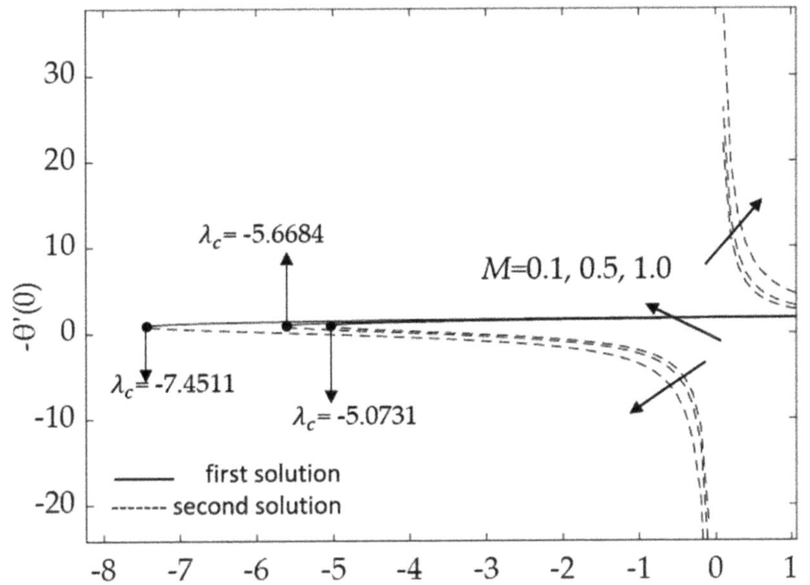

Figure 9. Deviations of $-\theta'(0)$ for some values of M.

Figures 2 and 3, respectively, show the effects of the curvature parameter K on the skin friction coefficient and the local Nusselt number (heat transfer rate at the surface). The figures illustrate that when the curvature parameter K increases from $K = 0$ (the surface of the cylinder behaves as a flat surface) to cylinders ($K = 0.2, 0.4$), both $f''(0)$ and $-\theta'(0)$

increase for the first solution. This is due to the decrement in the radius of the cylinder and the increment of the fluid motions. Less contact between the fluid and the cylinder surface causes a reduction in resistance to the fluid flow. Moreover, increasing K will delay the boundary layer separation. Using cylindrical surfaces in boundary layer problems is one approach to obtain a high skin friction and a better heat transfer rate. However, the second solution demonstrates the contrary, with $f''(0)$ values that reduce as K rises. The second solution presented in Figure 3 produces the opposite results for both flow regions. The heat transfer rate $-\theta'(0)$ increases for the phenomenon of buoyancy assisting flow ($\lambda > 0$); however, it decreases for the phenomenon of buoyancy opposing flow ($\lambda < 0$) due to the superior impacts of K.

Figures 4 and 5 demonstrate the influences of the hybrid nanoparticle volume fractions ϕ_1, ϕ_2 on the skin friction and heat transfer rate, respectively, when $M = 0.1$, $\Pr = 6.2$, $\omega = 30^0$ and $K = 0.2$. It was discovered that the bifurcation value λ_c for water ($\phi_1 = \phi_2 = 0.0$) was -4.6636; for Cu/H_2O nanofluid ($\phi_1 = 0.0$, $\phi_2 = 0.01$) it was -4.9360; and for Al_2O_3-Cu/H_2O hybrid nanofluid ($\phi_1 = \phi_2 = 0.01$), it was -5.0731. It was realized that the bifurcation values λ_c (in the opposing region) and the values of $f''(0)$ increase when the hybrid nanoparticle volume fractions increase. The results prove that the augmentation of the hybrid nanoparticle volume fraction improves fluid viscosity and thus enhances skin friction, leading to an increase in drag forces. Figure 5 illustrates the same dual solutions pattern as Figure 3. As the nanoparticle volume fractions ϕ_1, ϕ_2 increase, the heat transfer rate for the first solution also increases, which implies that the Al_2O_3-Cu/H_2O hybrid nanofluid transfers heat more efficiently than the Cu/H_2O nanofluid and pure base fluid. However, for the second solution, the heat transfer rate decreases for the opposing flow but rises for the assisting flow. It is worth noting that a hybrid nanofluid can have a higher heat transfer rate than a regular nanofluid as well as a pure fluid, as has been proved experimentally and numerically by numerous researchers, see for example [52,53].

Moreover, Figures 6 and 7 display the changes in the values of the skin friction coefficient and the heat transfer rate of the dual solutions for the numerous values of an inclination angle ω when $\Pr = 6.2$, $M = 0.1$, $\phi_1 = \phi_2 = 0.01$ and $K = 0.2$, respectively. The value of $f''(0)$ for opposing flow ($\lambda < 0$) increases as the inclination angle rises. Physically, as the value of ω increases relative to the posited x-axis and when the curvature parameter is present, the effect of buoyancy force due to gravity decreases, causing $f''(0)$ to increase but $-\theta'(0)$ to decline. The critical values $\lambda_c = -5.0731, -6.2111$, and -8.7722 for $\omega = 30°, 45°$, and $60°$, respectively, are shown in this illustration as well. The larger values of inclination angle augment the drag force, thus lowering the velocity within the boundary layer. The reason for this is the decreasing effect of gravity (see Rehman et al. [54]). Figure 7 illustrates how the increasing value of the inclination angle reduces the rate of heat transmission. As the temperature rises, the rate of heat transmission decreases. The influences of the drag force and heat transfer rate against λ for some values of magnetic parameter M when $\omega = 30°$, $\Pr = 6.2$, $K = 0.2$ and $\phi_1 = \phi_2 = 0.01$ are demonstrated in Figures 8 and 9. Because of the Lorentz force, the velocity and temperature gradients escalate with rising values of the magnetic parameter. This force is generated by the presence of a magnetic field in the fluid flow, which forms resistance to the movement of the fluid. As a result, the skin friction increases, raising the velocity gradient $f''(0)$ and the temperature gradient $-\theta'(0)$, which are proportional to the skin friction and the heat transfer rate, respectively (see Equation (12)). In addition, in the presence of a magnetic field ($M > 0$), the heat transfer rate is enhanced as the inclination angle increases. This is because the magnetic force counteracts the gravitational buoyancy force at $<90°$ but no longer counteracts it at $\omega = 90°$.

Figures 10–19 reveal the variations in velocity and temperature profiles for assisting flow ($\lambda > 0$) with several values of non-dimensional parameters. Figures 10 and 11 demonstrate the influences of K on the fluid velocity as well as the fluid temperature. At first, $f'(\eta)$ increases when the curvature parameter increases. This tendency is seen because of the larger impacts of the curvature parameter; the more it decreases the radius of the cylinder and then accelerates the fluid motions, the less contact there is with the cylinder surface, which causes a reduction in resistance to fluid flow. However, the opposite effect is seen when η_∞ is approximately greater than one. In contrast, the temperature field shows the opposite behavior to the curves of velocity. The temperature initially decreases ($\eta_\infty < 1$) and then rises when η_∞ is roughly greater than one.

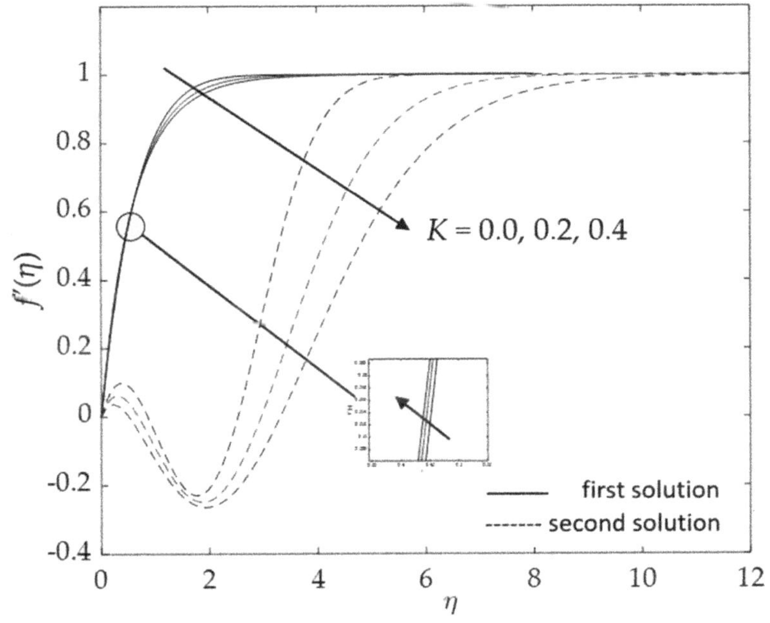

Figure 10. Influence of K on $f'(\eta)$ for assisting flow.

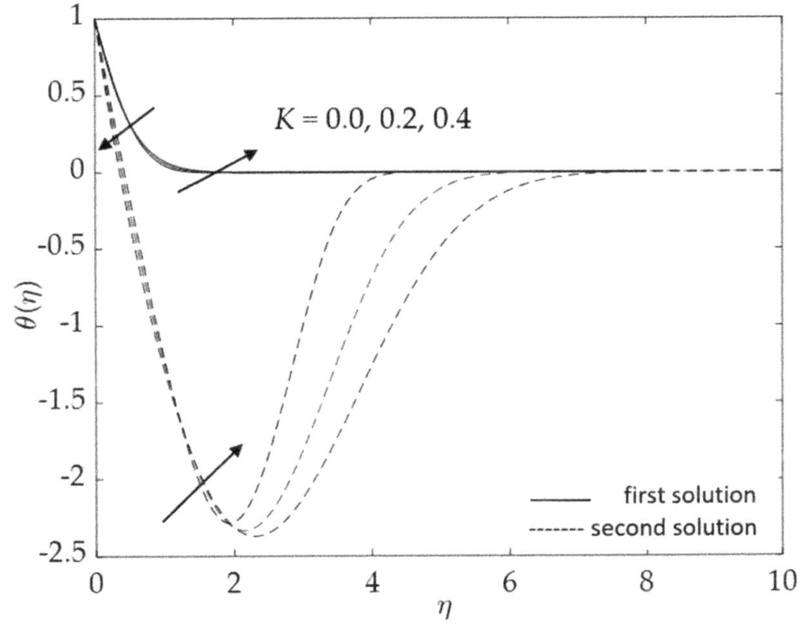

Figure 11. Influence of K on $\theta(\eta)$ for assisting flow.

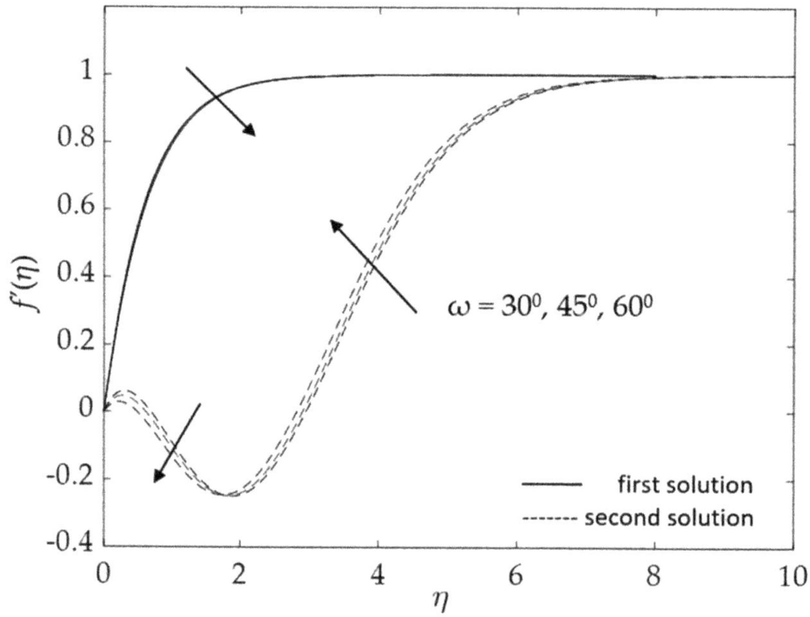

Figure 12. Influence of ω on $f'(\eta)$ for assisting flow.

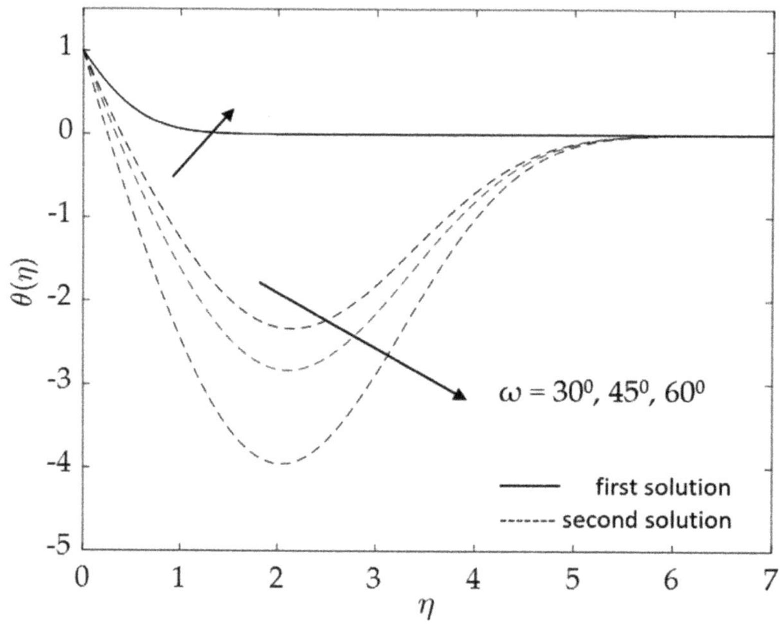

Figure 13. Influence of ω on $\theta(\eta)$ for assisting flow.

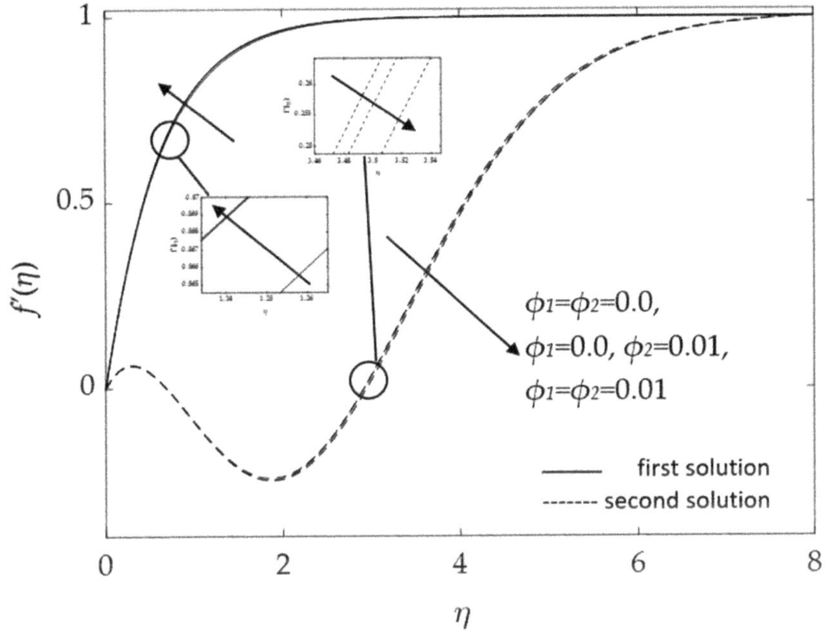

Figure 14. Influence of ϕ_1, ϕ_2 on $f'(\eta)$ for assisting flow.

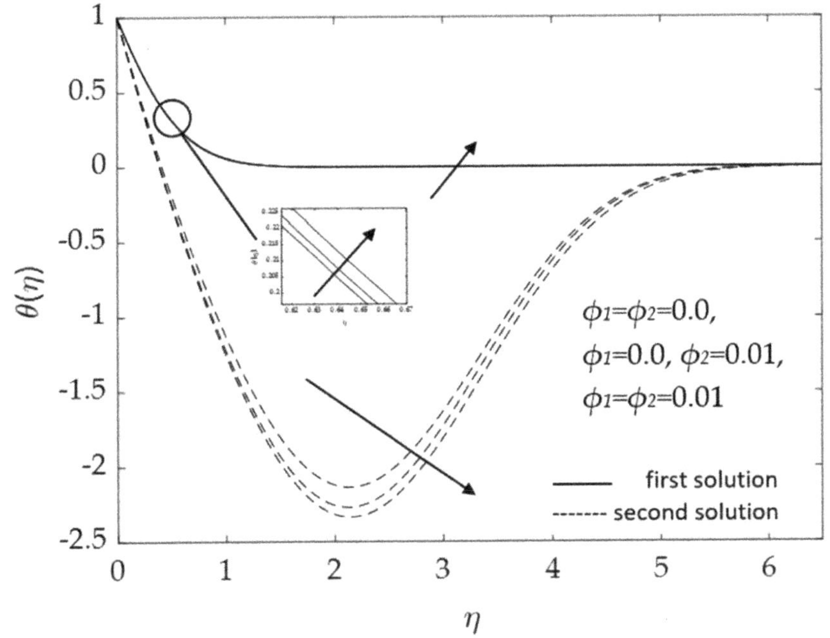

Figure 15. Influence of ϕ_1, ϕ_2 on $\theta(\eta)$ for assisting flow.

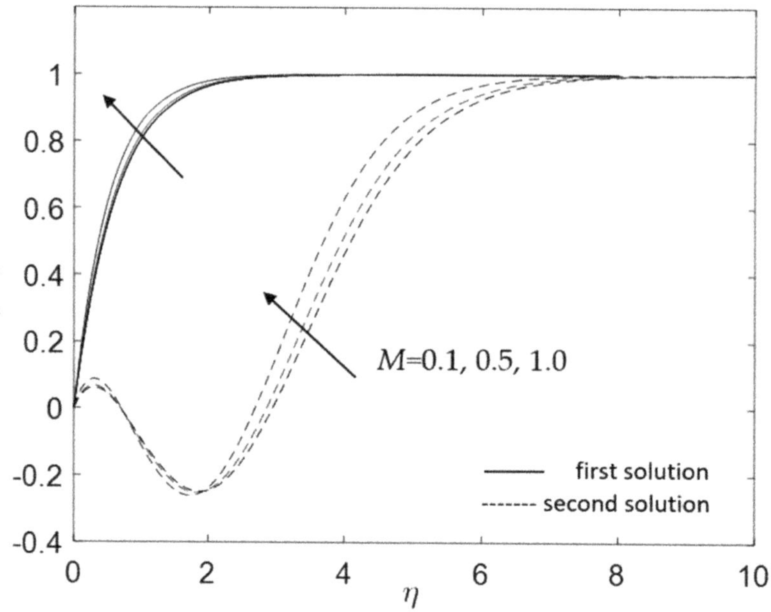

Figure 16. Influence of M on $f'(\eta)$ for assisting flow.

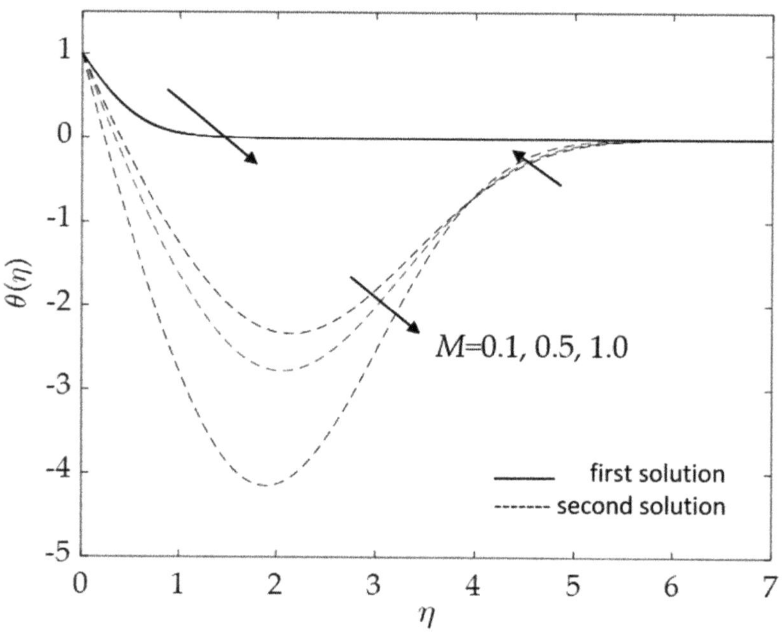

Figure 17. Influence of M on $\theta(\eta)$ for assisting flow.

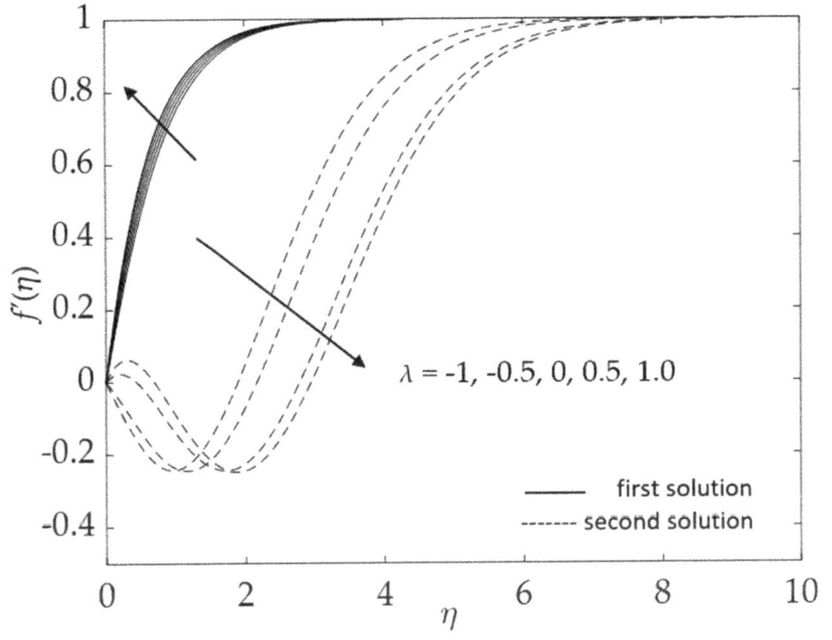

Figure 18. Influence of λ on $f'(\eta)$.

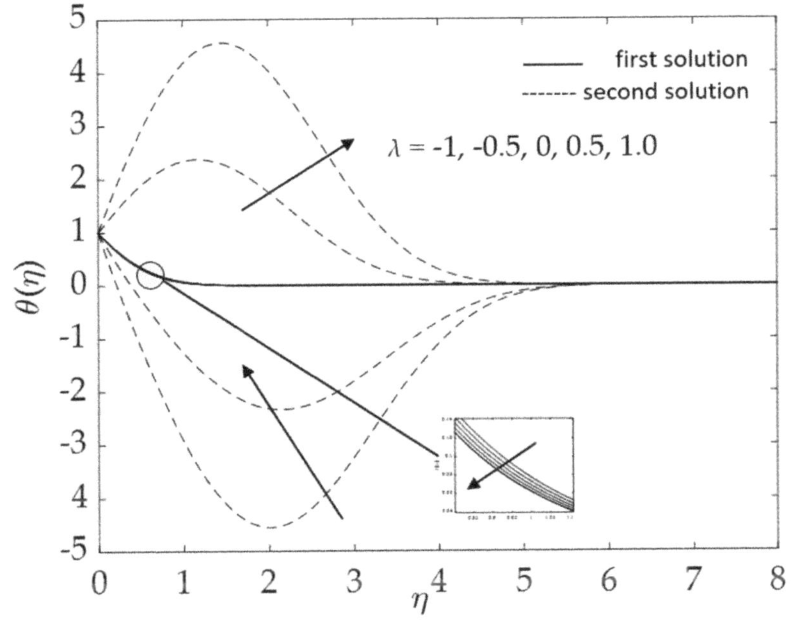

Figure 19. Influence of λ on $\theta(\eta)$.

Further, the effects of the inclination angle ω on the profiles are depicted in Figures 12 and 13. Physically, an increase in the inclination angle reduces the buoyancy effects, thus decreasing the velocity but enhancing the temperature in the boundary layers. The results in these figures correspond to the results shown in Figures 6 and 7. As can be seen in Figure 12, when the inclination angle rises, the velocity decreases (for both solutions). However, the second solution augments when the boundary layer thickness η_∞ is roughly greater than two. Meanwhile, Figure 13 shows a similar trend where the second solution displays the opposite behavior. We note that in the present study, there is a reduction to a vertical cylinder when $\omega = 0°$ and to a horizontal cylinder, when $\omega = 90°$. The buoyancy effects for the latter case are very small and thus can be neglected.

Figures 14 and 15 portray the influences of the hybrid nanoparticle volume fraction ϕ_1, ϕ_2 for the base fluid ($\phi_1 = \phi_2 = 0.0$), Cu/H_2O nanofluid ($\phi_1 = 0.0$, $\phi_2 = 0.01$) and Al_2O_3-Cu/H_2O hybrid nanofluid ($\phi_1 = \phi_2 = 0.01$). It was found that the velocity elevates when the volume of the nanoparticles augments, as presented in Figure 14. The thickness of the momentum boundary layer reduces as the concentration of nanoparticles uplifts. Hence, the velocity of the fluid flow increases. Figure 15 shows that as the hybrid nanoparticle rises, the temperature curves and the thermal boundary layer thickness also upsurges. The surface heat flux lowers as the thermal boundary layer thickens. These discoveries are according with the results illustrated in Figures 4 and 5. However, the second solutions for both figures show contradictory behavior, which is associated with the assisting flow ($\lambda = 1$).

Apart from that, Figures 16 and 17 present the effects of the magnetic parameter M on the velocity and temperature profiles when $\omega = 30°$, $\Pr = 6.2$, $K = 0.2$, $\phi_1 = \phi_2 = 0.01$ and $\lambda = 1$. By increasing the values of M, the velocity increases, as portrayed in Figure 16. The momentum boundary layer becomes thinner as the impact of M is boosted, causing an increase in the velocity gradient at the surface $f''(0)$. Thus, the skin friction coefficient increases with increasing M for the present case. This result is consistent with that shown in Figure 8. Additionally, Figure 17 shows that the temperature distribution reduces as the magnetic parameter rises. Deceleration in the thermal boundary layer thickness leads to the increment of the heat flux. This shows a similar trend to that shown in Figure 15. This finding agrees with the illustration in Figure 9.

Figures 18 and 19 show the influences of the buoyancy force on the velocity and temperature for numerous values of the mixed convection parameter λ when $\omega = 30°$, $\Pr = 6.2$, $K = 0.2$, $\phi_1 = \phi_2 = 0.01$ and $M = 0.1$. In Figure 18, the velocity increases for the first solution, whereas the second solution shows the opposite trend with the increment of λ (opposing to assisting flow). Nevertheless, the temperature for the first solution decreases when the mixed convection parameter increases, as shown in Figure 19. The second solution presents inconsistent behavior in the temperature profile.

Table 5 presents the smallest eigenvalues, γ, of the Al_2O_3–Cu/H_2O hybrid nanofluid for several values of K when $M = 0.1$, $\Pr = 6.2$, $\omega = 30°$ and $\phi_1 = \phi_2 = 0.01$. The obtained minimum eigenvalues are positive for the first solution, but negative for the second solution. Referring to Equation (19), positive values of γ indicate a decay of disturbance as time evolves $\tau \to \infty$. On the other hand, negative values of γ show an initial growth in disturbance, and thus the solutions obtained for this case are unstable in the long run. Moreover, for both solutions, the smallest eigenvalue γ approaches zero as λ approaches the bifurcation value λ_c. This means that, at turning points (from the first to the second solution), there is a change in the values of λ from stable (positive) to unstable (negative). Thus, this finding supports the stability and physical reliability of the first branch solutions. As shown by the temporal stability analysis, the second solution is unstable in the long run. Although such solutions may lack physical significance (in the long run), they are still of mathematical interest as far as the differential equations are concerned. The second solution is also a solution to the differential equations. Similar equations may arise in other situations where the corresponding solution can have a more realistic meaning.

Table 5. The eigenvalue γ for distinct values of K for Al_2O_3–Cu/H_2O when M = 0.1, Pr = 6.2, ω = 30° and $\phi_1 = \phi_2 = 0.01$.

K	λ	Smallest Eigenvalue	
		First Solution	**Second Solution**
0.0	−4.5	0.0972	−0.0963
	−4.51	0.0641	−0.0641
	−4.517	0.0199	−0.0218
	−4.5177	0.0104	−0.0103
0.2	−4.5	0.5486	−0.5091
	−5.0	0.1996	−0.1943
	−5.07	0.0709	−0.0703
	−5.073	0.0597	−0.0592
	−5.0731	0.0593	−0.0588
	−4.5	0.7357	−0.6675
0.4	−5.0	0.5353	−0.4991
	−5.5	0.2065	−0.2010
	−5.55	0.1385	−0.1361
	−5.552	0.1351	−0.1328
	−5.5521	0.1350	−0.1326

5. Conclusions

The present study investigated the behavior of the mixed convection flow of an Al_2O_3-Cu/H_2O hybrid nanofluid over an inclined cylindrical surface. The effects of the dimensionless parameters on the skin friction coefficient and the heat transfer rate at the surface were numerically computed and graphically presented. The summarized discoveries are listed as follows:

- Two solutions were obtained for both buoyancy assisting and opposing flows, whereas a unique solution was found in the absence of buoyancy force ($\lambda = 0$);
- Larger inclination angles ω lead to a lower gravitational force, increasing the friction factor $f''(0)$ but decreasing the heat transfer $-\theta'(0)$;
- The skin friction coefficient and the heat transfer rate of the Al_2O_3-Cu/H_2O hybrid nanofluid ($\phi_1 = \phi_2 = 0.01$) is higher than that of the nanofluid ($\phi_1 = 0\%$, $\phi_2 = 1\%$) and water ($\phi_1 = \phi_2 = 0.0$);
- Larger values of the curvature parameter K and mixed convection parameter λ result in a slower detachment of the boundary layer;
- Velocity increases with increasing values of the buoyancy parameter λ, magnetic parameter M, and hybrid nanofluid nanoparticle volume fraction ϕ_1, ϕ_2, whereas it declines for curvature parameter K and the inclined angle ω;
- The velocity and temperature gradients upsurge with a higher impact of the magnetic parameter;
- According to the temporal stability analysis, the first solution is physically stable as time evolves, whereas the second solution is not reliable in the long run.

Author Contributions: Conceptualization, F.H.S. and A.I.; methodology, F.H.S., U.K., S.K.S. and I.W.; software, F.H.S. and U.K.; validation, U.K., S.K.S., I.W. and A.I.; formal analysis, F.H.S. and A.I.; investigation, F.H.S. and A.I.; writing—original draft preparation, F.H.S. and U.K.; writing—review and editing, S.K.S., A.I. and I.W.; supervision, A.I., S.K.S. and I.W.; project administration, A.I.; funding acquisition, A.I. All authors have read and agreed to the published version of the manuscript.

Funding: Ministry of Education Malaysia (FRGS/1/2019/STG06/UKM/01/4).

Data Availability Statement: Not applicable.

Conflicts of Interest: The authors declare no conflict of interest.

References

1. Daungthongsuk, W.; Wongwises, S. A critical review of convective heat transfer of nanofluids. *Renew. Sustain. Energy Rev.* **2007**, *11*, 797–817. [CrossRef]
2. Buongiorno, J. Convective transport in nanofluids. *J. Heat Transf.* **2006**, *128*, 240–250. [CrossRef]
3. Tiwari, R.K.; Das, M.K. Heat transfer augmentation in a two-sided lid-driven differentially heated square cavity utilizing nanofluids. *Int. J. Heat Mass Transf.* **2007**, *50*, 2002–2018. [CrossRef]
4. Wang, R.; Wang, J.; Yuan, W. Analysis and optimization of a microchannel heat sink with V-ribs using nanofluids for micro solar cells. *Micromachines* **2019**, *10*, 620. [CrossRef] [PubMed]
5. Wang, R.; Chen, T.; Qi, J.; Du, J.; Pan, G.; Huang, L. Investigation on the heat transfer enhancement by nanofluid under electric field considering electrophorestic and thermophoretic effect. *Case Stud. Therm. Eng.* **2021**, *28*, 101498. [CrossRef]
6. Suresh, S.; Venkitaraj, K.P.; Selvakumar, P.; Chandrasekar, M. Synthesis of Al_2O_3-Cu/water hybrid nanofluids using two step method and its thermo physical properties. *Colloids Surf. A Physicochem. Eng. Asp.* **2011**, *388*, 41–48. [CrossRef]
7. Guan, H.; Su, Q.; Wang, R.; Huang, L.; Shao, C.; Zhu, Z. Why can hybrid nanofluid improve thermal conductivity more? A molecular dynamics simulation. *J. Mol. Liq.* **2023**, *372*, 121178. [CrossRef]
8. Waini, I.; Ishak, A.; Pop, I. Mixed convection flow over an exponentially stretching/shrinking vertical surface in a hybrid nanofluid. *Alex. Eng. J.* **2020**, *59*, 1881–1891. [CrossRef]
9. Devi, S.S.U.; Devi, S.P.A. Heat transfer enhancement of $Cu-Al_2O_3$/water hybrid nanofluid flow over a stretching sheet. *J. Niger. Math. Soc.* **2017**, *36*, 419–433.
10. Khan, M.R.; Pan, K.; Khan, A.U.; Nadeem, S. Dual solutions for mixed convection flow of $SiO_2-Al_2O_3$/water hybrid nanofluid near the stagnation point over a curved surface. *Phys. A Stat. Mech. Its Appl.* **2020**, *547*, 123959. [CrossRef]
11. Suresh, S.; Venkitaraj, K.P.; Selvakumar, P.; Chandrasekar, M. Effect of Al_2O_3-Cu/water hybrid nanofluid in heat transfer. *Exp. Therm. Fluid Sci.* **2012**, *38*, 54–60. [CrossRef]
12. Alshuhail, L.A.; Shaik, F.; Sundar, L.S. Thermal efficiency enhancement of mono and hybrid nanofluids in solar thermal applications–A review. *Alex. Eng. J.* **2023**, *68*, 365–404. [CrossRef]
13. Ranga Babu, J.A.; Kiran Kumar, K.; Srinivasa Rao, S. Thermodynamic analysis of hybrid nanofluid based solar flat plate collector. *World J. Eng.* **2018**, *15*, 27–39. [CrossRef]
14. Farajzadeh, E.; Movahed, S.; Hosseini, R. Experimental and numerical investigations on the effect of Al_2O_3/TiO_2H_2O nanofluids on thermal efficiency of the flat plate solar collector. *Renew. Energy* **2018**, *118*, 122–130. [CrossRef]
15. Steinheuer, J. Die lösungen der Blasiusschen grenzschichtdifferentialgleichung. *Proc. Wiss. Ges. Braunschw.* **1968**, *20*, 96–125.
16. Klemp, J.B.; Acrivos, A. A moving-wall boundary layer with reverse flow. *J. Fluid Mech.* **1976**, *76*, 363–381. [CrossRef]
17. Bognár, G. On similarity solutions of boundary layer problems with upstream moving wall in non-Newtonian power-law fluids. *IMA J. Appl. Math. Inst. Math. Its Appl.* **2012**, *77*, 546–562. [CrossRef]
18. Hussaini, M.Y.; Lakin, W.D. Existence and non-uniqueness of similarity solutions of a boundary-layer problem. *Q. J. Mech. Appl. Math.* **1986**, *39*, 15–24. [CrossRef]
19. Ishak, A.; Bachok, N. Power-law fluid flow on a moving wall. *Eur. J. Sci. Res. ISSN* **2009**, *1450*, 55–60.
20. Khashi'ie, N.S.; Arifin, N.M.; Pop, I. Mixed convective stagnation point flow towards a vertical riga plate in hybrid Cu-Al_2O_3/water nanofluid. *Mathematics* **2020**, *8*, 912. [CrossRef]
21. Waini, I.; Ishak, A.; Pop, I. Hybrid nanofluid flow and heat transfer past a vertical thin needle with prescribed surface heat flux. *Int. J. Numer. Methods Heat Fluid Flow* **2019**, *29*, 4875–4894. [CrossRef]
22. Sohut, F.H.; Ishak, A. MHD Stagnation Point of Blasius Flow for Micropolar Hybrid Nanofluid toward a Vertical Surface with Stability Analysis. *Symmetry* **2023**, *15*, 920. [CrossRef]
23. Hiemenz, K. Die Grenzschicht an einem inden gleichförmigen Flüssigkeitsstrom eingetauchten geraden Kreiszylinder. *Dinglers Polytech. J.* **1911**, *326*, 321–324.
24. Chiam, T.C. Stagnation-point flow towards a stretching plate. *J. Phys. Soc. Jpn.* **1994**, *63*, 2443–2444. [CrossRef]
25. Wang, C.Y. Stagnation flow towards a shrinking sheet. *Int. J. Non. Linear Mech.* **2008**, *43*, 377–382. [CrossRef]
26. Lok, Y.Y.; Ishak, A.; Pop, I. MHD stagnation-point flow towards a shrinking sheet. *Int. J. Numer. Methods Heat Fluid Flow* **2011**, *21*, 61–72. [CrossRef]
27. Awaludin, I.S.; Weidman, P.D.; Ishak, A. Stability analysis of stagnation-point flow over a stretching/shrinking sheet. *AIP Adv.* **2016**, *6*, 045308. [CrossRef]
28. Ishak, A.; Nazar, R.; Bachok, N.; Pop, I. MHD mixed convection flow near the stagnation-point on a vertical permeable surface. *Phys. A Stat. Mech. Its Appl.* **2010**, *389*, 40–46. [CrossRef]
29. Aman, F.; Ishak, A.; Pop, I. Mixed convection boundary layer flow near stagnation-point on vertical surface with slip. *Appl. Math. Mech.* **2011**, *32*, 1599–1606. [CrossRef]
30. Abbas, Z.; Mehdi, I.; Hasnain, J.; Alzahrani, A.K.; Asma, M. Homogeneous-heterogeneous reactions in MHD mixed convection fluid flow between concentric cylinders with heat generation and heat absorption. *Case Stud. Therm. Eng.* **2023**, *42*, 102718. [CrossRef]
31. Ullah, Z.; Ehsan, M.; Ahmad, H.; Ilyas, A. Combined effects of MHD and slip velocity on oscillatory mixed convective flow around a non-conducting circular cylinder embedded in a porous medium. *Case Stud. Therm. Eng.* **2022**, *38*, 102341. [CrossRef]

32. Zainal, N.A.; Nazar, R.; Naganthran, K.; Pop, I. MHD mixed convection stagnation point flow of a hybrid nanofluid past a vertical flat plate with convective boundary condition. *Chin. J. Phys.* **2020**, *66*, 630–644. [CrossRef]
33. Wang, G.; Zhang, Z.; Wang, R.; Zhu, Z. A review on heat transfer of nanofluids by applied electric field or magnetic field. *Nanomaterials* **2020**, *10*, 2386. [CrossRef] [PubMed]
34. Azmi, H.B.M.; Isa, S.S.P.M.; Arifin, N.M. The boundary layer flow, heat and mass transfer beyond an exponentially stretching/shrinking inclined sheet. *CFD Lett.* **2020**, *12*, 98–107. [CrossRef]
35. Hayat, T.; Farooq, M.; Alsaedi, A. Inclined magnetic field effect in stratified stagnation point flow over an inclined cylinder. *Z. Fur Naturforsch.–Sect. A J. Phys. Sci.* **2015**, *70*, 317–324. [CrossRef]
36. Bilal, M.; Khan, I.; Gul, T.; Tassaddiq, A.; Alghamdi, W.; Mukhtar, S.; Kumam, P. Darcy-forchheimer hybrid nano fluid flow with mixed convection past an inclined cylinder. *Comput. Mater. Contin.* **2020**, *66*, 2025–2039. [CrossRef]
37. Khalil-Ur-Rehman; Malik, M.Y.; Bilal, S.; Bibi, M.; Ali, U. Logarithmic and parabolic curve fitting analysis of dual stratified stagnation point MHD mixed convection flow of Eyring-Powell fluid induced by an inclined cylindrical stretching surface. *Results Phys.* **2017**, *7*, 544–552. [CrossRef]
38. Dhanai, R.; Rana, P.; Kumar, L. MHD mixed convection nanofluid flow and heat transfer over an inclined cylinder due to velocity and thermal slip effects: Buongiorno's model. *Powder Technol.* **2016**, *288*, 140–150. [CrossRef]
39. Gupta, S.; Sharma, K. Mixed convective MHD flow and heat transfer of uniformly conducting nanofluid past an inclined cylinder in presence of thermal radiation. *J. Nanofluids* **2017**, *6*, 1031–1045. [CrossRef]
40. Takabi, B.; Salehi, S. Augmentation of the heat transfer performance of a sinusoidal corrugated enclosure by employing hybrid nanofluid. *Adv. Mech. Eng.* **2014**, *2014*, 147059. [CrossRef]
41. Ganguly, S.; Sikdar, S.; Basu, S. Experimental investigation of the effective electrical conductivity of aluminum oxide nanofluids. *Powder Technol.* **2009**, *196*, 326–330. [CrossRef]
42. Waini, I.; Ishak, A.; Pop, I. Hybrid nanofluid flow over a permeable non-isothermal shrinking surface. *Mathematics* **2021**, *9*, 538. [CrossRef]
43. Oztop, H.F.; Abu-Nada, E. Numerical study of natural convection in partially heated rectangular enclosures filled with nanofluids. *Int. J. Heat Fluid Flow* **2008**, *29*, 1326–1336. [CrossRef]
44. Ishak, A. Mixed convection boundary layer flow over a vertical cylinder with prescribed surface heat flux. *J. Phys. A Math. Theor.* **2009**, *42*, 195501. [CrossRef]
45. Dzulkifli, N.F.; Bachok, N.; Yacob, N.A.; Arifin, N.M.; Rosali, H. Unsteady stagnation-point flow and heat transfer over a permeable exponential stretching/shrinking sheet in nanofluid with slip velocity effect: A stability analysis. *Appl. Sci.* **2018**, *8*, 2172. [CrossRef]
46. Lund, L.A.; Omar, Z.; Khan, I.; Sherif, E.-S.M. Dual Solutions and Stability Analysis of a Hybrid Nanofluid over a Stretching/Shrinking Sheet Executing MHD Flow. *Symmetry* **2020**, *12*, 276. [CrossRef]
47. Anuar, N.S.; Bachok, N. Double solutions and stability analysis of micropolar hybrid nanofluid with thermal radiation impact on unsteady stagnation point flow. *Mathematics* **2021**, *9*, 276. [CrossRef]
48. Merkin, J.H. On dual solutions occurring in mixed convection in a porous medium. *J. Eng. Math.* **1986**, *20*, 171–179. [CrossRef]
49. Weidman, P.D.; Kubitschek, D.G.; Davis, A.M.J. The effect of transpiration on self-similar boundary layer flow over moving surfaces. *Int. J. Eng. Sci.* **2006**, *44*, 730–737. [CrossRef]
50. Harris, S.D.; Ingham, D.B.; Pop, I. Mixed convection boundary-layer flow near the stagnation point on a vertical surface in a porous medium: Brinkman model with slip. *Transp. Porous Media* **2009**, *77*, 267–285. [CrossRef]
51. Lok, Y.Y.; Amin, N.; Campean, D.; Pop, I. Steady mixed convection flow of a micropolar fluid near the stagnation point on a vertical surface. *Int. J. Numer. Methods Heat Fluid Flow* **2005**, *15*, 654–670. [CrossRef]
52. Waini, I.; Ishak, A.; Pop, I. Hybrid nanofluid flow towards a stagnation point on a stretching/shrinking cylinder. *Sci. Rep.* **2020**, *10*, 9296. [CrossRef] [PubMed]
53. Jamaludin, A.; Naganthran, K.; Nazar, R.; Pop, I. MHD mixed convection stagnation-point flow of Cu-Al_2O_3/water hybrid nanofluid over a permeable stretching/shrinking surface with heat source/sink. *Eur. J. Mech. B/Fluids* **2020**, *84*, 71–80. [CrossRef]
54. Rehman, K.U.; Malik, M.Y.; Salahuddin, T.; Naseer, M. Dual stratified mixed convection flow of Eyring-Powell fluid over an inclined stretching cylinder with heat generation/absorption effect. *AIP Adv.* **2016**, *6*, 075112. [CrossRef]

www.ingramcontent.com/pod-product-compliance
Lightning Source LLC
Chambersburg PA
CBHW061810210926
51850CB00006B/664
* 9 7 8 3 7 2 5 8 7 8 2 5 3 *

MDPI AG
Grosspeteranlage 5
4052 Basel
Switzerland
Tel.: +41 61 683 77 34

Micromachines Editorial Office
E-mail: micromachines@mdpi.com
www.mdpi.com/journal/micromachines